SUPERCOLLIDER 2

SUPERCOLLIDER 2

Edited by
Michael McAshan

Superconducting Super Collider Laboratory
Dallas, Texas

Plenum Press • **New York and London**

Library of Congress Cataloging-in-Publication Data

International Industrial Symposium on the Supercollider (2nd : 1990 :
  Miami Beach, Fla.)
    Supercollider 2 / edited by Michael McAshan.
        p.   cm.
    "Proceedings of the Second International Industrial Symposium on
  the Supercollider held March 14-16, 1990, in Miami Beach, Florida"-
  -CIP t.p. verso.
    Includes bibliographical references and index.
    ISBN 0-306-43801-1
    1. Superconducting Super Collider--Congresses.   I. McAshan,
  Michael. II. Title. III. Title: Supercollider two.
  QC787.P7I57 1990
  539.7'3--dc20   90-49839
                                                                 CIP

Proceedings of the Second International Industrial Symposium on The Supercollider,
held March 14-16, 1990, in Miami Beach, Florida

© 1990 Plenum Press, New York
A Division of Plenum Publishing Corporation
233 Spring Street, New York, N.Y. 10013

Printed in the United States of America

Preface

The Second International Industrialization Symposium on the Supercollider, IISSC, was held in Miami Beach Florida on March 14-16, 1990. It was an even bigger and more successful meeting than our first in New Orleans in 1989. There were 691 attendees and 75 exhibitors. The enthusiasm shown by both the speakers and the audience was exhilarating for all attendees. The symposium again brought together the physicists and engineers designing the machine, the industrial organizations supporting the design and construction, the education community, and the governmental groups responsible for the funding and management of the SSC project. We believe it is this unique mix which makes this particular meeting so valuable. The theme of this symposium was "The SSC-Americas Research Partnership" and the varied presentations throughout the meeting high-lighted that theme.

The keynote speakers were:

Dr. Roy Schwitters, Director of the SSC
Mr. Paul F. Oreffice, Chairman of the Board of Dow Chemical Company
Honorable W. Hinson Moore, Deputy Secretary of Energy
Mr. Morton Meyerson, Chairman of the Texas National Research Laboratory Commission
Honorable Robert A. Roe Congressman from New Jersey and Chairman,
    House Science and Technology Committee
Honorable Tom Bevel, Representative from Alabama, Chairman House Energy and Water
    Development Appropriation Subcommittee

In addition there was a discussion of issues by a panel of four Congressmen:

Honorable Jim Chapman, Representative from Texas
Honorable Vic Fazio, Representative from California
Honorable James A. Hayes, Representative from Louisiana
Honorable Carl D.Purcell, Representative from Michigan

There were a total of 108 presentations on a wide variety of subjects related to the SSC, including reports on other U. S. accelerators, on non U. S. large accelerator projects and details on the design of the SSC accelerator, detectors and conventional facilities.

Representative Joe Barton of Texas received an award from the IISSC Board of Directors for his meritorious service to the SSC project.

The very able program committee for this symposium were:

Carl Rosner Chairman	Intermagnetics General Corporation
Owen Anglum	Armco, Inc.
David Berley	National Science Foundation
Tom Bush	SSC Laboratory
K. Wendel Chen	University of Texas at Arlington
Paul Gilbert	Parsons Brinkerhoff Quade and Douglas
M. G. D. Gilchriese	SSC Laboratory
Tom B. W. Kirk	Argonne National Laboratory
W. Arthur Porter	Houston Area Research Center
W. Parke Rohrer	Brookhaven National Laboratory

The 1990 International Industrial Symposium on the Supercollider was produced by IISSC Corporation a nonprofit company organized solely to conduct an annual symposium. The Board of Directors for the 1990 Symposium were:

Member	Affiliation
Mr. Charles E. Anderson	Air Products and Chemicals, Inc.
Mr. Robert Baldi	General Dynamics Space Systems
Dr. Edward Bingler	Texas National Laboratory Commission
Dr. K. Wendell Chen	University of Texas, Arlington
Mr. Owen Anglum	Armco, Inc.
Dr. David Berley	National Science Foundation
Dr. Tom Bush	SSC Laboratory
Mr. Paul Gilbert	Parsons Brinkerhoff, Inc.
Dr. M. G. D. Gilchriese	SSC Laboratory
Dr. Eric Gregory	IGC Advanced Superconductors, Inc.
Dr. Tom Kirk (*)	Argonne National Laboratory
Mr. Robert Marsh	Teledyne Wah Chang, Albany
Dr. W. Arthur Porter	Houston Area Research Center
Dr. Leonard M. Goldman (*)	Bechtel Corporation
Mr. A. J. Jarabak	Westinghouse Electric Corporation
Dr. Paul Mantsch (*)	Fermi National Accelerator Laboratory
Dr. Michael McAshan	SSC Laboratory
Mr. Paul Reardon (*)	Science Applications International Inc.
Mr. M. Parke Rohrer	Brookhaven National Laboratory
Dr. Giuseppe Scarfi	ANSALDO Componenti S.p.A.
Dr. Clyde Taylor	Lawrence Berkeley Laboratory
Dr. Carl H. Rosner	Intermagnetics General Corporation
Mr. Sven Svendsen	Daniel, Mann, Johnson and Mendenhall
Mr. Kuniyasu Toga	Hitachi Ltd.

(*) Designates Officers of the Corporation

The fact that the Symposium proceeded smoothly and effectively was due to the effective work carried out by members of the Board on the working committees. Equally important was the contribution of the Conference Manager, Ms. Pamela Patterson, who tended to all the myriad of details that must be addressed to produce a successful meeting.

The IISSC sincerely wishes to thank the Department of Energy for their $20,000 grant which helped to finance the initial costs for the Symposium. We also wish to thank the many companies and organizations which contributed money and effort to the IISSC for this meeting.

The following companies, organizations, societies, and agencies assisted us in producing the IISSC 1990:

ABB Technology Company
Air Products and Chemicals, Inc.
Ansaldo Componeneti S.p.A.
Argonne National Laboratory
Armco, Inc.
Bechtel Corporation
Brookhaven National Laboratory
CRS Sirrine
Daniel, Mann, Johnson, and Mendenhall
EG&G, Inc.
Fermi National Accelerator Laboratory
General Dynamics Space System Division
Hitachi Ltd.
Houston Area Research Center
IGC Advanced Superconductors, Inc.
Institute of Electric and Electronic Engineers
Intermagnetics General Corporation
Kawasaki Steel Corporation
Knight Architects, Engineers, Planners, Inc.
Koch Process Systems, Inc.
Lawrence Berkeley Laboratory
LeCroy Corporation
Lester B. Knight and Associates, Inc.
Martin Marietta Corporation
Morrison Knudsen
National Science Foundation
National Society of Professional Engineers
Parsons Brinkerhoff Quade and Douglas, Inc.
Science Applications International Corporation
SSC Laboratory
Teledyne Wah Chang Albany
Tempel Steel Company
Texas National Research Laboratory Commission
The University of Texas at Arlington
UNISTRUT Corporation
U. S. Department of Energy
Westinghouse Electric Corporation

The Board of Directors met in Miami Beach on March 14, 1990 and took the following actions to prepare for the 1991 Symposium. The officers of the IISSC Corporation for the period March 1990 to March 1991 elected by the Board are as follows:

President and Meeting Chairman	Paul Gilbert
Vice President	Andy Jarabak
Secretary	Chuck Anderson
Treasurer	Paul Mantsch

New members of the Board are:

Ms. Regina Borchard	Martin Marietta Corporation
Ms. Catherine Burns	TNRLC
Dr. Tony Favale	Grumman
Ms. Phyllis Hale	SSC Laboratory
Mr. John Nonte	Lockheed
Dr. Satoshi Ozaki	Brookhaven National Laboratory
Mr. Robert Tener	SSC Laboratory

The Board wishes to thank Dr. Mike McAshan and all the people at the SSC Laboratory who have assisted him in editing this volume. We believe it is a valuable record of progress in the development of the SSC.

I and the Board of Directors of the IISSC cordially invites all readers to the 3rd IISSC meeting to be held in Atlanta, Georgia in March of 1991.

Leonard M. Goldman
Chairman
IISSC '90

Acknowledgements

The editor is greatly indebted to Pam Patterson, conference manager, for her great organizational skills.

The success of this volume resulted on the dedication of Valerie Kelly and the Technical Information and Publications Group at the SSC Laboratory. Again it has made this project a great pleasure during the production process.

Contents

1. Worldwide Experience and Challenges

Chairman: W. N. Ness
U. S. Department of Energy

2. Accelerators

Co-Chairmans: H. Edwards and R. Stiening
SSC Laboratory

3. Computers/Controls

Co-Chairman: S. C. Loken
Lawrence Berkeley Laboratory

4. Conventional Construction

Co-Chairman: L. R. Smith
SSC Laboratory

5. Cryogenics

Co-Chairman: R. Byrns
Lawrence Berkeley Laboratory

6. Detectors I

Co-Chairman: T. B. W. Kirk
Argonne National Laboratory

7. Materials and Magnets I

Co-Chairman: J. Peoples
Fermi National Accelerator Laboratory

8. Systems and Controls

Co-Chairman: J. Nonte
Lockheed Engineering and Sciences Company

9. Poster Sessions

Co-Chairman: M. A. Green
Lawrence Berkeley Laboratory

10. Program Schedules and Challenges

Co-Chairman: R. Diebold
U. S. Department of Energy

11. Materials and Conductors II

Co-Chairman: W. Fietz
U. S. Department of Energy

12. Detectors II

Co-Chairman: M. Marx
SUNY, Stony Brook

13. Education

Co-Chairman: K. W. Chen
University of Texas at Arlington

14. Magnets II

Co-Chairman: P. Reardon
Science Applications International Corporation

15. Technology Transfer

Co-Chairman: D. Berley
National Science Foundation

1. Worldwide Experience and Challenges

PAST AND FUTURE OF THE US/JAPAN

COOPERATION IN HIGH ENERGY PHYSICS

Ken Kikuchi

National Laboratory for High Energy Physics
1-1 Oho, Tsukuba-shi
Ibaraki-ken, 305, Japan

ABSTRACT

The US/Japan cooperation in high energy physics has been in place
under an Implementing Arrangement signed on Nov. 11, 1979, between the US
Department of Energy and the Japanese Ministry of Education, Science &
Culture. This collaboration has been quite successful having produced
remarkable scientific results and a number of PhD's in both countries, and
has laid a firm foundation on which a future collaboration in an extended
scale including the SSC experiment can be built. Detector and accelerator
R&D for the SSC have already been included as cooperative programs in the
US/Japan cooperation and, in particular, a strong effort has been made in
the R&D of superconducting dipole magnets for the SSC accelerator.Based
upon the success of the collaboration in the last ten years, the US/Japan
cooperation should be continued in 1990's and expanded to include
significant participation in experiments at the SSC and continuing
activities of on-going programs.

HISTORY OF THE US/JAPAN COOPERATION BEFORE 1979

Between the U.S. and Japan there has been a big difference in
history, civilization and culture, and there were very few exchanges one
hundred years ago. Now we have many exchanges in cultural and scientific
areas, and particularly, in the collaborative program between our two
countries in the field of high energy physics.

After the World War II, a number of Japanese scientists and graduate
students got training or participated in research activities at US
universities and laboratories. In these days, this was almost the only
way of making high energy physics experiments for Japanese scientists,
since in Japan there was one small electron synchrotron which became
available in 1962.

In 1971, National Laboratory for High Energy Physics (KEK) was
established in Tsukuba Science City, which is located about 60 km
northeast of Tokyo, and the construction of a 12 GeV proton synchrotron
was started as a five years' project. It was completed in 1976 as was
scheduled and physics experiments started. But, the energy is limited to
12 GeV and then, in addition to the domestic research activities at KEK,

some university groups initiated to participate in physics experiments at US laboratories, such as bubble chamber experiment at BNL, CDF at Fermilab and LASS at SLAC. They wished to have a specific budget and a framework of international collaboration to make such international research activities. On the other hand, in the occasion of the Economic Summit Conference held in May, 1978, US/Japan cooperation in the R&D of new energy was discussed by the US President Carter and the Japanese prime minister Fukuda. Under the circumstances, the concept of cooperation was discussed by scientists of our two countries in the 19th High Energy Physics Conference held in Tokyo in August of the same year, and a conclusion was reached to make all efforts that the new agreement for R&D of energy will include high energy physics as one of collaborative areas. The earnest desire of scientists was accepted by the Governments of the US and Japan, the high energy physics was included in the new agreement of the US/Japan cooperation.

TEN YEARS OF THE US/JAPAN COOPERATION: 1979-1988

The new Agreement on Cooperation in Research and Development in Energy and Related Fields was signed between the Government of the USA and the Government of Japan on May 2, 1979. The formal scheme of the cooperation is shown in Fig. 1.

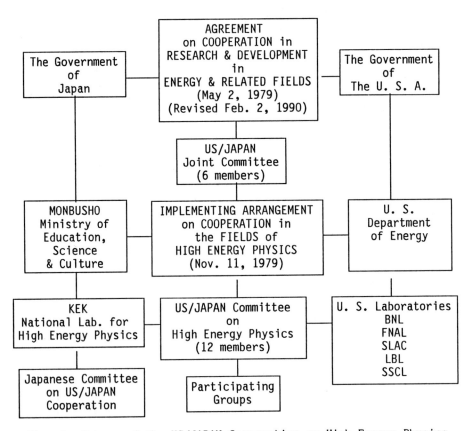

Fig. 1. Scheme of the US/JAPAN Cooperation on High Energy Physics

An Implementing Arrangement which falls under the Agreement was signed between the US Department of Energy and the Ministry of Education, Science & Culture, Monbusho, Japan, on Nov. 11, at the Stanford Linear Accelerator Center. According to the Arrangement, the US/Japan Joint Committee consisting of six members from both sides was formed to coordinate cooperative programs and has been held once a year alternatively in the US and Japan. The first meeting of the Committee which was held following the signing of the Implementing Arrangement authorized cooperative particle experiments at the US laboratories, i.e., BNL, FNAL, LBL and SLAC, including cooperative activities which had been already initiated.

In 1981, TRISTAN, a 30 GeV electron-positron collider, was approved by the Japanese Government and the construction started as a five years' program at KEK. In the 1982 Committee Meeting of the Cooperation, the Japanese side stated that participation and collaboration of US groups to the TRISTAN project were expected and welcome. In response to this statement, an international group AMY, consisting of US, Chinese, Korean and Japanese groups, was organized by the initiative of the US group and, in addition, a US group performed another experiment of TRISTAN, SHIP, which partly changed the pattern of the US/Japan cooperation from the one-way Japanese participation in the collaborative activities at US laboratories to an alternate exchange of groups of both countries.

As a whole, the US/Japan cooperation has been quite successful and produced a number of remarkable scientific papers and presentations, and Ph.D's on both sides, as is shown in Table 1. The annual amount of Japanese funding has been kept at a level of 1.5 BY and every year about 100 Japanese scientists and graduate students participated in the cooperative activities at the US laboratories. In Table 2 is shown also Japanese manpower dedicated to this cooperation in units of man-months, i.e., (number of participants) × (period of stay in the US, in unit of month).

Table 1. Product of the US/Japan Cooperation (1979-1989)

Laboratories	Project	Papers	Presentation	Ph.D. US	Japan
Experiments					
BNL	Neutrino/Heavy Ion	35	84	9	6
FNAL	CDF	31	25	17	7
	Fixed Target	26	93	24	11
LBL/SLAC	BEP-4	60	42	40	5
SLAC	LASS	24	6	3	5
SLAC/FNAL	Bubble Chamber	38	9	19	8
KEK	TRISTAN-AMY	24		10*	1
Accelerator R&D		43	59	−	1
Detector R&D		47	11	−	−
Total		328	329	122	44

* including US, China & Korea

Table 2. Japanese Funding & Participation in Cooperative Activities
at the US Laboratories

Lab. & Project		Funding (1979-1989) M¥	Participation (1979-1988) Man×Months
Experiments			
BNL	Neutrino/Heavy Ion	1398	635
FNAL	CDF	3825	949
	Fixed Target	3072	454
LBL/SLAC	PEP-4	1127	413
SLAC	Fixed Target	1013	711
	SLD	215	37
Accelerator R&D			
BNL/FNAL	SC Magnet	445	
SSC	SC Magnet	700	} 147
SLAC	Linear Collider etc.	1265	
Detector R&D			
SSC		320	
			} 142
Other Lab.		868	
Others		695	
TOTAL		14,943	3,488

APPROACH TO THE SSC

In 1982, the US/Japan Committee decided to add a new cooperative
program of "Detector R&D", reflecting the Japanese proposal. The aim of
the project was to make R&D associated with detectors for the TRISTAN
experiments. The efforts have brought fruitful results which gave various
useful impacts on the design and the construction of TRISTAN detectors.
The TRISTAN experiments started in 1986 as scheduled and the first purpose
of the Detector R&D program finished in 1986.

On the other hand, at that time there was a strong indication that a
new large hadron collider project, SSC, will be materialized in the near
future and discussion from various points of view were made in 1986
Committee meeting.   The Committee concluded that the potential
collaboration of the US and Japan in the creation of an SSC raises
exciting possibilities for a new level of internationalism in high energy
physics, and also noted that the scale and methodology needed for SSC
collaboration will require new modalities which the Committee should begin
to formulate. It was agreed that such discussions should be carried out
in a stepped approach with due consideration of the financial
circumstances and the concerns of the scientific communities involved.
Under these circumstances, the Committee decided to concentrate more
efforts on R&D of accelerator and detector technology associated with the
SSC.

The accelerator R&D of superconducting magnets and klystrons has
already been included in the cooperative program since 1981. However,

from 1987, the effort has been more concentrated on the R&D of superconducting magnets of SSC and the following themes have been undertaken, i.e.,
1. Development of a short dipole magnet with an inner coil diameter of 5 cm.
2. Improvement of properties of superconducting compacted strand wires.

The dipole magnet designed by the Central Design Group of SSC (CDG) had an inner diameter of 4 cm but we thought that the 4 cm aperture was too small to attain a high luminosity of the proton beam and started R&D of dipole magnets with an inner diameter of 5 cm.

In 1987, three 1 m dipole magnets with a cross section similar to the CDG design but having 5 cm aperture were constructed. The first dipole experienced almost ten quenchings to reach the design value of the current of 6500 A. The second and third ones reached the design current with only one quenching.

In 1988, a 5 m dipole magnet with a cross section similar to the BNL type magnet was constructed. This magnet worked successfully with a critical current of 7100A at a magnetic field of 7T and at a temperature of 4.2K.

In 1989, the SSC laboratory was established and the CDG design of the dipole magnet was re-examined from various points of view. We continued our efforts to improve the dipole magnet with an inner aperture of 5 cm and in addition to develop magnets which did not use wedges. From computer calculations it is believed that the wedge is needed to get a uniform magnetic field suited for the accelerator. However, if it is possible to avoid using wedges, the winding processing will be much simpler and a better reproducibility will be achieved. In fact two wedgeless 1 m dipoles, constructed in 1989, were found to be quite promising.

In 1990, this fiscal year, our plan is to construct, in cooperation with the SSC laboratory,
a. 1 m dipole magnet with wedges designed by the SSC laboratory,
b. 1 m dipole magnets without wedges designed by KEK,
c. full-size (13 m) dipole magnet designed by KEK.

Now that the dipole magnet for the SSC has changed in the aperture from 4 cm to 5 cm, the significant difference between the SSC and KEK designs is in the use of wedges.

FUTURE PERSPECTIVES OF THE US/JAPAN COOPERATION

The programs of the US/Japan Cooperation in 1990's will consist of an extension of the past ten years' of the cooperations and a new programs related to the SSC.

On-going cooperative experiments at BNL, FNAL and SLAC are expected to produce important scientific results and should be continued. In addition, Japanese groups have a strong interest to participate in experiments of the upgraded Tevatron and the heavy ion collider RHIC. As a next generation accelerator, a lepton linear collider is of primary importance and R&D efforts should be continued by the SLAC-KEK collaboration.

The SSC will raise a new aspect of the US/Japan cooperation in the scale and the methodology. In January, 1988, the Japanese MONBUSHO was asked informally by the US Department of Energy to cooperate with the SSC project and since then discussions have been made in considerations with scientific merits and financial situations. The most important aim of high energy physics in 1990's is to establish so-called standard model and further to explore the basic structure of elementary particles beyond the standard model. The main apparatus to attack these problems is high energy accelerators in the TeV region which is ten times higher than the energy so far we have attained. Frontier accelerators in this energy region are hadron colliders as UNK, SSC and LHC, and lepton linear colliders which are being studied by the world-wide collaboration of SLAC, CERN, Novosibilsk and KEK. Among these hadron colliders, the SSC is the most realistic project in which a sizeable Japanese group will be possible to participate.

By the completion of the TRISTAN accelerator, experimental high energy physics in Japan reached the energy frontier and we should continue to make particle experiments at the energy front by two modes: (1) participation in international collaborative experiments in hadron colliders, SSC or LHC, and (2) promotion of R&D efforts for a lepton linear collider as a possible future domestic plan, JLC (Japan Linear Collider). In fact, Japanese group has a strong intention to participate in the SSC experiment and already more than 100 Japanese scientists, who signed up in the participation in the SSC experiment, are working in R&D and design of a solenoidal detector.

On the other hand, the Japanese participation in the SSC accelerator construction is still uncertain. This matter has also been discussed from various points of view, with due considerations on the following issues.
  A. Relation with other domestic accelerator projects, particularly Japan Linear Collider. In addition, high intensity proton machine has been proposed by scientists of nuclear physics and neutron diffraction studies. Also several proposals have been made to construct electron storage rings for synchrotron radiation.
  B. International cooperations with Europe. Now University of Tokyo group is participating in the OPAL collaboration, one of the LEP experiments of CERN and this group may possibly participate in experiments of LEP II or LHC. INS (Institute for Nuclear Study) groups is participating in ZEUS collaboration of HERA experiments at DESY and will continue the activity in 1990's.
  C. Balancing between so-called "Small" and "Big" sciences, which is always a difficult problem probably in any country.

Taking account of these issues, we have reached a interium conclusion: R&D for a lepton linear collider in the energy region of several 100 GeV ~ 1 Tev should be promoted and the Japanese participation in the SSC experiment is recommended to make research activities at the energy front, with considerations on other international collaborations which should be continued in an appropriate scale. Participation in the accelerator construction of the SSC is a new type of international cooperation for Japan and, in order to avoid a financial influence to other field of science, it is requested to set up a new category of budget. At this time, the Japanese participation in the accelerator construction is uncertain. However, if it is needed, we will try to do our best efforts so that Japan may contribute to one of the most important scientific projects of the human being, SSC.

ACCELERATOR PROJECTS IN THE U.S.A.*

Nicholas P. Samios

Brookhaven National Laboratory
Upton, New York 11973

ABSTRACT

A review was presented of operating, planned and high energy
accelerators under construction in the U.S. It was pointed out that the
program is broad based, addressing a variety of frontiers and encompassing many of most fruitful projectiles, intensities, and energy ranges.
The U.S. program is also geographically diversified across the country in
major research centers. The physics productivity over the past years has
been excellent with a emphasis on the importance of sustaining the base
program at least until the SSC is operating and possibly beyond.

ACCELERATOR CENTERS

This paper will focus on accelerator projects in the United States,
concentrating on those that are operating, under construction or proposed. Such accelerators are located at the major centers for these
activities and are at Brookhaven National Laboratory, Cornell, Fermilab
and SLAC. The SSC endeavors which are of paramount importance are being
extensively discussed by others at this meeting. Therefore, I will not
make any comments on this subject, however for completeness, I will make
a few remarks on the CEBAF Project.

I will begin by noting the different types of accelerators, fixed
target and colliders, and the use of a variety of projectiles ($e^-, e^+, p, \bar{p}$,
A) that are utilized to investigate the fundamental properties of matter.
The electron e^-, is the only particle that we know of that has no
structure (is point like), is stable, and exists naturally. Its anti-
particle the positron, (e^+), is also stable in vacuum, however, it does
not exist naturally and must be artificially produced, thereby somewhat
limiting its intensity. The other naturally occurring stable particle is
the proton. It is just hydrogen gas stripped of its electrons. We now
know, however, that the proton is not fundamental, it is composed of
smaller units called quarks and the gluons which hold the quarks together. In our present understanding of the dynamics of these quarks,
they can never become free, as electrons do, but are forever destined to
be entrapped inside the proton (or other hadrons such as neutrons,
mesons, etc.). As such the energy of a proton (or its anti-particle the

---

*Work performed under the auspices of the U.S. DOE.

anti-proton) is distributed, in a continuous fashion, in the quarks and gluons. For completeness, we note a complex nucleus such as 0^{16} S^{28}, Au^{198} can be used as a projectile or a target, being viewed either as a collection of protons and neutrons or alternately as a large number of quarks or gluons.

With this assortment of possible projectiles $e^-, e^+, p, \bar{p}, A$ one can envisage all combinations but only a subset are considered to be useful. The two common types of interactions involve the so called fixed target mode where one projectile is very energetic, nearly always traveling close to the speed of light, hitting a stationary target; the second being the collider mode where two projectiles are made to collide head on. The former has the forte' of yielding a high intensity and rate of interactions, but more limited in energy while the latter (collider) allows access to the highest energies but are more limited in intensity. Again among the colliders, as noted earlier, in an e^-, e^+ collider since these particles are point-like all the e^-, e^+ energy is available for the production process while in the pp or $p\bar{p}$ case, the pp or $p\bar{p}$ energy is distributed among the quarks and gluons and therefore only a fraction of the total energy is available for the fundamental quark or gluon inter-actions. Nevertheless, the energy available to these quarks and gluons can be substantial and greater than is in the e^-, e^+ case because, up to now, it has been possible to build much higher energy proton than elec-tron machines (1,000 GeV vs. 100 GeV) due to the fact that electrons more readily radiate away their energy than protons. Below is listed the different types of machines and their locations.

Colliders:

e^-e^+	SLAC, Cornell
pp	Fermilab
pp	SSC
AA	BNL

Fixed Target:

PA	BNL, Fermilab
eA	SLAC, CEBAF

I now begin my survey of the activities at the major U.S. centers beginning with Fermilab. This laboratory near Chicago, houses an ac-celerator complex encompassing both fixed target and collider capability with protons and anti-protons as principle projectiles. A site overview is shown in Fig. 1. As in most hadron machines, the fixed target PA collision gives rise to a wide variety of intense secondary beams com-posed of $\pi's$, k's, p's, nu's, etc. Over the years there has been exten-sive work on the detailed dynamics of πN, pN, kN interactions as well as utilizing nu N interactions for the study of weak interactions and quark gluon energy distribution inside the nucleon. In addition, the relative-ly high energy available at the Tevatron accelerator at Fermilab has allowed for studies of charm particles and led to the discovery of the fifth quark, the b quark.

In developing the collider mode, $p\bar{p}$, this laboratory built a machine utilizing superconducting magnets for the first time. A view of the tunnel with both the warm and cold magnets is shown in Fig. 2. At present both machines are running well with the collider achieving luminosities $10^{30}/cm^2/sec$, at spec, with two large detectors CDF which has been operational for several years, and D0 which should begin taking data next year. The CDF has been studying the highest energy collisions at any particle accelerator. This allows precision studies of well known

10

Figure 1. Site overview of Fermilab.

Figure 2. View of the tunnel with both warm and cold magnets at Fermilab.

particles because they are produced in abundance, as well as sensitive searches for hitherto unobserved massive objects. Recently measurements have been made of the masses of the W,Z bosons the carriers of the weak force; $M_W = 80.0 \pm 0.6$ GeV and $M_z = 90.9 \pm 0.35$ GeV. A search has been made for the expected sixth quark, the t quark, with a limit for its mass greater than 77 GeV! The question of whether the quark itself is an elementary particle has also been addressed experimentally with the result that if there is structure within the quark, it is at a distance scale smaller than 2×10^{-17} cm. Presently this accelerator complex is being upgraded by increasing the energy of the proton linac from 200 MeV to 400 MeV. In addition, the construction of a new main injector is proposed, namely a 150 GeV proton synchrotron in a separate ring which will allow for a higher flux of $\bar{p}$ and a luminosity of $\sim 10^{31}/cm^2/sec$ for the TEVATRON. The main justification is for more reliable operation and extending the search for the top quark to ~ 200 GeV.

I now turn to the Stanford Linac complex at Stanford, California which encompasses both a fixed target and collider mode of operation with electrons. An overview of the site is shown in Fig. 3 with all its complexity. Among the earlier important findings uncovered at SLAC with the two-mile electron linac was the phenomenon of deep inelastic scattering, which indicated that there was structure to the proton, namely the quarks. The development and use of electron-positron colliders was pioneered at this laboratory culminating in a succession of three e^-e^+ machines: SPEAR 3 GeV x 3 GeV; PEP 15 GeV x 15 GeV and recently SLC 50 GeV x 50 GeV. The tau lepton, a third generation lepton analogous to the electron and the muon was discovered at this laboratory as well as the psi particle which with the J particle simultaneously found at BNL (to be discussed later) was the first indication of a new flavor of quark, namely, charm. This latter finding with the J received a Noble Prize in physics. More recently experimenters at the SLC and at the LEP accelerator at CERN established that there are only three flavors of neutrinos. The advent of successful operations of the SLC accelerator has also demonstrated the feasibility of building linear colliders. This new technique which accelerates electrons in a straight line eliminates one of the major difficulties of radiation loss experienced by electrons going in circles. As such it allows for the possibility of constructing high energy e^-e^+ colliders. The short term plan at SLAC is a high luminosity asymmetric B factory to be considered over the next year with a long term plan for an intermediate linear collider (ILC). A sketch of a layout of a ILC is shown in Fig. 4. It involves two linacs of ~ 3 km length, each with their accompanying sources, compressor rings and fine focus. Total energies of 0.5 - 1 TeV and luminosity of 10^{33}-$10^{34}/cm^2/sec$ are contemplated. All in all an ambitious program.

Of the major centers, Cornell is the only one that is funded by the NSF. Numerous electron accelerators have been built at this institution, each one superceding and improving on the previous version, culminating in the CESR e^-e^+ collider completed in 1979. An overview of the Cornell site is shown in Fig. 5. The CESR facility has operated for many years with an ever improving performance, now achieving a luminosity of $10^{32}/cm^2/sec$ a record for e^-e^+ colliding machines. There are two interaction regions where the CLEO and CUSB detectors have accumulated millions of interesting events. One of the beautiful results has been the unraveling and deciphering of the bb bound states, namely upsilon spectroscopy. This involved the upsilon (4S), upsilon (5S), upsilon (6S), the intermediate chi states and their numerous decay modes. This detailed study is shown in Fig. 6 where the complexity is illustrated by the numerous states and decay schemes. The other recent main research activity has involved the study of b decays, mainly into states containing charm quarks. This has involved the accumulation of a million $b\bar{b}$ pairs per

Figure 3. Site overview of the Stanford linac complex.

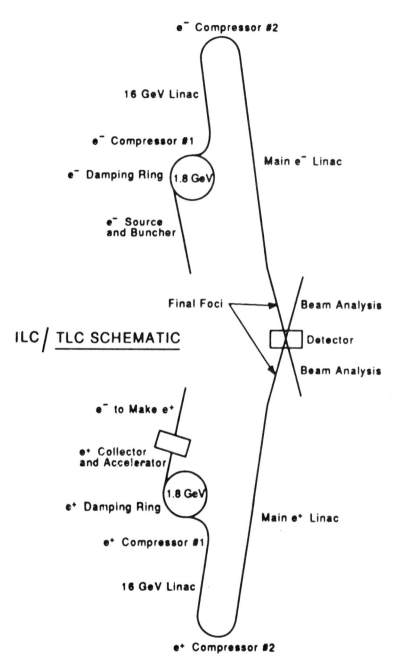

Figure 4. Sketch of SLAC's intermediate linear collider.

15

Figure 5. Site overview of Cornell.

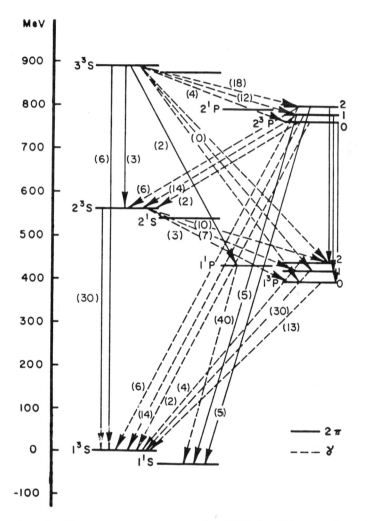

Figure 6. Upsilon spectroscopy at the CESR facility at Cornell.

17

Figure 7. Site overview of Brookhaven National Laboratory.

18

year. Since some of the more interesting decay modes have very small
rates $\sim 10^{-4}$, 10^{-5} upgrades of the accelerator are clearly desirable.
Already underway are activities to bring the luminosity to 5 x
$10^{32}/cm^2/sec$ by 1991. This will result in reducing the number of interacting regions to one, at this stage of the game, well worth the price.
A further upgrade, named CESR PLUS (i.e. a B factory) is presently being
studied. Its goal is a luminosity of $10^{34}/cm^2/sec$ at an energy of 10.6
GeV with a proposal for construction expected in approximately one year.

The program at Brookhaven National Laboratory on Long Island
centers around the AGS accelerator which with several major improvements,
has been productive for more than 25 years. Its main virture, beyond its
longevity, has been the production of high intensity protons ($\sim 10^{13}$
protons/second) that are accelerated to 30 GeV, and the concomitant
intense secondary beams of pions, kaons, antiprotons and neutrinos.
Experiments on this facility have resulted in three Nobel Prize Awards,
the previously noted discovery of the J particle, CP violation in K^0
decays and the discovery of the second type of neutrino, the μ neutrino.
An overview of the Brookhaven site complex is shown in Fig. 7. The
current physics program uses high intensity protons to search for new
physics in rare processes, and heavy ion beams to explore high density
nuclear matter. The study of the rare decay of K mesons probes the 100
TeV mass scale and physics beyond the standard model. One should recall
the example of μ decay, this particle of mass ~ 100 MeV which decays
weakly into an electron and two neutrinos. A measurement of its decay
rate and the knowledge that the electromagnetic and weak forces are
unified (the same) would have led a very smart person to deduce, years
ago, that the W mass was between 30-100 GeV. In an analogous fashion, by
searching for the rare K decay into a muon and an electron to a sensitivity of 10^{-12}, plus some knowledge would allow one to explore the mass
of an unknown new particle to the ~ 100 TeV level.

The recently completed and ongoing upgrades of the AGS complex
involve the acceleration of light ions O^{16}, Si^{28} and the construction of a
booster accelerator to be completed this year. This booster will increase the proton intensity by a further factor of 4, the polarized
proton intensity by a factor of 20 and will allow for the acceleration of
ion species up to $A \approx 200$. This accelerator infrastructure provides the
basis for RHIC (Relativistic Heavy Ion Collider) which is included for
start of construction in the President's 1991 budget. It has received
the enthusiastic support of the Nuclear Community over the past years and
NSAC gave it its highest priority for construction. The physics impact
is to heat up these large nuclei $A \approx 200$ by colliding them head on,
thereby creating conditions that existed during the first microseconds of
this universe. In effect, one is creating in the laboratory a quark-
gluon plasma, where these entities, thousands, are free to roam and
interact within a large nuclear volume. In effect one will be repeating
the creation of the universe--thousands if not millions of times and
seeing if it could come out different. A schematic of the RHIC accelerator is shown in Fig. 8 with the two concentric rings of superconducting magnets intersecting at several points around the ring where detectors would be placed to study the interesting interactions of this A-A
collider. An indication of the readiness of this project is given in
Fig. 9 where the quench history of 8 full scale dipole magnets is displayed. The first quench of all magnets is above that required for RHIC
operation, 3.5 Tesla, and the plateau field has ~25% margin of safety.
With the civil construction already in place and the R&D in good shape,
it is a project ready to go.

For completeness, I include a short discussion on CEBAF (Continuous
Electron Beam Accelerator Facility) at Newport News, VA, which is pre-

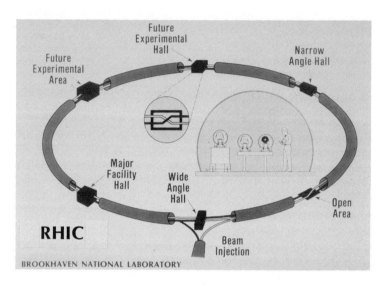

Figure 8.   Schematic of RHIC accelerator.

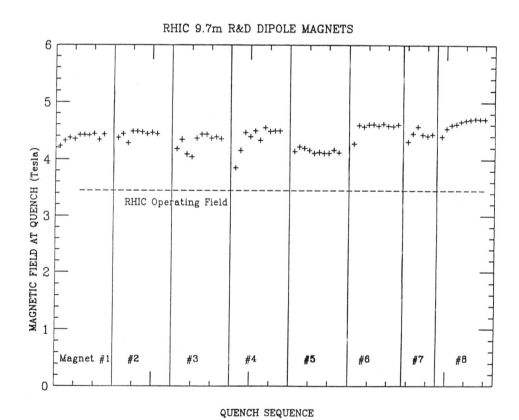

Figure 9.   Quench history of 8 RHIC magnets.

sently under construction. This is a nuclear physics facility utilizing high intensity electron beams at 4 GeV energy in a fixed target mode. The combination of energy, intensity and most important a continuous beam of electrons makes CEBAF a unique facility. Construction began in 1987 with completion scheduled for 1993. The project includes three instrumented experimental areas to service hundreds of experimenters in delving into the precise probing of nuclei by electrons.

I would conclude by stating that these major centers are providing a broad and exciting program in high energy physics. The emphasis at Fermilab will be on the pp collider, 2 TeV, $10^{31}/cm^2/sec$ exploring the 100 GeV mass scale with extensions to 200 GeV. At SLAC the focus is on the SLC collider with 100 GeV physics. The future possibilities include a B factory and a linear collider exploring the 1 TeV energy domain. Cornell will exploit its relatively high luminosity CESR with CLEO and if NSF is willing have a high luminosity B factory in its future. BNL will exploit the high intensity of its AGS accelerator complex taking a first look at the 100 TeV virtual mass range and other rare lower energy, 5 GeV, phenomena. This will be augmented by RHIC which should produce the quark gluon plasma, 200 MeV physics, which has the potential of uncovering hitherto unsuspected phenomena. All in all, the above totality of the programs should provide for a very exciting program in the years ahead.

2. Accelerators

STUDIES OF COLD PROTECTION DIODES

Ruben Carcagno

Superconducting Super Collider Laboratory*
2550 Beckleymeade Avenue
Dallas, TX 75237

John Zeigler

Texas Accelerator Center
4802 Research Forest Dr., Bldg #2
The Woodlands, TX 77381

Abstract: The feasibility of a passive quench protection system for the Superconducting Supercollider (SSC) main ring magnets depends on the radiation resistance and reliability of the diodes used as current bypass elements. These diodes would be located inside the magnet cryostat, subjecting them to liquid helium temperature and a relatively high radiation flux. Experimental and theoretical efforts have identified a commercially available diode which appears to be capable of surviving the cryogenic temperature and radiation environment of the accelerator. High current IV measurements indicate that the usable lifetime of this diode, based on an estimate of the peak junction temperature during a quench pulse, is an order of magnitude greater then than the expected lifetime of the SSC itself. However, an unexpected relationship was discovered between the diode turn-on voltage at 5 K and the most recent reverse voltage or temperature excursion. This turn-on voltage as a function of radiation exposure appears to be erratic and indicates a need for further investigation.

INTRODUCTION

A reliable quench protection system is essential for superconducting accelerator magnets. If a superconductor is warmed sufficiently, current no longer flows through it without resistance. The phenomenon of going from the superconductive state to the normal resistive state is called quenching. The current must be able to bypass a quenching magnet or the total stored energy of all the magnets in series can be deposited in this one magnet, producing joule heating that can easily overheat and damage the quenching magnet. Schemes

---

*Operated by the Universities Research Association, Inc., for the U.S. Department of Energy under Contract No. DE-AC02-89ER40486.

used to bypass the current around a quenching magnet are called quench protection schemes. Quench protection schemes can be active (the quench is detected and suitable switching is performed by elements in the protection circuit), or passive (no detection is required), with current shunting occurring automatically. The SSC Conceptual Design Report[1] specified an active quench protection system, in which room temperature diodes shunt a half-cell of six magnets. This type of active, heater-assisted quench protection with warm bypass loops is utilized in the Tevatron at Fermilab, and has proved to be very effective and reliable, and it was a conservative choice for the SSC conceptual design. However, it is also mentioned in the SSC Conceptual Design Report that an alternative passive system utilizing cold diodes would be an attractive option if uncertainties concerning the voltage distribution within a quenching magnet and radiation damage to cold diodes can be solved and if it can be shown that the magnets can be passively protected without quench heaters. A passive quench protection system with cold diodes would be an attractive option because of its inherent simplicity and reduced thermal loads to the cryogenic system. It was first proposed for the ISABELLE magnets,[2] and is now being implemented at HERA.[3] One of the problems of using cold diodes for the SSC is the relatively high radiation environment expected. The literature contains very little information relevant to this application of diodes which combines cryogenic temperatures, high currents, and neutron radiation. Therefore, in order to determine the feasibility of using a passive quench protection system with cold diodes for the SSC, the SSC Central Design Group, in cooperation with the Texas Accelerator Center, has performed a series of experimental and theoretical studies on the effects of low-temperature irradiation on several commercial semiconductor power diodes. In this paper we report the results obtained and we propose what further R&D is needed.

PASSIVE QUENCH PROTECTION SYSTEM WITH COLD DIODES

The main rings of the SSC are each divided into ten independently powered circuits, or sectors. For the latest SSC conceptual design,[4] each of these sectors is approximately 8.6 km long and contains a 300 V, 6500 A power supply connected in series with 480 dipoles and 96 quadrupoles. The total sector inductance is approximately 26 H, with 1.01 MJ of energy stored in each dipole and 0.145 MJ of energy stored in each quadrupole. As shown in the simplified schematic of Figure 1, each sector also includes four 0.27 ohm dump resistors which can be inserted in series with the power supply and the magnets. If a quench is detected anywhere in the sector, these resistors are used to extract the energy stored in the remaining superconducting magnets. The resulting exponential current decay has a time constant of approximately 24 sec, much longer than the 0.52 sec required to adequately limit joule heating in the quenched magnets.[5]

The passive quench protection system proposed for the SSC uses a diode installed in parallel with each magnet to provide an automatic, fast operating bypass path around a quenched magnet. The diodes are inside the cryostat to eliminate the need for external connections and take advantage of the voltage controlled switch behavior of a diode operating in liquid helium. This characteristic, shown in Figure 2 for an ABB DS6000 diode at 5 K, requires approximately 15 V of forward bias before any significant current is conducted. During normal charging and discharging of the magnet, the inductive voltage across the diode is sufficiently low that no current flows in the bypass path. During a quench, however, the resistive voltage developed inside the magnet would quickly exceed the "turn-on" voltage of the diode and commutate the current out of the magnet and into the bypass unit. If the magnet is self-protecting, or able to absorb its own stored energy without overheating under all quench conditions, this passive bypass circuit will safely protect the quenching magnet with no further actions such as firing heaters. After current begins to flow in the diode, the heat

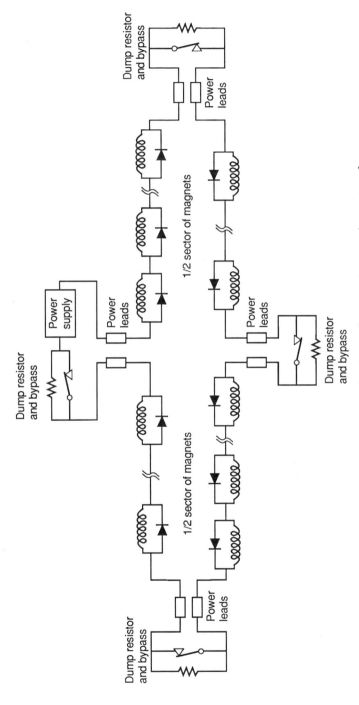

Figure 1. Simplified schematic of a passive quench protection system for one sector of the SSC.

generated at the diode junction is sufficient to raise its temperature and restore a relatively normal I-V characteristic.

The large number of cold diodes required in the SSC (about 10,000), and the long replacement time for a failed diode (about 1 week), make the feasibility of this type of system critically dependent on the diode's lifetime and reliability in the SSC environment. During a nominal SSC quench, the diodes would be required to conduct a 6500 A pulse with a 24 sec exponential decay.[1] Installing the diodes inside the magnet cryostat would maintain them in a liquid helium bath at a temperature of 4.35 K and subject them to a radiation level of approximately 3.2×10^{11} n/cm^2 per year.[6] This flux is produced by nuclear cascades created through beam losses and interactions with residual gas molecules in the bore tube.

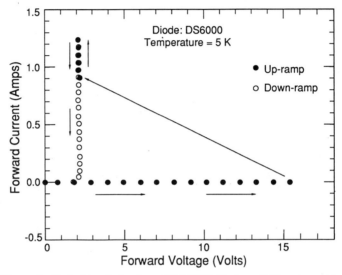

Figure 2. Switching behavior of DS6000 diode at 5 K, using a bias voltage step delay time of 1 second.

RADIATION HARDNESS OF COLD DIODES

Energetic neutrons displace atoms in the lattice of the diode, increasing the forward voltage drop during conduction and therefore increasing the energy dissipation during a quench pulse. If the junction temperature exceeds a specified limit (typically 200–300° C), the diode may be permanently damaged and subsequently degrade or prohibit proper operation of the accelerator.

The majority of published radiation damage studies, including those performed in the past specifically for quench protection systems,[3,7] have been based on room temperature irradiations of semiconductor devices. In these studies, a significant level of thermal annealing, or repairing, of the radiation damage occurred simultaneously with the radiation induced defect formation. Since the annealing rate is a nonlinear function of temperature and is very small at cryogenic temperatures,[8] these results cannot be used to directly extrapolate the performance of SSC cold bypass diodes.

A series of experimental and theoretical studies have therefore been performed to evaluate the radiation hardness of commercially available power diodes under simulated SSC operating conditions. The main results of these studies will be briefly discussed below; a more thorough treatment of portions of this information and results of additional tests has previously been reported. [5,9,10]

TEST SETUP

A variety of commercially available diodes with significantly different specifications were selected for inclusion in the test. Catalog specifications for the diodes are summarized in Table 1, although modifications were made to the packaging and passivation schemes for some of the devices.[10]

Table 1. Summary of Diode Specifications

Qty	Diode	Manufacturer	I_{FRMS}	V_{RRM}	Diameter
3	DS6000	ABB Asea Brown-Boveri	15600 A	200 V	50 mm
3	DSA1508	ABB Asea Brown-Boveri	5600 A	2000 V	50 mm
3	SSiRV60	Siemens	3930 A	1500 V	54 mm
1	RA20-A	Powerex	7535 A	1200 V	67 mm
2	RA20-D	Powerex	3920 A	3000 V	67 mm

The diodes were mounted in a cryostat designed to maintain all of the devices at the same ambient temperature while permitting electrical tests on individual diodes. Cooling was accomplished by attaching each diode mounting assembly directly to a central copper reservoir which was filled with liquid nitrogen or helium. This allowed single-sided conduction cooling to either 80 K or 5 K with the diodes located in the vacuum space surrounding the central reservoir. Electrical connections into the cryostat were made through a single pair of stainless steel safety leads. Pneumatically controlled G-10 rods operated copper switch contacts inside the vacuum space to connect a single diode to the power leads for each test.

The cryostat and diodes were then installed in the irradiation cell at the Texas A&M University Nuclear Science Center (NSC). The swimming pool type research reactor using FLIP TRIGA fuel[11] was operated at a steady state power level of 100 kW during each irradiation period. Boral and cadmium plates installed in the exposure window between the irradiation cell and the reactor core attenuated the lower energy neutron flux in the cell.

IRRADIATION TESTS AT 80 K

A preliminary irradiation test was performed at 80 K to evaluate the relative radiation hardness of the five diode types shown in Table 1. This operating temperature was chosen to minimize thermal annealing effects without requiring the additional complication and expense of a liquid helium coolant system. A variety of experimental procedures were used to monitor the forward and reverse performance characteristics of the diodes as a function of exposure.[10]

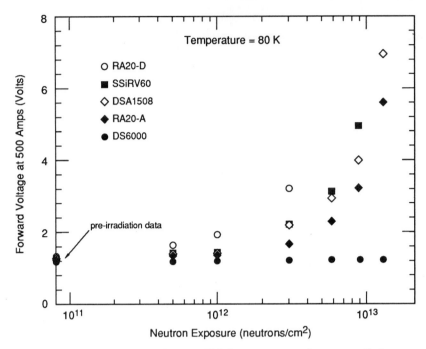

Figure 3. Forward voltage at 500 A vs. exposure for power diodes
irradiated at 80 K.

The data in Figure 3 shows the change in forward voltage at a current of 500 A as a function of exposure for a representative of each diode type. One of the diode types, the ABB DS6000, showed only a 5% increase in forward voltage after the full exposure of 1.2×10^{13} n/cm^2. The four remaining diode types showed sufficient degradation by the end of the irradiation that they would have failed if subjected to an SSC quench pulse.

Annealing

Figure 4 shows the effect of the irradiation and subsequent room temperature annealing cycle on the forward voltage at 80 K. Prior to the irradiation, the forward voltages exhibited only a small spread among the five diode types. After the irradiation, the DS6000 showed a 5% increase in forward voltage while the RA-20 and DSA1508 voltages increased by factors of five and six, respectively. The RA20-D and SSiRV60 had stopped conducting forward current by the end of the irradiation.

The forward voltages showed a significant effect from annealing after the diodes were warmed to room temperature and subsequently recooled to 80 K. The DS6000 showed a reduction of approximately 50% in the forward voltage increase caused by the radiation damage. The RA-20 and DSA1508 both showed a decrease in forward voltage by a factor of approximately three. The SSiRV60 exhibited even more dramatic improvement since it was once again able to conduct forward current. The RA20-D, however, never recovered forward conduction.

The significant effect of room temperature annealing shown in Figure 4 supports the need to perform radiation damage studies at the expected irradiation temperature of 5 K.

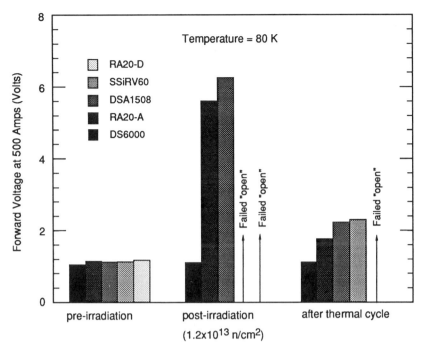

Figure 4. Forward voltage at 500 A before irradiation, after irradiation, and after thermal cycle.

IRRADIATION TESTS AT 5 K

The comparative irradiation test at 80 K showed the DS6000 to be the most radiation resistant of the diode types tested and indicated that it might be a suitable candidate for use as an SSC quench bypass diode. This diode type was therefore selected for further irradiation testing at 5 K to more accurately simulate the actual SSC environment.

Two reactor runs were made with a total of 11 diodes. High current IV characteristics were measured during both runs using 7 kA, 300 μsec sinusoidal current pulses. The forward turn-on voltage as well as the low current IV characteristics were measured during the second run using a 2 A, 0.75 sec triangular current pulse. These tests and the temperatures at which they were performed during each reactor run are summarized in Table 2.

Table 2. Tests and the temperature at which they were performed during each reactor run.

	RUN1 (6 Diodes)			RUN2 (5 Diodes)		
	Before irradiation	During irradiation	After irradiation	Before irradiation	During irradiation	After irradiation
High Current IV	300 K, 5 K	5 K	5 K, 300 K	f(300 K > T > 5 K)	5 K	f(5 K < T < 30)
Low Current IV	300 K, 5 K	5 K	5 K, 300 K	f(300 K > T > 5 K)	5 K	f(5 K < T < 30)
Turn-on voltage	——	——	——	5 K	5 K	5 K

Figure 5 shows the change in forward voltage at 7000 A versus exposure for each diode in the two experimental runs. The data for the two runs match very well, with both data sets exhibiting a moderate spread in the forward voltages at high fluences. A slightly nonuniform flux distribution across the test fixture contributed to this spread, with the highest voltages corresponding to the positions with the highest fluence. Manufacturing tolerances in the diodes also influenced the spread, with slight variations in the base region thickness probably being the most critical parameter. In either case, the spread in voltages does not appear to be significant until well beyond the expected fluence of 9.6×10^{12} n/cm^2 over the lifetime of the SSC.

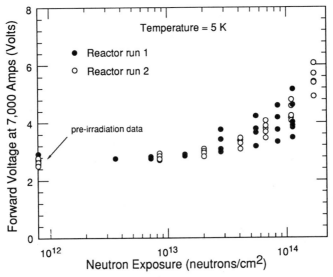

Figure 5. Forward voltage at 7000 A vs. exposure for DS6000 diodes irradiated at 5 K.

The switching behavior of a diode operating at liquid helium temperature, shown in Figure 2, was investigated in more detail during the second reactor run at 5 K. Pre-irradiation tests showed that the current through the diode in the "off" state was well below a microamp. Therefore, the turn-on phenomena appears to be related to the electric field established by the forward bias voltage rather than by heating of the junction caused by small leakage currents.

These tests also revealed that the magnitude of the forward turn-on voltage is dependent on the most recent reverse voltage or temperature cycle. Figure 6 shows the relationship between the turn-on voltage and the amplitude of the reverse voltage preceding the test. After the initial cooldown from 300 K to 5 K, the forward turn-on voltage was on the order of 10 V. Subsequent turn-on voltage measurements gave a value of approximately 2 V for all diodes. If a reverse voltage of 75 V or more was applied between measurements, however, the forward turn-on voltage was increased, even exceeding its original value for large reverse voltages. A similar recovery of the turn-on voltage was observed if the temperature of the diode was increased sufficiently between measurements of the turn-on voltage. Warming the diode to a temperature of 35 K was enough to restore the initial turn-on voltage of approximately 10 V.

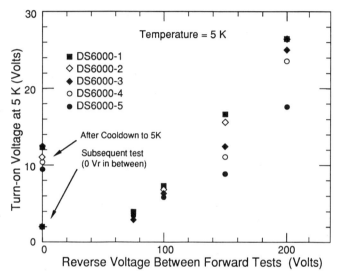

Figure 6. Turn-on voltage at 5 K vs. reverse voltage between forward tests.

During irradiation, a 200 V reverse bias was applied to the diodes before each measurement cycle to ensure full recovery of the turn-on voltage. The subsequent measurements of forward turn-on voltage as a function of exposure are shown in Figure 7. A minima of 1.8 V was observed immediately after the first irradiation period, which was equivalent to about 20 years in the SSC. This drastic decrease from the pre-irradiation value of 23 V for the same diode was unexpected and no consistent relationship was apparent between turn-on voltage and fluence.

THEORETICAL SIMULATIONS OF RADIATION EFFECTS

Computer simulations of the classical radiation damage mechanisms were performed and correlated to the damage observed in the diodes irradiated at 80 K.[12] These simulations were performed with SEDAN III, a one-dimensional program for the solution of differential equations governing the motion of carriers in a semiconductor material.

The theoretical simulations evaluated the relative contribution at cryogenic temperatures for the three major room temperature radiation damage mechanisms: reduction in carrier lifetime, carrier concentration, and carrier mobility. The results showed that the reduction in carrier lifetime remains the dominant damage mechanism even at cryogenic temperatures, with effects from reduced carrier concentrations only becoming significant after lifetime degradation has already rendered the diode unusable for the intended application.

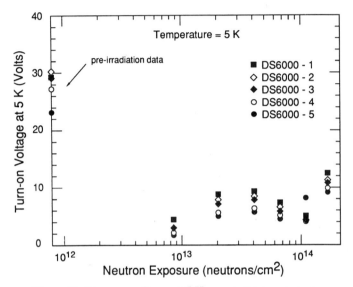

Figure 7. Turn-on voltage at 5 K vs. neutron exposure.

The results of simulations for diodes of varying base widths were then correlated to the experimental data as shown in Figure 8. The carrier lifetimes for the experimental data points were calculated from:

$$\frac{1}{\tau} = \frac{1}{\tau_0} + \frac{\Phi}{k_r}$$

where τ = post-irradiation carrier lifetime

τ_0 = pre-irradiation carrier lifetime

Φ = neutron fluence

k_r = carrier lifetime damage constant

A single value of $k_r = 2.7 \times 10^6$ n-sec/cm^2 provided reasonable agreement between the experimental and theoretical data sets over a wide range of base widths and exposure levels.

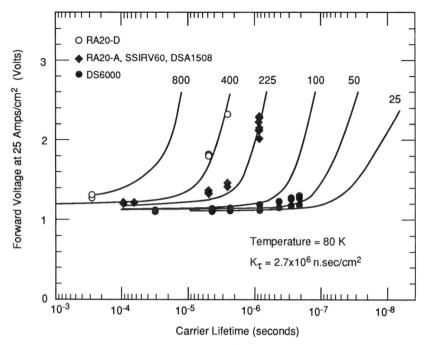

Figure 8. Forward voltage at 25 A/cm² vs. carrier lifetime for power
diodes irradiated at 80 K. Discrete symbols are measured
values; curved lines are computer simulations for diodes with
the base widths labelled in microns.

A strong relationship between base width and radiation hardness is apparent in both
the theoretical and experimental data. Proper diode performance in the high injection regime
requires that the majority of the injected carriers traverse the full width of the base region
before recombining. As radiation damage reduces the carrier lifetime, the number of carriers
traversing the base is reduced and the voltage required to support the current is increased.
Therefore, a diode with a narrow base width, and consequently shorter base transit times, will
appear more radiation resistant than a thicker diode. However, the reverse blocking voltage
capability of the diode is proportional to base width and may impose a severe constraint for a
radiation resistant diode if blocking voltages greater than 100-200 V are needed.

ESTIMATE OF DIODE LIFETIME IN THE SSC

For the purposes of this study, the lifetime of the DS6000 diode in the SSC
environment was considered to be limited solely by the peak junction temperature reached
during a quench pulse. This systematic failure mode was selected since the dominant effect
from radiation damage is an increase in the forward voltage, and therefore power dissipation,
in the diode. The catalog specification of 170°C was used as the maximum acceptable junction
temperature.

A Finite Element model was developed to predict the junction temperature of the diode
during an SSC quench pulse as a function of forward voltage.[13] This model includes the

35

temperature dependence of material properties and thermal contact resistances associated with the diode, its package, and its mounting assembly. Due to the low specific heat of these materials at liquid helium temperature, the junction temperature quickly exceeds 80 K during a quench pulse. Above this temperature, the magnitude of the temperature coefficient decreases and the junction temperature increases more slowly. Therefore, a conservative estimate of the peak junction temperature was obtained from the model by assuming a constant forward voltage throughout the quench pulse equal to the forward voltage drop at 80 K.

The result of this simplified model predicts a relatively linear relationship between the peak junction temperature and the forward voltage drop. This relationship can be approximated by:

$$T_j (max) = 170 \times VF(80 \text{ K}) - 221$$

where VF(80 K) is the value of the forward voltage drop at 80 K. From this equation, the maximum allowable junction temperature of 170°C will be reached when VF (80 K) is about 2.3 V, a 77% increase above the 1.3 V pre-irradiation value. From temperature coefficient measurements as a function of temperature,[9] a 77% increase in forward voltage at 80 K should roughly correspond to a 77% increase in forward voltage at 5 K. From Figure 5, this increase at 5 K occurs at a fluence of about 10^{14} n/cm^2, equivalent to more than 300 years of SSC operation.

RELIABILITY OF COLD DIODES

The only reliability information available for diodes actually operating under SSC type conditions is a result of experiments performed at DESY using standard DS6000 diodes in modified packages.[14] In these tests, 1250 of diodes were cooled to 4 K and subjected to 19 high current pulses very similar to those expected in the SSC. Only four diodes were destroyed during forward current tests.

CONCLUSIONS

The experimental results show that the ABB DS6000 diode is significantly more radiation resistant than the other four types of diodes tested, and theoretical simulations have shown that this diode is more radiation resistant than the others because of its thinner base width. The lifetime of the DS6000 in the SSC radiation environment has been estimated to be on the order of 300 years. This estimate was based on the peak junction temperature expected during a quench pulse after radiation damage has increased the forward voltage of the diode at high currents.

The relatively narrow base width of the DS6000 contributes to its radiation resistance. However, a narrow base width could be also a limitation if multiple diodes and internal taps in the magnets are required, because in this case the reverse voltage across individual diodes could exceed the rating of the DS6000. If a single diode is used to bypass each magnet, the DS6000 easily meets the maximum applied reverse voltage requirement of 20 V.

The turn-on voltage as a function of exposure exhibited erratic behavior and unexpectedly low values at moderate exposures. This behavior creates concern over the effects that a background level of neutron radiation, ionizing radiation, and other energetic particles will have on this critical parameter. Further theoretical and experimental studies should be pursued to evaluate these relationships and their potential impact on the proper operation of cold diodes in the SSC.

A strong relationship was discovered between the forward turn-on voltage of the diode and the most recent reverse voltage or temperature cycle. During a quench cycle, the power dissipation in the diode would raise its junction temperature well above the 35 K level which was found to be sufficient to restore the original turn-on voltage. Therefore, a diode which has conducted a quench pulse will automatically have its forward turn-on voltage restored as the junction temperature cools back down to the 5 K ambient temperature.

If the SSC magnets are self-protecting, a passive quench protection scheme using one cold bypass diode is an attractive option because of its inherent simplicity and reduced thermal loads to the cryogenic system. The theoretical and experimental results presented here suggest that using cold diodes is a feasible option for the SSC. However, further studies should be performed in the areas of optimum mechanical design of the diode holder, experimental and theoretical studies of turn-on voltage behavior, and reliability studies with statistically meaningful sample sizes of the DS6000 diode including all possible stress factors.

REFERENCES

1. "Conceptual Design of the Superconducting Super Collider," SSC-SR-2020, March 1986, SSC Central Design Group, Lawrence Berkeley Laboratory, Berkeley, CA.

2. K. Robbins, W. Sampson, and M. Thomas, "Superconducting Magnet Quench Protection for ISABELLE," IEEE Transactions on Nuclear Science, Vol. NS-24, No. 3, June 1977.

3. K. H. Mess, "Quench Protection at HERA," Proceedings of the 1987 IEEE Particle Accelerator Conference, March 1987, pp. 1474–1476.

4. "SSC Site-Specific Conceptual Design," December 1989, SSC Laboratory.

5. J. Zeigler and R. Carcagno, "Feasibility of Passive Quench Protection in the SSC," Proc. MIDCON/88 Technical Conference, Dallas, TX, Aug. 30–Sept. 1, 1988.

6. D. Groom, "Radiation in the SSC Main Ring Tunnel," Appendix 21 from "Report of the Task Force on Radiation Effects at the SSC," SSC-SR-1035 (1988).

7. A. Gosh, W. B. Sampson, G. Stenby, and A. J. Stevens, "Radiation Exposure of Bypass Diodes," SSC Technical Note No. 10, February 1984, Brookhaven National Laboratory.

8. H. J. Stein, "Electrical Studies of Neutron-Irradiated n-type Si: Defect Structure and Annealing," Physical Review, Vol. 163, No. 3, November 1967, pp. 801–808.

9. J. Zeigler and R. Carcagno, "Lifetime of Passive Quench Protection Diodes in the SSC," Proc. of the 1989 Particle Accelerator Conference, March 1989.

10. J. Zeigler, R. Carcagno, and M. Weichold, "Experimental Measurements of Radiation Damage to Power Diodes at Cryogenic Temperatures," Proc. of the 1987 Particle Accelerator Conference, March 1987.

11. G. Schlapper and J. Krohn, "Use of the Nuclear Science Center Reactor for Diode Response Testing," Workshop on Quench Protection Diodes, Texas Accelerator Center, The Woodlands, TX, July 14–15, 1986.

12. J. Zeigler, R. Carcagno, M. Weichold, and G. Welch, "Results of Neutron Irradiation of Power Diodes at 80 K," Technical Note TC1799, July 1987, Texas Accelerator Center.

13. R. Carcagno, "A Heat Sink Thermal Design for the SSC Passive Quench Protection Diodes," SSC-N-432, SSC Central Design Group, Lawrence Berkeley Laboratory, Berkeley, CA, December 1987.

14. K. Mess et al., "The Quench Protection Diodes for Superconducting Magnets for HERA," Proc. Thirteenth International Cryogenic Engineering Conference, Beijing, China, 1990 (to appear).

NUMERICAL STUDIES OF THE SSC INJECTION PROCESS:

LONG TERM TRACKING

T. Garavaglia, S.K. Kauffmann, R. Stiening, and
D.M. Ritson

Superconducting Super Collider Laboratory*
2550 Beckleymeade Avenue, Dallas, TX 75237

ABSTRACT

Results are presented for tracking protons within the Superconducting
Super Collider during the injection phase for up to seven million turns.
The results for the 4 cm and the 5 cm aperture machines are compared.
X and y invariant amplitude and phase space information are given
which are characteristic of bounded chaotic motion. The present long
term tracking results were obtained using SSCTRK for groups of four
particles started at four different equally spaced injection points on
eight different machines, each characterized by a different random seed.
A qualitative interpretation of the dynamical behavior of the protons
is given in terms of some simple models using nonlinear differential
equations.

INTRODUCTION

SSCTRK was developed by D. M. Ritson to address the question of the dynamical behavior of the proton beams in the Superconducting Super Collider
(SSC) during the injection period. Long term tracking studies where initiated
using SSCTRK in late 1989 in order to determine the suitability of the 4 cm
magnetic aperture design. As a result of these and related studies decisions
were made to recommend a change in the basic design of the Collider so that
the magnetic aperture would be increased to 5 cm. Defining bounded chaotic

---

* Operated by Universities Research Association, Inc., for the U.S. Department of Energy
  under Contract No. DE-AC02-89ER40486.

dynamical motion as dynamical behavior which is sensitive to small differences in the initial injection conditions and which results in chaotic motion within a restricted domain, one can see from the studies reported in this paper that the SSC is characterized by bounded chaotic behavior. The dynamic aperture is the spatial region throughout which the beam operates and beyond which particle motion becomes unstable and results in unacceptable particle loss. The agreed recommended safe dynamic aperture radius for the Collider is 4 mm. This represents a distance of 10 standard deviations of the rms beam radius at injection. It is seen from these studies that the 4 cm magnetic aperture design is unsafe with respect to plausible magnetic multipole errors which lead to erratic dynamical motion as a result of the nonlinearities they introduce into the magnetic field.

During the injection period into the Collider the proton beams must be stable for up to 7.0×10^6 turns. This corresponds to approximately 40 minutes. Numerical simulations of this process have been performed to observe the dynamical behavior of particles tracked through the Collider. In this paper, three long term tracking runs are described and their results are reported. A particle is determined to be lost if its distance from the closed orbit exceeds $\sqrt{2}$ 15 mm during a drift between thin elements. The number of turns before a particle is lost has been recorded against initial injection amplitude, and these graphs are shown. In addition, the maximum values of the x and y invariant amplitudes averaged over 10^4 turns are presented. The results have been obtained using the code SSCTRK. Plausible multipole errors have been introduced to simulate the expected realistic levels which can be measured. These errors and other assumptions regarding the plausible machine have been described in these proceedings by T. Garavaglia, K. Kauffmann and R. Stiening in Ref. 1.

SSCTRK

SSCTRK is a FORTRAN 77 code for the numerical simulation of particle tracking within the SSC lattice. There are versions which can be run on Sun workstations, IBM, Cray, and SX-2 supercomputers. The supercomputer version for the Cray is a vectorized code with inner DO loops written in assembly language to enhance the running speed. The Cray version can track up to 64 particles simultaneously. It has been used to track particles in a plausible machine of 4 cm aperture for seven million turns and in a plausible machine of 5 cm aperture for three million turns. Data on the $|i_x|$, $|i_y|$ values, and $(|i_x|^2 + |i_y|^2)^{1/2}$ values where $|i_x| = (x^2 + (\alpha_x + \beta_x \theta)^2)^{1/2}$ and where $|i_y| = (y^2 + (\alpha_y + \beta_y \theta)^2)^{1/2}$ for the usual α and β of betatron oscillations can be extracted at each element. Also information regarding the lifetime of a particle as well as smear information can be obtained. Typically data are taken and averaged over 10,000 turns for runs of more than 3 million turns. In addition, the values of x' and y' have been used to make phase space plots. In a typical long run, four particles are injected with equal x and y values on four different grid points for one of eight possible random seeds, each seed representing the simulation of a different machine, which represent the various random errors associated with the multipoles.

LONG TERM TRACKING

The results for three cases, Figures 1 through 5 (RUN50), Figures 6 through 10 (RUN58), and Figures 11 through 13 (RUN68), are presented. The first and last are simulations of the SSC with a 4 cm magnetic aperture, and the second is a simulation with a 5 cm magnetic aperture.

RUN50: 4 cm Magnet

Chromaticity corrections in x and y have been made; however, a 3.75% error is introduced in the setting of the chromaticity sextupoles. This is equivalent to an error of about 5 units. Systematic multipole errors appropriate for a 4 cm magnet have been assumed. The quadrupole persistent current duode-capole error has been suppressed. All multipoles have a systematic value of 0.05 units except the regular sextupole, the regular quadrupole, and the skew quadrupole. This level is viewed as the limit of measurement. No ripple has been introduced, and a momentum error of 5×10^{-4} has been used. The momentum error is achieved by injecting with no dp/p but with a 16.4 cm error in z_o. Four particles where started on a grid of 2, 3, 4, and 5 mm in eight different machines. This run had the same input information as a previous 10^5 turns simulation, RUN25 of Ref. 1; however, particles where tracked for 7×10^6 turns. As is usual, when a particle is lost another is started at a grid point that has been stepped in by one grid unit from the starting position of the one which was lost.

The results of this run can be seen in Figure 1 where a dynamic aperture of between 3 and 4 mm is seen. This is to be compared with the result of RUN68 in Figure 11 where a finer grid was used. This figure indicates an aperture of approximately 3.5 mm. Graphs for particles started at four different equally spaced grid points are given for the x and y invariant amplitudes as a function of turns for two representative random seeds. The results for RUN50 are given in Figures 2 through 5. It is clear that nonlinear dynamical behavior is seen at 3 mm and above. In addition, one can see in comparing the graphs for the x and y invariant amplitudes, e.g. Figures 2 and 3, a clear cross coupling where the y behavior is an approximate reflection symmetry of the x behavior; however, the x and y invariant amplitudes can be driven to different maximum values.

RUN58: 5 cm Magnet

This is an example where the multipoles have been chosen which are appropriate for a 5 cm magnet. The quadrupole persistent current duodecapole has been suppressed. All multipoles have a systematic value of 0.05 units except for the regular sextupole, regular quadrupole, and skew quadrupole. This level is viewed as the limit of measurement. The momentum error associated with

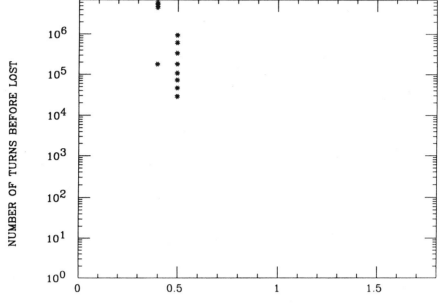

Figure 1. Particle Lost Run = 50.

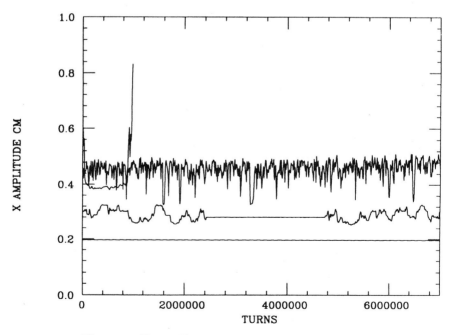

Figure 2. X vs. Turns Run = 50 4 cm Magnet.

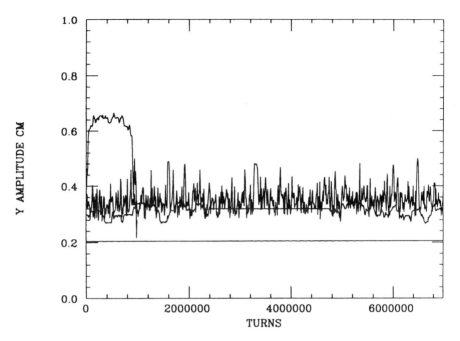

Figure 3. Y vs. Turns Run = 50 4 cm Magnet.

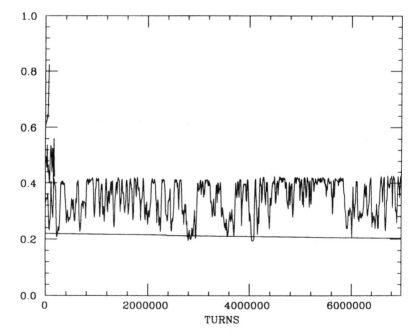

Figure 4. X vs. Turns Run = 50 4 cm Magnet.

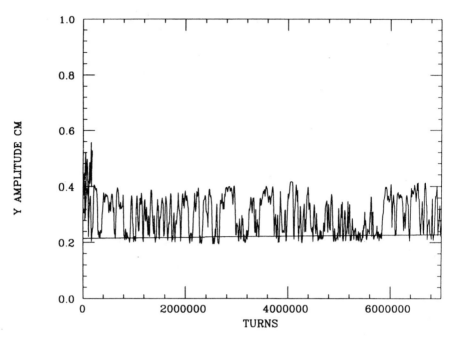

Figure 5. Y vs. Turns Run = 50 4 cm Magnet.

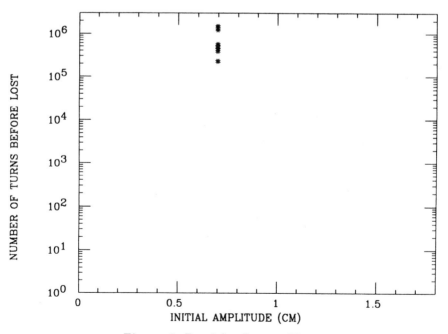

Figure 6. Particles Run = 58.

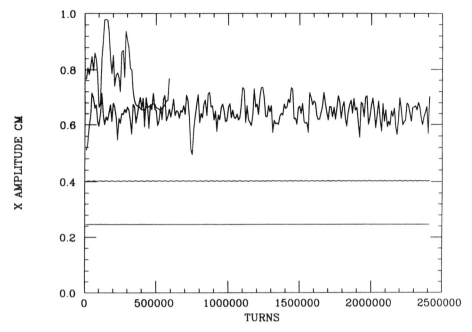

Figure 7. X vs. Turns Run = 58 5 cm Magnet.

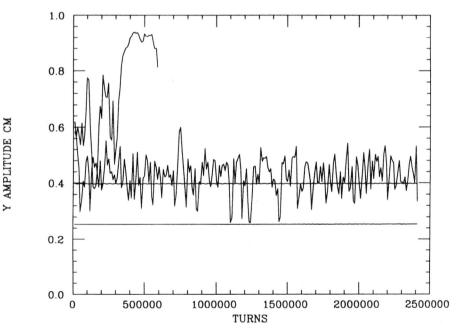

Figure 8. Y vs. Turns Run = 58 5 cm Magnet.

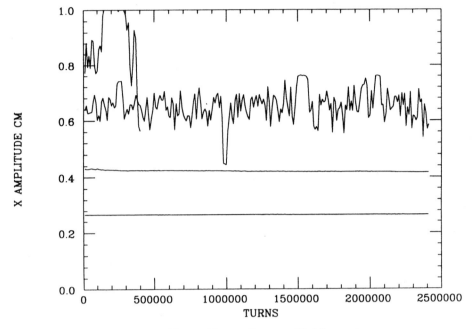

Figure 9. X vs. Turns Run = 58 Magnet.

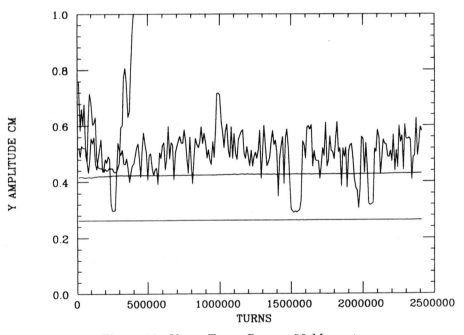

Figure 10. Y vs. Turns Run = 58 Magnet.

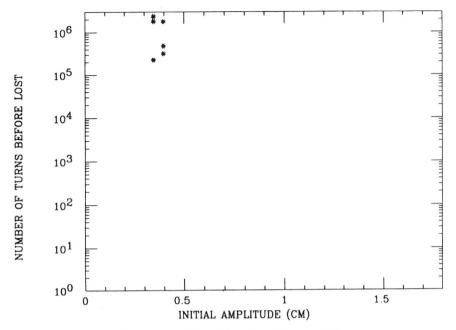

Figure 11. Particles Lost Run = 68.

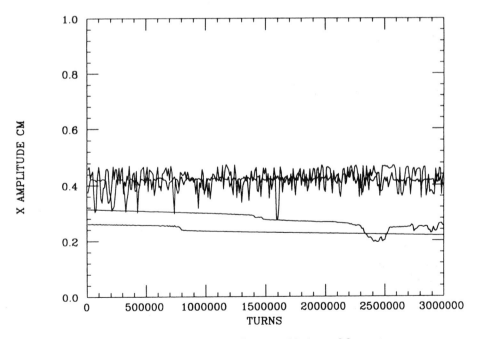

Figure 12. X vs. Turns Run = 68 4 cm Magnet.

synchrotron oscillations is 5.0×10^{-4}. This has been achieved with $dp/p = 0$ and a 16.4 cm z_0 error. This is the same as RUN36 which is discussed in Ref. 1; however, it has been run for 3×10^6 turns. Particles where injected at 2.5, 4.0, 5.5, and 7.0 mm into eight different machines. Figure 6 shows an aperture of approximately 5.2 mm. In addition, the x and y invariant amplitude graphs for this run, given in Figures 7 through 10, show particles running stably at 4 mm. Here one sees the usual xy coupling and the presence of nonlinear effects beyond 5.5 mm.

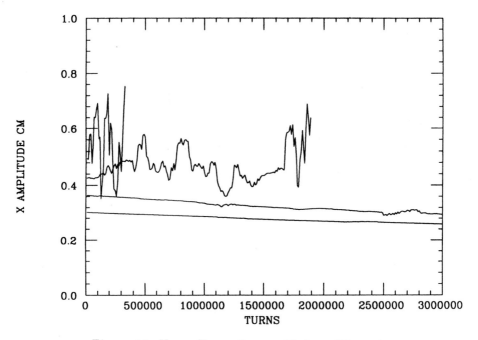

Figure 13. X vs. Turns Run = 68 4 cm Magnet.

RUN68: 4 cm Magnet

Particles where tracked in this run for 3×10^6 turns with the multipoles appropriate for the 4 cm magnet. This is basically the same as RUN50; however, particles were injected into the eight machines on a finer grid at the points 2.5, 3.0, 3.5, and 4.0 cm. It is seen from Figure 11 for the number of turns before a particle was lost that the dynamical aperture is approximately 3.5 mm. In addition, and inspection of the graphs for the x and y invariant amplitudes as a function of turns, Figures 12 and 13, demonstrates the usual irregular nonlinear behavior of the amplitudes.

BOUNDED CHAOTIC MOTION: Figures 14 through 16 (RUN681)

RUN681 has the same run card as Figures 14 through 16 (RUN68); however, two particles where started at nearby points and tracked using seed 1 for 10^6 turns on a Cray 2 machine. Particle one was started at 3.5 mm and particle two was started at $3.5 + 10^{-10}$ mm. Graphs where made of the value of the modulus, $(|i_x|^2 + |i_y|^2)^{1/2}$ of the two particles as a function of turn number. These are shown in Figures 14 and 15, respectively. The Euclidean distance between the two particles as a function of turn number is given in Figure 16. It is clear from Figure 14 that a bifurcation takes place after 15,000 turns. Figure 15 shows that bounded chaotic motion persists for the full 10^6 turns. Furthermore, Figure 16 shows that the long term growth of the Euclidean distance ·between the points, $\sqrt{(|i_{x2}| - |i_{x1}|)^2 + (|i_{y2}| - |i_{y1}|)^2}$, is neither Lyapunov exponential nor diffusive, i.e. growing as ($\sqrt{\text{time}}$).

In addition to the x and y invariant amplitude figures already discussed, one can see the nature of the chaotic behavior of the dynamical motion of the particle injected at 3.5 mm from phase space figures for both (x, x') and (y, y'). These are seen in Figures 17 and 18 where the phase space points are filling an annulus. The skew symmetry between these two figures is the result of x and y coupling.

QUALITATIVE BEHAVIOR

A certain qualitative understanding of the results presented above for the nonlinear dynamical behavior of the particles during the long term tracking simulations can be obtained from the study of a simple model using nonlinear differential equations of the type

$$x'' + \omega^2 x = d_3 x^2 + d_4 x^3 + e_{11}xy + e_{12}xy^2 + \text{etc.} + F_{random}$$
$$y'' + \omega^2 y = d_3 y^2 + d_4 y^3 + f_{11}xy + f_{12}xy^2 + \text{etc.} + F_{random}$$

which are solved numerically using the CERN Programme Library Runge-Kutta integration routine RKSTP. For this system, x space and phase space plots show how a particle can be lost when it is near the separatrix. Also one can see how a random force can cause particle loss. Furthermore, one can study the effects of coupling, and one can simulate the behavior seen in Figure 2 for the long term tracking study where the invariant amplitude can make transitions between large and small average values.

For stable motion, with $\omega = 2$ and with only d_4 nonzero, the amplitude as a function of time behaves as shown in Figure 19. The associated phase space portrait shown in Figure 20 determines the separatrix for this system. If the parameter in this system is changed from $d_4 = 8.0007910$ to the value $d_4 = 8.0007911$, the result can be seen in Figures 21 and 22 for the amplitude and phase space behavior where the particle becomes unstable upon crossing the

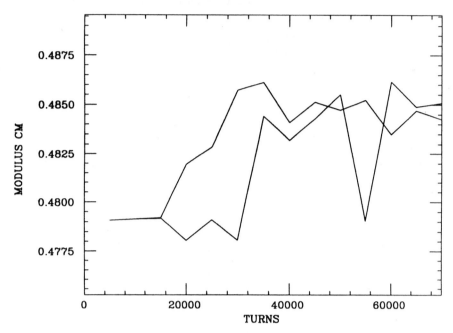

Figure 14. Mod 1 and 2 vs. Turns Run = 681 Chaos.

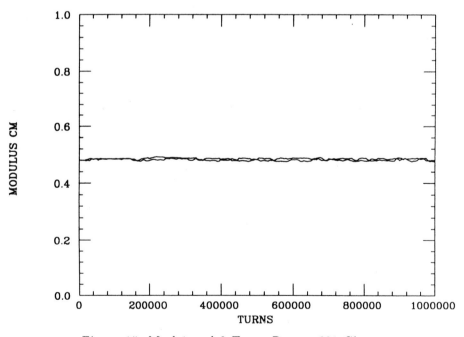

Figure 15. Mod 1 and 2 Turns Run = 681 Chaos.

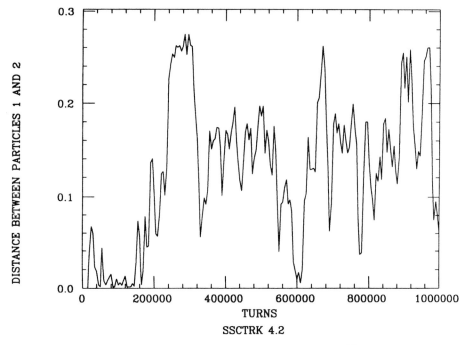

Figure 16. Initial Separation $1 \times 10E^{-10}$.

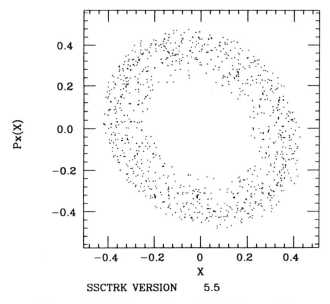

Figure 17. X Phase Space Plot Run = 68.

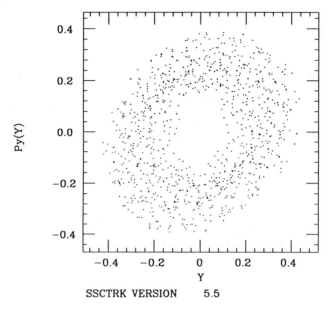

Figure 18. Y Phase Space Plot Run = 68.

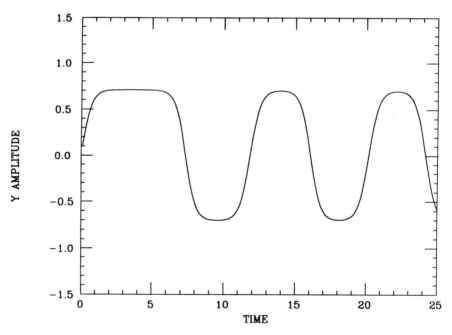

Figure 19. Amplitude vs. Time 1.

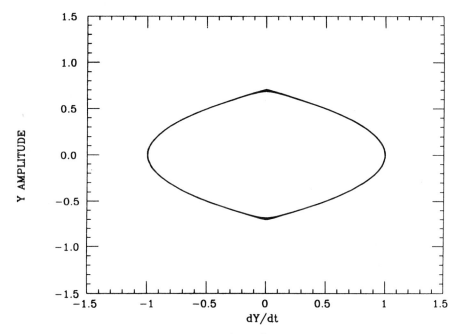

Figure 20. Phase Space 1.

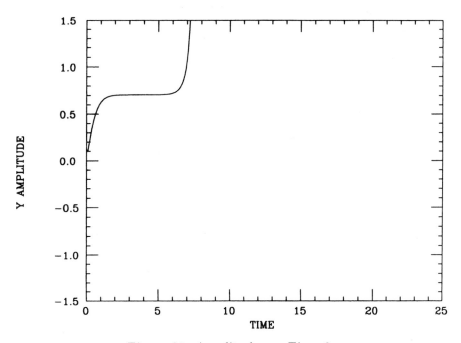

Figure 21. Amplitude vs. Time 2.

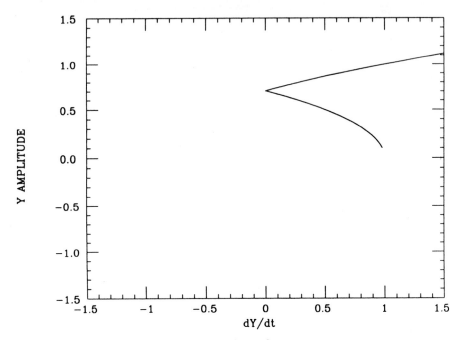

Figure 22. Phase Space 2.

separatrix. This represents a dynamical behavior that can account for particle loss. As a further illustration, a Gaussian random force has been applied to this equation and the results on stable motion can be seen in Figures 23 and 24 for the amplitude and phase space behavior respectively. Here one clearly sees how a random force can cause the particle to cross the separatrix and become lost. In addition, Figure 23 shows that the random force has an effect on the magnitude of the amplitude, and Figure 24 shows chaotic motion in phase space before the separatrix is crossed. It is possible to simulate the cross coupling behavior similar to that seen in the long term tracking results when the system of nonlinear differential equations for the x and y amplitudes are solved numerically and when this system is cross coupled with terms of the form xy, x^2y, y^2x, etc., which occur in multipole expansions. If both the x and y systems are started with the same initial conditions, one can see how a simple xy coupling introduced in the equation for x can drive the y amplitude. These results are represented in Figures 25 and 26 where $y1$ and $y2$ denote the x and y amplitudes, respectively. In this example there is no random force, and the nonzero parameters have the values $\omega = 2$, $d_3 = 0.5$, $d_4 = 7.000791$, and $f_{11} = 1$.

As a final remark, it is worth noting that behavior where the amplitude makes a transition between different values after a large number of turns is a characteristic of nonlinear driven differential equations. This results when the amplitude becomes a multivalued function of the driving frequency. This can result in a transition in the value of the amplitude. An example of this behavior is seen in Figures 2, 3, 27, and 28. The first two figures result from SSCTRK,

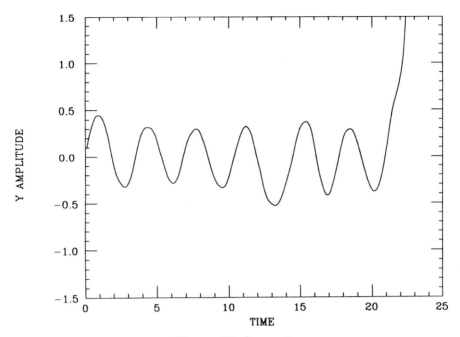

Figure 23. Space 3.

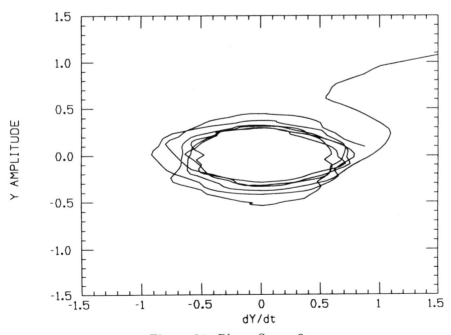

Figure 24. Phase Space 3.

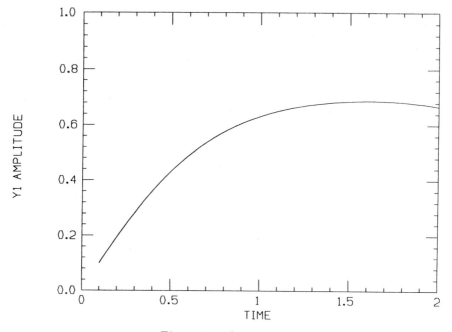

Figure 25. Space 4A.

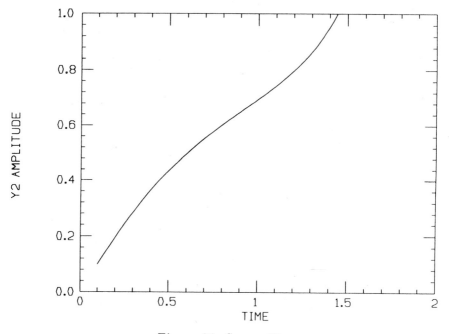

Figure 26. Space 4B.

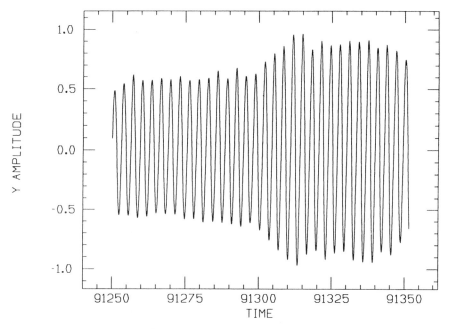

Figure 27. Space 7.FOR.

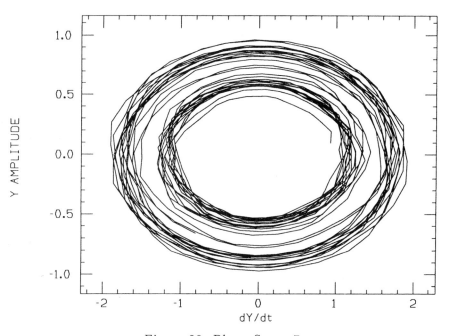

Figure 28. Phase Space 7.

and the last two figures are found from the first differential equation of the system already considered. The nonzero parameters have the values $\omega = 2$, and $d_4 = 0.1$, and the random force has been replaced with the driving force $0.909 \sin(4t + 0.08t^2)$. The amplitude transition seen in Figure 27 occurs after a large number of cycles, and the phase space graph in Figure 28 shows chaotic motion and transition regions. A further discussion of this phenomenon can be found in Refs. 3 and 4.

Although the system of differential equations used to produce the above results is much simpler than the tracking code SSCTRK, it does provide a qualitative understanding of the origin of the dynamical behavior seen in the long term tracking studies. Furthermore, the inclusion of random multipole errors in the tracking code does appear to have the effect of producing random forces characterized by mixtures of harmonics which can drive particles across a separatrix or produce transitions in the magnitude of oscillation amplitudes.

References

1. T. Garavaglia, S. K. Kauffmann, and R. Stiening, Collider Ring Particle Loss Tracking With SSCTRK, in: "International Industrialization for the Superconducting Super Collider (IISSC) Proceedings" (1990).
2. H. Mais, G. Ripken, and A. Wrulich, Particle Tracking, in: "CERN Accelerator School CERN 87-03," S. Turner, ed., CERN, Geneva (1987).
3. J. M. Jowett, M. Month, and S. Turner, Nonlinear Dynamics Aspects of Particle Accelerators, in: "Lecture Notes in Physics 247," Springer-Verlag, Heidelberg (1986).
4. E. A. Jackson, "Perspectives of Nonlinear Dynamics," Cambridge University Press, Cambridge (1986).

APPLICATION OF THE SSCTRK NUMERICAL SIMULATION

PROGRAM TO THE EVALUATION OF THE SSC MAGNET APERTURE

T. Garavaglia, K. Kauffmann, and R. Stiening

Superconducting Super Collider Laboratory[*]
2550 Beckleymeade Avenue
Dallas, Texas 75237

SUMMARY: The SSCTRK numerical simulation code has been used to estimate the benefit of increasing the SSC dipole aperture from 4 to 5 cm. The increase in maximum amplitude of stable betatron oscillations depends on the level to which systematic errors have been corrected. Two cases have been studied, a highly corrected ring and a ring with limited corrections. The maximum stable amplitude increase is approximately a factor of 1.3 in the case of the highly corrected ring and is approximately a factor of 1.6 in the case of the ring with limited systematic corrections. The aperture comparison has been made at 10^5 revolutions. Magnetic error assumptions are described in detail and a new table of errors suggested for future simulations is given.

INTRODUCTION

This report describes the application of the SSCTRK numerical simulation code developed by Ritson to the design of the Collider rings. Many principal features of the rings (magnet aperture, cell length, and injection energy) are determined by the requirement that the beam survive the long (40 minute) injection period without significant loss or emittance dilution. It is the injection process, that sets the design requirements, because at injection, both the beam width and the relative magnetic field errors in the dipoles are largest. Numerical simulation is the only method presently available for establishing the long term behavior of a beam.

SSCTRK became available as a tool in the Summer of 1989. At that time, analytic estimates of tune shifts caused by persistent current multipoles indicated that the cell length should be reduced and the injection energy increased. For this reason, SSCTRK was used mainly to study the related issues of magnet aperture and the levels of tuning required to establish stable beams with suitable margins. All of the studies reported here have been made with the assumption of the shorter, 90-m long, half-cell and the higher, 2 TeV, injection energy. A number of important considerations are not addressed by the SSCTRK simulation. The effects of time dependent persistent current multipoles, the requirement for careful control of superconductor magnetization, and the possible need for distributed correction of sextupole errors (if only for dipole alignment reasons) are examples of such considerations. Broadly speaking, the shorter cell length and higher injection energy removed these considerations from being first order concerns. SSCTRK has been used primarily to evaluate the benefit of increasing the magnet aperture from 4 to 5 cm and to establish the levels of tuning required to achieve the long term stability of betatron oscillations at a suitable amplitude.

[*]Operated by the Universities Research Association, Inc., for the U.S. Department of Energy under Contract No. DE-AC02-89ER40486.

At the time this work was carried out, 4 cm diameter dipoles had been produced in a variety of lengths. These dipoles were constructed in the course of studies undertaken to improve high field training and quench behavior. No production series had been made for the purpose of establishing distributions of random field errors nor had iteration of the magnet cross section to reduce allowed multipole errors been performed. For this reason, the magnetic field errors used in the simulations reported here do not use input data derived from measurements on SSC dipoles, but rely instead on the scaling of the magnetic measurements of dipoles produced in quantity for other accelerators.

MAGNETIC FIELD ERRORS (DIPOLE MAGNETS)

For the purposes of numerical simulation, the magnetic field errors present in an ensemble of magnets are described in a simplified form. For each magnet, a multipole expansion of the magnetic field is made about a reference beam trajectory, in the SSC case, the geometrical center of the magnet coils. The particular form of the multipole expansion used in SSC literature is as follows:

$$i B_x + B_y = B_o \Sigma \ (i A_n + B_n) \ (x + iy)^n$$

In this expression B_o is the central field of the dipole. The B_n terms are called regular multipoles. Regular multipoles cause tune shifts. The A_n terms are called skew multipoles. Skew multipoles couple horizontal and vertical modes of betatron oscillation.

In the simplified description used here, all information about the ensemble of magnets is condensed down to an average or "systematic" value for each multipole, $<A_n>$ or $<B_n>$ (the average is taken over all of the magnets in a ring), and a root mean square deviation σA_n or σB_n from the average. The rms deviation is frequently called a "random value" for a multipole. In making a numerical simulation, random numbers are chosen from Gaussian distributions having the appropriate average and rms values. The random numbers are appropriately assigned to each multipole in each magnet or block of magnets in the machine to be simulated. In order to avoid particularly fortunate or unfortunate combinations of errors from the random number generator, several machines are simulated, each from a different seed of the random number generator. In the work described here, the number of machines that have been simulated has been either eight or sixteen.

This description of an ensemble of magnets contains a potentially significant error because it has been assumed that the various multipole coefficients describing a magnet are uncorrelated. This assumption is incorrect. Future work should address this question.

If all magnets are measured, it is possible that ways of choosing the location of a magnet in the ring on the basis of its measurements may help to reduce the effects of random multipoles. In the work described here, magnet location has been chosen to reduce the effect of random regular sextupole $\sigma B2$. The rules for making use of magnetic measurement data are not established. It is possible that before the installation of magnets in the SSC tunnel begins, progress will have been made on a magnet placement strategy.

Random Multipoles

The value of a multipole varies from magnet to magnet. There are a number of effects that contribute to the variation. These include variations in the persistent current magnetization of the superconducting cables (M. A. Green, SSC-N-377), variations in the permeability of the collar and the yoke, and variations in the placement of the current carrying cable. Such placement errors are frequently called "geometrical" errors. The SSC injection energy has been chosen high enough (2 TeV) so that placement errors are dominant. If reasonable care is taken in the selection of materials the other errors should not be significant. Only placement errors have been considered in this report. Typical errors in the location of SSC magnet conductors are in the ±0.002 inch range.

60

In order to obtain expected multipole distributions for the 4 cm SSC dipole it is necessary to scale results from the production measurements of other accelerator dipoles. A number of different scaling rules have been described in the Preliminary Report of the Magnetic-Errors Working Group of the SSC Aperture Workshop (SSC-7). The authors of SSC-7 used the average of three different methods to estimate errors. Their recommended set of 4 cm magnet errors (listed on page 66 of SSC-7) has been used as input for SSCTRK simulations with one exception. The estimated value for random sextupole has been reduced from 2.01 units to 0.4 units. This was done to maintain consistency with previous Central Design Group recommendations. The effect of random sextupole can be reduced by arranging magnets in the collider rings on the basis of measurements of the sextupole field error in each magnet. The reduced value of random sextupole simulates the effect of this magnet arrangement procedure. Several investigations have been made using SSCTRK which suggest that there may be little benefit from the reduction of random sextupole by magnet arrangement procedures. These investigations will be described later in this report.

Multipole Distributions for a 5 cm Dipole Have Been Obtained by Scaling 4 cm Values

A somewhat simplified analysis suggests that bounds on the scaling law can be set by considering the following two plausible assumptions. The first is that the size of the construction errors is independent of the scale of the magnet. This assumption leads to the scaling law that the nth multipole varies as:

$$(B_n', A_n') = (B_n, A_n) R_s^{-(n+1)}$$

In this expression R_s is the ratio of the old aperture to the new aperture. An equally plausible assumption is that the construction errors are proportional to the size of the magnet. This second assumption leads to the scaling law that the nth multipole varies as

$$(B_n', A_n') = (B_n, A_n) R_s^{-n}$$

The multipole distributions for a 5 cm aperture dipole have been obtained by scaling the suggested values from SSC-7 using a scaling law that is intermediate between the two cases given previously.

$$(B_n', A_n') = (B_n, A_n) R_s^{-(n+1/2)}$$

The values of all multipoles with $n > 6$ were taken to be zero to speed the computation.

It is useful to compare these input data with actual magnetic measurements of production magnets. The data for the Tevatron have been published (H. Edwards, Proceedings of the 12th International Conference on High Energy Accelerators, page 7). Recent HERA data on the first 156 dipoles measured have also been made available (Peter von Handel, private communication). For the purpose of scaling to the SSC 4 cm aperture, the same $n+1/2$ scaling law used previously has been adopted:

$$(B_n', A_n') = (B_n, A_n) R_s^{-(n+1/2)}$$

The above table shows that the random skew quadrupole ($\sigma A1$) in the HERA dipole is much larger than the value used in the SSCTRK simulation. Skew quadrupole comes from an up-down asymmetry in the dipole. It is usually associated with differences in the dimensions of the upper coils and the lower coils. In the Tevatron dipole the iron yoke was not adjacent to the collared coil. The coil position could be adjusted to null the quadrupole components. The SSC dipole and the HERA dipole do not have the provision for such an adjustment. The manufacturers of the HERA dipoles attempted to reduce the random skew quadrupole by measurement and re-shimming of the coils. The value given in the table above can not easily be reduced. (Peter Schmuser, private communication) In future simulations the value of $\sigma A1$ should be increased from 0.7 units to 1.75 units.

Table 1.   4 cm and 5 cm dipole random multipole distributions used in SSCTRK simulations

	4 cm dipole	5 cm dipole	Notes
A1 B1	0.7 units 0.7 units	0.49 units 0.49 units	Probably too small
A2 B2	0.62 units 0.4 units	0.35 units 0.23 units	Obtained by sorting
A3 B3	0.69 units 0.34 units	0.32 units 0.16 units	
A4 B4	0.14 units 0.59 units	0.051 units 0.22 units	
A5 B5	0.16 units 0.059 units	0.047 units 0.017 units	
A6 B6	0.034 units 0.075 units	0.008 units 0.018 units	

Table 2.   HERA and Tevatron Magnetic Measurements Scaled to the SSC 4 cm Aperture (Random Multipoles)

Multipole	HERA	Tevatron	SSCTRK Simulation
A1	1.74	0.80	0.70
B1	0.68	0.57	0.70
A2	0.35	0.96	0.62
B2	2.00	2.80	2.01 Sorted to 0.4
A3	0.52	0.92	0.69
B3	0.16	0.49	0.34
A4	0.11	0.23	0.14
B4	0.36	0.58	0.59
A5	0.07	not available	0.16
B5	0.04	not available	0.06
A6	0.03	not available	0.034
B6	0.07	0.22	0.075

Note: The Tevatron measurements were made at 660 amps (injection) whereas the HERA measurements were made at 5000 amps (high energy).

Table 3.   4 cm SSC Dipole Persistent Current Systematic Multipoles

Multipole	1 TeV	2 TeV
B0	−15.06 units	−6.00 units
B2	−8.44 units	−3.29 units
B4	0.69 units	0.23 units
B6	−0.16 units	−0.06 units
B8	0.06 units	0.02 units

Systematic Multipoles

The average or systematic value of a multipole may be non-zero for a variety of reasons. All multipoles will have non-zero averages because a collider ring contains a finite number of magnets.

The rms width of this average is:

$$\sigma\!<\!A_n, B_n\!> \ = \frac{\sigma(A_n, B_n)}{\sqrt{N}}$$

In this expression N is the number of magnets in a collider ring. The collider rings contain 4800 magnets each. Random multipole sigmas are typically 0.5 units. The statistical value of systematic multipoles will be typically less than 0.01 units. These statistical systematics do not have a significant influence on the aperture.

Multipoles allowed by dipole symmetry, B2, B4, B6, etc. may have large values either caused by superconductor magnetization or by errors in the cross section. Superconductor magnetization multipoles have been calculated by Green for the C358 cross section (Progress on Magnetization Calculations for the SSC Magnets, M. A. Green, presentation to the MSIM, 17 January 1990). The values for 1 and 2-TeV field levels of the 4 cm SSC dipole are given in Table 3. The superconducting filament diameter used to obtain these results is 6 microns.

The scaling of persistent current multipoles to other magnet apertures depends on the way the magnet cross section is changed. If all components, including the dimensions of the superconducting coils are scaled together, persistent current multipoles scale as R_s^{-n}. It is assumed that the dimensions of the superconducting filament have not been changed when the magnet size is changed.

Transport current in the superconducting cable will generate allowed multipoles, the dipole for example, in a manner which depends on the location of the conductor. The magnet designer will attempt to make the B0, or dipole multipole, as large as possible while suppressing other allowed multipoles.

The SSC 358 dipole coil assembly contains many copper wedges. For this reason there are enough dimensions that can be adjusted so that all systematic allowed multipoles below B14 can be made to have, in principle at least, very small values (G. Morgan, CDG-N-342). The Tevatron magnet, in contrast, does not have wedges. The only adjustable dimensions in the Tevatron magnet are the pole angles of the inner and outer coils. These angles are chosen to null the sextupole and decapole. The Tevatron has large values of higher multipoles (see Table 5).

The cross section of a dipole coil as manufactured will differ from expectations based on the design. This is because actual dimensions differ from design dimensions. For example, at the present time the average value of B4 (other than that due to persistent current) in the 4 cm SSC dipole is 0.6 units. G. Morgan has given a prescription for reducing this value (G. Morgan, SSC-MD-241). The ends of the magnet coil may also contribute to allowed multipoles. It is reasonable to assume that the coil geometry contribution to the values of allowed multipoles, including the effects of ends, can be reduced to a value small in comparison with persistent current value.

Unallowed systematic multipoles occur when the magnet symmetry is broken. Symmetry is broken, for example, by buswork, coil ends, coil sagitta, and gravity. For example, a small length of unpaired conductor located at a distance of R_B cm from the center of the dipole produces a systematic multipole distribution as follows:

$$(B_n, A_n) = \left(\frac{l_B}{l_M}\right)\left(\frac{2}{R_B \text{ (meters)}}\right)\left(\frac{R_{ref}}{R_B}\right)^n$$

Table 4. Systematic multipoles in a 15-meter-long magnet caused by 5 cm of unpaired conductor located 2.5 cm from the dipole axis

B1, A1	0.107 units
B2, A2	0.043 units
B3, A3	0.017 units
B4, A4	0.007 units

In this expression, l_B is the length of unpaired buswork, l_M is the length of the magnet, and R_{ref} is the reference radius for magnetic measurements. The transfer function (1 Tesla/kA) for the SSC 4 cm dipole has been assumed. As an example, Table 4 shows the effect on a 15-meter-long magnet of 5 cm of unpaired conductor located at a distance of 2.5 cm from the axis.

The systematic multipoles which were used for the input to numerical simulations described in this report were chosen to be the persistent current value, if allowed, and 0.05 units, if not allowed. At the time when these values were selected, an attempt was being made to reduce the systematic decapole caused by a magnet cross section error. The value of 0.05 units was considered to be a difficult to achieve goal at that time. The signs of the unallowed systematics were all chosen to be the same to be consistent with the model of unpaired conductor. This assumption is not as important as it seems because even order systematics act only through momentum feed down. The tune shifts caused by momentum feed-down terms change sign every half synchrotron period.

Table 5 gives a comparison between the SSCTRK simulation assumptions and the measured data from HERA and Tevatron magnets. The HERA data have been taken at a high field level where persistent current multipoles are small.

The data of Table 5 suggest that the SSCTRK simulation values of high order unallowed multipoles may have been overestimated and that the values of low order multipoles may have been underestimated. The calculation given previously of the multipole errors caused by unpaired buswork support this conclusion..

Table 5. HERA and Tevatron Magnetic Measurements Scaled to the SSC 4 cm Aperture (Systematic Multipoles)

Multipole	HERA	Tevatron	SSCTRK Simulation
<A1>	−0.293	−0.007	0.00 (tuned to 0)
<B1>	−0.038	0.044	0.00 (tuned to 0)
<A2>	−0.164	0.060	0.05
<B2>	0.175	−2.567	−3.00 (tuned to 0)
<A3>	0.093	−0.016	0.05
<B3>	0.068	−0.093	0.05
<A4>	0.029	−0.012	0.05
<B4>	0.274	0.036	0.05 (tuned to 0.2)
<A5>	−0.017	not available	0.05
<B5>	−0.014	not available	0.05
<A6>	−0.002	not available	0.05
<B6>	0.047	0.887	−0.07

Note: The Tevatron measurements were made at 660 amps (injection) whereas the HERA measurements were made at 5000 amps (high energy).

SUGGESTED MULTIPOLES FOR FUTURE SIMULATIONS

A revised set of multipole expectations for 4 and 5 cm magnets of the SSC design is given in Table 6. The values for the 4 cm magnet were chosen in the following way:

1. Random multipole rms widths are those taken from SSC-7, page 66 except for the skew quadrupole (A1) which has been increased by a factor of 2.5 on account of a now different method of magnet construction.

2. The values in Table 6 do not assume any kind of magnet sorting. Future simulations should employ sorting only if it is established that it is beneficial.

3. Systematic allowed multipoles are equal to persistent current values. It has been assumed that iteration of the magnet cross section will reduce geometrical errors to levels that are considerably smaller than the persistent current values.

4. Systematic unallowed multipoles have been estimated by using Tevatron and HERA data in combination with the unpaired conductor model. A value for the systematic octupole of 0.06 units has been chosen. Other unallowed multipoles are obtained from the octupole value by assuming that the error occurs at a radius of 2.5 cm and scales according to the unpaired conductor model.

5. Table 6 does not take into account possible correction schemes. For example, the systematic regular sextupole <B2> obviously needs to be highly corrected.

Values for a 5 cm aperture dipole are obtained by scaling the 4 cm values in the following way:

1. Random multipoles scale as $(4/5)^{n+1/2}$.

2. Persistent current systematic multipoles scale as $(4/5)^{n}$.

3. Unallowed systematic multipoles scale as $(4/5)^{n}$.

MAGNET FIELD ERRORS (QUADRUPOLE MAGNETS)

The SSCTRK simulation results given in this report describe a collider ring constructed with error-free quadrupoles. Although the effects of quadrupole errors are estimated to be small, future simulation should include such errors. There is one systematic error that should be considered, <B5> caused by persistent currents. There are two random errors that should be considered, a quadrupole-to-quadrupole variation in length, and a random roll angle.

The persistent current <B5> has been estimated by M. A. Green (private communication) for an early version of the SSC 4 cm quadrupole. The SSCTRK code uses as input two parameters QB5F and QB5D, one for each of the two families of quadrupoles. These parameters are equal and of opposite sign since the same quadrupoles are used for both families. QB5F and QB5D are defined as follows:

$$QB5X = 10^4 \left(\frac{Lq}{Lb}\right) \frac{B_5 \, (1 \text{ cm})}{B_0 \, (\text{Dipole})} \, ,$$

Table 6.   Suggested Multipoles for Future Simulation of 2 TeV Injection

	4 cm	5 cm
$<A1>$	0.37	0.30
$\sigma A1$	1.75	1.25
$<B1>$	0.37	0.30
$\sigma B1$	0.70	0.50
$<A2>$	0.15	0.10
$\sigma A2$	0.62	0.35
$<B2>$	−3.29	−2.11
$\sigma B2$	2.01	1.15
$<A3>$	0.06	0.031
$\sigma A3$	0.69	0.32
$<B3>$	0.06	0.031
$\sigma B3$	0.34	0.16
$<A4>$	0.024	0.01
$\sigma A4$	0.14	0.05
$<B4>$	0.23	0.09
$\sigma B4$	0.59	0.22
$<A5>$	0.01	0.004
$\sigma A5$	0.16	0.047
$<B5>$	0.01	0.004
$\sigma B5$	0.059	0.017
$<A6>$	0.004	0.0012
$\sigma A6$	0.034	0.008
$<B6>$	−0.060	−0.016
$\sigma B6$	0.075	0.018

where Lq is the length of the quad and Lb is the total length of the bending magnets in a half cell. B5(1 cm) is the magnetic field due to persistent currents in the quadrupole at a radius of 1 cm. Green estimates that B5 (1 cm)/B1 (1 cm) $= 1.36 \times 10^{-4}$. This leads to the following values:

$$QB5F = -0.025$$

$$QB5D = +0.025$$

A rough estimate on the rms value of the quadrupole-to-quadrupole variation in gradient length product is 0.001. The rms value of the random roll angle is 0.0005 radians.

At the present time a prototype quadrupole is being constructed. Better estimates of the quadrupole field errors can be expected when the prototypes are measured.

NUMERICAL SIMULATION-OTHER INPUT CONDITIONS

Operating Tunes

The tune values chosen for all of the simulation results presented here are $Qx = XXX.425$, $Qy = XXX.410$. These values were chosen on the basis of Main Ring and Tevatron experience. This operating point is clear of all resonances below 9th order (see Figure 1).

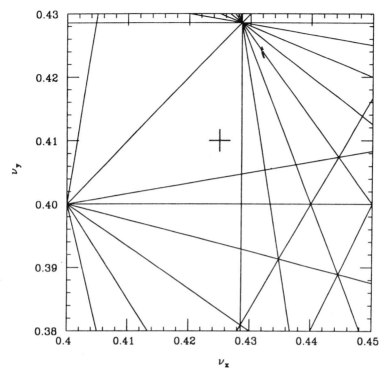

Figure 1. The cross marks the operating point chosen for SSCTRK simulation. All resonance lines below 9th order are shown.

Lattice Parameters

The half cell length is 90 meters and the betatron phase advance/cell is 95 degrees in all the simulations described in this report. The phase advance/cell differs from the Central Design Group recommendation of 90 degrees because it was thought that there might be a resonance between the beam and the lattice at 90 degrees/cell. Thus far simulations made to investigate this possibility have not shown a significant effect.

Alignment Errors

All magnetic elements have been assumed to be aligned with respect to the zero momentum error closed orbit of the beam with an rms deviation of 1.0 mm.

Radio Frequency Parameters and Momentum Spread

The RF frequency and voltage have been chosen to be 375 MHz and 20 MV. The maximum fractional momentum error of the tracked protons has been set at 5×10^{-4}. The corresponding value for the maximum longitudinal position error of the protons is 16.4 cm.

The momentum error is much larger than the minimum that might in principle be present in a low current bunch injected into the collider rings. Past experience with proton rings suggests that longitudinal stability of high current beams is enhanced by increasing the bunch length. Since the bunch length that accompanies a 5×10^{-4} fractional momentum error even at the injection energy is much smaller than the low beta value at the interaction point and since the beam size due to dispersion, $\eta(\delta P/P)$, is only 1 mm this choice of momentum error is reasonable.

Power Supply Ripple

These simulations are made with the assumption that power supply (or tune) ripple has been reduced to the point where it does not affect the results. Therefore, no power supply ripple has been used in these simulations.

RESULTS

The beam will need to circulate without loss for approximately 40 minutes or 10 revolutions. Ideally all tracking should be done for this number of revolutions. Tracking for 10^7 revolutions takes approximately 100 hours of CPU time on the Cray 2 or the NEC SX2, the two computers which were used for this work. One hour runs, 10^5 revolutions, have been used to quickly evaluate various error configurations, and longer runs have been made for interesting cases.

The injection conditions at the start of each simulation are as follows. Particles are injected such that their betatron amplitude in the two transverse directions is equal when measured at the maximum value of the appropriate beta function. Betatron amplitude is measured with respect to the closed orbit. All particles are injected with a momentum error of $(\delta P)/P = 5 \times 10^{-4}$.

In a typical run, 16 different machines are simulated simultaneously. (The machines differ only in the seed of the random number generator which is used to select the multipole coefficients for each magnet.) For each machine, particles are started with four different betatron amplitudes, typically 10, 10.33, 10.66, and 11 mm. Particles are circulated until they are lost. In these simulations a particle is considered to be lost if its distance from the closed orbit exceeds 21 mm. If a particle is lost, another particle is launched at the location of the loss at an initial amplitude which is typically 1 mm smaller than the initial amplitude of the particle that was lost. The results are given as a plot which shows the number of revolutions before loss as a function of initial betatron amplitude. For the purpose of estimating safety margins in the SSC design it should be noted that the rms betatron amplitude for the 2-TeV specification SSC emittance is 0.375 mm.

The SSCTRK code does not incorporate interaction regions. Interaction region beta values at injection are modest. It is not expected that this omission will significantly alter the results presented here.

All large accelerators employ magnetic correction systems to compensate the effects of errors. Two cases will be described in this report. The cases differ in the extent to which compensation is used to enlarge the aperture. The first case illustrates a highly corrected accelerator. In this case all systematic field errors have been corrected. The second case illustrates a moderately corrected accelerator. This case incorporates corrections similar (in concept, though not in detail) to those of the Fermilab Main Ring. These corrections are typical of past accelerator practice.

Details of diagnostic and correction systems are not contained within the SSCTRK simulation with one exception, the natural chromaticity sextupoles. The natural chromaticity of the quadrupole lattice is compensated by sextupoles located adjacent to the quadrupoles. All other errors are corrected essentially at the point of origin. This is accomplished in two different ways. The first way of correcting an error is simply to enter a reduced or zero value in the SSCTRK input data set. The second way is accomplished within the SSCTRK program. SSCTRK evaluates the strength of a multipole by making an appropriate computation using the multipole error in each magnet. The correction is made by subtracting a computed value from the value of each magnet.

The Highly Corrected Machine

Table 7 shows the corrections that have been made in this case.

Table 7.   The Highly Corrected Machine

<A1>:	0
<A2>:	Statistical value (approximately 0.005 units)
<A3>:	Statistical value
<A4>:	Statistical value
<A5>:	Statistical value
<A6>:	Statistical value
<B1>:	0
<B2>:	0
<B3>:	Statistical value
<B4>:	Statistical value
<B5>:	Statistical value
<B6>:	Statistical value
Chromaticity:	0
Quadrupole errors:	none

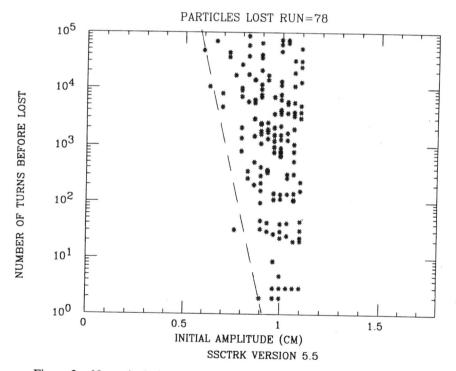

Figure 2. Numerical simulation of the collider ring with 4 cm dipole errors. This simulation is of the highly corrected ring. The dotted line is drawn for comparison with Figure 4, the 5 cm dipole error highly corrected ring.

Figure 2 shows the simulation of 10^5 turns of the highly corrected machine made with 4 cm diameter bore dipoles. In this report the left edge of the distribution of points is called the aperture. Figure 2 shows that the aperture for a few revolutions is 9 mm and that the aperture for 10^5 revolutions is 6 mm. Since the rms transverse size of an SSC spec emittance beam is 0.375 mm, this machine has ample aperture for 10^5 revolutions. This machine has not been simulated beyond 10^5 revolutions.

We have made a simulation identical to that of Figure 2 in all respects except for the value of random sextupole $\sigma B2$. Figure 3 shows the effect of increasing the value of the random sextupole a factor of five from 0.4 to 2.01 units. There is not a significant difference between the results of Figure 2 and those of Figure 3.

When the aperture of the dipole is increased from 4 to 5 cm, the rms values of random multipoles decrease according to the $n+1/2$ law discussed earlier. It is plausible that the amplitude at which betatron oscillations are stable would increase to the point where the field errors in the 5 cm magnet are identical to the field errors in the 4 cm magnet. This occurs according to the following ratio for the nth multipole:

$$R_5 = R_4 \left(\frac{5}{4}\right)^{1+(1/2n)}$$

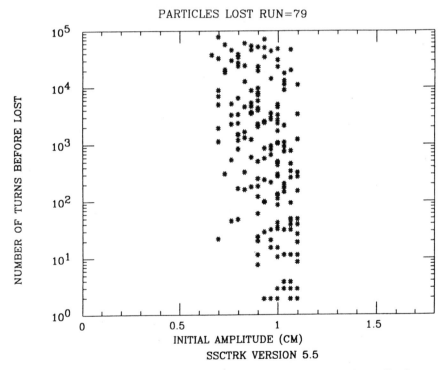

Figure 3. Numerical simulation of the collider ring with 4 cm dipole errors with the magnets not sorted on the basis of the value of the regular sextupole (B2). This simulation is of the highly corrected ring. It should be compared with Figure 2 where random value of regular sextupole has been reduced by a factor of 5 through sorting.

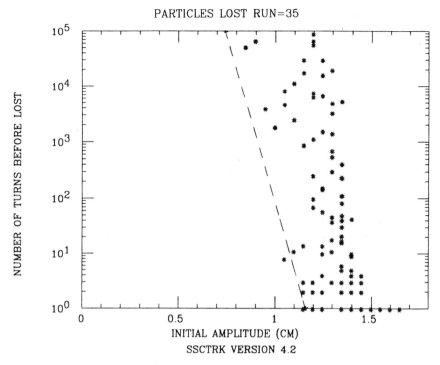

PARTICLES LOST RUN=35

Figure 4.   Numerical simulation of the collider ring with 5 cm dipole
errors. This simulation is of the highly corrected ring. The
dotted line is drawn at an amplitude 1.28 times larger than the
dotted line in Figure 2 to illustrate the scaling of stable betatron
aperture.

Table 8.   The Plausibly Corrected Machine

<A1>:	0
<A2>:	0.05 + Statistical value
<A3>:	0.05 + Statistical value
<A4>:	0.05 + Statistical value
<A5>:	0.05 + Statistical value
<A6>:	0.05 + Statistical value
<B1>:	0
<B2>:	See chromaticity error below
<B3>:	0.05 + Statistical value
<B4>:	0.05 + Statistical value
<B5>:	0.05 + Statistical value
<B6>:	−0.07 + Statistical value
X chromaticity:	+5 units
Y chromaticity:	+5 units
Quadrupole errors:	none

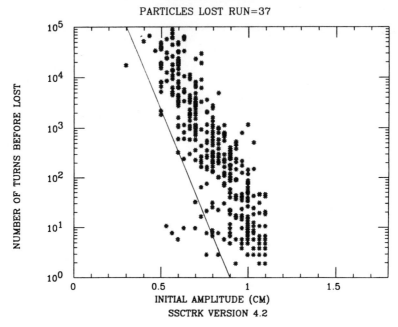

Figure 5. Numerical simulation of the collider ring with 4 cm dipole errors. This simulation is of the plausibly corrected ring. The line is drawn for comparison with Figure 7, the 5 cm dipole error plausibly corrected ring.

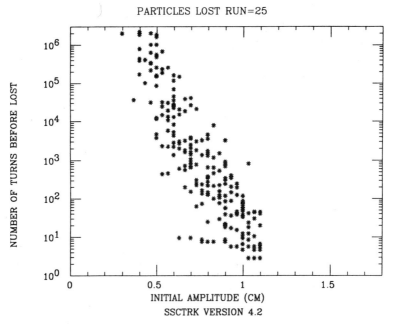

Figure 6. Numerical simulation of the collider ring with 4 cm dipole errors. This simulation is of the plausibly corrected ring. This simulation differs from that of Figure 5 in that only eight different rings are simulated. This simulation has been extended out to 3 million revolutions.

In this expression R_4 is some radius in the 4 cm magnet. R_5 is the radius in the 5 cm magnet at which the field error due to the nth multipole is equal to that in the 4 cm magnet at a radius R_4. If the betatron aperture is dominated by very high order multipoles the increase in aperture is 5/4. Since the aperture appears not to be influenced by random sextupole, the octupole ($n=3$) can be taken as the lowest multipole contributing to the aperture. The aperture should then scale in the ratio of 1.3 if it is octupole dominated or 1.25 if dominated by very high multipoles. This conclusion is supported by a simulation made with 5 cm magnet parameters. Figure 4 shows this result. The aperture is 8.5 mm at 100,000 revolutions. The dotted lines in Figures 2 and 4 are drawn according to a 1.28 radius ratio.

The Plausibly Corrected Machine

Table 8 shows the corrections that have been made in this case. These corrections are called plausible because it is likely that they could be made early in the commissioning of the machine.

The plausibly tuned machine differs from the highly corrected machine only in the values chosen for systematic multipoles. Figure 5 shows the results of 10^5 revolution tracking of the plausibly tuned machine. The presence of small, uncorrected, systematic multipoles reduces the aperture for stable betatron motion considerably. A longer term tracking (13 minutes of real time) of the plausibly tuned machine is shown in Figure 6. This tracking has been made on eight rather than sixteen machines in order to speed the computation.

The effect of reducing the random sextupole by magnet placement has been investigated by increasing the value of random sextupole from 0.4 units to 2.01 units. Figure 7 shows a simulation made with $\sigma B2 = 2.01$ units. It should be compared with Figure 5. There is practically no effect on the stable betatron aperture.

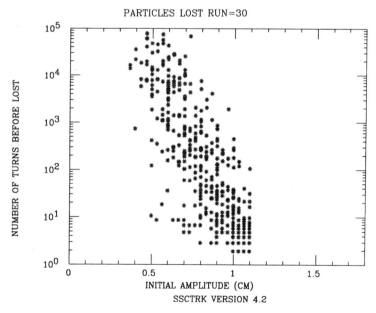

Figure 7. Numerical simulation of the collider ring with 4 cm dipole errors. This simulation is of the plausibly corrected ring. In this simulation the value of random sextupole has been increased by a factor of 5 to 2.01 units. It should be compared with Figure 5 where the value of random sextupole was 0.4 units.

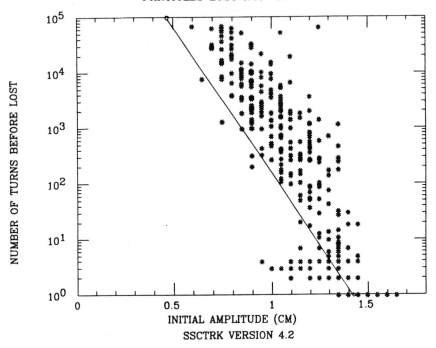

SSCTRK VERSION 4.2

Figure 8. Numerical simulation of the collider ring with 5 cm dipole errors. This simulation is of the plausibly corrected ring. The line is drawn at an amplitude 1.6 times larger than the line in Figure 5 to illustrate the scaling of stable betatron aperture.

When the aperture of the dipole is increased from 4 to 5 cm both the random and the systematic multipoles decrease. It has been assumed that the random multipoles decrease in the ratio $(4/5)^{n+1/2}$ and that the systematic multipoles decrease in the ratio $(4/5)^{n}$. Figure 8 shows the results of simulating a plausibly tuned machine made with 5 cm aperture dipoles. The fractional increase in stable betatron amplitude can be estimated by comparing Figure 8 (5 cm) with Figure 5 (4 cm). The lines in the figures are drawn in an amplitude ratio of 1.6.

The benefit of increasing the magnet aperture is clearly larger for the plausibly tuned machine than for the highly corrected machine. The ratio 1.6 suggests that the maximum amplitude of stable betatron motion is determined by the value of the product of systematic and random magnet errors. $(1.6 = 1.28 \times 1.25)$ It is clear that this scaling law cannot be valid for arbitrary increases in aperture because the maximum amplitude for stable betatron motion would increase beyond the magnet coils, an absurd notion. The large increase in stable betatron amplitude is a peculiarity of starting with a 4 cm diameter dipole aperture.

A future report will give simulation results based on a more complete and improved set of magnet errors. There is also clear evidence that splitting the tune values by one unit in order to avoid injection mismatch caused by feed-down systematic skew sextupole would be a desirable improvement. This improvement could be incorporated in the plausibly tuned machine with only a minor increase in the level of correction (a first harmonic skew quadrupole correction rather than a zeroth harmonic correction).

APPENDIX A

This appendix contains the job cards used for the SSCTRK simulations discussed in the report.

> Run 25: Same errors as run 37. Not shown.
>
> Run 30: A plausibly corrected 4 cm machine with five times the rms B2 value as run 37.

```
&inputs
      resume=0,
      run=30,
      turns=100000,
      every=5000,
      aplim=1.5d+0,
      e0=2.d+12,
      mudeg=95.d+0,
      mpoles=6,
      bends=6.0d+0,
      dx0=(1.d-1,1.d-1),
      oksmer=-1.,
      enoise=0.d-2,
      inx0=( 10.d-1, 10.d-1),
      maxdx0=(1.d-1,1.d-1),
      th0=(0.,0.),
      z0=0.d+0,

      dnux=4.25d-1,
      dnuy=4.10d-1,
      seed0=-1,
      sedcas=16,
      xerrab=10.d-2,
      xerr2=10.d-2,
      asys= 0.d+0,  5.d-2, 5.d-2, 5.d-2, 5.d-2, 5.d-2, 0.d+0, 0.d+0,
      bsys= 0.d+0,-30.d-1, 5.d-2, 5.d-2, 5.d-2,-7.d-2, 0.d+0, 0.d+0,
      asig= 7.d-1, 6.2d-1,6.9d-1,1.4d-1,1.6d-1,3.4d-2, 0.d+0, 0.d+0,
      bsig= 7.d-1, 2.0d-0,3.4d-1,5.9d-1,5.9d-2,7.5d-2, 0.d+0, 0.d+0,
      blks=5,
      blksig=0.d+0,
      avxy=.true.,
      avxy2=.true.,
      avquad=.true.,
      avsext=.true.,
      avoct=.false.,
      avdec=.false.,
      aset= 0.d+0, 3.d+0, 3.d+0, 3.d+0, 3.d+0, 3.d+0, 3.d+0, 3.d+0,
      bset= 0.d+0, 0.d+0, 3.d+0, 3.d+0, 3.d+0, 3.d+0, 3.d+0, 3.d+0,
      ngridx=4,
      ngridp=1,
      maxdp0=5.d-4,

      ramp0=0.d+0,
      rfv0=2.d+7,
      fdbtim=1.d+4,
      m720=0.d-2,
      ml0= 0.d-2,

      bumpel= 0, 0,
      bumpdx=(0.d+0,0.d+0),
      incsf=0.375d-1
      incsd=0.375d-1
```

```
modf=0.d+0,
ampf=0.d+0,

chekin=10,

ncas=64,
liscas=1,2,3,4,5,6,7,8,9,10,11,12,13,14,15,16,17,18,19,20,21,
        22,23,24,25,26,27,28,29,30,31,32,33,34,35,36,37,38,39,
        40,41,42,43,44,45,46,47,48,49,50,51,52,53,54,55,56,57,
        58,59,60,61,62,63,64,
neuff0=2.d+0,
qa5d=0.d+0,
qa5f=0.d+0,
```

COLLIDER RING PARTICLE LOSS TRACKING WITH SSCTRK

T. Garavaglia, S.K. Kauffmann and R. Stiening

Superconducting Super Collider Laboratory[*]
2550 Beckleymeade Ave, MS1047
Dallas, Texas 75237-3946

Abstract: The SSCTRK numerical simulation tracking code has been used to study the benefit of increasing the SSC dipole magnet aperture from 4 to 5 cm. This study has been carried out for both hypothetical highly corrected and plausibly corrected machines, the former having no systematic multipole errors and chromaticity identically zero. The choice of tune values, phase advance per cell, random multipole errors, systematic multipole errors and chromaticity (for the plausibly corrected machines), closed orbit error, the criterion for particle loss, etc. are set forth in detail. Runs of 10^5 turns and 3×10^6 turns are presented together with the approximate dynamic apertures they yield from their particle loss patterns.

This presentation describes the application of D.M. Ritson's numerical simulation particle tracking code SSCTRK to particle loss studies of the main SSC collider rings under injection conditions. The requirement that the beam survive the long (30+ minutes) injection dwell without significant loss is a central consideration in determining such basic collider ring parameters as cell length, injection energy, and magnet aperture. At the time this study was initiated, it appeared probable that the half cell length would be shortened to 90 m and the injection energy raised to 2 TeV. These parameters were incorporated into all the SSCTRK runs made.

The random multipole errors (Table 1) of 4 cm SSC dipoles (arising principally from random variations in the placement of the current carrying cable, often called "geometrical" errors) were put into SSCTRK from the values estimated in detail in the Preliminary Report of the Magnet-Errors Working Group of the SSC Aperture Workshop (SSC-7). (The method used in this document entailed appropriate scaling of results from production measurements of other superconducting accelerator dipoles.) The SSC-7 value for the random regular sextupole error was, however, reduced by a factor of five for input into SSCTRK, on the assumption that the dipoles would be sorted on this random error.

Random multipole errors for conceivable 5 cm SSC dipoles were put into SSCTRK by scaling the 4 cm values according to the formula

---

[*]Operated by the Universities Research Association, Inc., for the U.S. Department of Energy under Contract No. DE-AC02-89ER40486.

$$\sigma(B'_n, A'_n) = \sigma(B_n, A_n) \ (5/4)^{-\left(n+\frac{1}{2}\right)} ,$$

suggested by scaling considerations discussed in SSC-7.

TABLE 1
4 cm and 5 cm dipole random multipole distributions used
in SSCTRK simulations

Standards Deviations	4 cm dipole	5 cm dipole	Notes
$\sigma A1$ $\sigma B1$	0.7 units 0.7 units	0.49 units 0.49 units	
$\sigma A2$ $\sigma B2$	0.62 .units 0.4 units	0.35 units 0.23 units	Obtained by sorting
$\sigma A3$ $\sigma B3$	0.69 units 0.34 units	0.32 units 0.16 units	
$\sigma A4$ $\sigma B4$	0.14 units 0.59 units	0.051 units 0.22 units	
$\sigma A5$ $\sigma B5$	0.16 units 0.059 units	0.047 units 0.017 units	
$\sigma A6$ $\sigma B6$	0.034 units 0.075 units	0.008 units 0.018 units	

The values of all multipoles with $n > 6$ were taken to be zero to speed the computation.

The systematic multipole errors (Table 2) in these dipoles which are allowed by their symmetry are the even regular multipoles, B2, B4, B6, etc. These systematic errors can arise from superconductor persistent current magnetization or systematic variances in the dipole coil cross section from that designed. It is assumed that the latter will be negligible compared with the former, which, for the 4 cm dipoles, have been taken from the values given in "Progress on Magnetization Calculations for the SSC Magnets", presentation to the MSIM, 17 January 1990, by M.A. Green, for a superconducting filament diameter of 6 microns. Furthermore, in SSCTRK, the large (–3 units) systematic B2 value is subject to a correction which effectively eliminates it. The systematic B4 is not input into SSCTRK at its full value of 0.2 units, but at one quarter of that, an estimated reduction one might hope to achieve with preset decapole correction magnets.

Systematic multipole errors not allowed by the symmetry of the dipoles can occur if that symmetry is broken, for example, by unpaired buswork, by coil ends and Sagitta, or gravitational distortion. Reducing such unallowed systematic multipole errors to 0.05 units is considered to be a challenging task (several centimeters of unpaired buswork can already produce comparable values, all of the same sign), so this value was input into SSCTRK for the unallowed systematics.

TABLE 2
4 cm dipole systematic multipole errors
used in SSCTRK simulations

Multipole	SSCTRK Simulation Value	
<A1> <B1>	0.00 units 0.00 units	(tuned to 0) (tuned to 0)
<A2> <B2>	0.05 units −3.00 units	 (effective SSCTRK correction to 0)
<A3> <B3>	0.05 units 0.05 units	
<A4> <B4>	0.05 units 0.05 units	 (tuned from 0.2)
<A5> <B5>	0.05 units 0.05 units	
<A6> <B6>	0.05 units −0.07 units	

Exceptions were made for the systematic B1 and A1, which it was assumed are tuned to zero.

Systemic multipoles for a 5 cm aperture dipole were obtained by using the scaling formula

$$<B'_n, A'_n> = <B_n, A_n> (5/4)^{-n}$$

on the above Table 2.

We consider the systematic multipole values of Table 2 (and its scaling to the 5 cm aperture case) to be plausible for what might reasonably be achieved in the machine in the months after its commissioning. We also consider it plausible that chromaticity can be corrected to 5 units. In the longer term, it might be envisioned that all the systematic multipoles of Table 2 could be compensated by suitable delicate fine tuning of the correction magnets, and the chromaticity corrected to nearly zero. We have accordingly split our SSCTRK runs into two classes, the first being *plausibly corrected machines*, with systematic multipoles corresponding to Table 2, together with x and y chromaticity of +5 units, and the second being *highly corrected machines*, with all systematic multipoles set to zero (with the exception of <B2>, which is anyway effectively eliminated by the correction we select in SSCTRK) together with chromaticity zero.

The tune values for all the SSCTRK runs presented here is Qx=XXX. 425 and Qy=XXX. 410, chosen on the basis of Fermilab Main Ring and Tevatron experience. This operating point is clear of all resonances up to 8th order (see Figure 1).

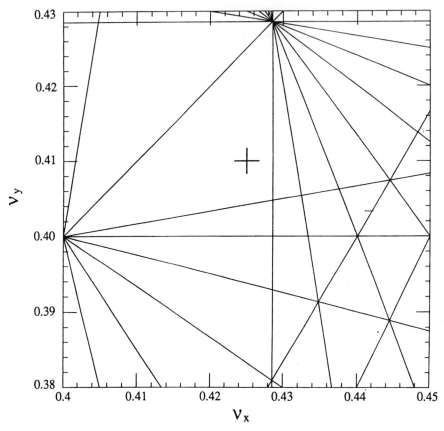

Figure 1. The cross marks the operating point chosen for SSCTRK simulation. All resonance lines up to 8th order are shown.

The half cell length chosen is 90 m, and the phase advance per cell is 95°. This is a little more than the design 90°, which it was feared might cause a damaging beam-lattice resonance (no convincing evidence for such an effect was seen when 90° phase advance runs were made).

Magnetic elements are taken to be misaligned with respect to the closed orbit by an rms deviation of 1 mm.

The RF voltage is taken as 20 MV and the frequency as 375 megahertz. The fractional momentum deviation of the particles in these runs is taken as 5×10^{-4}.

Power supply ripple is not included in these runs — other runs suggest that, at the tune operating point chosen, it is easily feasible to reduce it to a value which doesn't affect the results.

The SSCTRK code doesn't presently incorporate interaction regions. It is not expected that this omission will significantly alter the results presented here, as the interaction region beta values during the injection dwell are modest.

The SSCTRK code doesn't simulate the details of diagnostic and correction magnet systems. When a correction mode is selected, SSCTRK computes the mean value across the machine of the error multipole in question and then subtracts the appropriate offset value (often the mean itself) from that multipole's error values in each element. (These are "zeroth harmonic" corrections — recently a "first harmonic" correction capability has also been incorporated into the code.) In the runs presented here, this type of correction is performed for B1, B2, and A1.

The runs presented here are of a *particle loss* nature, with a circulating particle being considered lost if its distance from the closed orbit can possibly exceed $\sqrt{2} \times 15$ mm during a drift between thin elements (particles are checked for loss every two and half cells by evaluating the sum of the appropriate x and y Courant-Snyder emittance invariants). Particles are always launched with equal maximum x and y amplitudes with respect to the closed orbit, and angle equal to zero. When a particle is lost, a new particle (its "daughter") is launched at the same position in the machine where the "parent" particle was lost, but having initial maximum x and y amplitudes 1 mm less than the initial maximum amplitudes of the "parent".

For the runs presented here, the "first generation" of particles is always launched at the same defocusing quadrupole of the machine, with x and y maximum amplitude equal and gridded at the four values 10 mm, 10 1/3 mm, 10 2/3 mm and 11 mm. Particles running at such high amplitudes see a very bad field region and are rapidly lost. However, each subsequent generation is stepped in by 1 mm in initial maximum x and y, and within a few tens of thousands of turns, losses become quite infrequent, as the pattern of sufficiently high generation running particles edges on (indeed, defines) the dynamic aperture.

The random multipole errors for a machine are simulated using a random number generator. In order to avoid getting an especially fortunate or unfortunate sequence of random numbers defining the machine, many machines (typically 16) are simultaneously simulated by choosing different seeds for the random number generator. Each machine is then run with four particles and their subsequent stepped-in generations, as described above. On a vectorizing supercomputer, all the particles in all the simulated machines can be run in parallel. One may summarize the results of such a run with a particle loss plot, where each point represents the number of turns a particle completed before it was lost versus its initial maximum $x = y$ launch amplitude. The left hand boundary of the pattern of such particle loss points may be roughly identified as the dynamic aperture, at least for the number of turns computed.

The injection dwell of 30+ minutes corresponds to around 10^7 turns. Running 64 particles in parallel (4 particles each in 16 machines) for this many turns in SSCTRK requires of order 100 CPU hours of Cray 2 or SX2 supercomputer time. Running only 8 machines (32 particles) cuts down on Cray 2 CPU time by over 30%, so the really long runs were made with this option. However, it was observed that particle losses become very infrequent after 10^5 turns, which only require about one CPU hour of supercomputer time. Indeed the dynamic aperture values which are to be gleaned from 100,000 turn runs with 16 machines are usually not far from those which may be read from multi-million turn runs with 8 machines.

Below we present particle loss plots for 4 and 5 cm magnet aperture machines which are highly corrected and plausibly corrected.

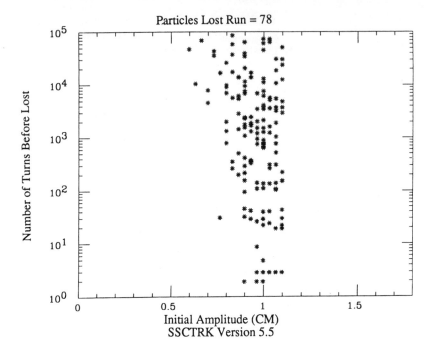

Particles Lost Run = 78

SSCTRK Version 5.5

Run 78 is 10^5 turns of 16 highly corrected machines of 4 cm magnet aperture. The particle loss plot shows a dynamic aperture of about 9 mm after a few turns and about 6 mm after 10^5 turns.

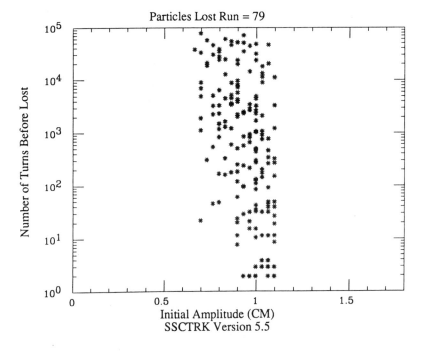

Particles Lost Run = 79

SSCTRK Version 5.5

Run 79 is the same as run 78, except that the random regular sextupole error was increased to its nominal SSC-7 value, i.e. dipole sorting on this random multipole was no longer assumed. There is no significant difference to run 78.

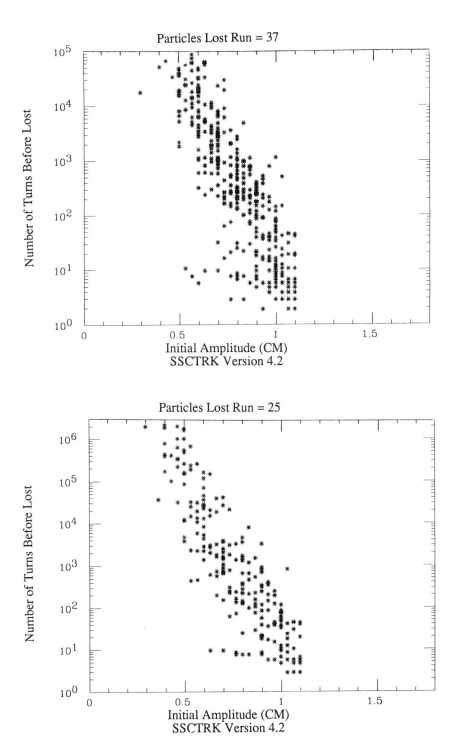

Particles Lost Run = 37

Initial Amplitude (CM)
SSCTRK Version 4.2

Particles Lost Run = 25

Initial Amplitude (CM)
SSCTRK Version 4.2

Run 37 as shown in previous run is 10^5 turns of 16 plausibly corrected machines of 4 cm magnet aperture (systematic errors from Table 2 together with +5 units of x and y chromaticity). The dynamic aperture has dropped to about 3 to 4 mm, in contrast with the 6 mm for the highly corrected machines of run 78. The same picture shows up in run 25, which has the same parameters as run 37, but with only eight machines run out to 3,000,000 turns.

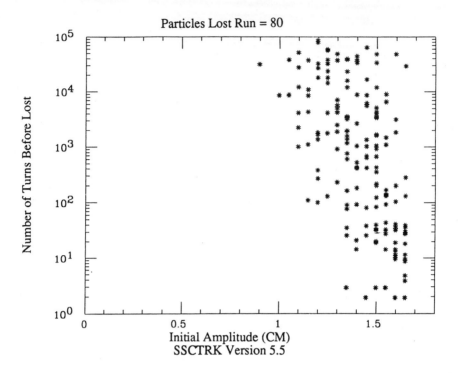

Particles Lost Run = 80

SSCTRK Version 5.5

Run 80 is 10^5 turns of 16 highly corrected machines of 5 cm magnet aperture. Here the four first generation particles are gridded at 15 mm, 15.5 mm, 16 mm and 16.5 mm, and subsequent generations are stepped in by 1.5 mm. The dynamic aperture for 100,000 urns is about 9 mm.

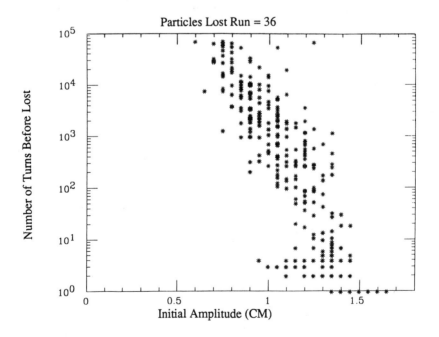

Particles Lost Run = 36

Run 36 is 10^5 turns of 16 plausibly corrected machines of 5 cm magnet aperture. The dynamic aperture for 100,000 turns is about 6 mm.

LOW ENERGY BOOSTER RESONANT POWER SUPPLY SYSTEM

Cezary Jach

Superconducting Super Collider Laboratory[*]
2550 Beckleymeade Ave.
Dallas, Texas 75237-3946

INTRODUCTION

During the particle-acceleration period, the magnet guide-field intensity, along the equilibrium orbit must rise from a minimum field (corresponding to injection energy) to the peak field, at which the protons attain their maximum energy. The field must then be reduced to its minimum value and the cycle repeated. This operation is achieved by excitation of the guide-field magnets with a 10 Hz sinusoidal waveform, which is biased by superimposing a dc component on the ac component excitation, giving a field variation of the form:

$$b(t) = B_{dc} - B_{ac} \sin (2\pi \, ft) \tag{1}$$

where f is the magnet excitation frequency (nominally 10 Hz). Aside from saturation effects in the magnets and chokes, b(t) may be considered as a linear function of the magnet current (design of network magnetic elements should not allow more than 1% saturation, i.e., peak field should not exceed 1.25 T). With this assumption, the magnet power supply generates a current of the form (Figure 1)

$$i_m (t) = I_{dc} - I_{ac} \sin (2\pi \, ft) \tag{2}$$

RESONANT MAGNET NETWORK

To ensure accurate tracking between dipoles and quadrupoles during the acceleration cycle, the magnet lengths and coil configurations are chosen so they can be driven from a common current supply. The coils have relatively few turns, to keep the magnet induced voltage within manageable bounds. The biased excitation requirements are detailed in Table 1.

The design of the magnet -network configuration is influenced by two factors:

- The need to avoid drawing a large reactive power from the magnet ac excitation excitation source,

- The maintenance of equal currents and hence a uniform field intensity in all dipole and quadrupole magnets.

---

[*]Operated by the Universities Research Association, Inc., for the U.S. Department of Energy under Contract No. DE-AC02-89ER40486.

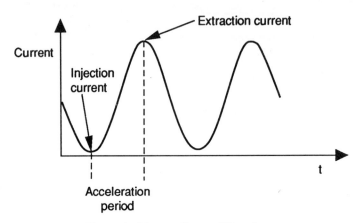

Figure 1. Magnet Current Waveform

Table 1. Power Supply System Parameters

Number of dipoles	84
Number of quadrupoles	108
Total winding resistance (@40°C)	0.390 Ω
Total bus work resistance(@40°C)	0.170 Ω
Total choke dc resistance (@40°C)	0.094 Ω
Maximum field at equilibrium orbit	1.23 T
Maximum field gradient	16.5 T/m
Dipole turns/pole	8
Quadrupole turns/pole	3
Dipole inductance at peak field	3.07 mH
Quadrupole inductance at peak field	0.26 mH
Magnet current, peak	3745 A
Magnet current, rms	2294 A
Magnet current, dc component	1872.5 A
Magnet current, ac component, peak	1872.5 A
Total peak magnet stored energy	2.0 MJ

The magnet reactive power of some 63 Mvar can be excluded from the excitation source by using a magnet circuit which is resonant at the accelerator operating frequency (10 Hz). This resonant circuit must be designed to provide a path for the dc bias and quadrupole magnets.

These basic requirements are satisfied by the simple circuit shown in Figure 2, in which the magnet inductance L_m forms a resonant circuit with the capacitor C_m and the dc bypass choke inductance L_{ch} forms a resonant circuit with the capacitor C_{ch}. The resonant frequency is given by:

$$f = \frac{1}{2\pi} \sqrt{\frac{L_{ch} + L_m}{L_{ch} L_m (C_{ch} + C_m)}} \tag{3}$$

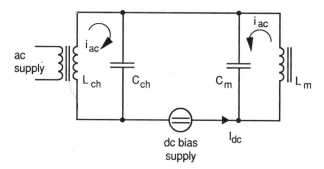

Figure 2. Simple Resonant Magnet Circuit

It is not possible to utilize such a system for the direct connection of 108 quadrupoles and 84 dipoles, since a simple series circuit, preferable for basic magnet current uniformity, would yield an excessively high induced voltage for the magnet ac component of:

$$\sum_1^{84} Vmd_i + \sum_1^{108} Vmq_i = 33.6 \text{ kV}!$$

DISTRIBUTED RESONANCE CIRCUIT

The dipoles and quadrupoles are connected as shown in Figure 3, in twelve series groups of seven dipoles and nine quadrupoles, together with associated resonant capacitors. Each magnet-capacitor group is then effectively connected in parallel with a secondary winding of the energy storage choke for dc bypass and a choke resonant capacitor. Every other energy storage choke secondary winding is split into two identical halves for the insertion of the dc bias power source. This arrangement assures that the dc bias power source does not see the 10 Hz component of the current. One half of the energy storage choke secondary winding and capacitor form a 10 Hz parallel resonant circuit. Power to make up for the ac losses of the network is fed via primary windings of the energy storage chokes. It can be seen from Fig. 3 that under symmetrical conditions, the ac voltage of each magnet-capacitor group is zero, i.e.

$$L_m \frac{di_m}{dt} + \frac{1}{C} \int i_C \, dt = 0 \qquad (4)$$

Therefore, ac voltage is not cumulative, as would be the case for a simple series connection of magnets based on the circuit in Fig. 2, but alternates around the ring, as shown in Figure 4.

Hence, except for the presence of the dc bias excitation voltage, which adds a low positive gradient to the distributed ring voltage, the center points of each capacitor, choke and magnet group are at earth potential (for normal operation). This situation allows the bias power supplies to be inserted into the magnet resonant system at multiple locations. With six such insertions, the maximum dc component of the voltage is about 205 V, lowering magnet dc insulation stress. A distributed ground system is used to detect ground faults. All terminals of the bias power supplies are tied to a common wire via 50 kΩ resistors. The common wire runs all the way around the ring and is terminated to ground via current sensing circuit. If the average voltage to ground of all the power supplies is zero, no current will flow from the common wire to ground. If a ground develops, the average voltage to ground will not be zero, resulting in ground current and a trip.

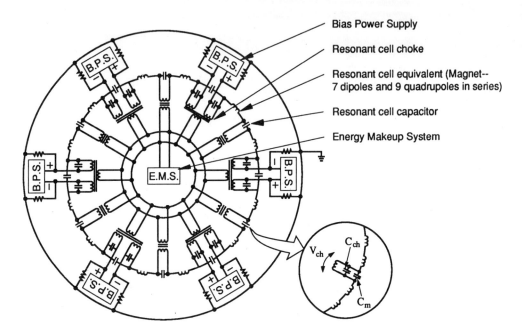

Figure 3. Twelve Cell Interconnection of Resonant Magnet Network.

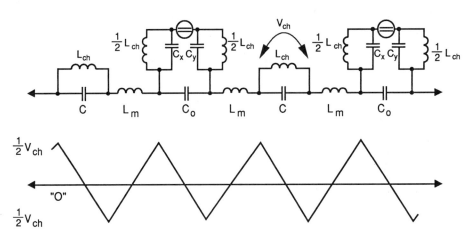

Figure 4. Resonant Magnet Network ac Voltage Distribution.

Owing to its series connection, the distributed resonance circuit provides a basic uniformity in magnet currents. The effect of leakage capacitance currents to earth can distort this equality to a degree that is significant in terms of guide field. Hence, since these leakage paths are distributed around the network in a generally uniform manner, it is necessary to ensure that the ac potentials of corresponding points in the network attain a similar identity. This is achieved by close tolerance in the permitted variations of the energy storage choke inductance and resonant capacitor capacitance.

CIRCUIT ALTERNATING VOLTAGE

The choice of the circuit ac voltage is dependent on the total inductance of the series connected magnets in a group.

A good compromise between insulation cost, leakage-current considerations, reliability and circuit complexity is achieved with seven dipoles and nine quadrupoles connected in series. This results in total of 12 resonant cells and total magnet group-induced voltage of 2800 V peak. Maximum operating voltage to ground is 1400 V peak and 990 V rms.

CIRCUIT DIRECT VOLTAGE

The total dc resistance of the ring at 40°C is 0.654 Ω, therefore an onload dc voltage of 1225 V is required for the maximum excitation condition of 1873 A. This is provided by six 12-pulse SCR type power supplies connected into the resonant magnet network as shown in Fig. 3.

CIRCUIT WAVEFORMS

During resonant operation, the total network stored energy is constant, except for losses, and the magnet stored energy is transferred to the choke and back again once per cycle, with capacitors storing a portion of the peak energy during the transfer.

The magnet resonant capacitance which is fixed by the magnet group inductance, the ratio of the choke inductance L_{ch} to the magnet group inductance L_m and hence, the choke resonant capacitance C_{ch}, can be selected to minimize the choke capacitor energy storage cost. The system cost has a minimum when the choke inductance is about 1.5 times larger than the magnet group inductance. Since under symmetrical conditions, the induced voltages across each choke secondary winding and magnet group are identical in amplitude (but reversed in phase), and the dc bias current is common to both, the resultant choke current is

$$i_{ch} = I_{dc} + \frac{L_m}{L_{ch}} I_{ac} \sin (2\pi ft) \tag{5}$$

and the peak choke stored energy for fully biased magnet current $(I_{ac} = I_{dc})$ is

$$W_{ch} = \frac{1}{2} L_{ch} (I_{ch})^2 = \frac{1}{2} (1.5 L_m) \left(\frac{5}{6} I_m\right)^2 = \frac{25}{26} W_m \tag{6}$$

The current and voltage waveforms of magnets, capacitors and chokes are shown in Figure 5 and parameters of the resonant network are summarized in Table 2.

DELAY-LINE MODES

The distributed L and C components exhibit resonant modes that when excited, disturb the uniformity of magnetic field around the accelerator. The particle beam can be either seriously disturbed or completely lost. In addition to the circuit components, the stray capacitances from the elements of the network to earth must be considered. In general, these capacitances are much smaller than the resonant capacitor banks, and resonate with the circuit inductances at a frequency that is appreciably higher than the network fundamental frequency (10 Hz). At these frequencies, the resonant capacitors have a very low impedance, leaving the magnet windings as the only appreciable impedance in the line. While the capacitance from the magnet coils to earth is distributed, there are also appreciable amounts of lumped capacitance from certain parts of the network to earth, particularly at the points where magnet, choke and

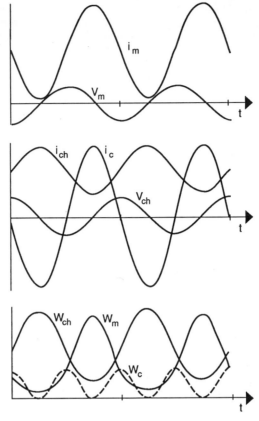

Figure 5. Waveforms of Resonant Magnet Network.

Table 2. Resonant Network Parameters

Number of Magnet Groups	12
Induced Voltage Across Each Magnet Group (rms)	1980 V
Alternating Voltage to Earth (rms)	990 V
Inductance Per Magnet Group, L_m	23.8 mH
Choke Inductance (Secondary), L_{ch}	35.7 mH
Choke Current, Peak	3121 A
Choke Induced Voltage/Capacitor Voltage, rms	1980 V
Choke Stored Energy, Peak	173.9 kJ
Capacitance Per Group (standard capacitor), C	17.74 mF
C_x, C_y	14.20 mF
C_o	10.64 mF
Peak Capacitor Stored Energy (standard capacitor)	69.54 kJ
Network ac Power Loss at Maximum Excitation, P_{ac}	1.68 MW

capacitor are connected, since these links involve long lengths of bus bar. Thus, when considering oscillations involving stray capacitance to earth, it is possible to simplify the circuit of Fig. 3 to that shown in Figure 6. When some disturbance occurs in one part of the magnet network, voltages are induced across the stray capacitance, and oscillations between these capacitors and the magnets result. The oscillating voltages, and associated currents in the magnets, form delay line modes of resonance, resulting in standing waves along the magnet network. The number of nodes and the frequency of the oscillations depend on the particular harmonic that is excited by the disturbance. Theoretically, an infinite number of harmonics are possible, but in practice it is usually only the mode with the lowest possible frequency (fundamental) together with some of the first few harmonics that are excited. For the circuit model shown in Fig. 6, the magnet current, associated with delay line modes of resonance, at any part of the network is given by:

$$i_m = \frac{\pi V}{\omega L} \cos\left(\frac{n\pi x}{l}\right) \sin (n\omega t) \tag{7}$$

where L is total inductance of the series magnets;
l is the total length of magnet windings;
x is the length from the network earth point to the point under consideration;
n is the harmonic number of the oscillation;
ω is the angular velocity of the fundamental mode;
V is the in standing wave peak voltage.

The frequency of the fundamental mode is given by:

$$f = \frac{1}{2\pi\sqrt{LC}} \tag{8}$$

where C is the total capacitance from the magnet system to earth.

Assuming $C = 1000$ nF, $L = 285$ mH, then $f = 300$ Hz, and it would take only 60 V of 300 Hz signal for the resultant current to exceed 0.01% of the peak excitation current (LEB field ripple requirement). Techniques for damping these modes are available and will be applied as appropriate.

ENERGY MAKEUP SYSTEM (PULSER)

Under symmetrical resonance conditions, the high Q-factor magnet network presents a purely resistive load to the ac power supply, and has a power requirement of 1.68 MW at the maximum magnet-excitation level. The ac power supply can be one of three basic types:

- A rectifier-inverter set with 10 Hz tuned filter, and continuous excitation (BESSY, DESY, Germany)

- A Pulse Width Modulator with GTO (Gate Turn Off) Thyristors, and continuous excitation (ESRF, France)

- A pulse-power supply in which power to make up the ac loss is supplied to the resonant magnet network in the form of an impulse, via the choke-primary windings (NINA, England, CEA and SSRL, USA).

The pulse power supply is favored because of its operational simplicity and reliable performance record.

PULSER

The major components of the pulser are shown in Figure 7. It is comprised of an energy storage capacitor and associated charging circuit, and a pulse discharge circuit triggered from the resonant magnet network. The phase-controlled 3-phase power supply charges the energy storage capacitor C_F via the filter choke L_F to twice the energy storage choke secondary peak voltage V'_{ch} (referred to primary). The thyristor is then triggered to discharge C_F via the pulse choke L_P and the energy storage choke primary, so that the resulting half cycle current pulse i_P occurs approximately symmetrically around the positive peak of the choke voltage V_{ch}. The cycle repeats itself with a 10 Hz frequency. It can be seen from the analysis of the pulser circuit that its wave forms are dependent on the values chosen for the frequencies f_p and f_F of the pulse and charging circuits. The most important considerations in this choice are:

- Minimum introduced disturbance to the magnet network during normal operation.

- Acceptable transient-fault levels.

- Economic circuit-component ratings.

- Limited fluctuations in mains-supply input power.

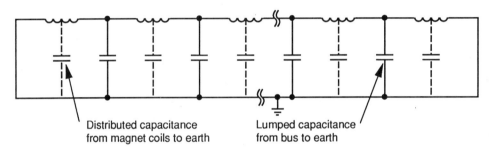

Distributed capacitance
from magnet coils to earth

Lumped capacitance
from bus to earth

Figure 6. Simplified Magnet Circuit for Delay-Line Mode Considerations.

On this basis, the selected values are $f_F = 2.5$ Hz and $f_p = 30$ Hz. Based on an analysis and given values for the frequencies, the coefficients of peak, minimum and rms currents can be calculated in terms of the average charging current i_F (av):

$$i_F \text{ (max)} = 1.108 \ i_F \text{ (av)} \qquad\qquad i_P \text{ (max)} = 9.026 \ i_F \text{ (av)}$$
$$i_F \text{ (min)} = 0.816 \ i_F \text{ (av)} \qquad\qquad i_P \text{ (av)} \ = \ i_F \text{ (av)}$$
$$i_F \text{ (rms)} = 1.004 \ i_F \text{ (av)} \qquad\qquad i_P \text{ (rms)} = 2.678 \ i_F \text{ (av)}$$

where i_F (av) $= P_{ac}/V_s$, and P_{ac} is the ac power loss of the resonant magnet network. The pulser parameters based on an energy storage choke turns ratio of 1:2 are shown in Table 3.

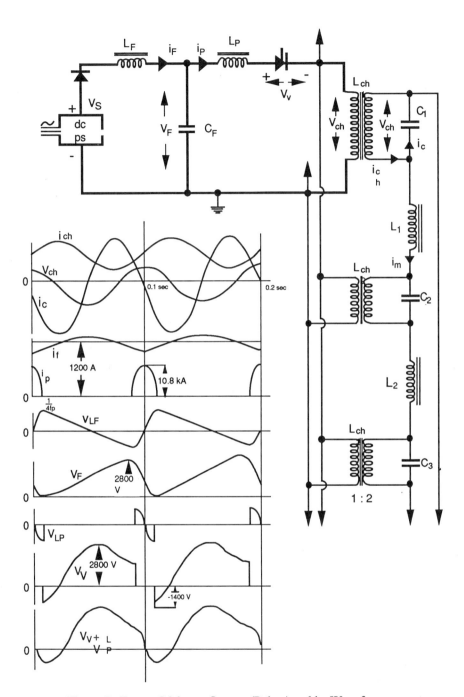

Figure 7. Energy Make-up System (Pulser) and its Waveforms.

Table 3. Pulser Parameters

Dc input	
On Load voltage, Vs	1400 V
Power, P_{ac}	1680 kW
Current, average	1200 A
Filter circuit	
Frequency, f_F	2.5 Hz
Inductance, L_F	11 mH
Current, rms	1200 A
Current, minimum	980 A
Energy storage capacitor	
Capacitance, C_F	36.52 mF
Peak voltage	2800 V
Peak stored energy	143 kJ
Pulse circuit	
Frequency, f_p	30 Hz
Inductance, L_p	0.77 mH
Current, peak	10.8 kA
Current, rms	3.2 kA
Current, average	1.2 kA
Thyristor switch	
Conduction period	16.6 ms
Duty cycle	16.6%
Peak forward voltage	2800 V
Peak inverse voltage	-1400 V

MAGNET FIELD CONTROL

Since saturation can be neglected and linear operation is assumed, the quantities B_{ac} and B_{dc} can be controlled by controlling I_{ac} and I_{dc}. The ac component of magnet current is controlled by the pulse power supply feedback system and consists of two principal closed loop systems:

- A magnet ac control loop of slow response, in which a reference is compared with the measured amplitude of oscillation (the amplitude error is characterized by measuring the elapsed time between downward and upward crossings of a field level just above the injection-time value of the field. These crossings are detected using bias peaking strips or other techniques such as NMR probes). The resultant error signal drives the SCR control circuits of the charging power supply.

- A thyristor firing control loop of slow response in which the phase position of the pulser discharge current is compared with the firing phase reference.

The dc component of the magnet current is controlled by the dc bias power supply feedback loop of slow response in which an adjustable reference is compared with a signal proportional to the magnet direct current. The resultant error signal drives the SCR control circuit of the bias power supplies.

ADJUSTMENT OF ACCELERATOR FREQUENCY

During normal operation, the resonant magnet network is excited at 10 Hz. If the pulse is adjusted to occur either in advance or retard of the optimum phase position (symmetrically about the resonant capacitor current zero) a corresponding increase or decrease in accelerator frequency is obtained. In this way, the magnet excitation frequency can be adjusted by ±0.1 Hz. Further frequency swing would result in excessive pulse current.

TUNING OF MAGNET RESONANT NETWORK

In order to reduce ac losses, the magnet resonant network is tuned to its resonant frequency. This is done by means of a trimming section of the resonant capacitor bank. Change in capacitance (due to temperature coefficient or aging) in the particular bank can be detected by measuring the primary current of the bank's energy storage choke which the bank is connected to. A change of capacitance of 1% of its nominal value, results in a change of a factor of two in the choke's primary current amplitude, which can be easily noticed.

THE SSC COLLIDER RING CORRECTION MAGNET SYSTEM

S.R. Stampke,[1] J.M. Peterson,[1] and D.V. Neuffer[2]

(1) Accelerator Division
    Superconducting Super Collider Laboratory*
    Dallas, Texas 75237
(2) AT-6, Los Alamos National Laboratory
    Los Alamos, NM 87545

Abstract

The correction system for the SSC Collider rings will have about 14,000 superconducting magnetic elements. Linear correctors (dipoles and quadrupoles) are located in spool pieces next to focusing (F) and defocusing (D) main quadrupoles. Systematic multipole correction utilizes nonlinear correctors (sextupoles, octupoles, and decapoles) located at positions (C) near half cell centers as well as in the F and D spools. The basic functions of correction magnets and the dynamics leading to the selected configuration are described. Strength requirements, the number and distribution of correction magnets, and initial prototype efforts at collaborating laboratories are outlined.

Introduction

The dominant features of the SSC Collider rings are roughly 8600 superconducting main dipoles for bending, nearly 2000 main quadrupoles for focusing, and about 1860 cryogenic spool pieces. Every cell of the SSC Collider arcs includes ten main bending dipoles, two main quadrupoles, and two cryogenic spool pieces. Much smaller in size and cost, but still essential to the machine are the superconducting corrector magnets. Table 1 outlines the major functions required of the corrector system.

The 1986 Conceptual Design Report[1] envisioned a correction system that utilized a standard dipole-quadrupole-sextupole primary corrector package located in each spool piece, plus powered bore tube trim coils (sextupole and decapole) for compensation of systematic errors in the main dipoles. (Octupole trim coils were added shortly thereafter.) Numerous secondary corrector packages were also located in spool pieces to compensate skew multipoles and augment the primary corrector packages.

The SSC Collider correction system has evolved significantly since the 1986 CDR. Requirements have changed due to a more complete knowledge of SSC main dipole and quadrupole design, deeper understanding of linear and dynamic aperture requirements,[2]

---

* Operated by the Universities Research Association, Inc., for the U.S. Department of Energy under Contract No. DE-AC02-89ER40486.

Table 1. Corrector functions and corresponding magnets.

Function	Magnets
Steering and Closed Orbit Correction	Dipoles
Tune Correction and Control	Quadrupoles
Linear Chromaticity Control	Sextupoles
Compensate x-y Coupling	Skew Quadrupoles
Augment Main Quadrupoles	Quadrupoles
Compensate Main Dipole Error Fields	Various Multipoles
Compensate Persistent Current Error Fields at Injection	Sextupoles and Decapoles
Compensate Saturation induced Error Fields at 20 TeV	Sextupoles and Decapoles
Second Order Control	mainly Octupole
Compensate Error Fields in IR Quads	Various Multipoles
Control IR Beam Crossing Angle and Separation	Dipoles

and resulting changes in the lattice (linear dynamics) design. New correction methods involving the use of nonlinear correction elements at intra-cell (C) positions have been developed for the Collider.[3] These methods, suitable to large synchrotrons in general, have eliminated the need for bore tube trim coils.

Beam stability demands highly linear motion and, therefore, linear fields. The greatly increased energy and circumference of the SSC compared to earlier machines magnifies the effects of linear (e.g. alignment and tuning) and nonlinear (e.g. uncorrected multipole) errors, while forcing the design towards small aperture, more nonlinear magnets. The correction magnet system must compensate for such problems both during a thirty minute 2 TeV injection period, when beam profiles are largest and some multipoles (b_2, b_4) are time dependent, and also at 20 TeV where strength demands on the correctors are highest. The correction system described in the SSCL Site-Specific Conceptual Design Report (SCDR)[4] and outlined here should meet these requirements. It also contains allowances for optimization, future refinements, Collider upgrades, and uncertainties in correction magnet technology.

The correction system, shown in Figure 1, starts with linear correctors (dipoles, and normal quadrupoles) in spool pieces adjacent to each main focusing (F) and defocusing (D) quadrupole. Orbit correction and control employs horizontally bending dipoles at each F spool and vertically bending dipoles at each D spool. Since the main quadrupoles are on the same current bus as the main dipoles, quadrupole control is through the corrector quadrupoles. These devices are responsible for maintaining precise control of the central tune (ν) of the machine while compensating for imperfect tracking or "differential saturation" between the main dipoles and quadrupoles.

Nonlinear correction elements control tune spreads ($\Delta\nu$) due to the collider's natural chromaticity (ξ_{nat}) and low order error multipoles. Chromaticity control is provided by sextupoles in the F,D spools. These sextupoles also contribute to compensation of normal sextupole (b_2) errors in the main dipoles. For reasons discussed below, compensation of normal sextupole (b_2), octupole (b_3) and decapole (b_4) errors in the main bending dipoles uses the "Neuffer-Simpson" (NS)[3] quasi-local method with elements located at F, C, and D positions within each half cell. (The NS corrector pattern is also referred to as an "FCD" pattern.)

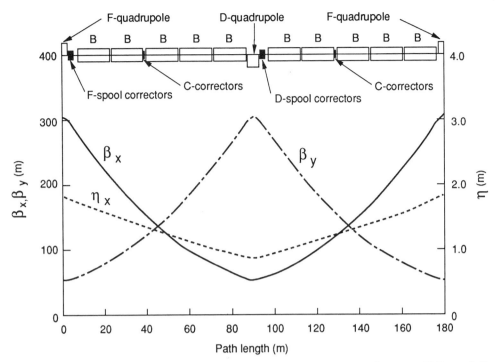

Figure 1. Normal cell of SSC Collider arc. Correction magnet packages at F,D, and C positions are indicated. Main dipoles, focusing and defocusing quadrupoles, as well as the amplitude (β_x, β_y) and dispersion (η) functions are also shown.

Thus each F and D spool contains a five element correction "package" consisting of dipole, quadrupole, sextupole, octupole, and decapole magnets. C-corrector packages contain sextupole, octupole, and decapole elements.

Physics of Multipole Correction and Control

The use of mid-cell (C) correctors for higher order nonlinearities is a significant change from previous correction systems. We outline here the physical basis for this change.

The magnetic fields in the dipoles may be represented by the complex expression

$$B_y + iB_x = B_0\{1 + \sum_{n=1}[b_n(s) + ia_n(s)](x + iy)^n\}$$

where B_0 is the bending field and $b_n(s)$ and $a_n(s)$ are the normal and skew multipole components. The transverse motion may be described by the Hamiltonian:

$$H = \frac{I_x}{\beta_x(s)} + \frac{I_y}{\beta_y(s)} + \Re\sum_n \frac{B_0}{B\rho}\frac{[b_n(s) + ia_n(s)](x + iy)^{n+1}}{n + 1}$$

where I_x and I_y are the action coordinates, $\beta_x(s)$ and $\beta_y(s)$ are the Courant-Snyder[5] betatron functions of the linear motion. $\Re$ implies taking the real part of the summation. The coordinates x and y of particle motion are represented to first order by the action-angle variables: $x = \sqrt{2\beta_x I_x}cos(\phi_x) + \eta\delta$ and $y = \sqrt{2\beta_y I_y}cos(\phi_y)$. The terms ϕ_x and ϕ_y are the angle variables (betatron phases), and the off-momentum orbit displacement, $\eta\delta$, determined by the dispersion function $\eta(s)$ at $\delta = dp/p$ is included.

99

Table 2. Tolerances to Dipole Multipole Strengths.[*]

Mult	Tolerance				Assumed Values			
	No Corr	At Quads Only	NS at 1:2	NS Opt.	4 cm Spec	4 cm Persist	5 cm Spec	5 cm Persist
b_2	0.022	4.0	5.7	10.2	1.0	-3.0	0.63	-1.9
b_3	0.042	0.051	3.25	7.25	0.1		0.05	
b_4	0.093	0.097	2.8	18.0	0.2	0.2	0.09	0.09
b_5	0.18	0.18	2.5		0.04		0.02	
b_6	0.34	0.34	2.7		0.07	-0.05	0.02	-0.017
b_7	0.63	0.63	3.5		0.1		0.03	
b_8	1.28	1.28	4.75		0.2		0.05	

[*] Collider tolerances to normal dipole systematic errors (b_n) with various correction schemes are given in "Units" (10^{-4} of dipole strength at 1 cm). A 90-deg, 180-m lattice with 2 TeV injection is used. Assumed multipole values for 4 cm and 5 cm dipoles from Ref. 2 are also listed.

In the SSC, the dominant first-order nonlinear effects are the nonlinear tune-shifts. These can be calculated by integrating the phase advance around the ring:

$$\Delta\nu_{x,y} = \frac{1}{2\pi} \int \frac{d\phi_{x,y}}{ds} = \left\langle \frac{dH}{dI_{x,y}} \right\rangle$$

To first order in the coefficients b_n and a_n, only systematic multipoles ($\overline{b_n}$) contribute. The integrals have been calculated for the SSC lattice.[6] Requiring adequate linearity within the SSC design aperture ($\Delta\nu \leq 0.005$ for $x, y < 0.5$ cm and $dp/p < \pm 0.001$) sets limits on the allowable $\overline{b_n}$, which can then be compared with expected $\overline{b_n}$ for SSC magnets, Table 2 and Figure 2. Some correction of b_2, b_3, and b_4 is desirable regardless of the magnet choice (i.e. "4" or "5" cm dipoles).

Previously, correctors have been placed near F and D quadrupoles, and such correctors were sufficient for dipole, quadrupole, and first-order sextupole (linear chromaticity) control, but are ineffective for higher multipole effects. This failure is directly related to the increased apparent complexity of the Hamiltonian which includes coupled motion terms, as well as separable horizontal and vertical terms. However, a great improvement is obtained by adding correctors to the center (C) location of each half cell. For the correction of constant systematic multipoles, the optimum corrector strengths are close to the Simpson's Rule derived values $(S_F, S_C, S_D) = -(\frac{1}{3}, \frac{2}{3}, \frac{1}{3})B_0 b_n L$, where $B_0 b_n L$ is the n'th systematic multipole error integrated over a half cell. This reduces all nonlinear effects by about two orders of magnitude.

The accuracy of the correction can be understood by noting that any $\Delta\nu$ term can be expressed as an integral over the lattice. For example, a b_3 term may be written as

$$\Delta\nu_3 = \int b_3 \beta_x^2 ds - S_{3,F}(\beta_x(0))^2 - S_{3,C}(\beta_x(L/2))^2 - S_{3,D}(\beta_x(L))^2.$$

where $S_{3,I}$ are octupole corrector strengths. All other nonlinearities, such as orbit distortions and higher-order $\Delta\nu$, can be expressed as similar integrals. The correction is equivalent to approximating a continuous integration by a sum over discrete points. Simpson's Rule is a generally valid solution. Its use corresponds to forming the FCD correctors into an optimal three-point quasi-local cancellation of the continuous mul-

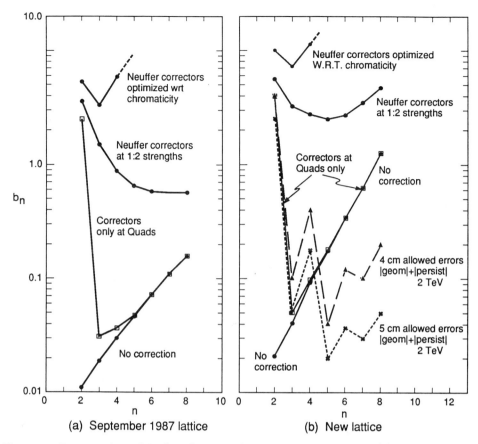

Figure 2. Systematic multipole tolerances from tune shift criteria. (a) For September 1987 lattice. (b) For current lattice. Also included are expected multipoles from Table 2.

tipole content of the dipoles. Optimization about that solution can reduce critical nonlinearities by another order of magnitude (Table 2).

For effects above first order sextupole, the Hamiltonian can be separated into horizontal, coupled, and vertical motion terms; for example x^4, $-6x^2y^2$, and y^4 terms in octupole order. The FCD correctors are at the optimal locations for control of these horizontal, coupled, and vertical motion parameters, and these are precisely the operational observables, as well as the separable terms in the Hamiltonian. This tunability can be used in improving correction from initial approximations. For instance, FCD octupoles are appropriate elements for control of all amplitude dependent and second order chromatic tune shifts. The FCD elements permit exact control of the motion through 10-pole order. The results of Table 2 and Figure 2 do not imply that one should allow dipoles with excessive multipoles: corrector magnet strengths and space for them are limited. However, they represent a compelling case for FCD correction, and illustrate the ability to optimize operation beyond simple multipole cancellation.

In the SSC cell with 5 dipoles per half-cell (Figure 1), the center corrector is slightly displaced from the optimal half-cell center. This shifts corrector values from the symmetric cell Simpson's Rule values of F:C = 1:2 to F:C = 22:50, yet correction capabilities are reduced only 10-20%.[6]

Primary Correctors

The correction magnet strengths were set assuming the Collider will use FCD correctors in each half cell of the regular lattice. A given family of primary correctors would be included in the system if linear aperture tune shift calculations and tracking studies demonstrated that it was necessary.[3,7] The 90 degree, 180 meter cell lattice with five dipoles per half cell (Fig. 1) is assumed. Injection energy is 2 TeV. The corrector strengths, however, are determined by requirements for 20 TeV magnetic fields. The present strength values were set assuming expected multipole errors for 4 cm dipoles as estimated in Table 2, and conservatively estimated dipole saturation and saturation sextupole moments at 20 TeV. Table 3 lists the strengths of primary correctors at F,C, and D locations within a cell.

Corrector dipole strengths are driven primarily by main quadrupole alignment uncertainties (0.5mm rms), with main dipole strength and roll errors contributing a few percent of the needed strength. The rms strength needed was found to be 0.60 T-m. To account for the statistical nature of the errors involved, the corrector dipole strength requirement is set at 2.50 T-m, a factor 4.2 above the rms need. This "safety factor" is consistent with Tevatron experience. Each corrector dipole is independently powered.

A corrector quadrupole is associated with each main quadrupole of the arcs and cluster region cells. Requirements for a central tune control range of $\Delta \nu = \pm 3$ while compensating up to 2% differential saturation between the main dipoles and quadrupoles dominate the integrated strength of GL = 53 Tesla (or BL = 0.53 T-m at r = 1 cm). The corrector quadrupoles are powered in two families, one family for focusing and one for defocusing.

Sextupole strengths were estimated by adding in absolute value the strengths needed to cancel a natural (from the linear lattice optics) chromaticity of $\xi_{nat} = -340$ with sextupoles at F and D locations, and the strengths to compensate a systematic sextupole (b_2) error of 2.6 "units" in the main dipoles. The b_2 compensation uses the NS arrangement with F:C strength ratio 22:50 appropriate to a 5 dipole half cell. The resulting integrated strengths are 0.13, 0.21, and 0.09 T-m at r = 1 cm for the F, D, and C corrector sextupoles. (Magnets made to the strength standards of the D location are likely to be used also at F positions instead of having two distinct types.) These are conservative choices reflecting significant uncertainties in 20 TeV dipole saturation b_2 and desires to allow luminosity upgrades through reducing interaction region β^* values. The sextupole strength is sufficient to consider binned, multi-cell applications of the Forest-Peterson[8] random error correction scheme using NS correctors. The F, D, and C sextupoles are powered as three separate families.

Although octupole field components do not occur in magnets with perfect dipole symmetry, extrapolation from Tevatron experience suggests a possible residual systematic an order of magnitude larger than desired. Effective octupole compensation requires NS correction (see Table 2 and Figure 2); spool-only octupoles have little effect. The FCD octupole corrector strengths of $BL_F = BL_D = 0.007$ T-m and $BL_C = 0.016$ T-m are able to compensate 0.46 units of systematic b_3. This allows the options of compensating random b_3 if necessary, of reducing the number of octupole correctors, and compensating skew (a_3) octupoles. Further, if the C corrector is separately powered, b_3 correctors can be useful for independent control of horizontal, vertical, and coupled amplitude dependent tune shifts, and for control of second order chromaticity. With the current strengths, second order tune shifts of ≈ 0.07 at 1 cm amplitudes and ≈ 0.04 at $(\Delta p/p)^2 = (0.002)^2$ can be controlled at 20 TeV.

Table 3. Primary Corrector Magnet Strengths*

Pole	F	C	D
Dipole	2.50	–	2.50
Quadrupole	0.53	–	0.53
Sextupole	0.13	0.09	0.21
Octupole	0.007	0.016	0.007
Decapole	0.004	0.009	0.004

* Values are full field (20 TeV) integrated strengths, BL, in Tesla-meters at a reference radius r = 1.00 cm.

Decapole field components are allowed by dipole magnet symmetry. While the random decapole is not considered a problem, compensation of systematic b_4 is required, especially at injection. As with octupoles, FCD correctors are required. The strengths to correct 0.24 units of b_4 are $BL_F = BL_D = 0.004$T-m and $BL_C = 0.009$T-m.

The strength requirements discussed above are for estimated 4 cm dipole error content. Use of larger aperture (5 cm) main dipoles, as currently planned, may reduce some multipole correction requirements. However, the requirements on corrector dipoles do not depend on main dipole aperture, and corrector quadrupole strength is dominated by tune adjustment ability and differential saturation. Also, the sextupole strength is dominated by the linear chromaticity of the lattice and saturation sextupole moments of the dipoles. These effects are not subject to simple aperture scaling. Only b_3 and b_4 may be dominated by scaling arguments. The actual multipole content of SSC dipoles is not precisely predictable, and will not be known until dipole production and measurement are underway. With that uncertainty, smaller effects have not yet been folded into the corrector strength estimations. Also, the degree of desirable tuning flexibility to be obtained with the correctors has not been fully evaluated. As these factors become more accurately known, corrector specifications will be appropriately revised.

Secondary Correctors

Much less reliance on secondary correctors is placed in the current system than in the 1986 CDR. However, with expectations of a possibly large a_1 skew multipole, correction of x-y coupling is essential. If global x-y coupling correction is sufficient, then skew quadrupoles might be placed only in cluster straight sections.[9] A more local correction would replace primary corrector packages with skew quadrupoles in pairs of adjacent C locations at a rate of one pair per half sector throughout the ring. This arrangement would make optimal use of the vertical and horizontal phase advances in a half cell.[10]

Cluster Region Correctors

The basic NS pattern of FCD primary correctors is repeated in each cell of the clusters where bending dipoles are present, with modifications for the changing optics. In empty cells, which include no main dipoles, the C correctors are deleted.

The interaction regions (IR's) will demand special attention. Correction dipoles will be used for closed orbit correction and control of beam separation and crossing angle. Their strength requirements will be dominated by the need to maintain (separated

Table 4. Collider Corrector Magnet Totals.

Region	D_H $+D_V$	Q_F $+Q_D$	S_F $+S_D$	S_C	O_F $+O_D$	O_C	De_F $+De_D$	De_C
North Arc	392	392	392	372	392	372	392	372
South Arc	392	392	392	372	392	372	392	372
Clus. W	12	12	12	12	12	12	12	12
Clus. E	12	12	12	12	12	12	12	12
Skew		40	40			40		
Clus. Type	124	124	80	50	56	50	56	50
Subtotal	932	972	928	818	864	858	864	818
I.R.	176	24						
Total	2040	1968	1856	1636	1728	1716	1728	1636

beam) injection optics during acceleration to 20 TeV. The problem is complicated by the need for local control at each IR. In one study,[11] a minimum of 8 horizontal/vertical dipole pairs per beam were required and rather strong (over 10 T-m) dipoles were needed. Space constraints imply that special dipoles will be needed.

Non-linear error fields in the IR quadrupoles can severely restrict the Collider linear aperture because of large amplitudes in the beam motion and high gradients in the long quadrupole triplet magnets. Local corrections of all a_n and b_n through $n = 5$ were found necessary in the 1986 CDR and bore tube correctors were proposed for compensating all these errors. Since then, it has been found that discrete correctors within the quadrupole triplets can also provide sufficient compensation, provided quadrupole units are arranged to allow corrector placement. The details of IR region design and correction are important remaining problems for the SSC Laboratory.

Magnet Totals

Each arc of each ring in the Collider contains 192 cells. When account is taken also of the cluster regions, there are roughly 3900 corrector packages in about 1800 F/D spools, 1700 C locations, and roughly 360 other assemblies. Including skew secondary correctors and about 250 IR region corrector dipoles, the corrector system will consist of about 14,000 magnetic elements, exclusive of IR quadrupole triplet correctors. This assumes that each bending half cell has a full corrector package of dipole through decapole, and FCD correction for the nonlinear elements. These elements are summarized by region in Table 4. In the table, we list the elements as if each were an independent magnet. Totals for each Arc, and two sets of standard "Arc type" cells within the clusters hold to a regular pattern. This is broken in the other cluster region cells ("Clus. Type") where both empty cells and dispersion suppressor cells are included. The "subtotal" line accounts for one of the two Collider rings excluding interaction regions. Both rings and the IR corrector dipoles are included in the final "total" line.

Magnet Development

The correction magnets needed to implement the corrector system are relatively small, but numerous. They need to be cost effective, but strong in comparison to previous superconducting correction magnets. The preliminary specification requires the magnets to reach full strength at or below 100 Amps, and to do so without training. The field quality is currently 1% field error at r = 1.0 cm. Mechanical tolerances must

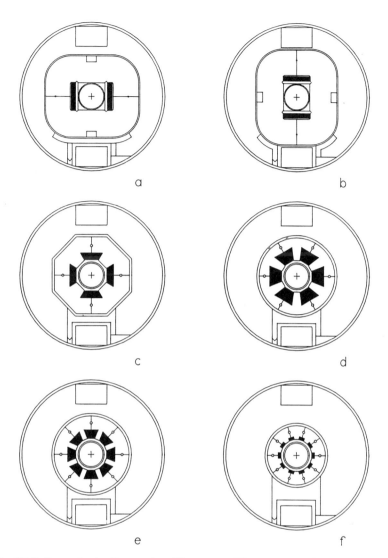

Figure 3. TAC Superferric Correction Magnets. The outer circles represent the cold
mass pipe. Rectangles above and below magnets are main dipole current busses
and signal busses. a) Dipole oriented for F spool, b) Dipole at D spool, c)
Quadrupole, d) Sextupole, e) Octupole, and f) Decapole.

permit alignment of the F/D sextupoles and beam position monitors to within 0.1 mm.

The space in which the correctors must fit is limited. The current length allocation
for the primary spool package of dipole through decapole is only 2.25 meters. The
corrector cold mass diameter is about 18 cm (in the spools). Main dipole current
busses and signal busses, both about 25 mm by 42 mm, also fit within the cold mass
above and below the correction magnets. The length available for C region correctors
is 0.50 meters. Present thinking is that at least the dipole corrector will be a separate
magnet. While the others might be a radially nested package, a sequentially distributed
package of magnets has received the most study.

In addition to developing specifications for correction magnet strengths and performance, SSCL needs to explore promising designs and methods of corrector fabrication.

Given the magnitude of the development and production tasks, close collaboration with existing labs and good working relations with industry are essential. We intend to have correction magnet development facilities at SSCL by the end of this year. Corrector design studies and prototyping are already in progress at Lawrence Berkeley Laboratory (LBL), the Texas Accelerator Center (TAC), and Brookhaven National Laboratory.

Development of "random wound" magnets at LBL has demonstrated successful operation of dipoles at over 2.5 T on first quench.[12] These magnets are based on the technique used for Tevatron correctors, although fields needed at the SSC are significantly higher. Careful control of the winding and potting process, as well as attention to the insulation have raised first quench fields from approximately 1.0 T to over 3.0 T.

TAC has designed a series of superferric magnets which meets the above criteria and integrated strengths of Table 3, assuming a beam tube diameter of 34 mm (as for the 4 cm dipole). The B fields for these magnets (evaluated at r = 1.00 cm) are Dipole: 2.50 T, Quadrupole: 1.44 T, Sextupole: 0.61 T, Octupole: 0.25 T, and Decapole: 0.072 T. Allowing 2 cm intermagnet spacing, and total lengths 4 cm longer than magnetic lengths, the D spool package requires 2.10 meters, leaving 0.15 m contingency. Figure 3 illustrates the cross sections of the TAC magnets.[4,13] TAC is currently working on prototype quadrupoles of the this design. They will also test application of Multiwire[14] methods for coil winding without splices.

Conclusion

The SSC Collider correction magnet system reflects the evolution of the machine and recent advances in correction theory. While future modifications and adjustments to the correction system can be expected, we feel we have the broad outline of the new system in hand. Initial test and prototype facilities at SSCL will begin operation this year. Prototype development is currently underway at LBL and TAC, and we hope that industry will join in the development process. Developing cost effective magnets to satisfy the strength, space, and reliability requirements will be a challenging task.

References

1. J.D. Jackson, ed., Superconducting Super Collider Conceptual Design Report, SSC Central Design Group, SSC-SR-2020, 1986.

2. D. Bintinger *et al.*, Report of the Correction Element Working Group, SSC-SR-1038, 1989.

3. D. Neuffer, "Correction of the Multipole Content of Synchrotrons," NIM **A274**, 400 (1989); D. Neuffer, "Multipole Correction in Large Synchrotrons," in Proceedings of the Second Advanced ICFA Beam Dynamics Workshop, J. Hagel and E. Kiel, eds., CERN 88-104, p.159, 1988; D. Neuffer and E. Forest, "A General Formalism for Quasi-Local Correction of Multipole Distortions," Phys. Lett. **A135**, 197 (1989).

4. J. Sanford, ed., SSCL Site-Specific Conceptual Design Report (SCDR), 1989.

5. E.D. Courant and H.S. Snyder, Annals of Physics, **3**, 1-48 (1958).

6. D. Neuffer, "Asymmetric Nonlinear Field Correction with 5-Dipole Half Cells," SSC-N-673, 1989.

7. T. Garavaglia, S.K. Kauffmann, R. Stiening, and D.M. Ritson; several papers on SSCTRK in these proceedings.

8. E. Forest and J. Peterson, "Correction of Random Multipole Errors with Lumped Correctors", SSC-N-383, September 1987, and Proc. of the European Particle Accelerator Conference, June 1988, p.827, S. Tazzari, ed.

9. L. Schachinger, "Interactive Global Decoupling of the SSC Injection Lattice", Proc. of the European Particle Accelerator Conference, June 1988, p.857, S. Tazzari, ed.

10. R. Talman, private communication.

11. A.A. Garren, and D.E. Johnson, "Controlling the Crossing Angle in the SSC", SSC-213, April 1989.

12. D. Bintinger, P. Bish, K. Franck, and M. West, "SSC Superconducting Dipole Prototypes using a Random-Wound Potted Coil Technique", poster session, this conference.

13. Figure 3 courtesy R. Huson, Texas Accelerator Center.

14. "Multiwire" is a registered trademark of the Kollmorgen Corporation.

EARLY INSTRUMENTATION PROJECTS AT THE SSC

D. J. Martin, L. K. Mestha, S. A. Miller and R. Talman

Superconducting Super Collider Laboratory
2250 Beckleymeade-MS 1046
Dallas, TX 75237

Abstract: Conceptual designs for some SSC instrumentation is given. Stripline beam position monitors, appropriate for cryogenic operation are described, along with plans for their A/D conversion and recording. A global timing system based on fibre optics is described; it is to be capable of ±100 psec accuracy over many tens of kilometers. Stabilization is patterned after a scheme in use at CERN. Timing pulses (roughly 60 MHz) as well as pulses synchronized to the various frequency modulated RF systems are distributed and scaled by digital clocks situated at those locations where accurate timing is required. Finally, a digital control circuit to be used for synchronizing beam transfer from the Low Energy Booster to the Medium Energy Booster is described. It is based on controlling the relative phases of the two RF systems even though one of the frequencies is variable.

INTRODUCTION

In this paper, three related instrumentation projects at the SSC are described. These projects are still at the conceptual design level, with acquisition and prototype development either just beginning or not yet started. The first report describes beam position measurement. The second describes a distributed timing system capable of providing timing signals with accuracy ± 0.1 nsec anywhere on the site. The third report describes the Low Energy Booster (LEB) to Medium Energy Booster (MEB) beam transfer synchronizer.

SSCL BEAM POSITION MONITORING SYSTEM

The BPM system contains the most important instrumentation for beam control. Position pick-ups, each with four electrodes, will be located at most quadrupoles in all the

---

*Operated by the Universities Research Association, Inc., for the U.S. Department of Energy under Contract No. DE-AC02-89ER40486.

accelerators. Doublet pulses of 1 ns duration and 16 ns separation are produced as the beam traverses the pickups. The peak-to-peak amplitude of pulses varies from 0.5 V at machine commissioning to 20 V at operating intensity. The detector signals will be processed and digitized in the niches, and the digitized data stored in registers for readout by the control system. At least three methods of analog signal processing are being considered to meet accelerator requirements.

Each sensing device is a detector consisting of four 15-cm, 50 ohm strip transmission lines placed above, below, and to both sides of the beam as shown in Figure 1 and Figure 2. To maximize the signal to noise ratio, each electrode subtends most of one quadrant. So that the electrodes will not be aperture defining elements, the beam tube bulges out around them. The electrodes are also recessed 2 mm outside the aperture so that synchrotron radiation cannot strike them. The characteristic impedance of the four electrodes must be matched within $\pm$ 0.25% to hold the electrical to mechanical center difference to 0.005 in., and to be within $\pm$ 0.5% of 50 ohms to control reflections at the cable interface. The detectors are rigidly welded to the spool pieces in each half-cell for alignment purposes and therefore operate at 4 K. The 8000 vacuum feedthroughs used in the collider must isolate beam vacuum from liquid helium. Because of their great quantity and inherent difficulty of replacement they must be very reliable (MTBF $> 870 \times 10^6$ hrs). The feedthrough and cable to the outside of the cryostat form an integral assembly and are composed of 316 Stainless Steel (S.S.) and Al_2O_3 ceramic. The cable dielectric is SiO_2. The 0.142 in. dia. S.S. jackets of the four cables form a part of the hermetically sealed cryostat which confines the liquid helium. The cables will hold off 20 atm. LHe, be radiation resistant, tolerate welding, and be very rugged.

Since beam detector directionality is not required in the collider, one end of each stripline is shorted. This measure saves construction cost, reduces heat leak, and improves reliability. The integral cable assemblies mate to SMA bulkhead vacuum feedthroughs which isolate guard vacuum from atmosphere. These feedthroughs are at the outside surface of the cryostat and are readily replaceable if damaged. All BPM signals will be brought to processor racks in the tunnel niches from the adjacent three consecutive upstream and downstream half-cells. The longest cable runs are 270 m.

The BPM electronics must provide position and intensity signals under various operating conditions: during machine commissioning; at full intensity collider operation; in fault diagnosis; and during specialized accelerator studies. Optimal performance in a variety of applications requires front-end processing tailored to the various modes and ring locations.

For maximum sensitivity and maximum dynamic range in sensing trains of bunches separated by 5 m intervals, down-converted amplitude-to-phase conversion is used[1]. Signal processing is done using only one harmonic of the RF bunching frequency. With 15 kHz wide bandpass filters in the IF section and 15 dB noise figure limiters, resolution of 100 nm should be achieved. The good sensitivity of these channels at low beam current will be valuable in steering the beam through the first turn commissioning. For this purpose, it is not necessary to instrument every half-cell, and for that reason, AM/PM processing will be used at the BPM stations adjacent to the niches, every three cells. These locations use 30 m of 7/8 in. solid copper jacketed corrugated cable, making the niche BPM's the most sensitive locations in the accelerator. Position monitoring near the IR's will also use this system.

At the BPM's fartherests distance, from the niches, there exists skin effect losses even in high quality, and low loss cable remains excessive for the pulse risetime to be preserved.

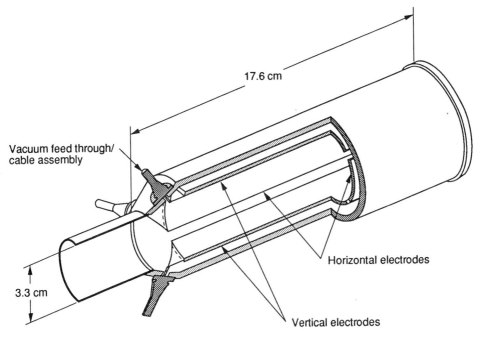

Figure 1. Isometric view of beam position monitor.

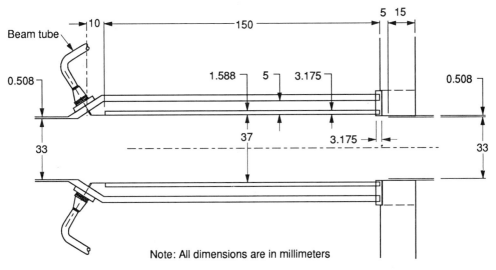

Note: All dimensions are in millimeters

Figure 2. Plan view of beam position monitor.

Therefore, located close to the pickups, will be peak detection circuits which effectively down-convert the high frequency components. The diode detectors that provide the bulk of the front-end BPM processing, are the least costly of the methods, and use lower quality RG-213 cable to transmit rectified pulses to the niches. The diode circuits have limited dynamic range, about 35 dB, and must be hardened against radiation. To measure the transverse position of a particular bunch, adjustable timing signals are used to gate the 12-bit ADC. These timing signals are derived from the global timing system described in the next section. Each station will be equipped with First In First Out (FIFO) memories that retain the most recent ten thousand turns of data. These will be used for occasional readout, for example: for post mortem after a beam abort to reconstruct the beam orbit. Under special conditions, all position monitors must be able to measure orbit distortions of about 10 microns peak amplitude to locate orbit cusps.

For specialized beam studies, bunch-by-bunch and turn-by-turn processing will be implemented in a few places in each accelerator. Such a system will be able to measure the position of any bunch, any sequence of bunches, the same bunch on every beam orbit (turn), or any other definable bunch pattern. The electronics will employ 8-bit 100 MHz flash A/D converters, digital intensity normalizing, and FIFO memories to record position histories. To obtain reasonable resolution, wideband sum and difference signals are obtained using hybrid transformers integral to each detector. The flash converters clocked at 60 MHz take one sample of the vertical position, horizontal position, and intensity of each bunch. The ratios of differences to sums are taken digitally and are written into fast memories, which can be read by the control system. A 100Kbyte memory could store the position data of 6 full turns, or the position of a single bunch on 10^5 turns. The electronics would be similar to the front-end processing done in the Tevatron beam dampers[2].

SSCL PRECISION TIMING SYSTEMS

Introduction

The timing systems of the SSCL generate the individual timing triggers in the accelerator complex. They are used by the synchronization of beam transfer, beam dump, and beam position data acquisition systems, which require a jitter of ≤ 100 psec. Their use by the ramping and corrector magnets require a jitter of ≤ 1 msec. Finally, the systems provide a phase compensated RF reference. It is no small task to distribute a timing signal with a jitter and precision of ≤100 psec over a geographic area of approximately 200 square miles. The collider itself will contain 161 niches, spaced evenly 540 meters apart. Each of these niches must receive the high precision timing signal.

The precision timing system incorporates features from the Tevatron, CERN, and the SLC project at SLAC. An SSCL timing system consists of a distribution network which carries RF reference signals and timing modules which count the number of cycles of the RF reference signal. At predetermined counts, timing modules produce trigger signals. For more precise timing, individual RF cycles can be subdivided.

There are 7 different timing systems: the Linac, LEB, MEB, HEB, SSC Top, and SSC Bottom Beam Syncs, and Global Timing. The Beam Sync systems are distributed to their respective accelerators, while the Global Timing is distributed to all accelerators. The systems perform the same functions but have different master clocks. The master clock of a system is the origin of the RF reference signal of a system. For the Global Timing, the master clock is phaselocked to the 1,000,000th harmonic 60 Hz line frequency, or 60 MHz. For the Beam Syncs, the master clock is phaselocked to the RF in the accelerator cavities.

During acceleration, the Beam Sync systems are approximately 60 MHz, but do sweep a significant frequency range (47 to 60 MHz for the LEB). Other than the effects of the FM signal, the systems may be considered identical in function and construction. We will confine our observations to the Global Timing System (GTS).

Goals of the GTS

We require that timing modules trigger simultaneously to ±100 psec anywhere in the accelerator complex. Jitter and repeatability of individual modules should be better than 100 psec. A high mean time between replacement (MTBR) is required, as the GTS must be operating properly for the collider to operate. MTBR should exceed 100,000 hours for the individual modules. Self diagnostics should allow faults to be determined during production and as an aid to troubleshooting during installation and operation. If possible, the system should give indications of how soon various failures are predicted for modules.

Network Topology

The precision, cost and reliability of the timing system is strongly influenced by how the signals are actually routed - the distribution network topology. The intention of the distribution network topology is to minimize jitter, uncertainty, drift and cost, while maximizing reliability.

The simplest distribution network, a horseshoe (Figure 3a) would have one fiber strung from niche to niche, with a repeater at each niche. This would minimize cost of fiber used to $262,500, but would result in a probable increase in jitter, uncertainty and drift of 22 times that of a single repeater, and a worst case increase of 484 times that of a single repeater. The loss of a single repeater would cripple all the down stream timing circuits, or as much as 1/2 of the collider.

The opposite extreme, as shown in Figure 3b (star net), would be to string a separate fiber to each niche, which would require bundles of as many as 484 fibers to be strung in the cable conduit. It would minimize the jitter, uncertainty and drift to that of one repeater, but would require $63,381,000 of fiber. Loss of any repeater would affect only one niche.

A compromise star-star network (Figure 3c) would distribute one fiber to each sector, which would then subdistribute to each niche. This results in bundles of no more than 13 fibers. The total cost of fiber is $982,200. The probable jitter, uncertainty and drift are 1.41 times that of a single repeater, and the worst case is 2 times that of a single repeater. Loss of a subdistribution repeater would affect only one niche, while loss of a distribution repeater would affect at most one sector. The star-star network will require 10 more repeaters than the other systems. The system is shown in more detail in Figure 4.

Installation of the network is no small task. There is some question of the long term radiation resistance of the fiber, as silica becomes more opaque as it is exposed to neutrons. The fibers are shielded underneath a minimum of 18 inches of rubble and concrete. The

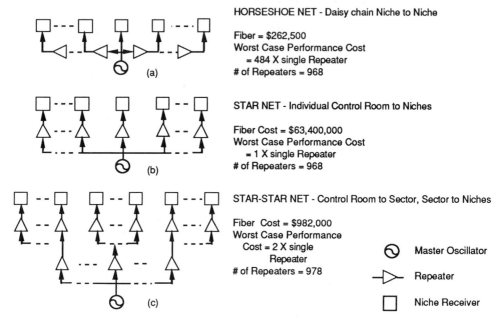

HORSESHOE NET - Daisy chain Niche to Niche

Fiber = $262,500
Worst Case Performance Cost
    = 484 X single Repeater
# of Repeaters = 968

(a)

STAR NET - Individual Control Room to Niches

Fiber Cost = $63,400,000
Worst Case Performance Cost
    = 1 X single Repeater
# of Repeaters = 968

(b)

STAR-STAR NET - Control Room to Sector, Sector to Niches

Fiber Cost = $982,000
Worst Case Performance
    Cost = 2 X single
        Repeater
# of Repeaters = 978

(c)

Master Oscillator

Repeater

Niche Receiver

Figure 3. Timing Distribution Network

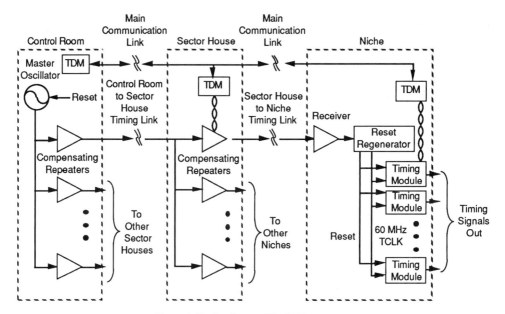

Figure 4. Timing System Block Diagram

114

niches are 540 meters apart, and contain the pull boxes for the fiber. To reduce the number of splices required, the fiber bundle is pulled from one niche and respooled at the next. This process is repeated, with appropriate breakouts being made, until the entire network is strung.

High Accuracy Timing Signal Transmission

Either coaxial or fiber optic cables could be used to distribute the timing signal. Fiber optics are superior to coax in their cost, attenuation, variation of propagation velocity with temperature, and common mode noise rejection characteristics. The radiation resistance of fiber optics is comparable to coax[3]. Fiber optic links have also been used in the Tevatron and LEP timing systems.

Variations in temperature of ±5 °C will make variations of as much as ±13 nsec in triggering time from one side of the collider to the other due to the variation in propagation velocity with temperature. A technique originally developed by Peschardt and Sladen[4], enhanced this application which allows the temperature variation to be compensated (Figure 5). It does this by measuring the phase shift through the fiber round trip and compensating for any variation.

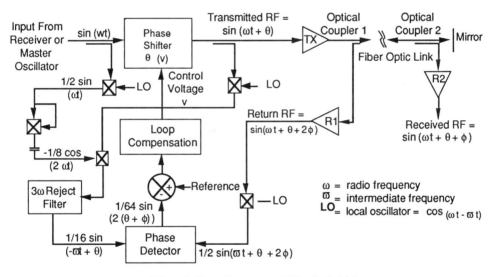

Figure 5. Phase-Compensated Fiber Optic Link
(Adapted from Peschardt and Sladen)

As seen in Figure 5, an input signal, designated "From Receiver/Signal Processor", is assumed to be the phase reference V_i.

$$V_i(t) = \sin(\omega t) \tag{1}$$

This is passed through the phase shifter, and results in V_t

$$V_t(t) = \sin(\omega t + \theta(v)), \tag{2}$$

where the phase shift θ is a function of the control voltage v. V_t is then transmitted over a single mode optical fiber. The media adds an additional phase shift of $\phi = \omega\tau$, where τ is the one way time delay of the media. Optical Coupler 2 picks off a fraction of the signal, which is converted into V_{rec} by Receiver RX2.

$$V_{rec}(t) = \sin(\omega t + \theta(v) + \phi) \qquad (3)$$

The signal which is not absorbed by RX2 is reflected back to the transmitter, where it is picked off by Optical Coupler 1 and converted into V_{ret}.

$$V_{ret}(t) = \sin(\omega t + \theta(v) + 2\phi) \qquad (4)$$

V_i is down converted by a local oscillator LO. The down converted frequency is represented by ϖ. The frequency of LO is set so ϖ is about 100 KHz.

$$V_{LO}(t) = \cos((\omega-\varpi)t) \qquad (5)$$

$$V_i(t)\,V_{LO}(t)\Big|_{LowPass} = \sin(\omega t)\cos((\omega-\varpi)t)\Big|_{LowPass} = \frac{1}{2}\sin(\varpi t) \qquad (6)$$

This down converted reference is then multiplied by itself and the DC component removed, resulting in $-\frac{1}{8}\cos(2\varpi t)$. This is multiplied by a down converted image of V_t and the 3rd harmonic removed:

$$-\frac{1}{8}\cos(2\varpi t)\,\sin(\varpi t + \theta(v))\Big|_{\varpi\,BandPass} = \frac{1}{16}\sin(-\varpi t + \theta(v)). \qquad (7)$$

V_{ret} is also down converted, resulting in

$$V_{ret}(t)\,V_{LO}(t)\Big|_{LowPass} = \sin(\omega t + \theta(v) + 2\phi)\cos((\omega-\varpi)t)\Big|_{LowPass}$$
$$= \frac{1}{2}\sin(\varpi t + \theta(v) + 2\phi). \qquad (8)$$

When these last two signals are multiplied together in the phase detector and harmonics removed, the result is V_{PD}:

$$V_{PD}(t) = \frac{1}{64}\cos(2\theta(v) + 2\phi)). \qquad (9)$$

If the error amplifier and the loop filter feedback to the phase shifter, they will cause

$$V_{PD}(t) = \frac{1}{64}\cos(2(\theta(v) + \phi)) = constant = V_{ref} \qquad (10)$$

or

$$\theta(v) + \phi = constant, \qquad (11)$$

116

provided that θ(v) is approximately linear and has a positive slope, and the loop filter stabilizes the loop. But this last equation is merely the condition that the phase shift from the input to the receiver output be constant. Therefore any changes in φ with time, temperature, etc., will result in a change in θ(v) which cancels it out.

At CERN, this resulted in a reduction of the variation of the delay of the LEP RF reference line, the longest of which is 7.9 Km, from ±8 nsec with the feedback loop off to ±19 psec with the feedback loop on, over a 20 °C temperature range. Extrapolation to the requirements of the SSC show that ±100 psec stability should be achievable.

Two additional modifications are required for the SSC. The system must be modified when used as a Beam Sync for the lower energy boosters. This is because of the wide variation in operation frequency of these accelerators. Since the phase compensated line keeps the total phase shift through a line constant, the net time delay of the system is a function of frequency:

$$\tau = (\theta(v) + \phi)/\omega = \text{constant} / \omega \qquad (12)$$

This may be handled by using a TDC which keeps track of the total transit time. This number is then used to control a variable delay line, rather than a variable phase shifter. Under this control scheme, the system keeps the time delay, rather than the phase, constant (Figure 6). An alternative may be to account for the time delay variation with frequency in the database of the central control computer, and to redistribute new trigger times as required. The hardware solution is perhaps better, since it requires about the same level of complexity and would greatly reduce the amount of communication required between the timing modules and the central computer.

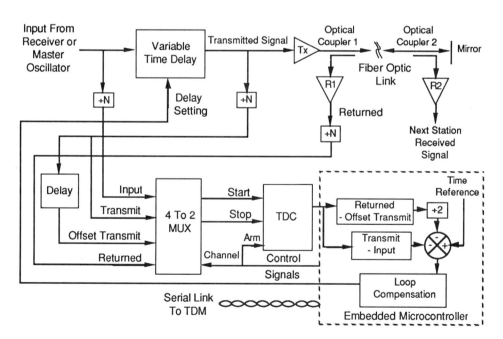

Figure 6. Time-Compensated Fiber Optic Link

A reset mechanism must also be provided which allows the timing modules to be set back to a zero count (Figure 4). This is done by removing a pulse from the transmitted pulse train. The missing pulse is detected and generates a reset signal. The reset signal is used by timing modules to reset their internal counters to zero. The RF timing frequency itself is regenerated using a phase locked loop (PLL) in a process similar to that used in high speed serial transmissions (Figure 7). This regeneration process greatly enhances the phase stability and noise margin of the signal. The signal is then distributed locally to the niche or recombined with the reset signal for subdistribution.

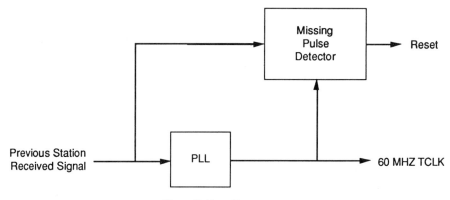

Figure 7. Reset Regenerator

A diagnostic signal is returned from the loop filter to the central computer. This diagnostic measures the required compensation and hence is a direct measurement of the phase shift of the line. Each repeater's nominal phase shift can be recorded in a database in the central computer and periodically compared with the current value. A large shift in compensation would indicate that some component in the system is shifting drastically and will need to be replaced at the next maintenance cycle.

Timing Module Operation

The timing modules themselves are elementary (See Figure 8). The number of requested time clock cycles required is down-loaded from the central computer to an individual timing module via a dedicated serial link to a Time Domain Multiplexer/Demultiplexer. This is received and stored by a dedicated microcontroller unit (MCU). The MCU holds the least significant bits of the requested time word (RTW) on the input of an equality detector.

The RF reference signal TCLK enters the High Speed Counter, which simply tallies the number of clock cycles since the counter was last reset, or since overflow. The size of the High Speed Counter is considerably smaller than the RTW, but gives the MCU enough time to anticipate the instant when the RTW is equal to the total number of cycles since reset . When the number of cycles counted by the High Speed Counter is equal to the least significant bits of the RTW, a pulse is generated which alerts the High Speed Arming Circuit, which in turn alerts the MCU. The MCU then compares the rest of the RTW with the number of alerts received from the High Speed Arming Circuit. Once the rest of the RTW minus one matches with number of alerts, the High Speed Arming circuit is armed by the MCU. When the next equality is detected, the High Speed Arming Circuit produces an output pulse. This pulse is delayed by a fraction of a cycle by a programmable delay line, which acts as a vernier on the RTW.

The MCU clock is slaved to TCLK, assuring that its software counters are synchronized to TCLK at least to the accuracy required to properly anticipate and arm the High Speed Arming circuit. More than one timing module can be serviced by one MCU, which allows a reduction in cost and complexity, especially in the serial communication link to the central computer. For timing applications which do not require very high accuracy, the vernier delay line can be left out. For applications which require only crude timing, most of the high speed circuitry may be omitted.

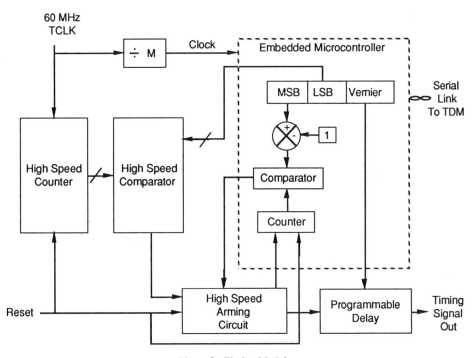

Figure 8. Timing Module

Timing System Calibration

The central computer provides RTW's which are compensated for delay time from niche to niche, FM effects in the lower energy boosters, etc. Each of the timing systems must be calibrated and the calibration stored in the central database. For example, the difference in phase between two clocks separated by a communication link of time delay τ is given by

$$\Delta\phi = \int_{t_0}^{t_0+\tau} \omega(t)\,dt$$

$$\cong \tau\,\omega(t_0). \qquad\qquad (13)$$

It is clear that the time delay from the master oscillator to each niche must be known with high accuracy. This calibration procedure is done in two steps. The first, a crude calibration, allows the SSCL to be commissioned, but does not provide the ultimate level of accuracy of the system to be achieved. The technique is simply to replace the master oscillator with a high stability atomic clock, and then to synchronize a second atomic clock to the temporary master. This atomic clock is portable and connected to a timing module . Once this is done, the second atomic clock and its attendant timing module may be moved quickly from niche to niche including the mobile timing module, compared with the count achieved by the timing modules at each niche.

The accuracy of this system is limited by the maximum speed with which a precision atomic clock may be moved from point to point in the system and by the phase stability of state-of-the-art atomic clocks. The best accuracy requires a hydrogen maser clock. The speed of a clock through the collider is limited to about 15 Km/hour. An overland route might speed this up considerably, but might unduly vibrate the atomic clock.

The second calibration step can proceed once protons can circulate through the collider. The protons travel at a known velocity, for a known distance, and their time of flight from niche to niche can be determined with arbitrary accuracy and precision. By using the timing modules to note the exact time of arrival of a single bunch of protons, the time delays of the system may be calibrated out to an accuracy of better than 100 psec at all locations in the accelerator complex.

SSCL BEAM TRANSFER SYNCHRONIZATION

Introduction

At present there are two schemes for achieving synchronous transfer of bunched beams from one circular accelerator to another in various laboratories around the world[5,6]. In the Booster to Main Ring transfer at Fermilab, a phase-locking scheme is in use. In this approach, at a suitable time before injection to the Main ring, the phase and frequencies of the Booster RF are locked to the Main Ring RF system. When the RF phase error is zero synchronous transfer is triggered. In the corresponding transfer at KEK in Japan the phase matching is done by using a phase-slippage scheme. In this scheme, at a predetermined time before the transfer, the Booster RF frequency is offset by 10Khz relative to the Main Ring RF frequency. As a result, since the Main Ring RF frequency is constant, the RF phase of the Booster ring slips relative to the Main Ring. Phase coincidence is detected and then a kicker system is fired to transfer the beam. Some emittance dilution occurs owing to the mismatch of bunch spacing; that is, an inherent drawback of the system.

The phase-locking scheme gives good results when there is sufficient time, first to achieve the RF lock, and then to perform the "cogging" operation in which the bunch pattern is rotated into its correct orientation. In the SSC, especially for the beam transfer between the

Low Energy Booster (LEB) and the Medium Energy Booster (MEB), this scheme is not suitable because at top energy the LEB is operating very close to transition which makes the cogging time too large[7,8] and the peak B-field does not have a long flat top since it is derived from the biased sinusoidal resonant power source. (It has been judged impractical to derive a long flat top by switching an auxiliary power supply). However, for all other transfers at SSC, this scheme can be applied since the B-field can be held constant until the transfer process is completed.

The ideal scheme for synchronizing beam transfer from the LEB to MEB is to keep control of the relative beam phase throughout the acceleration. Since the frequency of the LEB is modulated by about 20% during acceleration, the concept of relative phase must be generalized and for that purpose we introduce a new term "synchronizing phase" for a phase that is controlled through the cycle and acquires its standard meaning when the two frequencies become equal. In this method, transfer is triggered at the instant the synchronizing phase vanishes. This avoids emittance dilution due to mismatch of bunch spacing as in the phase-slippage scheme, even with large bunch separation (e.g., 90m instead of the nominal 5m). The scheme has two essential elements. First, it involves the accurate detection of the synchronizing phase error by comparing the measured values with the known "trip plan" stored in the computer. The trip plan and the synchronizing phase are related terms with the trip plan being ideal values of the synchronizing phase at discrete intervals through the cycle. Second, the measured phase error is passed through a feedback controller digitally, so that a compensating RF frequency shift is generated to minimize the error. In this way, the synchronizing phase is controlled to a pre-programmed trajectory. In the following section, we discuss the method qualitatively and show how it will be possible to detect the synchronizing phase at regular intervals. The feedback controller is not explained. A detailed analysis of the scheme is contained in reference[9].

Principle of the Phase-Control Scheme

To begin with, let us assume that there is no frequency error in the LEB and MEB RF systems. Transfer line delays and measurement errors in beam position (due for example, to coupled bunch oscillations) are ignored. Also, to understand the basic principle, let us assume that the synchronous phase of the reference particle in the beam bunch is constant throughout the acceleration cycle. With these assumptions in mind, we can write an expression for the path length covered by the beam for a time duration of τ as follows.

$$L_{LEB}(\tau) = \frac{2\pi R_L}{h_L} \int_0^\tau f_L(t)dt + L_{LEB}(0)$$

(14)

where, τ = Time interval
R_L = Radius of the orbit
h_L = Harmonic number for the LEB
$f_L(t)$ = RF frequency of the LEB

(Numerically L_{LEB} is roughly 10^7m at transfer time. Since a longitudinal uncertainty of less than 10^{-2}m is required at that time, it is clear that absolute control of all parameters to the corresponding precision of one part in 10^9 is an impractical way of synchronizing transfer).

Taking advantage of the constancy of the MEB frequency, the path length in the MEB can be written:

$$L_{MEB}(\tau) = \frac{2\pi R_M}{h_M} f_M \tau + L_{MEB}(0).$$

(15)

The difference in path lengths, $L_\psi = L_{MEB} - L_{LEB}$, is equal to the synchronizing phase. The phase when defined in this form has units of length. To solve for L_ψ, the following equation is used for the LEB RF frequency with usual notations.

$$f_L(t) = \frac{h_L c}{2\pi R_L \sqrt{1 + \left(\frac{M_p c^2 / e}{\rho c B(t)}\right)^2}}$$

(16)

Where ρ is equal to the effective bending radius. The accelerator guide field $B(t)$ varies with time according to the following expression:

$$B(t) = B_{min} + \frac{B_{max} - B_{min}}{2}(1 - \cos \omega t).$$

(17)

Using Eq. 15 in Eq. 13 the path length, L_{LEB}, is calculated numerically with a fine time step of 1ns. We assumed $L_{LEB}(0)=0=L_{MEB}(0)$, i.e., the beam bunch in the LEB was assumed to have started at the same time as the MEB beam bunch from the reference points. The reference points were chosen at the beam pick up points nearest to the beginning and the end of the transfer line. For synchronous transfer, we would like the synchronizing phase to be equal to zero. Since we are dealing with a circular machine, when the MEB beam bunch has completed one single turn, the LEB beam bunch will have completed a few whole turns plus a semi-turn. The semi-turn is due to the difference in the two RF frequencies throughout acceleration and the fact that the circumference ratio between the MEB and the LEB is not a whole integer. Hence L_ψ can be rewritten in the following form.

$$L_\psi = L_{MEB} - L_{LEB} = (N_{MEB} + \gamma_M)2\pi R_M - (N_{LEB} + \gamma_L)2\pi R_L$$

(18)

Where N_{MEB}, γ_M and N_{MEB}, γ_L, represents the whole and the semi-turn completed by the MEB and the LEB reference beam bunch respectively. The whole turn is of no significance for synchronous transfer for obvious reasons. Hence we simply drop it and consider only the fractional part representing the incomplete turn.

For synchronization, at the time of transfer, the fractional part γ_M and γ_L, must be equal to zero. In Figure 9, L_ψ is plotted each time the MEB beam bunch returns to the reference point (i.e. when γ_M becomes zero). It can be seen clearly from this figure that after about 47ms, the LEB beam bunch tends to come back to a fixed point in the orbit more frequently than before. We see three curves that are approaching constancy. (Three curves are due to the fraction "1/3" in the MEB to LEB circumference ratio: when the ratio is a whole integer, then we would have only one curve.) The decay is due to the fact that the difference between the two RF frequencies is narrowing as the time approaches the nominal 50 msec transfer time at which B(t) is maximum. These curves settle down eventually to a fixed value; for example, 68.38 m is the final settling value for curve 2. The final settling value can be moved anywhere in the orbit if we set the initial relative beam position, $L_{MEB}(0) - L_{LEB}(0)$, appropriately. For example, to make the decay to zero, we have to set $L_{MEB}(0) - L_{LEB}(0)$ to be equal to 68.38m; then curve 2 will exponentially decay to zero. Experimentally this can be achieved by controlling the injection time from the Linac into the LEB and the LEB RF turn-on time. That is, by knowing the time when the first pulse in the Linac has been injected, the RF switch-on time can be adjusted to a little later time so that there is enough time to arrange the MEB phase relative to the LEB phase to a desired point in the ring. By doing this initial phase adjustment we can achieve correct phase at the instant of transfer provided we have kept control of the phase to an adequate degree of accuracy throughout the cycle. This path length, L_ψ, becomes our trip plan for the LEB beam bunch. We set up a detection scheme digitally, to enable the deviation of the synchronizing phase from the designated trip plan for every MEB turn and then use this error information to carefully adjust the LEB RF frequency such that it is forced to take up a desired value sufficiently in advance within a reasonable time before extraction. We have plans to "feed forward" by refreshing the trip plan every cycle so that a pseudo-adaptive loop is arranged to adjust itself without human interference.

The detection of phase error will fail if the B-field fluctuation is too large since it may lead to a very large phase error. If it is greater than the LEB circumference within the settling time of the feedback controller, then we will lose control over the phase error. We evaluated this case for a fractional B-field error of 10^{-3}. It showed that the phase error exceeded one LEB circumference in 6 milliseconds. This is quite tolerable since we can design a feedback controller that adjusts the LEB RF frequency to control the phase error within 6 milliseconds. Since the B-field error is expected to be less than 4×10^{-4} we anticipate acceptable behavior.

In Figure 10, we have sketched the block diagram of the complete digital control system. In this scheme, we have shown two inputs for the 'Synchronizing Phase Error Detector.' These can be either from the beam position monitors located at the reference points or from the two RF signals. The trip plan for the LEB such as the one shown in Figure 9 is calculated in the Trip Plan Generator. Apart from the computed phase, it also includes subtle information such as (1) the synchronous phase of the LEB and MEB reference particles when the two RF signals are used as the input for the Phase Error Detector, (2) transfer line delays, (3) the kicker rise time, and (4) the steady delay in the electronics associated with synchronization. The phase error is computed for each MEB turn on a real time basis in the detector by time-slicing and then time-tagging the arrival time of the LEB beam bunch to the reference point. This information is then subtracted appropriately with the inexorable trip plan. The error in the Synchronizing phase is then processed in the Sliding-Mode Controller block[10]. The output of the Sliding-Mode Controller is the required frequency modulation which would compensate for the error in the synchronizing phase. The main RF frequency for normal acceleration of the beam is obtained from the B-field using high resolution A/D converters. Using modern high speed logic, we believe that it is possible to compute the phase error and also develop the control signal within one MEB turn which is about 13.2us. Details of the processor and the controller design are more complicated and hence are not shown here. It is however described in reference 9.

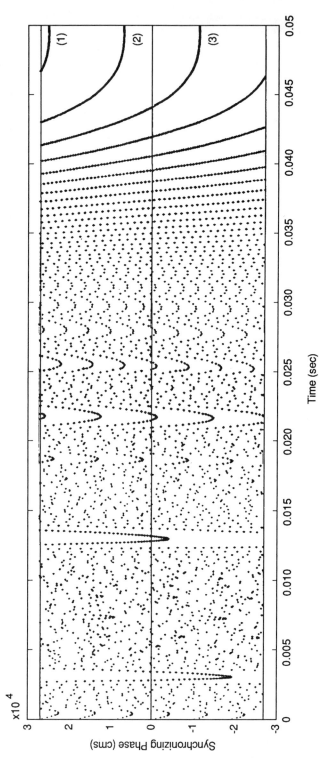

Figure 9. Synchronizing Phase During Acceleration In The LEB

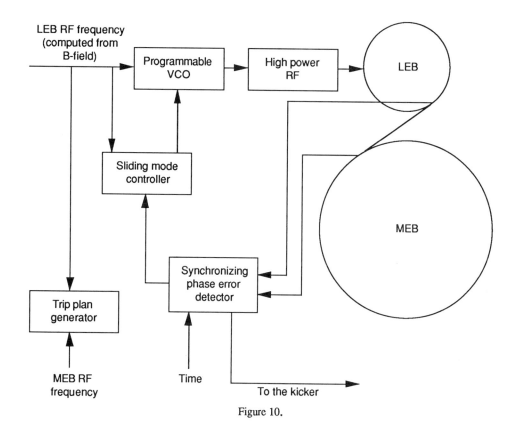

Figure 10.

REFERENCES

1. F. D. Wells and S. P. Jachim, "A Technique for Improving the Accuracy and Dynamic Range of Beam Position - Detection Equipment," IEEE PAC, 89CH2669-0, Proc. 1989.

2. J. Crisp et al., "A Programmable High Power Beam Damper for the Tevatron," IEEE Trans. Nucl. Sci., Vol. NS-32, No.5, October 1985.

3. C. E. Dickey, "Irradiation of Fiber Optics in the SSC Tunnel," private communication to be published in 1990.

4. E. Peschardt and J.P.J. Sladen, "Phase Compensated Fibre-Optic Links for the LEP RF Reference Distribution," IEEE Particle Accelerator Conference, Chicago, IL, p. 1960, 1989.

5. J. A. Dinkel et al., "Synchronous transfer of beam from the NAL fast cycling booster synchrotron to the NAL main ring system," IEEE Trans. Nucl. Sci., Vol. NS-20, No. 3, June 1973.

6. Y. Kimura et al., "Synchronous transfer of beam from the booster to the main ring in the KEK proton synchrotron," IEEE Trans. Nucl. Sci., Vol. NS-24, No. 3, June 1977.

7. Site-Specific Conceptual Design of the SSC, Technical volume 1, December 20, 1989.

8. Griffin, J. E.: 'Private technical discussions, FNAL, Chicago, Illinois, February 1990.

9. Mestha, L.K.: 'Phase-control scheme for synchronous beam transfer from the Low Energy Booster to the Medium Energy Booster', SSC Laboratory Report, 1990 (To be issued).

10. K. S. Yeung, Private technical discussions, Department of Electrical Engineering, University of Texas at Arlington, Arlington, TX, 1990.

3. Computers/Controls

VERSATILE COMPUTER PROGRAM FOR SOLENOID
DESIGN AND ANALYSIS

M. P. Lakshminarayanan and R. Rajaram

MHD Centre
Bharat Heavy Electricals Limited
Tiruchirapalli. 620 014
INDIA

ABSTRACT

Solenoid magnets have the simplest possible winding configuration and are used in various areas like (1). Magnetic Resonance Imaging (2). High Gradient Magnetic Separation (3). Superconducting Quantum Interference Devices (SQUIDS) etc. This paper describes SOLY, a highly interactive PC based computer program for Design and Field Analysis of Solenoid magnets, developed at MHD Center. The features of the program and also the principles and computation methodology employed are described. The analytical expressions have been derived exploiting the axi-symmetry of solenoid magnets. The computations are performed using simple numerical integration methods without the necessity to use Finite Element methods.

THE PROGRAM

SOLY can analyze and design solenoid magnets with homogeneous windings. It can also analyze solenoids with graded windings. It uses simple numerical techniques to provide fast and accurate results on Personal Computers. It employs highly interactive data editors for data entry and variation. There is a main menu which leads to design and analysis parts.

THE DESIGN PROCESS

The design of a solenoid for given magnetic flux density and homogeneity is performed by a set of routines to give a configuration with minimum usage of conductor. The specifications to be given are (1) inner diameter of solenoid (2) required magnetic flux density (3) size of wire and operating current and (4) minimum homogeneity in range. The out put of the program consist of (1) the length of the solenoid (2) number of turns per layer and (3) number of layers.

The basic equation used in this part is:

$$B_o = \mathrm{J.a.F}(\alpha, \beta)$$

where

B_o	= Required magnetic flux density,
J	= Current density in the winding,
a	= Inner radius of the solenoid and
$F(\alpha, \beta)$	= Fabri's constant

and

$$F(\alpha, \beta) = \mu_o \cdot \beta \cdot \ln \left[\frac{\alpha + (\alpha^2 + \beta^2)^{1/2}}{1 + (1 + \beta^2)^{1/2}} \right]$$

where

$$\alpha = \frac{\text{Outer Radius of Solenoid.}}{\text{Inner Radius of Solenoid.}}$$

$$\beta = \frac{\text{Half length of Solenoid.}}{\text{Inner radius of Solenoid.}}$$

Figure 1 gives the program logic of the design process. The first step is input of the specifications from the terminal through the data editor, then the value of the Fabry's constant $F(\alpha, \beta)$ is computed for the given flux density, inner diameter and current density. Now, the value of α and β for the given Fabry's constant are obtained from the minimum volume curve. Then the magnetic flux density values within the range specified are computed and checked for given homogeneity. If the required homogeneity is achieved the length, outer diameter, number of turns per layer and number of layers are computed and output on the terminal. If the required homogeneity is not achieved, the value of β is increased (in effect increasing the length of solenoid) and the corresponding value of α calculated and the new configuration is checked for given homogeneity. This process is repeated until given homogeneity is achieved. The output of the design process is automatically used by the analysis part if invoked subsequently.

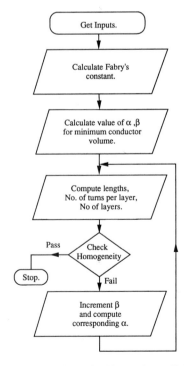

Fig. 1 Program logic for Designing a Solenoid.

The various routines used for performing the above computations can be easily used along with an optimization routine to design superconductor solenoid magnets with homogeneous as well as graded windings.

THE ANALYSIS PROGRAM

The analysis part of the program performs magnetic flux density computations (Axial and radial components) at any point due to a solenoid magnet. This part can analyze solenoids with simple homogeneous winding configurations as well as those with graded windings. The value of the magnetic flux density is given in the rectangular as well as the polar forms.

The inputs to this program are (1) Length of the solenoid (2) Inner diameter (3) Wire size (4) Operating current (5) Number of layers (6) Z Range for analysis (7) R Range for analysis.

THE COMPUTATION

Methods Available

Computation of the magnetic flux density at any point along the axis of a solenoid magnet is simple due to axi-symmetry and can be obtained, after integration of the basic equations, as a simple algebraic expression. For points off the axis the computation becomes complicated, since a closed form solution cannot be obtained. The solution can be obtained by reducing the integrand suitably in terms of elliptic integrals or Legendre polynomials. The evaluation of the elliptic integrals or the Legendre polynomials, as the case may be, gives the value of the magnetic flux density. In this paper we present a method of calculating off-axis fields using numerical integration.

Method employed

The equations for the magnetic field at a point due to a solenoid is arrived at by starting from the field due to a infinitesimal current element. Integrating this expression along a circle of given diameter and relative position, we get the expression for the field due to a circular current carrying loop. Now a single layer of a solenoid may be considered to comprise of a series of elemental loops. Therefore integrating the expression for the field due to an elemental loop along the length of the solenoid we get the expression for the field due to a single layer of a solenoid. Now if we consider a multi-layer solenoid to comprise of a stack of elemental layers and integrate the expression for the elemental layers from the inner radius to the outer radius of the solenoid, we get the expression for the field due to complete solenoid at the specified point.

The Expression

Consider Figure 2 which shows circular current carrying loop of a conductor. The point of interest P is defined by Z and D, where Z is the axial distance from the plane of the loop and D is the radial distance from the axis. Now, let us consider a current element $I.d\ell$ at point C.

By Biot-Savart's law, the magnetic flux density due to $I.d\ell$ at P is given by

$$d\vec{B} = \frac{\mu_o.I.\vec{d\ell} \times \vec{CP}}{4.\pi.|\vec{CP}|^3}$$

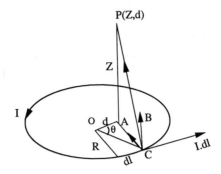

Fig. 2. Field due to a single current carrying loop.

This value of dB has an axial component dB and radial component dB (dB_A) and radial component dB (dB_R). Now, if we resolve CP along CA on the plane of the loop and CB perpendicular to the plane of the loop we have

$$|d\vec{B}_A| = \left| \frac{\mu_o.I.\vec{dl} \times \vec{CA}}{4.\pi.|\vec{CP}|^3} \right| = \frac{\mu_o.I.|\vec{dl}|.|\vec{CA}|.\sin\beta}{4.\pi.|\vec{CP}|^3}$$

and

$$|d\vec{B}_R| = \left| \frac{\mu_o.I.\vec{dl} \times \vec{CB}}{4.\pi.|\vec{CP}|^3} \right| = \frac{\mu_o.I.|\vec{dl}|.|\vec{CB}|}{4.\pi.|\vec{CP}|^3}$$

Now, substituting for CA, CB and CP in terms of θ, we have

$$|d\vec{B}_A| = \frac{\mu_o.I.R.\cos[\tan^{-1}(z/r)].\sin[\sin^{-1}(d.\sin\theta/r)].d\theta}{4.\pi.(r^2 + z^2)}$$

$$|d\vec{B}_R| = \frac{\mu_o I.R.\sin[\tan^{-1}(z/r)].\cos\theta.d\theta}{4.\pi.(r^2 + z^2)}$$

where $r = \sqrt{R^2 + d^2 + R.d.\cos\theta}$

Integrating the equations above between the limits 0 and 2π we get the magnetic flux density at P due to the loop.

Now, let us consider Figure 3 which shows a single layer of a solenoid which comprises of a series of elemental loops of thickness dx. Now, the ampere turns in section dx is given by

$$\frac{I.N.dx}{L}$$

where N = no. of turns per layer
L = length of solenoid.

Substituting this value for I in the expression for the loop and integrating along the length of the solenoid we get the expression for the field values due to a layer of the solenoid. Now, referring to Figure 4 showing an elemental layer of solenoid of thickness dr. Now, the ampere turns to be considered becomes

$$\frac{I.N.N_L.dx.dr}{L.H}$$

132

where N_L = No. of layers in the solenoid and

H = Height of the winding.

Substituting this value·in the equation for the magnetic due to a single layer of a solenoid and integrating wrt to r within the limits inner radius to the outer radius of the winding, we get the field value due to the complete solenoid.

The expression so obtained for the magnetic flux density has to be evaluated using numerical methods since the integrations do not yield a closed form solution. In SOLY, Gaussian Quadrature method using four (eight) points has been used for the numerical integration.

In case the solenoid has graded windings, the effective field components are calculated as the sum of those due to two concentric solenoids with homogeneous windings. Likewise, the routines can also be used to analyze the magnetic field due to Helmholtz coils.

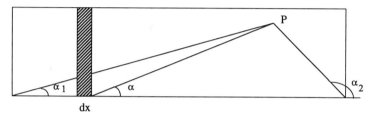

Fig. 3. Field computation for a single layer solenoid.

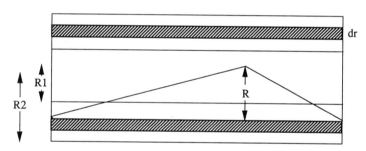

Fig. 4. Field computation for a complete solenoid.

CONCLUSION

The package has been implemented with the following features:-

1. Ease of problem definition.
2. High speed of computation on personal computers. The computation time is dependent only on the number of points at which magnetic field has to be computed.
3. Accuracy: The accuracy has been found to be within ±2%.
4. Easy adaptability for force computations.

References

1. Engineering Electromagnetics - W. Hayt
        (McGraw-Hill Kogakusha)
2. Superconducting Magnet Systems - H. Brechna
        (Springer-Verlag, Berlin)
3. Superconducting Magnets - H. Wilson
        (Clarendon Press - Oxford)
4. Designing Particular Magnetic Profiles by Computer
        - Jozsef Bankuti, Tamas Karman, Istvan Kirschner
        and Tamas Porjesz. (IEEE Trans. on Magnetics
        Vol. 24 Sept. 1988)

TRENDS IN DATA ACQUISITION INSTRUMENTATION

George J. Blanar

LeCroy Corporation
700 Chestnut Ridge Road
Chestnut Ridge, New York 10977-6499

ABSTRACT: Particle physics research demands unique data acquisition instrumentation in terms of speed, size, cost and architecture. This paper will focus on principal issues related to trends in high-speed, large-scale, economical, sophisticated instrumentation for high energy physics, heavy ion, nuclear and atomic physics as well as large scale astronomical experiments. Examples will be taken from experiments at many national laboratories including BNL, FNAL, CERN, SLAC, etc. as well as LeCroy Corporation's 26 year history in the field of physics research instrumentation.

Finally, instrumentation needs for the next generation of high energy, hadron colliders including the Superconducting Super Collider (SSC) and the Large Hadron Collider (LHC) at CERN will be reviewed and compared to current technologies.

INTRODUCTION

Exponential Problems of Data Acquisition

Physics research electronic instrumentation requirements for data acquisition, triggering and support functions have grown dramatically in the last quarter century. The introduction of higher energy, higher beam intensity facilities has forced the number of channels of data acquisition to expand exponentially. Figure 1 shows this growth with respect to the number of ADC or TDC channels for a particular detector of a high energy physics experiment (for example, hadron calorimeter or muon spectrometer), an entire heavy ion experiment, or an entire nuclear physics facility. In all cases, the trends are clear with only a scale change necessary to actually compare disciplines.

Since experimental particle physics has not had an exponentially expanding budget, this growth has been fortunately balanced by an exponential suppression in the price per channel of the data acquisition electronics. Figure 2 plots the cost per channel of many of the most popular commercial LeCroy ADCs and TDCs versus their period of peak use in experimental physics programs.

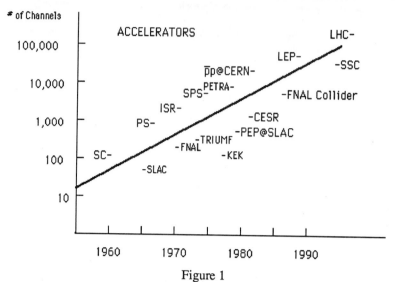

Figure 1
Start-up dates for several Particle Physics Research
Facilities vs. Number of Channels of Data Acquisition Electionics

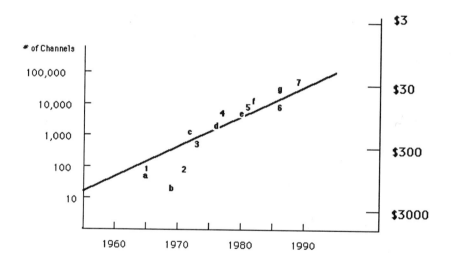

LeCroy Instrument Notation

ADCs	Model #	TDCs	Model #
a	143A	1	108H
b	243	2	226
c	2248	3	2226
d	2249	4	2228
e	4300	5	4303
f	2282	6	1879
g	1882	7	HTD161

Figure 2
Number of Channels/Detector/Experiment
and Cost/ADC or TDC Channel vs. Time

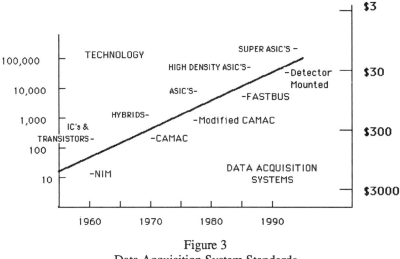

Figure 3
Data Acquisition System Standards
and the technology they use

Until now, the physics community has achieved this cost reduction with the introduction of standards to the data acquisition problem. Standards mean that instrumentation may be configured and constructed based on money saving principles including economies of scale, common crates, power supplies and interfaces, common software libraries etc. Therefore, the same FASTBUS IEEE-960 multi-hit TDC for Drift Chambers can be used by a heavy ion experiment at BNL, or a hadron collider group at FNAL, or an electron collider group at SLAC, or CERN.

The primary reason for the balance in density and cost has been the tailoring of the technology to the fixed parameters represented by both the standard itself and the scope of the instrumentational needs it must cover. Figure 3 reminds us of the exponential trends described above but notes the dates of the introduction of a number of data acquisition standards and the technology they exploited.

THE EVOLUTION OF STANDARDS

NIM, the First Standard

NIM standard instrumentation which was introduced in the 60's, used discrete components and standard TTL logic integrated circuits to provide the first true international instrumentation standard. It included mechanical standards for crates and modules as well as power supplies. NIM, used primarily for trigger logic, was never considered as a mainstream data acquisition system since it had no standardized computer interface to pipe the acquired data through.

CAMAC, the first C is for Computer

CAMAC (IEEE-583) represented a huge step forward in the 70's. It defined several architectures for control and data flow. The data flow was optimized to the standard data acquisition computer of the 70's, usually the "mini" such as Digital Equipment Corporation PDPs, Norsk Data NORDs or Data General NOVAs. With its common instrumentation control language ("CNAF"), standardized modules and similar computers, laboratories like CERN and FNAL could develop sophisticated software libraries.

CAMAC also supported the idea of multi-channel data acquisition modules with over 300 channels per crate. This increase in density was made possible through the use of hybridized circuits. Thick film hybrids eventually became the preferred manufacturing technique and permitted not only an enormous savings in the "real estate" that a circuit occupied, but also high performance in terms of noise isolation and speed.

Modified CAMAC, a Standard of Necessity

By the end of the 70's many of the limitations of CAMAC began to affect its use in the higher performing experiments. The most restrictive was the limit on the number of crates that could be run off of one parallel branch (7) and the speed of the transfers (1μsec/ word). Performance was also lost to the 24-bit data format that did not fit well to either the dynamic range of the instrumentation or the word size of the popular acquisition computers (DEC VAX).

In the absence of an IEEE or ESONE standard, many laboratories and commercial companies offered modifications of CAMAC that expanded both the architecture and the data transfer speed. With these changes, Modified CAMAC also benefited from the introduction of Application Specific Integrated Circuits (ASICs) to increase the number of channels of data acquisition that could be contained in a single slot of CAMAC; up to 32 TDCs and even 48 channels of ADCs for example. These single and dual channel ASICs were originally done in semi-custom bipolar and CMOS processes with feature sizes starting at 10 μm and eventually coming down to 4 μm.

FASTBUS, the Physics Community's Own Standard

In the beginning of the '80's, the IEEE-960 FASTBUS standard was established to "tame" the problems stired by CAMAC. Primarily driven by the labs and the particle physics community in general, problems of architecture, speed, flexibility and in-line computer processing were all accommodated. The cost per channel question was addressed by the unpublished goal of packing 10,000 channels of ADC or TDC in a single rack of electronics. This was roughly the number of channels in the initial design of a CERN LEP or FNAL Tevatron detector subsystem. Therefore, a data acquisition system had to have 4 or 5 crates per rack, 20 to 25 data acquisition modules per crate and finally about 100 channels per slot.

Single or dual channel ASICs were insufficient for the FASTBUS requirements. Higher density (4 channels or more per die) ASICs were now required. This push also meant a change from semi-custom processes to full custom designs. This requirement had been felt earlier since the problems of testing these semi-custom, high performance (15-bit dynamic range, 12-bit resolution, 5 psec jitter) devices was difficult using VLSI industry tools. In addition, migrating the design from an engineering run to production often was more difficult than the initial design.

Custom ASICs Were the Key

By the time CDF at the Tevatron and the four LEP experiments started installing their detectors and pulling cables, the FASTBUS standard with 96 channel ADCs and TDCs had already proven itself in several high rate experiments. These instruments were designed around higher density, 4-channel completely custom ASICs, as shown in Photograph 1.

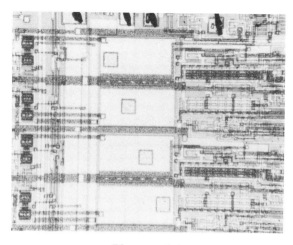

Photograph 1
Micro-photograph of the 4-Input Charge Multiplexer
used in the 15-bit, 96-channel LeCroy FASTBUS ADCs.

With time, the various roles of different semiconductor processes were more clearly understood. CMOS worked well for high-speed digital designs like scalers, digital delays, data compression, etc. On the other hand, bipolar lent itself to high performance analog designs including amplifiers, comparators, charge converters, multiplexers, etc. Experiments were tried with exotic processes (for the time) like SoS and GaAs with mixed results. Sometimes it was found that the exotics (SoS for example) were not that well understood by the semiconductor foundries themselves, and their intensive application in physics instrumentation was premature. Others were more expensive than originally foreseen with lower yields (GaAs).

DENSITY IS NOT THE ONLY STORY

Speed and High Rate Trends

Before turning to the next generation of experiments and their requirements, two other performance parameters must be traced for their trends and directions: Triggering and Data Compaction. The former is critical since the integration of the trigger into the data acquisition system is a natural consequence of both the increase in background reactions that the trigger must suppress and the increase in the number of detector channels. Other aspects of the problem include the disappearance of separate detector elements to form the trigger as well as the need to get both the data and the trigger functions out of the tightly packed detector systems.

The Integration of Trigger Systems

Figure 4 illustrates the merging of the trigger systems into deeper levels within the data acquisition system. This trend is driven by the increases in the background rate from the higher energy beams, the higher luminosity of the new accelerators and the corresponding lower relative cross section of interesting physics phenomena.

NIM systems were usually completely separate from the data acquisition system with a single cable running between them signaling a COMMON START to the TDCs or a

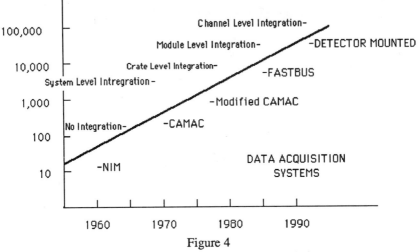

Figure 4
Triggering and Data Compression applied to
finer and finer levels of the data acquisition system.

GATE to the ADCs. The situation was similar at the start of the CAMAC era but eventually provisions were made to pick off some data at the point where the data acquisition system met the computer (programmable branch drivers) in order to make a third level trigger decision. Deadtime in the computer could then be avoided if the trigger conditions were not met.

With the increases in rates and sizes that were possible with Modified CAMAC, many systems inserted a trigger port at the custom crate controller level to supply data to second level triggers. The LeCroy PCOS III Multi-Wire Proportional Chamber System is probably the most successful example of this implementation and is in widespread use even today.

Finally, FASTBUS provided a backplane wide enough (especially with the auxiliary backplane) to allow data from individual modules and even groups of channels within the modules to contribute to second level triggers.

Data Compression, the Key to High Throughput

Figure 4 is also a record of the integration of data compression applied to finer and finer levels of the data acquisition system. This solution was dictated by higher rates and larger amounts of data that could not be simply handled by increasing the width and speed of the data acquisition backplanes. In addition, the speed and bulk of the data had to be reconciled with only a limited amount of computer archive capability.

NIM and early CAMAC systems had relatively few detector channels to read out. Therefore, no compression was necessary. However, with larger CAMAC systems the data acquisition system/computer interface was given the job of compressing data by excluding un-hit TDC channels and ADCs below pedestal.

Once again, Modified CAMAC used the crate controllers to encode the data so that while all the data was read out of the individual modules, only interesting data was passed on the branch. FASTBUS originally used the same approach but recent instruments including the LeCroy Model 1871 Rapid Encoding TDC and others compact data at the module level. Therefore, the backplane bandwidth of the crate is not compromised.

The next generation of data acquisition electronics will have to provide trigger information and compress data on an individual channel bases in order to cope with the demands of both interaction rates and data volumes.

ELECTRONICS FOR FUTURE COLLIDERS

The Case for Detector Mounted Electronics

The next generation of hadron colliders presents us with a critical problem. The late 90's will see the need for several hundred thousand channels per detector device at a cost requirement of less than $10 per channel. Additional restrictions are detailed below, but these two are enough to seriously abandon the idea of a branch/crate/module type CAMAC, VME, FASTBUS or VXI system. The problem is that such standard systems carry an overhead for cooling, power, mechanics, support services, etc. of approximately $10 per channel. With a $10 total average budget for the million SSC/LHC channels, this overhead is not acceptable. The most direct solution is to mount the electronics directly on the instrument and integrate the entire electronic chain into the logical and physical structure of the detector.

SSC/LHC, Additional Restrictions and Problems

The list of additional problems and constraints that have to be accommodated is impressive even if compared to the most sophisticated experiments running at LEP or the Tevatron: Beam Crossings every 16 nsec as opposed to several μseconds, Flight times of a ß=1 particle through the detector of up to 30 nanoseconds, Timing accuracies of multi-hit electronics for drift chambers to 0.5 nsec, ADCs to measure calorimeter information with 20 bits of dynamic range and 14 bits resolution, Interaction rates a thousand times higher than presently handled, Data per event ten times larger, etc.

The "zeroth" level trigger is always provided by the beam crossing. In order to create time to make a first level trigger, all the possibly useful data must be stored in an analog or digital pipeline for approximately 1 or 2 μseconds. Note that CCDs will not work for this storage element since the clocking currents for them would require the equivalent of an FM radio transmitter in the middle of the detector. Switch capacitor arrays are a more likely solution.

Trigger information must then follow its own routing to be used in a first level trigger decision. This pipelined process determines if the data in the storage pipelines should be routed to the next level of buffering or flushed. If the original storage pipelines are analog, data has to be converted. The second level trigger uses more of the data itself as opposed the trigger signals. This next level of buffering will probably be organized as a FIFO with the capacity of holding an event for about 5 μsec.

There have been a number of studies on both the intensity and type of radiation background the electronics will have to tolerate. It is not clear that the data used by the defense industry to set their standards of radiation hardness are relevant. However, one conclusion can be drawn: it will be hotter then anything that has been instrumented in elementary particle physics experiments to date.

Estimates range from 10MR for the inner detector to 1 MR by the muon chambers, for 10 year exposures. This eliminates several integrated circuit processes including simple bipolar, MOS and others. CMOS, special bipolar, GaAs and even SoS are better candidates.

Power will be critical. With a million channels of electronics mounted on or near the detector, power will have to be cut by a factor of 20 from the past levels. The use of appropriate technologies (CMOS) and separate, specific valued power lines will help. Fiber Optic transmission with the LED transmitters on the chambers will not be acceptable. However, an interesting idea was presented at the 1990 IISSC conference by Rykatzewski, et. al. for the L* proposal with off-chamber mounted light sources and an individual shutter integrated in each channel.

All future collider proposals call for hermetic detectors, with as close to 4π coverage as possible for each layer of detector elements. There is little or no room left for the electronics much less the cables or heat/cooling plumbing.

Therefore, data acquisition electronics will have to use new techniques like flexible circuit boards. Trigger, data compression, data signal processing and multiplexing of the data on the detector will help to keep the cabling to a minimum. The use of glass fiber optic links at this point will be highly desirable to get the data out of the detector and into a computer "farm" for further in-line computing and higher level trigger decisions.

Architectures for capturing data and techniques for designing and implementing electronics will have to change to accommodate new levels of reliability and fault tolerance. Studies of techniques used in the space program, avionics and military should help us get fresh ideas.

Because of space restrictions, power, noise sensitivity and reliability, we will have to continue the trend of designing higher channel density ASICs. Eight to 64 channel designs are now discussed and experience with some of the 128 channel Si Strip preamps must be integrated.

A Couple of Models to Consider

Practical solutions to at least some of these severe restrictions can be found in several older experiments. Proton decay and underground astrophysics detectors have been using detector mounted electronics for many years to read out their hundreds of thousands of streamer tube channels. For example, the LeCroy STOS (Streamer Tube Operating System shown in Figure 5) used on the Mt. Blanc experiment featured on-chamber mounted, limited overhead, data acquisition using bipolar based ASICs to keep the channel price and power low. In addition, the trigger is distributed so that all the channels can contribute to a fast global trigger decision with minimum deadtime. However, there was no data compaction used because the trigger rates for these experiments tended to be very low (proton decay is a very rare process!).

Another example comes from the mid '80's. The BNL Multi-Particle Spectrometer needed to instrument approximately 20,000 channels of image chamber detector with a multi-hit TDC but with a density problem similar to some SSC designs. This group used chamber mounted 16-channel, thick film hybrids (LeCroy HTD161) to amplify, discriminate, time encode and compress the signals and interface to an output bus. The hybrids (Photograph 2) have a footprint of 38 mm by 90 mm and cost about $25 per channel and could be pipelined.

However, this hybrid/ASIC system had no trigger outputs and dissapated at least 250 mW per channel. The ASICs were not particularly radiation hard and the hybrid used several different incompatible processes including CMOS and three different types of bipolar including ECL. Therefore, it would be impossible to make this design directly in one high density multi-process ASIC.

CONCLUSIONS

Trends from 25 years and Projections for the next 10...

The table below summarizes the critical points developed in this paper. It reviews the changes in density, standards, technology, sensitivity and speed over the last 25 years. We have seen the field go through at least 4 or 5 clearly identifiable stages. For the next generation of hadron colliders, we will have to go through yet another.

Decade	Channels	Systems	Technology	Trigger & Data Conversion
1960	50-500	NIM & DMA	Transistors & ICs	Separate
1970	500-2,000	CAMAC & MOD. CAMAC	Hybrids & ASICs	Controller Level
1980	2,000-50,000	FASTBUS	High Density ASICs	Module Level
1990	>100,000	Detector Mounted	Super ASICs	Channel Level

The data acquisition instrumentation problems of the 1990's are very difficult. Solving them will be critical to the success of the SSC/LHC programs. We must learn to use multi-process, high density, low power ASICs or be able to compensate for single process ASICs by using advanced techniques like Silicon hybrids. We must study radiation hardness as it applies to a 40 TeV hadron collision environment as opposed to a nuclear holocaust. We must develop architectures that include several levels of pipelines, that are self sparsifying, perform some digital signal processing and allow for the distribution of trigger information.

The last 25 years have taught us a great deal about data acquisition. By studying this discipline we can see the directions we have to go in for the SSC/LHC. Experienced commercial, laboratory and university groups working together are following these directions and meeting these challenges today so that by the end of 1990, the next generation of experiments will be ready to make their contributions to physics research.

ACKNOWLEDGEMENT

The author gratefully acknowledges the assistance in the preparation of this manuscript by Linda Nelson and the careful editing and critique by John Hoftiezer, both of LeCroy Corporation, New York.

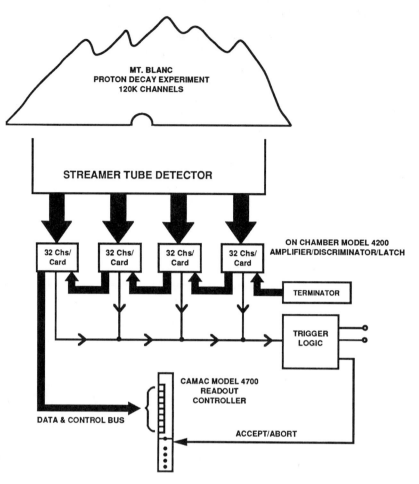

Figure 5
Streamer Tube Operating System Block Diagram

Photograph 2
LeCroy's 16-Input On-Chamber Multi Hit TDC Hybrid
with amplifier, discriminator, time digitizers and control logic.

THE MAGNET COMPONENTS DATABASE SYSTEM*

M.J. Baggett, Magnet Division, Bldg. 902B
Brookhaven National Laboratory, Upton, NY 11973

R. Leedy, C. Saltmarsh, and J.C. Tompkins
Superconducting Super Collider Laboratory
2550 Beckleymeade Avenue, Dallas, TX 75237

ABSTRACT

The philosophy, structure, and usage of MagCom, the SSC magnet components database, are described. The database has been implemented in Sybase (a powerful relational database management system) on a UNIX-based workstation at the Superconducting Super Collider Laboratory (SSCL); magnet project collaborators can access the database via network connections. The database was designed to contain the specifications and measured values of important properties for major materials, plus configuration information (specifying which individual items were used in each cable, coil, and magnet) and the test results on completed magnets. These data will facilitate the tracking and control of the production process as well as the correlation of magnet performance with the properties of its constituents.

PURPOSE

From the earliest R&D days of the SSC project, the magnet test groups at the labs have recognized the need for a complete and up-to-date database of configuration and test data pertaining to the components of the magnets. Such a database would enable magnet scientists to correlate performance properties of any magnet with its structural raw materials.

For instance, magnet testers predict the likely magnet quench currents from the short sample measurements on the cable. Magnet designers compare the measured field with properties of the cable and coil assembly. In an operating environment, the accelerator controls system will need to know the performance features of each individual magnet and relate these to its lattice location.

The Magnet Components (MagCom) database is intended to serve these scientific needs. Its design includes some aspects of purchase tracking, quality

---

*Work supported by U.S. Department of Energy.

control, and configuration control, so that scientists know the serial numbers for the major components of each magnet, the vendors that supplied the materials, and references to the backup quality assurance tests and documents. However, MagCom does not eliminate the need for complete bill of materials and purchase tracking systems. Ultimately it should have access to such tracking information by direct links to databases maintained by appropriate groups.

The goal for the MagCom database is to have all magnet data, spanning all phases of construction and testing, accessible from a central, user–friendly system. This system should allow both quick access to summary data of broad interest and retrieval of detail data for in–depth studies.

SOFTWARE & HARDWARE

The MagCom database is implemented using the SYBASE relational database management system (RDBMS) in a UNIX environment on a SUN 4/280 server which is connected to Internet and DecNet. It can be accessed via SUN workstations or vt100–type terminals. Some introductory documentation and copies of general reports are also available on the SSC central VAX computer. (See Figure 1)

SYBASE was selected as the vehicle for SSC scientific databases in December 1987 by an SSC Central Design Group (SSC/CDG) task force.[1] Initial experience, including development of a general data export system and graphics utilities, was concentrated in the accelerator controls area.[2] From the beginning, it was considered essential to have a "hub and spoke" data organization, with a central database to specify the interrelationship of databases maintained by various groups. (See Figure 2.)

DATABASE STRUCTURE

A relational database consists of independent tables with specific column names and unlimited rows for data. For useful retrievals, a table should have one or more key columns that provide a unique identifier for each row. These key columns also allow data from different tables to be related when they have the same key values. This provides the power and flexibility in a relational database system.

The first pass at specifying the desired contents of a magnet components database was made by a "commission on cable data" that was appointed by Tom Kirk, the head of the SSC/CDG Magnet Division, in fall 1988. The commission included representatives from all the laboratories involved in SSC magnet R&D.[3]

The commission quickly recognized that although wire and cable test data were the items of prime interest initially, they were only part of the data needed for a full study of the magnets. The proposed database contains tables in eleven general sections, spanning the magnet building cycle from specifications on major materials through performance measurements on magnets. (See Figure 3)

IMPLEMENTATION & CURRENT STATUS

Going from design to a functioning database operation involved work on several fronts. In the first phase, a program was developed that automatically

ACCESS INFORMATION FOR
MAGNET COMPONENTS (MAGCOM) DATABASE

Information and Current Summary Reports

DecNet: SSCVX1::User 2:(MagCom)

Database

Internet: grumpy.ssc.gov 134.3.10.10
DecNet: SSCSUN 41.375
Server: SUN 4/280 with SYBASE 4.0
Public Account: mdbuser
Access Command: magcom

E-Mail for MagCom Staff

Bitnet: MagCom @ SSCVX1
DecNet: SSCVX1::MagCom
Internet: magcom @ sscvx1.ssc.gov

Figure 1.   Database Access

TOP LEVEL "HUB AND SPOKE" DATA ORGANIZATION

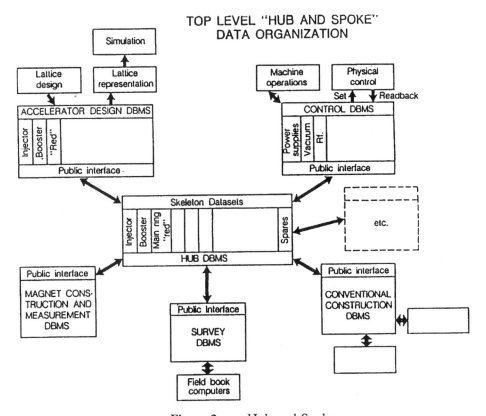

Figure 2.   Hub and Spoke

produced all the SYBASE table creation commands from the design document text file. Cable measurement data and magnet configuration information were obtained from Lawrence Berkeley Labortory (LBL) and Brookhaven National Laboratory (BNL) collaborators via ASCII files, and a summary report of cable properties by magnet was created. (See Figure 4.)

Later, import/export utilities were developed to convert tables between SYBASE format and the formats for the PC database systems of R:BASE or dBASE, which are in use by the BNL and LBL groups. Recently, a SSCL Magnet Systems Division quality assurance engineer has developed a dBASE application that will be run on a portable computer for collection of cable production data at the cabling site.

MAJOR TABLE CATEGORIES FOR
SSC MAGNET COMPONENTS DATABASE

1. Vendor, purchasing, and general information
2. Specifications for strand and cable
3. Ingot and billet production and tests; strand production and receiving
4. Strand tests
5. Cable production and receiving
6. Cable tests
7. Coil information
8. Cold mass assembly information
9. Cryostat assembly
10. Magnet data
11. Comments on tables; gossip; formats and naming conventions

Figure 3. Major table categories for SSC Magnet Components Database

The MagCom staff works with the staff of the engineering groups to identify the prime data that should be available to the full project and to set up the procedures for collecting and reporting that data in a timely way. The initial focus has been on summary data of broad interest with references to further sources (paper or computer files) for backup details.

Summary and detail reports have been written for the items of general interest, and a menu program that runs directly from UNIX allows user-friendly access to these reports. After logging in to the UNIX system, the user simply types the word "magcom" and then follows the directions. All reports offer the option of being displayed on the screen or written to a file for printing.

Magnet Cable Data Summary2 01   * - denotes new data

MAGNET	CABLE	Cu/SC	Jc(5T) amps/mm2	Ic(5T) amps	spec Tesla	Jc(sp) amps/mm2	Ic(sp) amps	R295K uOhm/cm	R10K uOhm/cm	ratio RRR	ramp mT/sec	field Tesla	fil+eddy mTesla	eddy mT
DD-0010	SO-0080	1.65	2672	9,970	5.6	2346	8,757	28.7	0.41	70	21.0	0.30	18.5	1.3
DD-0011	SI-0052	1.19	2686	14,430	7.0	1627	8,740	27.2	0.20	136	20.5	0.30	21.4	1.3
	SO-0087	1.64	2639	9,890	5.6	2317	8,689	28.8	0.43	67	21.0	0.30	19.2	2.0
DD-0012	SI-0054	1.57	2337	10,730	7.0	1420	6,530	24.5	0.31	79	21.0	0.30	53.6	1.0
	SI-0055	1.60	2380	10,790	7.0	1447	6,560	24.3	0.31	78	21.0	0.30	54.4	1.0
	SO-0098	1.65	2565	9,570	5.6	2248	8,387	28.5	0.21	136	21.0	0.30	20.4	3.7
	SO-0099	1.67	2576	9,550	5.6	2257	8,367	28.4	0.21	135	21.0	0.30	21.1	3.6
DD-0013	SI-0063	1.49	2700	12,789	7.0	1576	7,463	25.1	0.40	63	21.0	0.30	19.8	1.8
	SO-0102	1.69	2508	9,220	5.6	2198	8,079	28.3	0.25	113	21.0	0.30	18.9	3.4
	SO-0103	1.69	2511	9,230	5.6	2201	8,089	28.3	0.25	113	21.0	0.30	19.7	3.8
DD-0014	SI-0059	1.24	2541	13,380	7.0	1502	7,910	26.8	0.24	112	21.0	0.30	26.6	3.1
	SO-0104	1.68	2524	9,300	5.6	2211	8,147	28.3	0.21	135	21.0	0.30	19.7	
DD-0015	SI-0060	1.24	2551	13,440	7.0	1508	7,940	26.8	0.23	117	21.0	0.30	27.1	2.8
	SI-0058	1.24	2651	13,980	7.0	1567	8,260	26.8	0.22	122	21.0	0.30	27.0	
	SO-0074	1.80	2446	8,615	5.6	2143	7,549	27.9	0.50	56	20.8	0.30	13.9	0.6
	SO-0105	1.65	2515	9,370	5.6	2202	8,211	28.5	0.22	130	20.0	0.30	18.7	3.7
DD-0016	SI-0064	1.47	2775	13,265	7.0	1619	7,741	25.2	0.40	63		0.30	20.7	1.3
	SI-0065	1.37	2580	12,807	7.0	1548	7,684	25.8	0.36	72		0.30	21.0	1.8
	SO-0122	1.65	2453	9,164	5.6	2159	8,064	28.7	0.39	74	21.0	0.30	16.3	2.0
DD-0017	SC11-0394	1.59	2823	12,848	7.0	1662	7,562	24.4	0.35	70	21.0	0.30	26.6	2.6
	SC22-0391	1.72	2435	8,860	5.6	2135	7,770	28.3	0.41	69	21.0	0.30	14.4	1.3
	SC22-0001	1.64	2467	9,321	5.6	2171	8,203	28.9	0.40	72				
DD-0018	SC11-0398	1.42	2730	13,284	7.0	1586	7,718	25.5	0.37	69	21.0	0.30	20.4	2.4
	SC21-0396	1.69	2422	8,910	5.6	2132	7,841	28.5	0.44	65				
DD-0019	SC12-00001	1.44	2499	12,076	7.0	1545	7,465	25.3	0.73	35				
	SC12-00001	1.44	2591	12,515	7.0	1602	7,736	25.7	0.34	74				
	SC12-00005	1.44	2512	12,137	7.0	1533	7,407	25.7	0.73	35				
	SC22-00004	1.75	2646	9,516	5.6	2318	8,337	28.6	0.49	58				
DD-0026	SC12-00001	1.44	2499	12,076	7.0	1545	7,465	25.3	0.73	35				
	SC12-00001	1.44	2591	12,515	7.0	1602	7,736	25.7	0.34	74				
	SC22-00004	1.75	2646	9,516	5.6	2318	8,337	28.6	0.49	58				
	SC22-00005	1.75	2715	9,770	5.6	2386	8,586	28.6	0.48	60				
	SC22-00005	1.76	2747	9,829	5.6	2412	8,630	28.5	0.47	61				

Page 2

Figure 4. Magnet cable data summary report

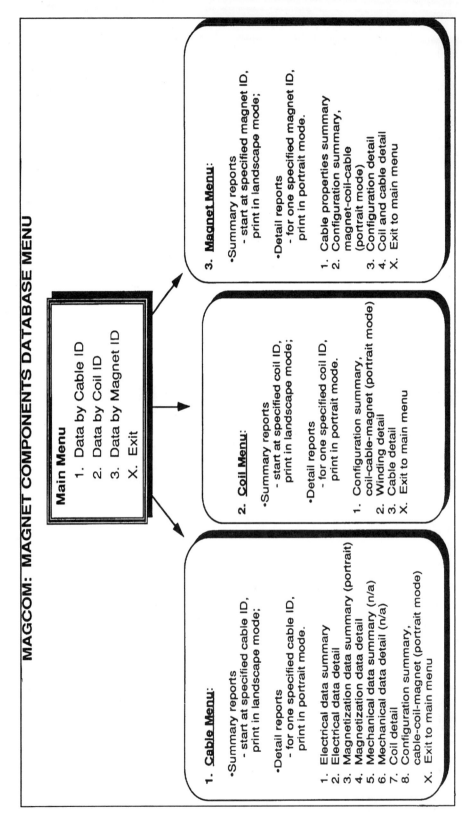

Figure 5. Magnet components database menu

The menu program allows retrieval of full detail data for one specified item (such as a cable, coil, or magnet) or summary data for a group of items starting from a specified ID. Figures 5 through 10 show the menu, several versions of configuration reports, and sample summary and detail reports. At present, the data is still limited to the cable measurement and magnet configuration tables.

Over the course of the past year, the ID schemes for cable, wire, and magnets have been revised, based on experience. Descriptions of the current schemes are available in the online documentation files. Also, the structure and contents of various tables have been changed. For instance, the cable measurement data tables were changed to a paired master and detail design to better reflect the structure of the data.

FUTURE PLANS

So far, all requests for data retrievals have been handled by members of the MagCom staff because the system has been undergoing development. However, the basic user features are now in place and the database will be available to the SSCL engineers to run their own retrievals if they wish.

The data collection will be expanded to new areas, beginning with summaries of quench and multipole results from magnet testing.

```
            MAGNET -> COIL -> CABLE SUMMARY
            ----------------------------------

        MAGNET              COIL                CABLE
        -------             ----                -----
        DD-0018             LLNI-28             SC11-0398
                            LLNI-29             SC11-0398
                            LLNO-27             SC21-0396
                            LLNO-28             SC21-0396

        DD-0019             DDI-31              SC12-00001
                            DDI-32              SC12-00005
                            DDO-30              SC22-00004
                            DDO-31              SC22-00004

        DD-0026             DDI-30              SC12-00001
                            DDI-34              SC12-00001
                            DDO-32              SC22-00004
                            DDO-33              SC22-00005

        DD-0027             DDI-35              SC12-00007
                            DDI-36              SC12-00007
                            DDO-34              SC22-00005
                            DDO-35              SC22-00005

        DSS-001             DSSI-03             SI-0044
                            DSSI-04             SI-0044
                            DSSO-03             SO-0070
                            DSSO-04             SO-0070

        DSS-002             DSSI-05             SI-0047
                            DSSI-06             SI-0047
                            DSSO-01             SO-0070
                            DSSO-02             SO-0070
```

Figure 6. Magnet configuration summary report

```
                    CONFIGURATION FOR

                  MAGNET    DD-0027

═══════════════════════════════════════════════════════
                                  |
            COIL                  |          CABLE
                                  |
═══════════════════════════════════════════════════════
                                  |
INNER:          DDI-35            |        SC12-00007
    (UPPER)                       |
──────────────────────────────── | ────────────────────
                                  |
INNER:          DDI-36            |        SC12-00007
    (LOWER)                       |
──────────────────────────────── | ────────────────────
                                  |
OUTER:          DDO-34            |        SC22-00005
    (UPPER)                       |
──────────────────────────────── | ────────────────────
                                  |
OUTER:          DDO-35            |        SC22-00005
    (LOWER)                       |
```

Figure 7. Magnet configuration detail report

```
                    MAGNET -> COIL -> CABLE
                    ───────────────────────

MAGNET ID
DD-0027
                         COIL INFORMATION
                         ────────────────

      LOCATION: INNER              UPPER
      COIL ID: DDI-35              CABLE SPOOL: SC12-00007

                    CABLE INFORMATION
                    ─────────────────

        CABLE SPOOL: SC12-00007      VENDOR: SCN
            RUN #:    2318      RUN DATE: May 15 1989

B FIELD     Ic       Jc       R295      R10      RRR    Cu/Sc
(tesla)   (amps)  (amps/mm2)    (uOhm/cm)
  5.00     13037    2740     25.40     0.66      38     1.48
  5.60     11473    2411
  6.00     10430    2192
  7.00      7822    1644
No obvious surface damage as seen in coil conductor.  No wire samples suppl
ied.  SCN Billet 2300-1

                         COIL INFORMATION
                         ────────────────

      LOCATION: INNER              LOWER
      COIL ID: DDI-36              CABLE SPOOL: SC12-00007

                    CABLE INFORMATION
                    ─────────────────

        CABLE SPOOL: SC12-00007      VENDOR: SCN
            RUN #:    2318      RUN DATE: May 15 1989



              :
              :              etc.
              :
              :
```

Figure 8. Magnet coil and cable detail report

CABLE ELECTRICAL DATA SUMMARY

CABLE SPOOL	CABLE SEGMENT	RUN #	RUN DATE	R(295) uOhm/cm	R(10) uOhm/cm	RRR	Cu/Sc	B(T)	Ic amps	Jc amps/mm2
SC12-00005		2247	Mar 3 1989	25.70	0.73	35	1.44	5.00	12137	2512
								5.60	10718	2218
								6.00	9772	2022
								7.00	7407	1533
SC12-00006	END	2264	Mar 10 1989	24.70	0.70	35	1.60	5.00	10938	2411
								5.60	9659	2129
								6.00	8806	1941
								7.00	6675	1471
SC12-00007		2318	May 15 1989	25.40	0.66	38	1.48	5.00	13037	2740
								5.60	11473	2411
								6.00	10430	2192
								7.00	7822	1644
SC12-0001		2071	Oct 31 1988	26.90	0.42	64	1.25	5.00	12156	2320
								5.60	10726	2047
								6.00	9772	1865
								7.00	7388	1410
SC13-00001		2185	Jan 23 1989	25.40	0.33	77	1.43	5.00	13281	2733
								5.60	11687	2405
								6.00	10625	2187
								7.00	7968	1640
		2434	Nov 29 1989	25.30	0.33	77	1.45	5.00	13235	2746
								5.60	11617	2410
								6.00	10538	2186
								7.00	7840	1627

SCN 6 um. Front. Use same I5/I6 as 12-00004.

SCN cable - Back. 6 um.

No obvious surface damage as seen in coil conductor. No wire samples supplied. SCN Billet 2300-1

SCN 364E-2, 6um fils.

IGC 5275-1009, 6 um. NO WIRES: ASSUME I5/I6=1.25

IGC 5275-1009. Rerun of R2185. Old pts. included.

Figure 9. Cable electrical data summary report.

153

```
                    CABLE ELECTRICAL DATA
                    --------------------

CABLE SPOOL: SC12-00007                    VENDOR: SCN


RUN #:    2318          RUN DATE: May 15 1989

  Measurement Data:    B(T)     T       It      Iq       n
                              (deg K)  (amps)  (amps)
                       ----   -------  ------  ------    -
                       5.37    4.36    11496   12198     28
                       5.38    4.36    11584   12198     33
                       5.79    4.40    10194   10832     26
                               4.40    10202   10832     27
                       6.15    4.42     9097    9779     25
                               4.42     9218    9779     32


  Summary Data:        R295          R10       RRR    CU/SC
                         (uOhm/cm)
                       -----         ---       ---    -----
                       25.40         0.66       38     1.48


                        B(T)        Ic          Jc
                                  (amps)     (amps/mm2)
                        ----      ------     ----------
                        5.00      13037        2740
                        5.60      11473        2411
                        6.00      10430        2192
                        7.00       7822        1644

PROCEDURE DOCUMENT:                   REVISION:
No obvious surface damage as seen in coil conductor.  No wire samples
supplied.  SCN Billet 2300-1
```

Figure 10. Cable electrical data detail report.

REFERENCES

1.   R. Leedy, S. Peggs, C. Saltmarsh, L. Schachinger, R. Talman, T. Toohig, J. Tompkins, "Report of the Database Definition Task Force", SSC/CDG unpublished report, December 7, 1987

2.   E. Barr, S. Peggs, and C. Saltmarsh, "Relational Databases for SSC Design and Control", SSC-N-606; to be published in Proceedings of the Particle Accelerator Conference, Chicago, March 1989

3.   M.J. Baggett, B.C. Brown, J. Carson, T. Ferbel, R. Leedy, R.M. Scanlan, R.H. Remsbottom and P.Wanderer, "General Remarks on the Organization and Contents of the Proposed Cable Data Base", SSC-N-582, Rev.No.1, March 24, 1989

COMPUTING REQUIREMENTS FOR THE SSC MAGNET TEST LABORATORY

C. Day, D. Balaban, W. Greiman, and D. Hall

Lawrence Berkeley Laboratory
Berkeley, CA

C. Saltmarsh and J. Tompkins

SSC Laboratory
DeSoto, TX

1   Background

The superconducting magnets of the SSC are the most expensive technical component of the project. There will be 10,000 magnets at a cost of roughly $220,000 each, or a total of $2.2 billion. For the SSC to achieve 96% availability, there can be no more than 3 magnet failures per year. Over an estimated twenty year lifetime, the magnets must survive 10,000 acceleration cycles, 20 thermal warmup cycles, 50 quenches and a radiation exposure of 1 MGy (a.k.a., MegaRad). Field quality requirements are typically a few parts per 10,000.

The purpose of the Magnet Test Lab (MTL) is to certify that SSC magnets meet all these requirements, to qualify magnets within the SSC construction schedule, and to provide rapid and effective magnet repairs during SSC operation.

The MTL will monitor the industrial production of the SSC magnets by running thorough tests on approximately 10% of the magnets. It is scheduled to begin operations in July 1991 and test one magnet a day for three years. MTL is crucial to the success of the SSC. LBL's goal is to provide working software for data acquisition, hardware control and test analysis within this time scale.

2   Software Problem

We do not anticipate major technical problems with the software needed by the MTL. Data rates and volumes are modest compared to the experiments. The major problem, as with many software projects, is management. To meet our goal, we must be able to ensure that the software will do what is needed and that the software development is proceeding as scheduled. To aid in meeting these management goals, we are using various evolving documents designed to make the software development activites more visible and more accountable.

Table 1. Generic Software Development Schedule

	Pre Dev	Requir. Analysis	Prelim. Design	Detail Design	Implement & Unit Test	Software Integration & Test	Acceptance Testing
SDP		P	U	U	U	U	U
STP		P	U	U	U	U	U
URS	F						
SRS		F					
SDD			P	F			
TD			P	F			
TP					P	F	
Doc			P	U	U	F	

Keys:

SDP	-	Software Development Plan	F	-	Final
STP	-	Software Test Plan	P	-	Preliminiary
URS	-	User Requirements Statement	U	-	Updated
SRS	-	Software Requirements Specification			
SDD	-	Software Design Document			
TD	-	Test Description			
TP	-	Test Procedures			
Doc	-	Documentation as appropriate			

3 Monitoring Progress

The progress of software development will be tracked for each software subsystem or module. For each identified software module, there is a fairly standard set of development stages, *i.e.*,

- Predevelopment Analysis

- Requirements Analysis

- Preliminary Design

- Detailed Design

- Implementation and Unit Test

- Acceptance Testing

Progress through these stages is monitored by the evolution of a standard set of documents. See Table 1.

These documents provide timely and concrete detail about specifications (What is to be done), requirements (Why it is to be done) and testing (How to determine if it has been done). While the detailed form of the documents is adjusted to the particular project, there are relatively few outlines which handle many cases. Table 2 is an example taken from [1].

As long as these documents can be kept approximately in step with the development work, there is a basis for judging how far the project has come and estimating how much more time and effort will be needed for completion.

Table 2. Prototype Outline for SRS

Table of Contents
1. Introduction
    1.1 Purpose
    1.2 Scope
    1.3 Definitions, Acronyms, & Abbreviations
    1.4 References
    1.5 Overview
2. General Description
    2.1 Product Perspective
    2.2 Product Functions
    2.3 User Characteristics
    2.4 General Constraints
    2.5 Assumptions & Dependencies
3. Specific Requirements
    3.1 Functional Requirements
        3.1.1 Functional Requirement 1
            3.1.1.1 Introduction
            3.1.1.2 Inputs
            3.1.1.3 Processing
            3.1.1.4 Outputs
        3.1.2 Functional Requirement 2
        . . .
        3.1.n Functional Requirement n
    3.2 External Interface Requirements
        3.2.1 User Interfaces
        3.2.2 Hardware Interfaces
        3.2.3 Software Interfaces
        3.2.4 Communications Interfaces
    3.3 Performance Requirements
    3.4 Design Constraints
        3.4.1 Standards Compliance
        3.4.2 Hardware Limitations
        . . .
    3.5 Attributes
        3.5.1 Security
        3.5.2 Maintainability
        . . .
    3.6 Other Requirements
        3.6.1 Data Base
        3.6.2 Operations
        3.6.3 Site Adaptation
        . . .
Appendices
Index

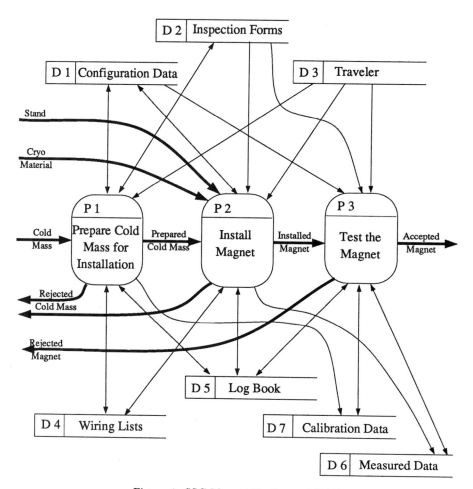

Figure 1. SSC Magnet Testing at FNAL

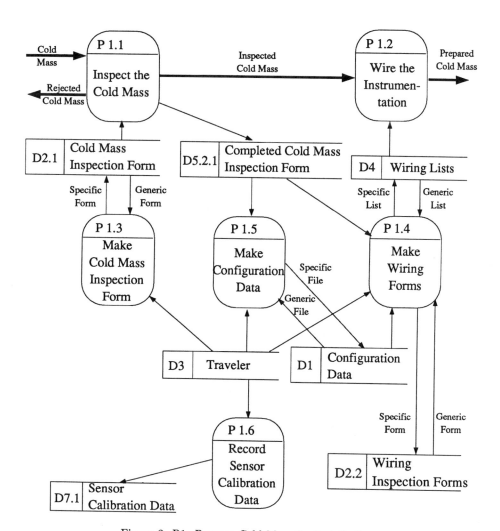

Figure 2. P1: Prepare Cold Mass for Installation

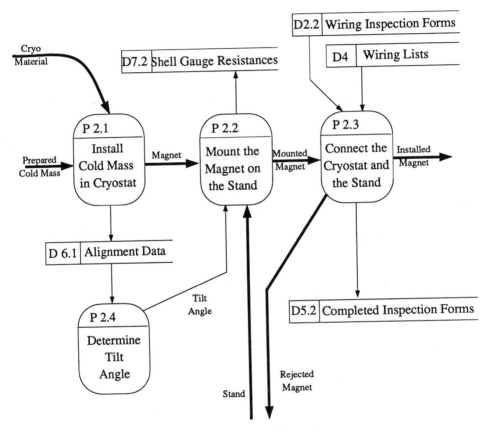

Figure 3. P2: Install Cold Mass

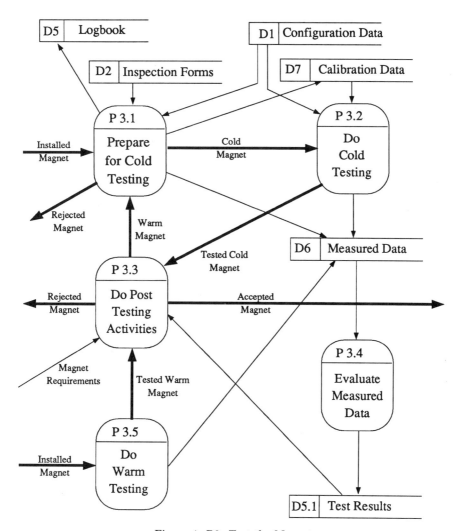

Figure 4. P3: Test the Magnet

4 Identifying Modules

These monitoring techniques work well only if coherent, well-defined software modules have been identified. The commercial world is more used to this process and often can provide the initial system breakdown into modules. Clearly defining modules at an early stage of design of the complete system is less common in HEP.

The MTL provides an example of the process used to define the modules. First, we have worked to understand the overall operations of the lab by observing the operations of the existing, comparable facility, Fermilab's Magnet Test Facility, interviewing as many experts as we could, and collecting and analysing whatever relevent documentation we could find. We have organized our understanding by building a Dataflow Model of MTF operations, supplemented by some Entity-Relation Models and State/Sequence Models. As an example, Figures 1–4 are the top and first refinement level of our Data Flow Diagrams for SSC magnet testing at FNAL.

These models have been iterated with reviews by potential users and experts. The next step, currently underway, is to make a strawman sketch of the MTL's toplevel conceptual design. This will also be iterated with reviews by users and experts. From the resulting sketch, we will try to identify distinct software modules and write preliminary User Requirements Statements. Once these have been reviewed and approved, we will be able to proceed with the software development proper.

5 Benefits

At the completion of the development cycle, the management documents provide a thorough description of the system which will be of great help to the operational and maintenance teams to follow. They will also be highly valuable for estimating time and resources required for similar projects in the future.

The system level understanding gained during the identification of software modules can benefit the whole project by encouraging an early consensus on the overall design and opening communications channels among all involved in the project's construction.

6 Reference

[1] ANSI/IEEE Std 830-1984, "IEEE Guide to Software Requirements Specifications", in *Software Engineering Standards*, IEEE, New York (1987).

REAL-TIME DATA ACQUISITION AND COMPUTATION FOR THE SSC USING OPTICAL AND ELECTRONIC TECHNOLOGIES

C. D. Cantrell

Center for Applied Optics
University of Texas at Dallas, Richardson, TX 75083-0688

E. J. Fenyves

Programs in Physics
University of Texas at Dallas, Richardson, TX 75083-0688

Bill Wallace

CONVEX Computer Corporation
3000 Waterview Parkway, Richardson, TX 75083-3851

Abstract

We discuss combinations of optical and electronic technologies that may be able to address major data-filtering and data-analysis problems at the SSC. Novel scintillation detectors and optical readout may permit the use of optical processing techniques for trigger decisions and particle tracking. Very-high-speed fiberoptic local-area networks will be necessary to pipeline data from the detectors to the triggers and from the triggers to computers. High-speed, few-processor MIMD supercomputers with advanced fiberoptic I/O technology offer a usable, cost-effective alternative to the microprocessor farms currently proposed for event selection and analysis for the SSC. The use of a real-time operating system that provides standard programming tools will facilitate all tasks, from reprogramming the detectors' event-selection criteria to detector simulation and event analysis.

Introduction

The high-energy physics community hopes that the Superconducting Super Collider (SSC) will be a copious source of Higgs bosons, which are a key component of the Standard Model of the elementary particles but which have so far not been observed. Physicists also hope that the SSC will reveal new physics beyond the Standard Model. However, much of the importance of the SSC outside the world of elementary-particle physics may come from the systems that must be developed to meet the greatest real-time computing challenge yet posed: to detect the events physicists seek amid a background that is up to 10^{13} times greater than the "signal" by filtering a data stream of the order of 100 terabytes per second. To solve this problem with today's technology would be difficult, perhaps impossible; yet there is every reason for confidence that problems in data acquisition and processing will not prevent the SSC from reaching its objectives.

Computing and data processing in the elementary-particle community have largely evolved within the context of proprietary computer systems and specialized data-acquisition instrumentation over the past twenty-five years. The magnitude and difficulty of the problem of SSC data acquisition, the open-systems revolution in the computer industry during the past decade, and the recent historically high rate rate of growth of the performance per dollar of commercial computer systems make an intersection of SSC instrumentation with mainstream computing not only inevitable, but desirable from all points of view. In this paper we outline new ideas for SSC data-acquisition devices and an approach to SSC data-processing systems based upon conservative estimates of the likely effects of rapidly evolving semiconductor technology on the development of electronic and optical processors and the design of computer systems over the next five years.

The influence of new technology on the design of hardware triggers for the SSC

An early design for data filtering and analysis at the SSC identified three levels of data acquisition, as shown in Fig. 1.[1] The most difficult Level I data-acquisition problems will occur in a general-purpose 4π detector, in which one will try to analyze potentially interesting events that produce large numbers of hadrons as well, perhaps, as relatively rare events such as the four-muon decay of the Higgs boson.

According to the 1985 plan and its subsequent modifications, a Level I (analog) trigger decision will not be available until 1 μs after an event,[1] even in a data-flow architecture in which the data stream moves in a pipeline through segmented trigger electronics.[2] The high rate of beam crossings for which the SSC has been designed (62.5 MHz), the expected event rate of approximately 1.6 per crossing, and the long delay between an event and a Level I decision expected from the 1985 plan would imply a need for massive buffering of the full data stream from all 10^6 channels of a general-purpose 4π detector. Signal pileup[3] and the deconvolution of signals from successive events[4] would be major problems given the slow detectors envisioned in 1985.

Level I trigger decisions will almost certainly be based upon signals received from calorimeter towers.[3] Recent calorimeter designs employing media with a fast response, such as a liquid Xe scintillator[5], suggest that it is reasonable to assume that liquid-noble-gas calorimeters deployed at the SSC will be able to furnish signals to the analog trigger well within the 16 ns time delay between successive bunches. The Level II triggers will probably employ preliminary tracking information as well as spatial information from the calorimeters. Novel tracking detectors such as those employing scintillating fibers[6,7] in which the scintillation decay time is as short as 3 ns[8] are expected to evolve within a few years to a point at which most of the light is emitted in less than 2 ns. Proper segmentation of the scintillating elements will be necessary to ensure that the width of the statistical distribution of arrival times of optical signals when fiber propagation delays are compensated for is less than 2 ns, so that hits can be located with a resolution of less than 2 ns. The overall trend in detector design appears to be towards the use of media that produce prompt optical signals and away from slow traditional detectors such as drift chambers.

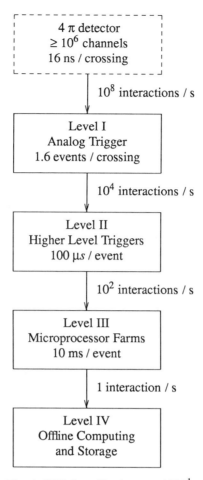

Fig. 1. SSC data filtering, ca. 1985[1]

We assume, therefore, that all signals needed for a Level I trigger decision will be available in less than 10 ns. The actual time required for a decision will depend on the technology used for the Level I logic. If the raw detector data is processed in a highly parallel pipeline[2] in which each stage contains the data from a single crossing, then the duration of the time interval in which a decision must be made is constrained in principle only by the feasible physical length and number of parallel channels of the pipeline. In practice, considerations of reliability, power consumption, geometrical volume and cost will probably limit the number of stages of data conversion or logic in the Level I trigger.

A partial list of reasonable requirements for the various levels of data acquisition and filtering in the light of new detector technology is as follows:

Level I triggers

- Make a decision within 500 ns after calorimeter signals enter the pipeline
- Reduce the raw event rate by a factor of $\approx 10^3$

Level II triggers

- Utilize partial tracking analysis plus segmented calorimeter data[9,10]
- Reduce the Level I data rate below the maximum capacity of an affordable software trigger system plus associated data interconnections (taken as 10^3 s^{-1})[1]

Level III triggers

- Accept the full data stream remaining after Level II decisions
- Provide jet reconstruction[11] and further tracking analysis
- Handle events at a rate not greater than 10^2 s^{-1}

Level IV (offline)

- Provide full event reconstruction, analysis and storage
- Provide for general physics computation

The boundaries between Levels II, III and IV can be expected to shift somewhat as the choice of technology becomes better defined.

The combined requirements of speed and radiation hardness at Level I bode poorly for MOS and bipolar semiconductor devices. Optical data transport, optical processing and CMOS or gallium arsenide semiconductor devices are strong candidates for inclusion in the Level I trigger hardware, some or all of which may be mounted in or on the detector itself. Optical data transport and processing are especially attractive given that modern detector technology relies increasingly upon scintillation detectors and optical readout.

For the SSC, fiberoptic data transport between major devices such as detectors, trigger hardware and computers offers the advantages of much lower crosstalk, lower electromagnetic interference, high data bandwidth over longer distances, and, for glass fibers, lower attenuation than electrical cabling. The key requirement of detector hermeticity[12] makes the thinness of optical fibers attractive in comparison with electrical cables. However, fiberoptic cables are currently more difficult to connect than electrical cables. Most of the development effort in fiber optics has been expended for the sake of long-distance telecommunications, where developers of semiconductor lasers and detectors have worked hardest at the wavelengths at which the attenuation of glass optical fibers is a minimum, namely, 1.3 μm and 1.55 μm. Laser sources at these wavelengths generally depend upon InP semiconductor technology. However, semiconductor lasers and light-emitting diodes based upon the older (Al, Ga)As technology are very highly developed by now, in considerable measure because of the market for compact disk players. For distances of interest for the SSC, the attenuation of modern glass optical fiber is negligible even at wavelengths in the emission range of (Al, Ga)As semiconductor lasers.

Optical data links may also be useful at the SSC within the trigger hardware (for example, as part of a data pipeline) or as part of a controller interfacing the readout from a detector or trigger to a data processor. Fiber and free-space optical interconnections are a topic of serious discussion in the VLSI community[13] and in other communities, such as high-performance computing, where the overall performance of a system is often limited by memory and I/O bandwidths rather than by CPU speed. The interconnection of gates in a VLSI array, the feasible number of pins on a gate-array package, or the interconnection of processors in a massively parallel computer can become limiting factors in both the design and the performance of complex systems.[13,14] Optical interconnections may also be useful in eliminating the access conflicts that are a characteristic problem of electronic bus architectures. Reconfigurable optical interconnections may be useful in general, as well as at the SSC, for changing the interconnection graph of the functional subunits of a computer to provide nearly optimal performance on different types of jobs that currently run optimally on quite different computer architectures. By the late 1990s the SSC will probably be making use of at least one commercially manufactured computer in which optical interconnections play a role.

At the SSC it will be necessary to switch high-volume data streams selected by a trigger decision. In current technology optical pulses carrying data over optical fibers must still be electrically detected, switched, and amplified before being retransmitted in optical form. The search for workable technology for a fully photonic switch is a major effort in optical device research because of the wide applicability of such a device in telecommunications and computing. Photonic switching devices generally depend upon the coherence, narrow bandwidth or polarization of a laser input beam. However, electrically activated optical switches are also of interest for switching high-bandwidth streams of data without the need for electrical detection or retransmission. A micromechanical deformable mirror device (DMD), which was intended for use as a spatial light modulator and is now being made in experimental quantities by Texas Instruments, Inc,[15] may be applicable at several stages in the SSC data-filtration process. The DMD consists of up to 128×128 micromechanical mirrors suspended over the cells of what is in effect a dynamic RAM. Charging one of the cells deflects the mirror above it by electrostatic forces. Several versions of the DMD exist. In the version that is relevant to data switching, the micromechanical mirrors are suspended by torsion bars that permit rotation about a fixed axis. The switching time of the present generation of DMDs is 10 μs.

Since fiberoptic data rates can easily exceed 1 Gbit/s, the DMD is an attractive candidate for switching data streams that have been selected by a trigger decision, especially between Levels II and III, where the full stream of data produced by a large 4π detector must be handled. Spatial light modulators such as the DMD have been proposed as optical crossbar switches in multiprocessor computers,[16,17] and might also find applications in a Level II track processor. One advantage of using an easily reprogrammed crossbar such as a DMD at the SSC would be that the logic or functions of a trigger could be reconfigured in software.

The direct optical processing of information, although currently much more speculative than optical data transmission or optical interconnections, offers potential advantages that are too great to be ignored in designing data-filtering hardware that will be fully deployed only in the second half of this decade. Much of the literature on information processing via nonlinear-optical devices assumes coherent illumination and is therefore not useful for Level I signals produced by SSC detectors. However, combinational logic functions can also be carried out with incoherent light if one is willing to interpose photodetectors at some point in the data pipeline, thereby partially sacrificing the advantage of avoiding conversions of energy between photons and electrons. The main advantages of optical processing, speed and parallelism, were effectively demonstrated over a decade ago in an apparatus for performing discrete Fourier transforms using incoherent light.[18]

More recently, a *pnpn* optical switch has been developed with some very interesting characteristics from the point of view of implementing optical logic without depending upon the coherence or polarization of light.[19] The *pnpn* switch is an optical amplifier (with a gain of 10^4 or more) and has the electrical properties of a thyristor whose switching characteristics can be controlled by incident light. The rise and fall times of the emitted optical signal were observed to be less than 10 ns, instrument-limited. The device can be operated in a threshold mode, and (depending on the threshold) can implement the "and" or (non-exclusive) "or" of as many channels as can be input to the active junction. Since the junction region may be of the same size as a single optical fiber, it may be necessary to concentrate the signals from different channels using a lenslet array.[20] It is also possible to operate *pnpn* optical switches in parallel as an optical comparator, i.e., a device in which only the switch receiving the highest power fires.[21] The *pnpn* optical switch might be invaluable in the Level I trigger circuits, since it could accomplish simultaneously the functions of a threshold detector and an optical signal generator with sufficient power to drive a logic gate connected with the *pnpn* switch via optical fiber. The logic gate could then turn "on" the appropriate channels of a DMD-based crossbar switch, letting data flow through to Level II. Also, *pnpn* optical switches operated as threshold-limited optical amplifiers could be useful in conjunction with scintillating-fiber track detectors, since the pulses generated by the *pnpn* switch could be powerful enough to be transmitted some distance away from the detector and short enough to avoid pileup of analog optical signals.

The *pnpn* devices may also be useful in analog as well as digital mode. Since linearity is an important property of the calorimeter,[12] it will be necessary to investigate the dynamic range over which a *pnpn* switch can be operated linearly. Unless the device is operated as a laser or at such high gains that the gain per unit length saturates for strong signals, there should be few problems in obtaining linear amplification.

While no one area of optical technology apart from fiberoptic data transmission can yet be considered as indispensable for the SSC, it is almost certain given the current vigorous pace of research and development in optical switching that some aspects of optical interconnection or optical processing will be considered to be standard technology by the time the SSC reaches full luminosity.

In fact, the SSC Laboratory is in a position to hasten technological developments in optical interconnections and processing by supporting experimental detector projects that use optical technology as well as selected optical-device-fabrication projects in academic and industrial laboratories. Technological spinoff that is widely perceived as useful would help provide convincing responses to criticisms that the SSC is unlikely to produce any results that will have a near-term effect on the competitiveness of U.S. industry.

A philosophy for SSC software trigger systems and offline computing

The past five years have been a period of rapid and dramatic change in the computer systems available to the technical computing community. Today, desktop computer systems have higher performance than the mainframes of only a few years ago. Engineering workstations, with integrated graphics capabilities and fast scalar processors, have caused basic changes in the way we perform many technical tasks. The use of leading-edge workstations has substantially increased the productivity of users from mechanical engineers to molecular biologists. One hopes that physicists will soon join this trend. The last five years have also seen the introduction of minisupercomputers. These systems, with supercomputer architectures and minicomputer prices, have made it possible for technical organizations to perform computationally intensive tasks much more economically and with much better response to users' needs than in the days of remote, centralized computing. Finally, the emergence of widely accepted standards for operating systems, networking protocols and I/O interfaces has made the computer marketplace truly competitive, freeing users from being chained to a single vendor through their investment in software and training.

As impressive as the performance and price/performance gains in technical computing over the last five years have been, the performance improvements anticipated for the next five years promise even more dramatic changes. The expected increases in computer performance levels are so great that simple extrapolations based on today's methods and practices are inappropriate. Much of our experience on older, slower computer systems with proprietary operating systems is not useful in establishing future computing needs. For a new installation such as the SSC we must re-examine our "common-sense" assumptions about technical computing environments and consider how the new higher-performance systems can be utilized effectively.

We argue that to take maximum advantage of the commercially driven revolution in computer performance, one should design the SSC data-acquisition and data-processing systems with an emphasis on heterogeneous multiprocessing, modularity, expandability and adherence to industry standards.

Careful attention to standards can directly benefit the physicists who use the SSC by providing a common command language for all users, from those who reconfigure the triggers to those who analyze reconstructed events. UNIX is already the de facto standard operating system for high-performance computing. Serious attempts to modify the UNIX kernel to support the efficient, deterministic interrupt handling required for real-time computing are now being made by several vendors.[22] The next five years will probably see several computer vendors offer operating systems whose user commands are the same as those of standard UNIX and which make the maximum performance of the computer available for the acquisition and on-line processing of a data stream generated by real-time events.

Taking advantage of commercial offerings in high-performance computing whose existence could not have been foreseen by the physics community five years ago should make it possible to confine reliance upon custom data-handling hardware to the analog and higher-level detector trigger systems and the interface of the latter to the software trigger.

A Glimpse of Future Technical Computing Systems

Several different trends in high-performance computer hardware and software are combining to create major changes in technical computing environments. Multiprocessors are now commercially available from a number of vendors; vector processing, once considered a feature found exclusively on the most expensive supercomputers, is now available on a broad range of systems; and software technology required to effectively utilize vector-parallel processors on programs written in traditional programming languages has been developed by several computer vendors. Heterogeneous multiprocessing, in which processors that are optimized for specific functions such as vector processing, superscalar processing, or digital signal processing share a common memory, is now being discussed actively in the computing community. Current experiments in heterogeneous multicomputing, in which different processors have local memories and communicate via message passing, may in the future lead to new types of supercomputers or to network supercomputing. Fast local-area networks based on fiberoptic technology are now commercially available,[23–25] providing us with the ability to connect multiple systems together well enough that the burden of improving the efficiency of multiprocessing and multicomputing for programs written in higher-level languages has temporarily shifted from interconnections to compilers and network queue-managing software.

However, the most dramatic improvements in the next few years will come in scalar processing performance. As the result of continued advances in semiconductor technology, scalar processors no longer cover several printed circuit boards; they consist of one or two integrated circuits. The widespread implementation of reduced instruction set computer (RISC) concepts has changed the rate of improvement in scalar processing performance from approximately 30% each year to well over 50% each year by greatly reducing the average number of CPU cycles per instruction. This rate of performance improvement cannot continue forever. However, we can expect a compounded annual growth rate of 50% or more in scalar performance over the next 4 to 5 years. These increases will come from improvements in semiconductor process technology and super-scalar or superpipelined implementations of scalar architectures. [26] Superscalar processors, which are capable of issuing more than one instruction per clock period, have recently come to market and are expected to have a strong influence on the evolution of the computer industry over the next several years. On selected jobs that cannot take advantage of the hardware concurrency of vector processors and that perform small or moderate amounts of I/O, relatively inexpensive superscalar processors have outperformed single processors of well-known vector supercomputers with substantially faster clock rates.[26,27] Such superscalar processors have been dubbed "killer micros".[27]

The estimated performance of single-chip mid-range scalar processors utilizing RISC, superscalar and superpipelining concepts is shown in Table 1. These estimates are conservative, since GaAs microprocessors with 200 MHz clock rates are already in an advanced stage of design. [28,29] Further, the estimates shown do not allow for issuing many instructions per clock period.

Predictions of the widespread use of GaAs in future computing systems have occasionally drawn the criticism that Ga is a much scarcer element than Si, so that GaAs can hardly be expected to to replace Si in the semiconductor industry. In the forecast in Table 1 we do not forsee the use of GaAs in commodity products such as dynamic RAMs, but only in relatively specialized products where speed is at a premium.

Table 1. Estimated Mid-Range Scientific Processor Performance

Property	1988[30]	1990[30]	1992[30]	1994
Performance (MIPS)	16	50	100	160
Performance (d.p. MFLOPS)	25	100	200	250
Clock Rate (MHz)	25	80	160	250
Technology	CMOS	ECL	ECL	GaAs

(1 MIP = 1 VAX 11/780)

The number of computer vendors supporting vector processing has grown to the point that vector processing has become a standard feature of scientific computers. With their recent product announcements, the Digital Equipment Corporation has joined the ranks of vendors supporting vector-parallel machines for the technical computing market. Clock-rate improvements for computers will permit higher performance in vector processing as well as scalar processing. However, improvements in vector performance are dependent on increased memory-system bandwidth, not just processor clock rates.

The chips that implement scalar and vector processing units have a relatively small effect on system costs, and will have an even smaller effect in the future. Even today, the performance of the memory system, I/O system, and packaging have the greatest influence on the cost of computer systems. We can expect a division of computer systems based on price classes in which each system class will provide a different level of capabilities in the memory system, I/O system, and overall processing power.

Future minisupercomputers will ride the cost curves produced by the semiconductor industry, just as workstations have done. However, they will provide greater overall processing power by integrating multiple heterogeneous processors into coherent computer systems with high memory and I/O bandwidth. Multiprocessors composed of several processor types will become the standard in the minisupercomputer class. Specialized processors, high memory bandwidth and capacity, and expandable I/O systems will provide higher system performance than in smaller systems with nominally similar processor clock rates.

Networking technologies are also improving dramatically. New fiberoptic local area networks have recently been introduced that can support significantly higher speeds than in the past, as well as an expanded geographical distribution of computers. If we are to capitalize fully on faster networks, memory and I/O systems must provide sufficient bandwidth to meet the greater demands of the new network interfaces.

Hardware is not the only area of improvement in the computing industry. In the past few years, system software has advanced significantly, and additional improvements can be expected. Compilers and operating systems for future computer systems will shield the general user from the increased architectural complexity of the underlying hardware implementation. In addition to traditional optimizations, compilers will perform interprocedural analysis (permitting the exploitation of fine-grained parallelism in user programs), static processor resource scheduling, and profitability analysis to determine the most efficient method of implementing a given computation in a heterogeneous multiprocessing environment. In addition to traditional batch and timesharing capabilities, operating systems for minisupercomputers will be capable of supporting real-time operations. One of the principal objectives of manufacturers in the minisupercomputer price class will be to provide software products that make high-performance systems easy to use and program.

Computing alternatives for the SSC

Computer technology has advanced rapidly since the original proposals by the Offline Computing/Networking Working Group. [31,32] Advances in optical processing technology and computer systems now provide, or soon will provide, alternatives in data acquisition and computing systems that would have been difficult to imagine even a few years ago.

In 1985 a large, custom-built array of small processors, called a processor farm, was proposed for reducing the raw data from the SSC detectors, for Monte Carlo simulation and possibly for lattice gauge calculations. At the time it was conceived, the proposed SSC microprocessor farm was a cost-effective, high-performance alternative to the systems offered by the major vendors of scientific computers. The aggregate performance of the proposed processor array was 10^4 MIPS at a cost of $14 million in 1986 dollars. [32] More recent proposals call for 10^6 MIPS. [33] Additional computing systems were proposed in 1985 for communications, printing, program development, and general administrative needs.

We question whether it makes sense for physicists to be in the computer business in the 1990s. It is not clear that scientists, however able, can compete in a cost-effective manner with a computer industry that has become extraordinarily good at delivering a wide variety of leading-edge systems at prices per unit performance that seem astonishingly low even by the standards of five years ago. Maintaining an operating system that can provide the same command language and as productive a user environment as the technical computing systems to which scientists are becoming accustomed will be an even greater undertaking than building and maintaining custom hardware. Computer vendors typically have hundreds of programmers working on the version of UNIX that runs on their hardware, despite the widespread perception that UNIX is a "standard" operating system. The important objective of providing a common software environment for all users could be met more economically than in the 1985 proposal by using the same general-purpose computer system for the Level III trigger and for offline simulation and analysis.

A possibly cost-effective alternative to the proposed processor farm is a network of superscalar workstations. However, such a network would have disadvantages in a multiple-use environment. The so-called "killer micros" have a performance on fully vectorizable programs that is of the order of 10^2 times worse than a vector supercomputer.[26] One of the reasons for the low cost of some superscalar machines is their lack of support for high memory and I/O bandwidth, which are crucial determinants of performance in a multiuser environment. Finally, multicomputer operating systems such as Mach[34] have not yet reached a degree of maturity that would permit real productivity for any users other than the most computer-wise.

An attractive alternative to the processor farm proposed in 1985 is a small number of general-purpose, shared-memory MIMD (multiple-instruction, multiple-data) systems that will be typical of the minisupercomputer class in the future. Today, systems of 4 to 8 processors sharing a large random-access memory are readily available. Over the next five years, we expect the number of processors to increase by a factor of 4. Using the performance estimates from Table 1 for 1994, one sees that a performance equivalent to that of the processor farm proposed in 1985 could be achieved using 4 systems containing sixteen processors each. Even if we drop back to Bell's performance estimates for 1992, a small number of general purpose minisupercomputer systems would be able to achieve the 1985 performance requirements at or below the projected budget. We regard the production by 1995 of a room-sized computer system that can actually sustain 10^6 MIPS in a general-purpose environment with a wide variety of users, job types and I/O requirements to be unlikely, regardless of architecture.

A minisupercomputer solution would have the additional advantages of increased reliability, lower maintenance cost and greater usability than a microprocessor farm. An array of low-cost, low-performance micros could result in lower system reliability and higher maintenance costs than with a smaller number of more powerful processors sharing a large common memory, especially if one considers the personnel and other costs of using physicists as computer designers and technicians. Using fewer components to achieve the required level of performance should improve system reliability and provide higher system availability. More time will be available for data collection and technical computation. Also, annual hardware maintenance expenses should be lower for a more reliable system.

Another significant advantage of the minisupercomputer solution comes in software development and support. These systems have rich software environments today, and they will certainly improve in the future. Vendor-supported operating systems, with real-time capabilities, will be available to support real-time operation concurrently with technical timesharing. Software tools, such as compilers, symbolic debuggers, and performance monitors, will provide an excellent software development environment for the physicist/programmer. In addition, a wide variety of other software will be available from the computer vendors and third parties, such as database managers, scientific visualization software, and data analysis packages.

Providing general-purpose technical systems capable of supporting more than raw data reduction and simulations is an attractive alternative to a specialized processor array. When desired, these general-purpose systems could support program development and other general technical computing requirements. Therefore, a portion of the expenditures planned for computers to satisfy general processing requirements could be used for additional workstations, thereby increasing the usability of the technical computing environment and making its users as productive as possible.

Acknowledgment

This work was partially supported by the Texas Advanced Technology Program under Grant No. 009741-082 and by the University of Texas System Center for High Performance Computing.

References

1. A. J. Lankford and G. P. Dubois, "Overview of data filtering/acquisition for a 4π detector at the SSC", in *Proceedings of the Workshop on Triggering, Data Acquisition and Computing for High Energy/High Luminosity Hadron-Hadron Colliders*, Fermilab, November 11-14, 1985, pp. 185-199.
   Hereafter these Proceedings will be referred to as *Workshop 85*.

2. W. Sippach, G. Benenson and B. Knapp, "Real time processing of detector data", IEEE Transactions on Nuclear Science **NS–27**, 578-581 (1980).

3. Paolo Franzini, "Lowest level trigger for SSC general purpose detectors", in *Workshop 85*, pp. 93-105.

4. P. S. Cooper, J. Dunlea, H. Kasha, S. Klein, W. M. Morse, L. Paffrath and M. Sheaff, "Data filtering-acquisition group: report of the hardware subgroup", in *Workshop 85*, pp. 200-210.

5. M. Chen, C. Dionisi, Yu. Galaktinov, G. Herten, P. LeCoultre, Yu. Kamyshkov, K. Luebelsmeyer, W. Walraff and R. K. Yamamoto, "The xenon olive detector: a fast and precise calorimeter with high granularity for SSC", in *Proceedings of the Workshop on Experiments, Detectors and Experimental Areas for the Supercollider*, Berkeley, CA, July 7-17, 1987 (Singapore, World Scientific Publishing Co., 1988), pp. 574-582.
   Hereafter these Proceedings will be referred to as *Workshop 87*.

6. M. Atac, D. B. Cline, M. Cheng, J. Park, M. Zhou, R. C. Chaney, E. J. Fenyves and H. Hammack, "A high resolution scintillating fiber gamma ray telescope", in *Supercollider I: Proceedings of the First International Symposium on the Supercollider*, New Orleans, LA, February 1989 (New York, Plenum Publishing Co.), pp. 699-707.

7. M. D. Petroff and M. Atac, "High energy particle tracking using scintillating fibers and solid state photomultipliers", IEEE Transactions on Nuclear Science **NS–36**, 163-164 (1989).

8. J. Alitti, A. Baracat, P. Bareyre, H. Blumenfeld, P. Bonamy, M. Bourdinand, J. Crittenden, J. P. Meyer, P. Perrin, A. Stirling, J. C. Thevenin and H. Zaccone, "The design and construction of the scintillating fiber detector for the UA2 experiment", IEEE Transactions on Nuclear Science **NS–36**, 29-34 (1989).

9. J. Dorenbosch, "Triggers in UA2 and in UA1", in *Workshop 85*, pp. 134-151.

10. L. D. Gladney, N. S. Lockyer and R. Van Berg, "The CDF track processor - prospects for the SSC", in *Workshop 85*, pp. 152-162.

11. J. T. Carroll, "Level 3 filters at CDF", in *Workshop 85*, pp. 254-260.

12. R. N. Cahn, M. Chanowitz, M. Golden, M. J. Herrero, I. Hinchcliffe, E. M. Wang, F. E. Paige, J. F. Gunion and M. G. D. Gilchriese, "Detecting the heavy Higgs boson at the SSC", in *Workshop 87*, pp. 20-67.

13. Joseph W. Goodman, Frederick J. Leonberger, Sun-Yuan Kung and Ravindra A. Athale, "Optical interconnections for VLSI systems", Proceedings of the IEEE 72^7, 850-866 (1984).

14. Mehdi Hatamian, Lawrence A. Hornak, Trevor E. Little, Stuart K. Tewksbury and Paul Franzon, "Fundamental interconnection issues", AT&T Technical Journal 66^4, 13-30 (1987).

15. Larry J. Hornbeck, "Deformable-mirror spatial light modulators", in *Spatial Light Modulators and Applications III*, Proceedings of the SPIE **1150** (1989), and references cited therein.

16. Alastair D. MacAulay, "Optical crossbar signal processor", in *Real Time Signal Processing VIII*, Proceedings of the SPIE **564**, 131-138 (1985).

17. Alfred Hartmann and Steve Redfield, "Design sketches for optical crossbar switches intended for large-scale parallel processing applications", Optical Engineering **28**, 315-327 (1989).

18. J. W. Goodman, A. R. Dias and L. M. Woody, "Fully parallel, high-speed incoherent optical method for performing discrete Fourier transforms", Optics Letters 2^1, 1-3 (1978).

19. J. I. Pankove, R. Hayes, A. Majerfeld, M. Hanna, E. G. Oh, D. M. Szmyd, D. Suda, S. Asher, R. Matson, D. J. Arent, G. Borghs and M. G. Harvey, "A pnpn optical switch", in *Optical Computing*, Proceedings of the SPIE **963**, 191-197 (1988).

20. I. Glaser, "Applications of the lenslet array processor", in *Real Time Signal Processing VIII*, Proceedings of the SPIE **564**, 180-185 (1985).

21. K. Hara, K. Kojima, K. Mitsunaga and K. Kyuma, "Differential optical comparator using parallel connected AlGaAs *pnpn* optical switches", Electronics Letters 25^7, 433-434 (1989).

22. See, for example, the work being performed by the IEEE POSIX committee on real-time standards: "Realtime extension for portable operating systems", P1003.4/D9, Work Item: JTC1.22.21.2, IEEE Computer Society, December 1, 1989.

23. Howard W. Johnson, "Effective performance in high speed networking", in *Digest of Papers, IEEE Compcon Spring 1989* (IEEE Computer Society Press), pp. 306-310.

    Hereafter this Digest will be referred to as *Compcon Spring 89*.

24. D. Deel, "VectorNet", in *Compcon Spring 89*, pp. 311-313.

25. Don E. Tolmis, "The high-speed channel (HSC) standard", in *Compcon Spring 89*, pp. 314-317.

26. John Hennessey, "Beyond RISC", Unix Review 7^9, 48-54 (1989).

27. Eugene Brooks, Lawrence Livermore National Laboratory, unpublished communication (1990).

28. E. R. Fox, R. W. Heemeyer, K. J. Kiefer, R. F. Vangen and S.P. Whalen, "A 32-Bit GaAs Microprocessor by CDC", in *Microprocessor Design for GaAs Technology*, Veljko Milutinović, editor (Englewood Cliffs, NJ, Prentice-Hall, 1990), pp. 154-247.

29. William A. Geideman, "A 32-Bit GaAs Microprocessor Implemented in GaAs JFET", in *Microprocessor Design for GaAs Technology*, Veljko Milutinović, editor (Englewood Cliffs, NJ, Prentice-Hall, 1990), pp. 248-293.

30. Gordon Bell, "The Future of High Performance Computers in Science and Engineering," Communications of the ACM 32^9, 1091-1101 (1989).

31. J.A. Appel, P. Avery, G. Chartrand, C. T. Day, T. Gaines, C. H. Georgiopoulos, M. G. D. Gilchriese, H. Goldman, J. Hoftun, D. Linglin, J. A. Linnemann, S. C. Loken, E. May, H. Montgomery, J. Pfister, M. D. Shapiro and W. Zajc, "Offline Computing and Networking", in *Workshop 85*, pp. 269-282.

32. "Cost Estimate of Initial SSC Experimental Equipment", SSC Central Design Group, Report SSC-SR-1023, June 1986.

33. A. J. Lankford, IISSC (1990).

34. Richard Rashid, Daniel Julin, Douglas Orr, Richard Sanzi, Robert Baron, Alessandro Florin, David Golub and Michael Jones, "Mach: A system software kernel", *Compcon Spring 89*, pp. 176-178.

4. Conventional Construction

SCHEDULING THE SSC CONSTRUCTION

Ted Kozman

Project Management Division
Superconducting Super Collider Laboratory *
2550 Beckleymeade Avenue, Dallas, TX 75237

ABSTRACT

The scheduling of the construction activities for the SSC Project involves two major concurrent emphases on design, construction, installation, and test. These two major activities include the completion of the construction project by the end of FY98 (108 months construction time) and industrially produced magnet tests by the end of FY92 (30-1/2 months from this point in time). Scheduling of the construction activities is further complicated by the fact that the baseline cost and schedule is still under review and negotiation with the Department of Energy and the currently anticipated funding for FY91 is somewhat less than originally requested. However, with the above limitations, the schedules presented herein are the most current at this time.

NEAR TERM—MAGNET/ACCELERATOR SYSTEM TESTS

The highest priority objective of the near term schedule is the demonstration of a collider magnet system test in an underground section of tunnel by the end of FY92, preceeded by an aboveground system test. The construction of the facilities to perform these tests and fabrication of the technical components is based on the following assumptions:

---

* Operated by Universities Research Association, Inc., for the U.S. Department of Energy under Contract No. DE-AC02-89ER40486.

1. Award AE/CM Contract–5/90,
2. Supplemental Environmental Impact Statement Record of Decision–11/90,
3. Award Cryogenics Systems Contract–7/90, and
4. Award Collider Dipole Magnet Contract–8/90.

The overall near term schedule is shown in Figure 1.

The current ongoing activities shown are: bid preparation for the first cryogenics plants, supplemental environmental impact statement work, AE/CM negotiation and contract review, and preparation of the collider dipole magnet RFP. Additional ongoing activities (not shown) include: preliminary design of accelerator components (power supplies, quench protection, controls, spool pieces, correctors, etc.) and ongoing magnet design and development at SSCL, FNAL, BNL, and LBL. It is our intent that both the aboveground and belowground test will use industrial fabricated 50 mm magnets. In our present planning, the collider dipole magnet subcontractor will fabricate these magnets using FNAL (or other) tooling after witnessing FNAL construction and testing of their 50 mm full length magnet. During the same time period, the collider dipole magnet subcontractor will start the acquisition and setup of this prototype tooling. If successful, magnets can be built and tested using these facilities, they will also be used in the two aforementioned tests.

It is our desire that these tests provide an early string test demonstration of magnet operation along with other prototypical accelerator components. Additionally, these tests and their facilities will provide us with an early understanding of the behavior and interaction of the various accelerator components. With this understanding, the time required for sector tests, several years later, should be reduced.

LONG TERM—PROJECT CONSTRUCTION SCHEDULE

The long term schedule was presented to DOE in January 1990. This schedule is based upon 108 month construction time starting in FY90 (October 1, 1989). The overall construction schedule is shown in Figure 2. Three levels of milestones were reported at the same time to DOE for control purposes. These milestones and schedules are to be considered *preliminary*, pending DOE approval of the project baseline. Category 1 milestones are intended for project control by the Office of the Superconducting Super Collider (DOE/Headquarters), Category 2 milestones are intended for project control by the On-Site Project Office (DOE/Texas), and Category 3 milestones are intended for control by the SSC Laboratory. All of the project control milestones are currently being reviewed internally for consistency with low level WBS schedules. While the schedule and milestones should be considered preliminary, the general times and approximate dates are what we feel will be the project baseline approved later this year.

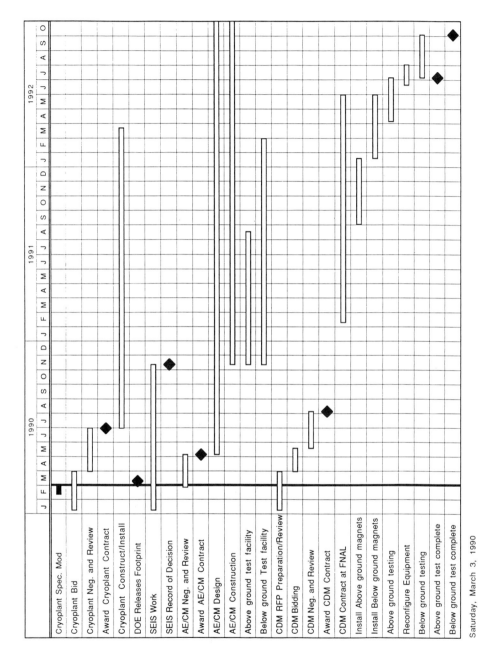

Figure 1. Overall near construction term schedule.

Saturday, March 3, 1990

179

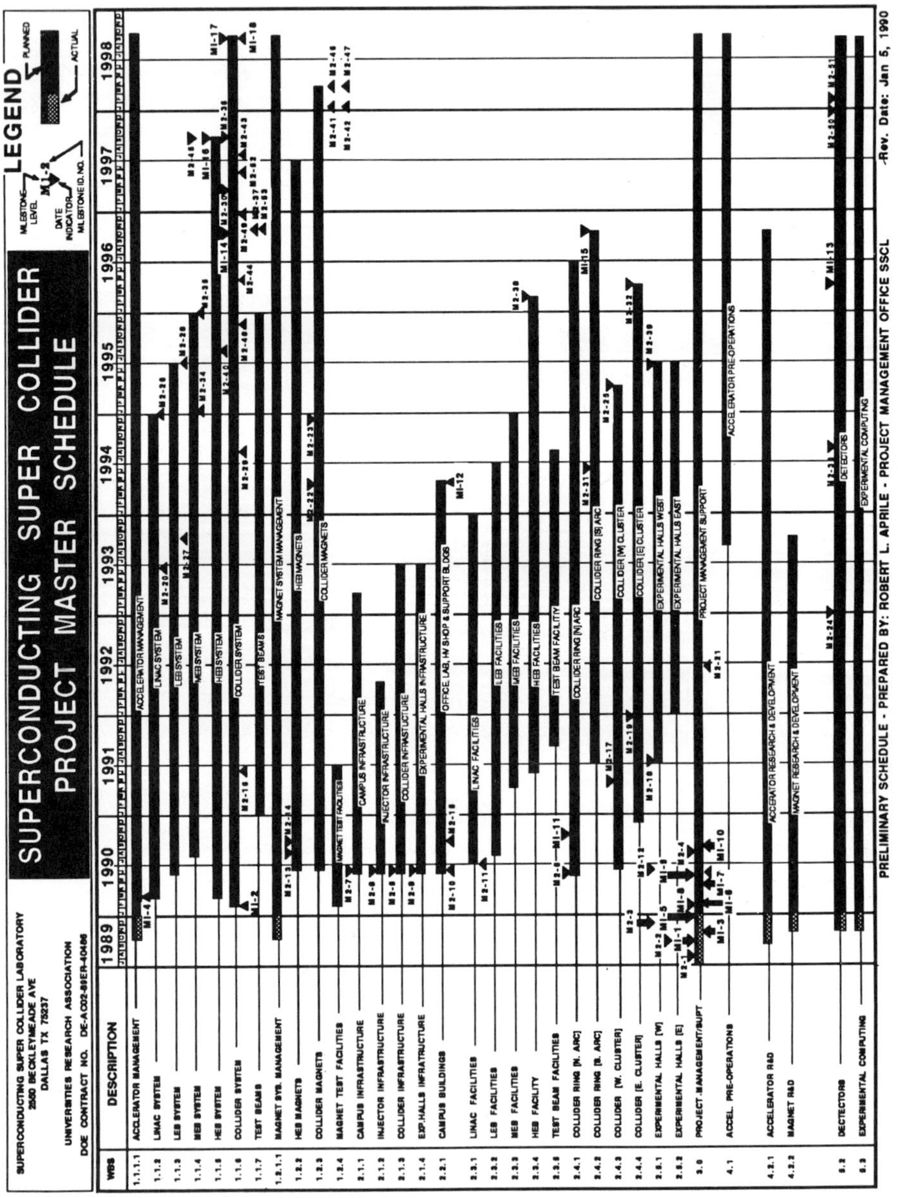

Figure 2. Overall long term construction schedule.

Table 1 shows the Category 1 milestones and Table 2 shows the Category 2 milestones. From these tables, the following observations of the near term activities can be made at this time:

- M1-1 — Completed on the original date,

- M1-2 — Completed for some activities on the original date,

- M1-3 — Completed on the original date,

- M1-4 — Conceptual design started,

- M1-5 — Preliminary draft issued as schedule,

- M1-6 — Now anticipated in March,

- M1-7 — No change in schedule,

- M1-8 — Now anticipated in May 1990,

- M1-9 — No change in schedule,

- M1-10 — Now anticipated in November 1990,

- M1-11 — Currently under review, will be later,

- M2-1 — Completed on schedule,

- M2-2 — Completed on schedule,

- M2-3 — Completed on schedule,

- M2-4 — No change in schedule,

- M2-5 — No change in schedule,

- M2-6 — No change in schedule,

- M2-7 — No change in schedule,

- M2-8 — No change in schedule,

- M2-9 — No change in schedule,

- M2-10 — No change in schedule,

- M2-11 — No change in schedule,

- M2-12 — No change in schedule,

- M2-13 — No change in schedule,

- M2-14 — No change in schedule, and

- M2-15 — No change in schedule.

Table 1. Master Milestones/Category 1

No.	WBS No.	Title		Milestone Date
M1-1	3.1.1	PM	First DOE Semiannual Review	Sep-89
M1-2	1.1.6	Collider System	Start Design	Feb-90
M1-3	3.1.1	PM	Start Construction Project	Oct-89
M1-4	1.1	Injection System	Start Design	Mar-90
M1-5	3.1.1	PM	SCDR Issued	Dec-89
M1-6	3.0	Footprint	DOE Approval	Feb-90
M1-7	3.1.1	PM	Baseline Validation Complete	May-90
M1-8	3.1.1	PM	PMP Approved By DOE	Feb-90
M1-9	3.0	AE/CM	Award of Contract	May-90
M1-10	3.0	SEIS	Record of Decision	Sep-90
M1-11	2.4	Collider Ring	Start First Tunnel Construction	Oct-90
M1-12	2.2	Campus Structures	Complete	May-94
M1-14	1.1.6	Collider System	Complete Component Design	Oct-96
M1-15	2.4	Collider Ring	Complete Conventional Construction	Oct-96
M1-16	1.1	Injection Systems Operational		Sep-97
M1-17	1.1.0	Collider System	Complete Acceptance Tests	Sep-98
M1-18	1.0	SSC Operational		Sep-98

Table 2. Master Milestones/Category 2

No.	WBS No.	Title		Milestone Date
M2-1	3.1.1	PM	Issue First Draft PMP	Aug-89
M2-2	3.1.1	PM	Award SE&I Contract	Oct-89
M2-3	3.1.1	PM	First C/SCSC Test Report Issued	Nov-89
M2-4	3.1.1	PM	First Land Tract Available	Aug-90
M2-5	2.1.3	Collider Infrastructure	Start Design	Jun-90
M2-6	2.4	Collider Ring	Start AE/Design	Jun-90
M2-7	2.1.1	Infrastructure	Start Campus Infrastructure Design	Jun-90
M2-8	2.1.2	Infrastructure	Start Injector Infrastructure Design	Jun-90
M2-9	2.1.4	Infrastructure	Start Exper Halls Infrastructure Design	Jun-90
M2-10	2.2.1	Campus Labs/Offices	Start AE/Design	Jun-90
M2-11	2.3	Injector Facilities	Start Design	Jul-90
M2-12	3.1.1	PM	Ready for C/SCSC Validation	Jun-90
M2-13	1.2.3.1	Collider Dipole Magnets	Award Preproduction Contract	Aug-90
M2-14	1.2.3.2	Collider Quad Magnets	Award Preproduction Contract	Sep-90
M2-15	2.2	Camp Hv Wks/Shops/Sup. Bldgs	Start AE/Design	Sep-90
M2-16	1.1.6	Collider Components	Start Fabrication	Jun-91
M2-17	2.4.3	Collider Ring	Start Construction West Cluster Tunnel	May-91
M2-18	2.5	Experimental Facilities	Start AE/Design	Jul-91
M2-19	2.4.4	Collider Ring	Start Construction East Cluster Tunnel	Dec-91
M2-20	1.1.2	LINAC	Complete Fabrication	Jun-93
M2-21	3.1.1	PM	Land Acquisition Complete	Jun-92
M2-22	1.2.3.1	Collider Dipole Magnets	Start Production	Apr-94
M2-23	1.2.3.2	Collider Quad Magnets	Start Production	Nov-94
M2-25	2.4.3	Collider Ring	Complete Construction West Cluster Tunnel	Apr-95
M2-26	1.1.2	LINAC Operational		Jan-95
M2-27	1.1.3	LEB	Complete Fabrication	Oct-94
M2-28	1.1.3	LEB Operational		Jun-95
M2-29	1.1.6	Collider System	Complete Acceptance Test Sector A	Aug-94
M2-30	1.1.6	Collider System	Complete Acceptance Test Sector B	Mar-97
M2-31	2.4	Collider Ring	Complete AE/Design	Jun-94
M2-32	2.4.4	Collider Ring	Complete Construction East Cluster Tunnel	Mar-96
M2-34	1.1.4	MEB	Complete Fabrication	Jan-95
M2-35	1.1.4	MEB Operational		Jan-96
M2-36	1.1.6	Collider System	Complete Acceptance Test Sector K	Sep-97
M2-37	1.1.6	Collider System	Complete Acceptance Test Sector C	Oct-96
M2-38	2.3	Injector Facilities	Complete Construction	Feb-96
M2-39	2.5.2	Experimental Facilities	Complete Construction	Jun-95
M2-40	1.1.5	HEB	Complete Components Fabrication	Aug-95
M2-41	1.2.3.1	Collider Dipole Magnets	Complete Fabrication	Jan-98
M2-42	1.2.3.2	Collider Quad Magnets	Complete Fabrication	Jan-98
M2-43	1.1.6	Collider System	Complete Acceptance Test Sector J	Jul-97
M2-44	1.1.6	Collider System	Complete Acceptance Test Sector H	Apr-96
M2-45	1.1.5	HEB Operational		Sep-97
M2-46	1.2.3.1	Collider Dipole Magnets	Complete Installation	Mar-98
M2-47	1.2.3.2	Collider Quad Magnets	Complete Installation	Mar-98
M2-48	1.1.6	Collider System	Complete Acceptance Test Sector D	Nov-95
M2-49	1.1.6	Collider System	Complete Acceptance Test Sector F	Dec-96
M2-52	1.1.6	Collider System	Complete Acceptance Test Sector E	May-97
M2-53	1.1.6	Collider System	Complete Acceptance Test Sector G	Oct-96

SUPER COLLIDER EGRESS SPACING AND LIFE SAFETY*

P. L. Clemens and R. R. Mohr

Sverdrup Technology, Inc.
600 William Northern Boulevard
Tullahoma, Tennessee 37388

Abstract: For the 53-mile tunnel of the Superconducting Super Collider (SSC), egress points are planned at an unprecedented spacing of five miles. In studying egress spacing with regard to life safety, no codes were found dealing specifically with accelerator tunnels. "General" codes were found which specify egress spacing not greater than several hundred feet. However, these "general" codes are neither for occupancies like the SSC nor do they credit the many overlapping safety features found in SSC conceptual design. A search for standards in counterpart underground activities disclosed a safety code for non-coal mines which prescribes egress requirements that the SSC surpasses. Egress-related risks, for SSC hazards were cataloged and found less profound than those for mining. Thus, the SSC egress-related injury rate should be less than that currently accepted in the mining industry. (The overall injury rate in mining is below the national average for industry at large.) As a further check, Fermi National Accelerator Laboratory was used as an extrapolation model for an egress-related injury rate study. Fermilab, with a statistically relevant database, has an injury rate below the national average. Of all accelerator tunnel injuries over its 18-year history, there were none for which probability of occurrence or severity of outcome would have been affected had egress spacing been either reduced from the present 800 ft. or increased by several miles. If the injury history experienced at Fermilab is acceptable, and if the extent to which the SSC surpasses non-coal mine requirements is satsfactory, there should be no reason to alter SSC egress spacing from the five-mile intervals now planned.

PURPOSE

The principal feature of the Superconducting Super Collider (SSC) laboratory will be the main collider ring, 53 miles in circumference, located in a tunnel with a diameter of approximately 12 feet at an average depth of 150 feet. Egress points for the tunnel are planned at intervals of 5 miles (Ref. 2).

---

*Acknowledgement: This paper reports the results of work carried out by an *ad hoc* team. Team members were Larry Coulson, Robert E. DeHart, Charles P. Lazzara, Robert J. Powers, and Ralph J. Vernon. The authors gratefully acknowledge their work The work of the team is described in more complete detail in Ref.1.

There are no safety codes specific to egress spacing for particle accelerators. However, there are precedents for egress spacing at various accelerators, as shown in Table 1. Spacing for the SSC, shown for comparison, exceeds that of the others by a factor of more than two.

Table 1 — Scale and Egress Spacing Precedents

FACILITY	CONST.	DEPTH	CIRCUM.	EXITS	EGRESS ΔX
Fermilab	1972	8 m	6.3 km	30	0.2 - 0.4 km
Tevatron	1983	8 m	6.3 km	30	0.2 - 0.4 km
SPS	1975	70 m	6.9 km	6	1.2 km
SLC	1986	10 m	2.8 km	4	0.3 - 1.1 km
HERA	1990	15 m	6.3 km	4	1.1 km
LEP	1989	70 m	27 km	8	3.4 km
UNK	1991	50 m	21 km	15	0.7 - 2.2 km
SSC	1994	>10 m	86 km	10	8 km

The purpose of the study reported here was to evaluate the adequacy of this unprecedented egress spacing with regard to life safety for the SSC work force.

CHARACTERISTICS OF THE PROBLEM

The task of evaluating life safety risks related to planned egress spacing for the SSC main collider tunnel poses an unusual challenge. Three chief factors contribute to this challenge: uniqueness of the facility, problems of codes applicability, and the presence of multiple, overlapping countermeasures.

Uniqueness

A number of underground particle accelerator facilities exist. (Some are listed in Table 1 above.) A few have accumulated several decades of operating experience. Thus, many hazards to be anticipated in the SSC are identifiable, based on experience elsewhere. Risk for many of these hazards can be assessed in the light of that same experience. (For the SSC, both the anticipated hazards and the countermeasures against them are well described in Ref. 2, based on the conceptual design.) Of the existing facilities, however, none are built to the large physical scale nor operated at the high energy levels planned for the SSC. As a result, for those egress-related hazards for which either the severity or the probability components of risk may be related to the size of the facility or to the energy level, risk assessment must be carried out with special care.

Extraordinary safeguards will be in place to prevent occupancy of the main collider tunnel during accelerator operation (Ref. 2). Consequently, risk for egress-related hazards associated with the high energy level is deemed to be under appropriate control, and hazards that are related to facility scale become the issue of concern.

The planned spacing of means of egress is greater for the SSC than for any existing accelerator facility, and it would be expected that life safety risk is related to the spacing between egress points. The questions raised by this consideration include determining the amount by which risk should be considered to be elevated over that found elsewhere by the increase in egress spacing, and the amount by which that increase might be offset by other features of the facility and its operating plan.

Codes Applicability

Safety codes prescribe measures that, if adopted, control levels of risk for hazards that may be anticipated in particular classes of facilities or operations. After the measures prescribed by the codes are imposed, the remaining residual risk must lie at a level deemed by the code-making body to be *de minimis*, a concept widely discussed in the current literature (e. g., Ref. 3). Codes can succeed at risk control only to the degree to which the code-making body is able to anticipate the hazard peculiarities of the facilities and operations. The process can be relied on especially well to ensure the acceptability of residual risk in facilities of ordinary nature, well known to the code-making body, and having hazards that are well understood.

Alternatively, codes that are inflexibly drafted to deal in general ways with broad ranges of facilities and operations suffer a substantial disadvantage. Because they cannot anticipate the individual hazard peculiarities of specialized cases, they may underprescribe countermeasures for some cases and overprescribe for others. For a novel facility, adherence to such a code may therefore result in unknowing acceptance of excessive residual risk in cases of the former kind, or in needless squandering of risk control resources without commensurate benefit in cases of the latter kind. This disadvantage becomes increasingly acute as the characteristics of the facility treated by the general code become increasingly novel.

As indicated above, the SSC is unquestionably a novel facility, if owing only to its physical scale. Imposing a "general" code — i. e., forcing conformance to an unequivocal code drafted for a broad range of general cases — therefore introduces the potential disadvantage of unwittingly accepting excessive risk, or of ineffectively deploying unnecessarily extravagant risk control resources. Nonetheless, the practice is often attractive to the unenlightened facility proprietor. The steadfast adherence to "a code" does confer the appearance of earnest concern for responsibly managing risk. It may also provide the alluring appearance of reduced liability.

Multiple Countermeasures

For many SSC hazards multiple layers of mitigation measures will be imposed to suppress risk, further complicating risk assessment. (Ref. 2 describes the measures.) These mitigation measures control both the severity and the probability components of risk to personnel. The multiple countermeasures include these examples:

- design to ensure the flammable burden within the tunnel will be low, and administrative controls to ensure it will remain low
- restricting electrical cable insulation to materials with low smoke properties and without halogenated hydrocarbons (Cable insulation is the major source of combustible material in most accelerator tunnels.)
- a system of redundant detectors to sense oxygen deficiency, temperature rise, smoke, and flooding
- emergency annunciation/evacuation alarms (activated automatically by the detection systems described above, or manually) at work locations, in the Control Room, and at emergency response crew locations, to ensure rapid emergency response

- full-time, instant, two-way communication between work crews, Control Room, and emergency response personnel to coordinate ongoing emergency response
- multi-skill emergency response crew makeup, including both trained, drilled firefighters and emergency medical technicians
- full-time, forced ventilation with greater-than-adequate throughput rate and with redundant, separately powered backup to operate during tunnel occupancy
- means for reversing forced ventilation flow, if necessary, to supply personnel undepleted, smoke-free air in the event of fire or a release of inert gas
- ventilation outage detection/alarm features to warn of failures and initiate evacuation procedures if backup systems fail
- screening employees who will work in the tunnel to ensure conformance to pre-established levels of physical fitness, and training and drills to ensure emergency preparedness
- limiting tunnel occupancy to two-man crews for two-day, fortnightly inspections and seven-man crews at widely spaced locations for major repairs, thereby reducing exposure to levels below those for which building egress codes are customarily drafted
- provisions to prevent energizing accelerator equipment during tunnel occupancy to preclude personnel exposures (Accelerator system equipment rests in a passive "storage state" during tunnel occupancy.)

These risk reduction measures do provide reassurances. However, because many of them combine to control a single class of hazard — i. e., fire — they also complicate assessing the true level to which residual risk has been suppressed. And it is that combined effect of these numerous measures, acting together to reduce risk, that must be considered rather than the effect of individual measures acting singly.

This concept is expressed in NFPA 550 (Ref. 4): "Firesafety aspects such as construction features, combustibility of contents, protection devices, and characteristics of occupants have traditionally been considered independently of each other. This can lead to unnecessary duplication of protection." This declaration embraces the principle of equivalent safety by alternate means ..That is, risk reduction to the *de minimis* level, as achieved by pursuing specific measures prescribed by necessarily unequivocal codes, can also be achieved at that same *de minimis* level by adopting suitably combined, multiple, alternative countermeasures.

APPROACHES AND RESULTS

The challenge of evaluating egress spacing is now narrowed. It may be resolved by (1) finding an appropriate industry-specific code for the unique SSC accelerator tunnel facility, or (2) otherwise ensuring that the combined effects of the multiple countermeasures planned for the SSC are appropriately accounted for in assessing risk related to the adequacy of egress spacing. Three approaches were adopted:

- Codes Search — A search was carried out for codes, standards, and regulations — within other domains of safety engineering practice — that might relate to identifying hazards and mitigating risks within the SSC.
- Preliminary Hazard Analysis — SSC hazards and their subjectively evaluated risks were inventoried and compared with those of a counterpart underground industry for which egress requirements have been codified.
- Use of a "Model" Facility — A "model" accelerator tunnel facility was sought and adopted as a reference standard from which to project evaluations of SSC risks.

The following sections describe the method used and the results achieved for each of these approaches.

Codes Search

Method. Because there are no safety codes specific to accelerator operations, a search was carried out to identify *non*-accelerator-specific codes that may exist for underground facilities having similar physical characteristics and employee activities, and to identify egress spacing requirements in those codes for cases in which parallels could be drawn with SSC hazards and their risks. The code search included:

- Codes cited in prior SSC documentation (e. g., Refs. 2 and 5)
- The Codified Federal Register
- The Library of the Board of Certified Safety Professionals

In examining codes found in the search, an attempt was made to apply comparison criteria to gage the extent of applicability in hazard and risk comparisons with the SSC. Chief among the comparison criteria were these:

- Restricted Access — Access to the SSC accelerator tunnel will be limited to authorized personnel and to selected, individually escorted visitors. The public at large will not be admitted. (Thus, codes applicable to tunnels for public transport systems and private vehicle traffic are useful in identifying hazards, but they have little value in addressing specific SSC risk considerations.)
- Limited Occupancy — Work crew sizes will not be less than two nor more than seven individuals, at widely spaced locations.
- Controlled Occupancy — SSC accelerator tunnel occupancy will be controlled to ensure that: only authorized crews are admitted, locations of all work crews are known at all times by Control Room monitors, and each work crew has immediate communication contact with Control Room monitors.
- Employee Training/Certification — Employees will be trained and certified in such areas as first aid, cardiopulmonary resuscitation, and emergency escape and rescue.
- Occupant Equipment — Each SSC accelerator tunnel occupant will be outfitted with a monitoring device to warn of exposure to oxygen deficient atmospheres and a supply of oxygen sufficient to permit escape.
- Low Combustible Burden — SSC design, construction, and operating criteria will impose rigid strictures to limit the quantity of combustible materials within the accelerator tunnel.

Results. Two codes were found which warranted detailed study when screened against the review criteria listed above: NFPA 101 and CFR 30, Part 57.

- NFPA 101 — *The Life Safety Code* (Ref. 6) — This is a general code as opposed to an industry-specific code. NFPA 101 seeks to prescribe countermeasures against risks that might be encountered in a wide spectrum of architectural configurations. It addresses broadly diverse occupancies and activities. Taken singly, many aspects of the SSC tunnel configuration and of the activities to be conducted within it were foreseen by the code-making body. However, certain of those aspects could not have been anticipated as existing in combination, within a single tunnel, containing only equipment supporting the operation of a laboratory facility. Among those aspects are:

  — EQUIPMENT — A high-energy particle accelerator, to which there is access by authorized personnel only during non-operating periods. All accelerator equipment is electrically de-energized during tunnel occupancy.

— OCCUPANCY/ACTIVITY — Occupancy is discontinuous, infrequent, and at low density. There are no fixed work locations to which personnel are assigned.

— COMBUSTIBLE BURDEN — Combustible materials making up the tunnel and installed within it are minimal. Electrical cable insulation constitutes the principal source of combustible material. The insulation selected has low smoke properties, and it contains no halogenated hydrocarbons. All major electrical power circuits — the chief source of ignition — are de-energized during tunnel occupancy.

Because this unusual combination of aspects could not have been foreseen by the code-making body, the egress spacing provisions of this broadly general code were not judged as specifically applicable in the unique setting of the SSC accelerator tunnel. Instead, an industry-specific code was sought. One was found which specifically relates to underground activity by a work force exposed to numerous hazards comparable to those foreseen for the SSC:

- CFR 30, Part 57 — *Mineral Resources / Safety and Health Standards — Underground Metal and Nonmetal Mines* (Ref. 7) — This industry-specific code was studied closely on the supposition that egress-related risks for personnel in an underground mine are comparable to those to which SSC accelerator tunnel workers will be exposed. Summarized briefly, this code requires, within the non-coal mining industry to which it applies, that two unobstructed directions of egress to the surface must be available at all times to each worker within the facility and that exiting must be possible from any work location by normal means within one hour, unless an internal method of refuge is provided. It also requires that damage to one exit route must not lessen the effectiveness of others.

A review of SSC background documents (Refs. 2 and 5) shows that the SSC conceptual design and planned operating procedures surpass these requirements. Moreover, within the industry to which this code applies, the recent-year recordable injury rate is 7.4 cases per 2×10^5 manhours (Ref. 8). This is lower than the overall U. S. average rate of 7.9 cases per 2×10^5 manhours for private sector industry at large, as also found in Ref. 8. Thus, a detailed comparison of the principal risks found in mines with those found in the SSC becomes relevant. That comparison was carried out using the Preliminary Hazard Analysis technique, below.

Preliminary Hazard Analysis

Method. The method used for the Preliminary Hazard Analysis (PHA) is described in Ref. 9. This method follows a technique that has become widely accepted in system safety practice in the aerospace and the chemical processing industries.

The PHA produces an inventory of hazards and assesses the risk of each hazard according to a method that, although subjective, is tightly disciplined. Both components of risk (severity and probability) are evaluated separately for each hazard in the inventory. For each hazard, the severity level for which risk is assessed is that of the worst credible consequence. The probability value expresses the likelihood of that consequence occurring, despite the presence of countermeasures. (SSC countermeasures are described in detail in Ref. 2.) Because probability of occurrence cannot be judged unless a particular interval of exposure is presumed, a 25-year facility lifetime was used for the SSC exposure interval in the analysis. A widely used risk assessment matrix (also found in Ref. 9) then guides the process of combining probability and severity evaluations to assess risk for each hazard.

The PHA technique suffers the obvious limitation that not all hazards can be foreseen by the analyst. To overcome this shortcoming, several sources were used to identify hazards for non-coal mines and for the SSC:

- Prior documentation — e. g., Refs. 2, and 10 through 18
- The OSHA *Policy for Systems Safety Evaluations of Operations with Catastrophic Potential* (Ref. 19)
- Consultations with operators of existing accelerator facilities and reviews of operating experiences
- Various proprietary hazards checklists
- Reliance on perceptive understanding of subtle hazards by experienced, prudent professionals
- Realistic scenario development — i. e., "what-iffing"

Results. Table 2 lists major headings, each representing a group of related hazards for which risk was evaluated in each of the two settings: the SSC and underground non-coal mines treated under CFR 30, Part 57. Under the heading "FIRE," hazards of *Electrical, Welding/Brazing, Solvent Misuse, Engine Heat, Spontaneous Combustion, Friction,* and *Explosion* were evaluated for the probability and the severity of risk they posed in each setting. The other major headings were broken down to assess risk for their respective hazards. Risk for all hazard ensembles shown in Table 2 is deemed to be egress-related.

Table 2 — Hazard Ensembles for the SSC and for Non-Coal Mines

FIRE	MATERIAL HANDLING
WALL/ROOF COLLAPSE	SLIP/TRIP/FALL
EXPLOSIVE BLAST	UTILITY INTERRUPTION
FLOODING	MACHINERY
RADIATION	PERSONNEL TRANSPORT
AIRBORNE DUST	ELECTRICAL
O_2 DEFICIENCY	CRYOGEN INJURY

In carrying out the mine risk assessment, severity and probability for each hazard were subjectively based on a population of more than 200 mines for which injury and fatality data have been examined over a period representing more than 1,500 mine-years of exposure. A representative non-coal mine was then subjectively postulated that would be comparable to the SSC accelerator tunnel, both in physical extent and in extent of personnel exposure.

The SSC risk assessment was based on study of a representative facility, scaled to represent the physical extent of the SSC and the personnel exposures anticipated in it. The Fermi National Accelerator Laboratory was used for this purpose, and was also used as a "model" facility, as described later.

Risk for the ensemble of hazards found in the mining case was found to be substantially more profound than that expected in the SSC. Of the 14 major hazard categories listed, all fell at the *de minimis* level for the SSC. For mines, many fell at more severe levels.

No SSC cases were found for which the residual risk — i. e., risk in the presence of documented countermeasures — was assessed to exceed the *de minimis* level. This analysis outcome is not unusual. In high-tech industries and state-of-the-art, high-energy systems, PHA inventories often list many hazards for which severity components of risk fall at the catastrophic level. For such high-severity cases, it is also customary that system designers will have given great attention to the need for multiple, independent countermeasures to establish adequate control over the probability component of risk. Risk control at the *de minimis* level is the result.

The injury rate of 7.4 cases per 2×10^5 manhours for underground non-coal mines (Ref. 8) may be taken as an index of the overall degree of risk prevalent within the industry to which CFR 30, Part 57 applies. It cannot be determined to what degree the egress provision found in the code maintains the rate. Neither is it known to what degree contraventions of that provision may contribute to the rate. However, because the industry is a mature one, it may be assumed that the principle of homeostasis applies. That is, for the many requirements within the mining code (e. g., ventilation, equipment, egress) none is outstandingly more or less effective at controlling overall risk than is any other.

This injury rate of 7.4 cases per 2×10^5 manhours for non-coal mines characterizes an industry having egress requirements that SSC conceptual design surpasses. That is not to say that those egress requirements for the mines would produce the same injury rate in the SSC case. As the PHA demonstrated, differences between risks for the principal hazards in mining operations and in accelerator tunnel activities are substantial. The risks for hazards ensembles shown in Table 2 are appreciably more profound for mining operations than for the SSC. It should be expected that conformance of the SSC to the egress spacing requirements of the mine code would produce a lower injury rate for the SSC case than the rate now prevalent and accepted in the mature mining industry. This presumes, of course, that egress provisions do have an effect upon the rate in mining, and that they would also have an effect on the SSC rate.

Use of a "Model" Facility

Method. An existing accelerator tunnel facility was sought as a representative reference "model." Such a model is useful in identifying potential SSC hazards and in evaluating their risks, as was done for the Preliminary Hazard Analysis, above. In seeking a facility to adopt as a model, the intention was not to ascribe to it any claim of merit nor to imply that its hazards, risks, and injury experience should be viewed as acceptable. Instead, the purpose was to adopt a baseline standard from which to make extrapolations of risk to the SSC case. Ideally, the model facility would have these characteristics:

- Physical dimensions, equipment configuration, and employee activities similar in all major, risk-related features to those of the SSC
- Egress spacing similar to that of the SSC
- A statistically substantial, trustworthy, actuarial database representing injury/fatality history over a lengthy period of facility operation

The Fermi National Accelerator Laboratory (Fermilab) was chosen to serve as the model facility. Fermilab satisfies the criteria listed above, with the obvious exception of egress spacing, which is 800 ft. (Table 1 provides dimensional comparison data.)

Fermilab employee injury data have been maintained in a way that makes it possible to examine the subset of data representing only those injuries experienced in accelerator tunnel work (as opposed to, and not including, support and operational work performed in all other locations of the facility). Manhour support has been similarly differentiated in the Fermilab database. Thus, accelerator tunnel injury rates that can be normalized to actual manhours of exposure can be computed. Moreover, the information has been gathered and maintained in a sufficiently descriptive way to enable determining — on a case-by-case basis — which injuries, if any, might have had either greater probability of occurrence/recurrence or more severe consequences had egress spacing been greater. Actual operating risk to personnel, in other words, can be related to egress spacing.

Results. The Fermilab injury record was reviewed for all accelerator tunnel operations from January 1982, when cryogenic technology was adopted, through June 1989. All documented injury cases for the period were examined, both OSHA-recordable injuries and less severe cases. During that period, there were 26 injuries and no fatalities. Of those 26 cases, 6 qualified as OSHA-recordable, based on severity. None of the cases involved more than a single employee — that is, there were no multiple-injury mishaps. The OSHA-recordable injury rate for the period was 7.1 per $2x10^5$ manhours.

The Fermilab injury rate of 7.1 per $2x10^5$ manhours is 11 percent below the recent-year nationwide, all-industry rate of 7.9 per $2x10^5$ manhours (Ref. 8). The Fermilab rate compares favorably with other modest-rate segments of industry — the aircraft and parts industry, for example, which has an injury rate of 7.0 per $2x10^5$ manhours (Ref. 8). The population of only 6 instances of OSHA-recordable injuries on which the Fermilab rate is based might appear to be too small for statistical relevance. However, total exposure for the period during which the 26 injuries occurred approaches 100 manyears, a statistically relevant database.

A case-by-case review of the Fermilab injuries has shown none for January 1982 through June 1989 for which either probability of occurrence/recurrence or severity of outcome would have been influenced by changes in egress spacing. Even very substantial increases in spacing (i. e., to several miles) would have had no effect. Interviews with Fermilab emergency response personnel for cases over the 1971-1981 period produced the same conclusion, based on anecdotal recall. Thus, there have been no injury cases in which increased egress spacing would have altered the mishap record during the 18-year Fermilab operating history — more than 250 manyears of accelerator tunnel work exposure. The Fermilab actuarial record shows relative independence of injury probability/severity from increases in egress spacing to distances of several miles.

Injury scenarios can indeed be postulated in which greater egress spacing could have an influence on the severity of outcome. However, the probability of such cases is envisioned as extraordinarily low. Current first aid and rescue philosophy bears on this. It was once the case that, in removing injury victims to remote emergency treatment facilities, haste was given priority over other considerations. Now, however, except in urban settings where specially-equipped trauma centers are quite nearby, it is customary to stabilize the victim at the scene of the injury before attempting transportation. Thus, the length of the egress path and rescue time have come to be less important.

CONCLUSIONS

Egress spacing planned for the SSC conforms to requirements imposed by an industry-specific code for underground mining operations, where a more profound ensemble of risks to personnel exists. While other codes can be found that claim universal applicability to a broad range of facilities, their specific applicability to the specialized character of the SSC is questionable. Those "general codes" could not have anticipated the unique configuration of the SSC nor the array of multiple countermeasures planned for the facility. Consequently, those codes stipulate egress spacing at very close intervals, as they must to ensure their applicability to the control of risk in other facilities posing greater threats to life safety.

The Fermilab accelerator tunnel serves as a reasonable hazard evaluation/risk assessment model for the SSC tunnel in all important risk-determining features except egress spacing. The Fermilab facility has a well-documented injury record and a statistically reasonable exposure interval. There have been no injuries of kinds for which either probability of occurrence or severity of outcome would have been worsened had egress locations been appreciably more widely spaced — e. g., by several miles. Although scenarios can be postulated in which such cases might occur, the probability of such cases is envisioned to be extraordinarily low. The actual Fermilab injury rate is below the national average and is comparable to that of other modest-rate segments of industry.

If risk for SSC operations is deemed acceptable at these levels, then egress spacing planned in the conceptual design for the SSC accelerator tunnel should also be judged acceptable. If risk is deemed unacceptable at these levels, additional countermeasures should be considered. But that can be done only after deciding the level of risk to be tolerated

In the work reported here, the goal has been to assess risk, i. e. to take its measure, not to address its acceptability. This work has sought to assess the level of SSC risk and to compare it to risks of other activities, specifically without addressing the acceptability of risk. Risk acceptance is in the domain of the facility proprietor. Further, this work has considered only matters of life safety, only as affected by egress spacing, and only during the post-construction period of facility use.

The Preliminary Hazard Analysis technique used as a part of this effort can now be extended to serve other SSC hazard identification and risk assessment purposes — e. g., risk to equipment, to the experimental data, and risk to the environment. And the PHA and the other facets of the study (codes examination and facility comparison) can provide continuing hazard identification and assessment of risk throughout future detail design work.

REFERENCES

1. (_____) — *Study and Findings Concerning Egress Spacing and Life Safety* — SSC-SR-1042, November 1989.
2. Toohig, T. E., Editor — *The Superconducting Super Collider / SSC Safety Review Document* — SSC-SR-1037, November 1988.
3. Whipple, Chris, Editor — *De Minimis Risk* — Plenum Press, 1987.
4. (_____) — *Firesafety Concepts Tree* — NFPA 550, 1988.
5. (_____) — *Preliminary Analysis of the Conceptual Design of the Superconducting Super Collider Main Collider Tunnel Operational Safety with Respect to Access, Egress and Ventilation* — *Preliminary Draft Report* (Agency / Publisher not listed), 8 February 1989.
6. (_____) — *Life Safety Code* — NFPA 101, 1988.
7. (_____) — *Mineral Resources / Safety and Health Standards —Underground Metal and Nonmetal Mines* — CFR 30, Part 57, July 1988.
8. (_____) — *Accident Facts* — National Safety Council, 1988.
9. (_____) — *Military Standard / System Safety Program Requirements* — MIL-STD-882-B, March 1984.
10. Casebolt, H. — *ODH Analysis for the Main Ring Tunnel* — 21 February 1984.
11. Brown, D. P. and J. H. Sondericker — *Oxygen Deficiency Hazard Induced by Helium Release in Accelerator Tunnel*.
12. Heyman, A. —*Tunnel Access and Egress Study* — 24 October 1985.

13. Casebolt, H. and P. J. Limon — *Simulation of Accident Involving the Release of Liquid Helium into the Main Ring Tunnel.*
14. Cryogenic Consultants, Inc. — *Spillage of 80K Shield Line Contents into the Tunnel of the SSC* — 25 November 1987.
15. Cryogenic Consultants, Inc. — *Spilling of Helium Gas from the 20K Shield into the Tunnel* — 1 December 1987.
16. Cryogenic Consultants, Inc. — *Break in 4K Vapor and/or 4K Liquid Helium Line* — December 1987.
17. Gilbert, Paul H. — *A Review of Current Practices for Providing Integrated Systems for Life Safety of Tunnel Occupants — Application to SSC Tunnel Requirements* — Parsons, Brinckerhoff, Quade & Douglas, Inc., August 1988.
18. Rode, C. H. — *S.S.C. Experiment #2: He Pumping of a Vertical Shaft* — Internal Memo; 29 May 1984.
19. (_____) — *Systems Safety Evaluations of Operations with Catastrophic Potential* — OSHA Policy Notice No. 50, 13 July 1987.

5. Cryogenics

A REFRIGERATION PLANT CONCEPT FOR THE SSC

R. Powell

H. Quack

Koch Process Systems, Inc.

Gebr. Sulzer AG

Westborough, MA, U.S.A.

Winterthur, Switzerland

INTRODUCTION

Determination of the optimum refrigeration plant arrangement for the SSC will require consideration of many factors. Starting from the latest information on the SSC cooling requirements and the given number of cooling sectors, an arrangement is proposed which will provide an optimum combination of low investment cost, low power consumption and simplicity of operation, while providing an adequate degree of redundancy.

The proposed arrangement consists of two identical helium refrigerator/liquefiers at each station. In normal operation, each of these plants will provide the refrigeration, liquefaction and 20K shield cooling for half a sector. In case of a scheduled or unscheduled shutdown of one of these plants, the operation of its twin plant will be changed to provide the 4K refrigeration load for the whole sector. The liquefaction and 20K cooling loads will be supplied by the plant adjacent to each half sector.

Design considerations for the required refrigerator/liquefier will be presented. Also, it will be shown that these systems can be built with present day technology.

THE OVERALL SYSTEM

The starting point of this study is the paper by M.S. McAshan presented at last year's IISSC, "SSC Refrigeration System Design Studies" and amended in the 12/20/89 revision of the SSC Central Design Study Report.[1] The layout of the SSC is described as having ten cryogenic sectors with the nominal total heat load for each sector established as:

5.2 kW at 4.15K, plus
28 g/s liquid helium, plus
8.335 kW at 20K

In this paper, we want to discuss different schemes for the cryogenic refrigerators focusing on the two aspects of system availability and first cost.

Our aim is not to propose any special type of equipment but to give indications from the viewpoint of the refrigerator industry where there exist degrees of freedom for the total cryogenic system designer.

For this study, the details of the duty cycle, system dynamics, use of stored refrigeration to cope with upsets, cooldown and quench recovery were not considered. These design aspects are very important. However, the primary emphasis was to look at the system as needing a certain amount of refrigeration, liquefaction and shield cooling and how these requirements may be met.

Figure 1 shows the SSC ring with its ten sectors. The distance between two neighboring cryogenic stations is about 8 km. System availability studies have set a goal for the cryogenic system of 98%.

To simplify our analysis, we will focus on three of these sectors. Figure 2 graphically depicts the cryogenic systems in three neighboring sectors, each providing refrigeration (solid lines) and liquefaction (dotted lines).

The simplest approach to providing the SSC cryogenic needs would be to provide one helium system for each of the ten sectors, each of which is capable of handling the entire load of one sector. While this is the most cost effective solution, the problem exists that if one of the refrigerators is lost, the whole SSC is out of operation. Therefore, to achieve the required total system availability of 98%, each single refrigerator must have a 99.8%

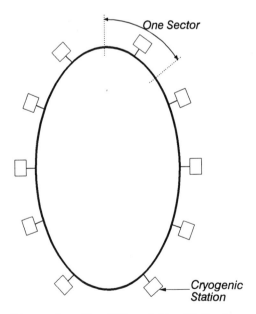

Figure 1. The SSC and its 10 Sectors.

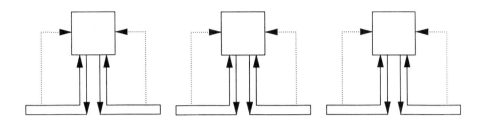

Relative Cost: 1

Required Unit Availablity: 99.8%

Figure 2.   Single Sector Scheme A (1 x 100%).

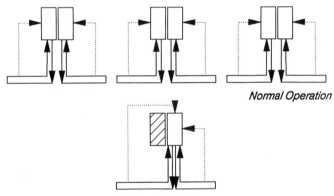

Normal Operation

"Emergency" Operation

Relative Cost: 2

Required Unit Availablity: 95%

Figure 3.   Single Sector Scheme B (2 x 100%).

availability (this translates to a down time of only 18 hours per year). To guarantee this level of performance would require a lot of courage from the manufacturer.

The first alternative to consider for reducing the availability required by each individual cryogenic system is to install two refrigerators for each sector, each having 100% capacity. This is shown in Figure 3. To obtain a total system availability of 98%, each refrigerator then needs to have an availability of only 95.5%. However, if the refrigerators are of equivalent design, the cost of this scheme is about twice as high as that of the 1 x 100% configuration. This cost picture does not improve if one considers a larger number of smaller plants per sector, e.g., three 50% plants.

The rule of thumb which we are using here to establish the relative cost is that the price of a refrigerator is proportional to the 0.5 power of its maximum capacity. This factor takes into account the fact that larger plants usually have a higher efficiency than smaller ones.

ASSISTANCE FROM NEIGHBOR SECTORS

It is obvious from the above that when each sector tries to solve the redundancy problem in isolation, it becomes very expensive. The next step is to ask how neighbor sections can assist each other in case of an emergency.

One solution, shown in Figure 4, is to provide each sector with a 150% unit. If one refrigerator fails in any way, the two neighbor units can provide refrigeration to the adjacent half sections. This results in a relatively cost effective system (relative cost of 1.22) and a realizable required unit availability of 97%.

The main problem with this configuration, however, is the effect on magnet temperature. The magnet temperature is determined by the pressure drop in the cold vapor line. If one compares the normal situation in a half sector with the situation of two half sectors in series, twice as much mass flow has to flow over twice the distance to achieve the required refrigeration. Since the pressure drop varies with the square of flow rate and linearly with distance, the pressure drop for the "emergency" operation is eight times the normal value. There could also exist in some of the sectors an additional pressure drop due to gravity effects since the SSC ring is not level. (We assume the sector length of 8 km has been chosen with these limitations in mind.)

The net result of all of this is that the temperature variation in such a system would undoubtedly exclude the idea that neighboring systems can assist each other in the refrigeration aspect of the cooling duty. However, the situation is different for the liquefaction and also the 20K shield cooling duties because the associated flows are not as pressure drop limited as the refrigeration duty.

One can then consider a refrigerator scheme where each half sector has two identical cryogenic plants which can provide refrigeration, liquefaction and 20K shield cooling for each sector. If one of these two plants fails, the remaining one will then provide the refrigeration load for both half sectors. On the other hand, the liquefaction load and the 20K shield load will be provided by two plants from the neighboring sectors. If two 67% systems are used, as shown in Figure 5, the relative cost factor

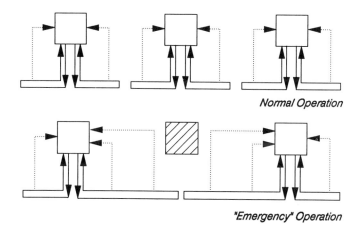

Normal Operation

"Emergency" Operation

Relative Cost: 1.22

Required Unit Availablity: 97%

Figure 4. Assisting Neighbors Scheme A (1 x 150%).

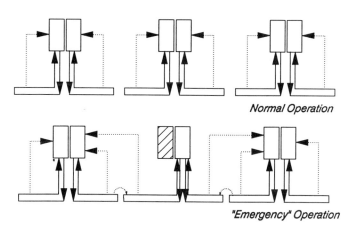

Normal Operation

"Emergency" Operation

Relative Cost: 1.63

Required Unit Availablity: 95%

Figure 5. Assisting Neighbors Scheme B (2 x 67%).

becomes 1.63 and the required unit availability is only 95%. The necessary design characteristic for this redundancy scheme is that the involved plants be able to change their internal distribution between refrigeration and liquefaction. The remaining operational single unit becomes a "pure refrigerator", whereas the assisting units from the neighboring sectors increase their liquefaction duty. Therefore, all the plants have to be flexible in their refrigeration-liquefaction performance.

DESIGN CHARACTERISTICS OF REFRIGERATOR/LIQUEFIERS

For the sake of simplicity, we will neglect the 20K shield cooling and investigate how one and the same plant handles refrigeration and liquefaction loads. In both cases, the plant delivers liquid helium. In the refrigeration mode, this helium returns to the plant as cold vapor near the saturation temperature. In the liquefaction mode, the helium returns at ambient temperature and atmospheric pressure. This is shown schematically in Figure 6.

In a diagram which shows refrigeration as the ordinate and liquefaction as abscissa, one can draw a straight line which constitutes a "thermodynamic equivalence" (shown in Figure 7). For a system without LN_2 precooling, the equivalence factor is 120 Watts of refrigeration per g/s of liquefaction capacity. When liquid nitrogen precooling is used, the liquefaction mode operation is improved more than the refrigeration mode operation, and the equivalence factor is reduced to about 80 Watts of refrigeration per g/s of liquefaction capacity.

The characteristic of real plants is seldom such a straight line. The plant designer has many degrees of freedom but must choose where he wants to put his priorities. Figure 7 also schematically depicts the characteristics of a typical refrigerator, a typical liquefier and a plant which has been optimized for a certain "mixed mode".

It should be noted that the plant characteristic may be extended into the "negative liquefaction" range. During this mode of operation, stored liquid helium is used to increase the refrigeration output beyond the plant's own capacity. A well known example for this kind of operation is the "satellite mode" of the FNAL Tevatron cryogenic system.

DESIGN CHARACTERISTICS REQUIRED FROM THE SSC REFRIGERATOR/LIQUEFIER

We return now to the case where each sector has two identical units installed in each cryogenic station. Again, for the sake of simplicity, we neglect the 20K shield cooling. In normal operation, each plant provides the duties needed for a half sector, i.e., 2.6 kW refrigeration plus 14 g/s liquefaction (shown as the "50% Point" in Figure 8).

To examine the case if one plant fails, we assume its direct partner would provide the total refrigeration requirements of both half sectors (i.e., 5.2 kW) and none of the liquefaction. The two assisting plants from the neighbor sectors would then need to provide the liquefaction duty of the half sector adjacent to each, in addition to their own sector requirements. Their total load becomes 2.6 kW plus 28 g/s each.

If these new operating points are shown on a refrigeration /liquefaction diagram (see squares on Figure 8), the connection

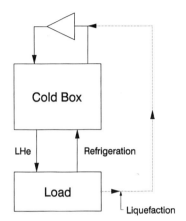

Figure 6.   Refrigeration and Liquefaction Load.

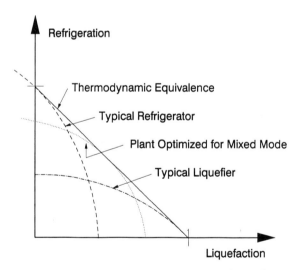

Figure 7.   Refrigeration-Liquefaction Diagram.

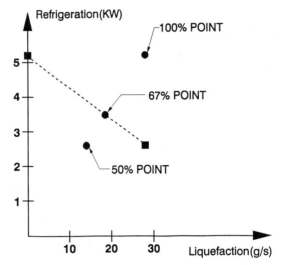

Figure 8.   Possible Characteristics of a 67% Plant.

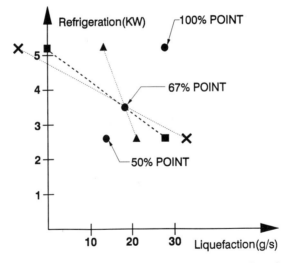

Figure 9.   Possible Characteristics of a 67% Plant.

line between the two "emergency" operating points is the required characteristic for the SSC refrigerator/liquefier.

It now turns out that this example is not a unique solution. In Figure 9, two other possible characteristics are shown. In one design case where the single operating unit is able to provide some of the liquefaction in addition to the 5.2 kW refrigeration loads, the operating characteristic is shown by the triangles. In this case, the helping units have to produce less liquefaction. In another design case where the operating unit needs some "liquid helium precooling" to produce the required 5.2 kW of refrigeration, the operating characteristic is shown by the X's. The liquid helium for precooling would be produced by the assisting units in adjacent sectors.

It should be noted that all three possible characteristics described have one point in common. This point is 133% of the normal operating point for that half sector, or, in other words, 67% of a full sector load, a case that was discussed earlier.

Using this approach, the refrigeration system designer would choose the plant characteristic which has the least overall cost.

SIZE OF THE SSC REFRIGERATOR/LIQUEFIER

In Figure 10, we have compared the proposed plant size of the SSC units with other plants built during the last 20 years. It is not easy to choose a correct scale for such a comparison. What should one take as measure — input power, mass flow, or duty? How should one value the use of liquid nitrogen precooling? Therefore, the comparison can only be a rather qualitative one. However, as can be seen, this approximation shows that the proposed SSC refrigerators are rather small compared with many previously built systems and are well within the range of demonstrated operation.

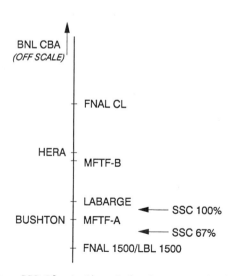

Figure 10. SSC Plant Size Relative to Existing Plants.

SUMMARY

This paper has attempted to shed some light on the thought process necessary to develop the best cryogenic system design to meet the needs of the SSC. Decisions which specify redundant systems to meet overall system availability requirements must take into consideration allowable operating characteristics of refrigerator/liquefiers. Two cryogenic plants, each with a capacity of meeting 67% of each sector cryogenic requirements, are shown to be one solution to this requirement.

In addition, a comparison of cryogenic system sizes currently under consideration for the SSC were shown to be easily within the range of present industry practice.

REFERENCES

1. M.S. McAshan: "SSC Refrigeration System Design Studies", Supercollider 1, Plenum Press, 1989, p. 287, and amended by Table 4.2.3.1-9, p. 430 of the 12/20/89 revision of the SSC Central Design Study Report.

2. R.A. Byrns: "Large Helium Refrigeration and Liquefaction Systems" in "Cryogenic Engineering" ed. by R.A. Hands, Academic Press, 1986, p. 357.

TRANSVERSE COOLING IN SSC MAGNETS*

R.P. Shutt and M.L. Rehak

Brookhaven National Laboratory
Accelerator Development Department
Upton, New York 11973

ABSTRACT

A new and more efficient cooling method for SSC magnets is presented. The simple cooling–by–diffusion scheme provides adequate cooling only when the synchrotron radiation heat is limited to $S_r = 2$ W. The mass flow that can pass through the coil passage cannot be increased without an unduly high pressure drop along the magnet. The transverse flow cooling described here allows for large increases in the heat load. This scheme lets cold helium from the upper bypass mix with the lower bypass helium at intervals of 30 cm for instance. This is achieved by introducing channels in the yoke laminations at given intervals using existing gaps between collar packs and using pickup notches in the collars.

INTRODUCTION

The purpose of this paper is to outline the principles behind the transverse cooling method. Only the general ideas are discussed. More details on the theory, refinements and an in depth parametric study will be found in a forthcoming report.

SSC magnets must be kept at liquid helium temperature in order to keep the coils operating as superconductors. The magnets are subjected to a heat load due to synchrotron radiation (S_R) localized along the bore tube at the center of the magnet. Additional heat comes from the exterior which is at room temperature. The aim of the cooling scheme is to keep the main coil and trim coil temperatures (T, T_t) and pressure drop Δp_m across a magnet at relatively low values.

REQUIREMENTS AND LIMITATIONS

The temperature of trim coil and the main coil which are closest to the major heat source, the synchrotron radiation, must not exceed $\Delta T = 0.15$ K and $\Delta T_t = 0.5$ K, respectively. In addition the pressure drop across the refrigeration system is

---

*This work is supported by the U.S. Department of Energy.

limited to $\Delta p = 0.5$ atm, including about 200 dipoles and 30 quadrupoles. Some of the cooling problems the 4 cm design faces are a narrow coil cooling passage cross section (radial width $d_c = 1.3$ mm), very long dipoles (≈ 17 m), and the coils are well insulated electrically, therefore thermally. A total mass flow of $M = 100$ g/sec is available per magnet, the average operating temperature is 4.4 K, the operating pressure is 5 atm, and recoolers to restore the operating temperature of the helium are positioned periodically along the magnet rings.

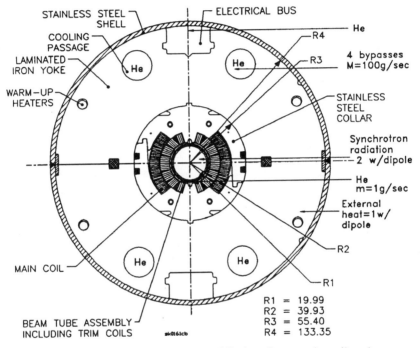

Figure 1.   Coil cooling by heat diffusion from main coil to bypass.

DIFFUSION COOLING METHOD

The simple cooling–by–diffusion method for SSC magnets consists of letting supercritical helium at 4.4 K temperature flow axially along four bypass holes in the iron yokes and along the coil cooling passage between the main coils and the bore tube which carries the trim coil. Under these conditions there is m=1 g/sec of helium flowing through the coil cooling passage mostly for heat transfer from the bore tube to the main coil. Subsequently heat diffuses radially from the main coil to the four bypasses through the yoke laminations. Fig.1.

210

This design was adequate for originally proposed operating conditions and material properties. For a synchrotron radiation heat load of S_r = 2 W per dipole a maximum temperature increase along the trim coil of 0.09 K and along the main coil of 0.06 K was found. Radial heat diffusion between inner coil and bypasses is sufficient to obtain these small temperature increases. The question of an eventual upgrading of the beam intensity has been raised recently. An increase of S_r by a factor of five, from 2 to 10 W, would result in temperature increases of 0.43 K for the trim coil and 0.29 K for the main coil which might still be acceptable. However the heat conductivity of Kapton (k_p = 6 x 10^{-4}) is not known precisely and could be one order of magnitude less (k_p = 6 x 10^{-5}) than the current value in which case the trim coil and main coil temperature increases would still be acceptable at 2 W but not at 10 W. Summarizing, the simple cooling by diffusion is acceptable for synchrotron radiation heat of 2 W but offers little or no margin – or is insufficient – at higher heat loads.

TRANSVERSE COOLING METHOD

This method, originally proposed for SSC magnets, is reconsidered as it would allow a large increase in heat load and accommodate the lower value for heat conductivity of Kapton (which will be assumed until confirmed by new measurements). It combines transverse flow with radial diffusion and axial coil cooling. Here, cold helium from the upper bypasses periodically reaches the coil cooling passage, mixes with the incoming helium, removes heat from a given section of magnet, say 30 cm, then exits to reach the lower bypass. Fig.2. Orifices at the ends of the bypasses produce the transverse pressure difference regulating the magnitude and axial distribution of the transverse mass flow. As in the previous scheme, axial helium flow finally mixes with the bypass flow after exiting into the magnet interconnection.

Every 30 cm a yoke lamination with channels is positioned in the upper half of the assembly while a similar yoke lamination with channels is also introduced into the lower half every 30 cm but offset by 15 cm from the upper one. Gaps between collar packs are provided every 15 cm. These bypass–to–collar gap openings exist in present magnets for quench pressure relief but they do not alternate. Since the yoke channels do not line up with the gaps between collar packs, helium arriving from the channels travels along collar pickup notches which are periodically blocked, presently by inserting indium blocks, to direct the flow to the coil cooling passage (Fig. 3). In addition leak paths due to gaps between collars and yoke at the pole and the midplane are blocked. (At present, and for test purposes only, this is done by applying a caulk–like compound called RTV). Fig.3.

MASS FLOW

The unknowns in the system of equations are M1, M2 the upper and lower bypass mass flows, m1 the transverse mass flow from the upper bypass to the coil cooling passage, m2 the transverse mass flow from the coil passage to the lower bypass, q1 the pressure differences between upper bypass and coil cooling passage, q2 the pressure difference between coil cooling passage and lower bypass (Fig.4). The change in pressure along the magnet length is proportional to the square of the mass flow and to a friction coefficient, also dependent on the mass flow. This is valid only for turbulent flow. For laminar (slow) flow, the pressure drop is a linear

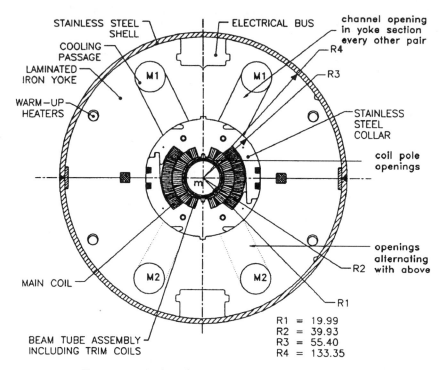

Figure 2. Transverse helium flow cooling (flow from top M1 to bottom M2).

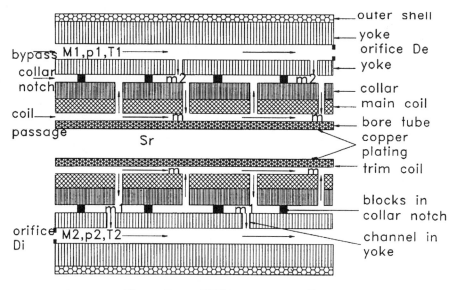

Figure 3. SSC transverse cooling.

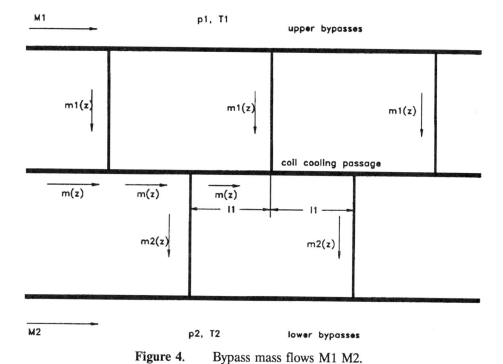

Figure 4. Bypass mass flows M1 M2.

function of the mass flow. These laws are used for each of the bypasses and for the coil cooling passage. Three more equations are derived by considering helium mass flow conservation. The boundary conditions provide relations between mass flows and also include dimensions of the orifices. Since the boundary conditions depend on the solution this is a boundary value problem solved by an iterative procedure using the relaxation method.

TEMPERATURE

The unknowns in this part of the problem are the main coil temperature T, the trim coil temperature Tt, the helium temperatures in the upper and lower bypasses T1, T2 and that of the coil cooling passage T0. Main coil and trim coil temperatures T, T_t (conductor temperatures) are given by diffusion equations. Using the fact that the diffusion lengths for both coils are much smaller than the magnet length, the diffusion equations can be set equal to 0. This decreases the complexity of the problem since T and Tt are now given as linear functions of T1, T2 and T0. Expressions for T0, T1, T2 (cooling passage temperatures) are established by considering heat exchanged between all the various components of the system.

The problem is reduced to integrating two first order differential equations defining T1 and T2. Due to the simple nature of the boundary conditions this is an initial value problem solved with one matrix inversion.

RESULTS

The aim is to minimize T (main coil), Tt (trim coil), Δ_p (pressure drop). The parameters that one can adjust to optimize the results are the gap sizes between collar packs (a_g = 0.16 cm), the lengths of the yoke and collar packs (l_1 = 15 cm), the end orifice sizes (D_i, D_e). A representative set of curves is shown with the following values: S_r = 10 W, D_i = $D_e \approx 0$. (Here it is assumed that the ends of the cooling passage are closed.) Figure 5. shows the upper and lower bypass mass flows along the magnet length. Starting at 100 g/sec M1 reaches 0 at the end of the magnet. Figure 6 shows the transverse mass flow (here m1=m2) which has a parabolic shape as a result of the mathematics of the problem. The goal of an

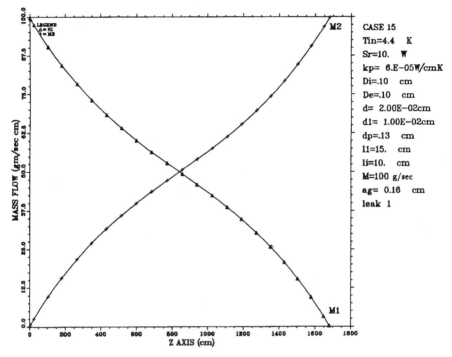

Figure 5. Bypass mass flows M1 M2.

optimization has been to flatten this shape as much as possible for more even cooling. Figure 7 shows how the upper and lower bypass helium temperatures vary, the increase from 4.4 K at the magnet end is 0.035 K for the lower bypass helium. Finally Fig. 8 indicates that there is a maximum increase in the trim coil of 0.24 K and in the main coil of 0.04 K. These values would be 1.44 K and 0.76 K, respectively, with the simple diffusion cooling method. Transverse cooling has been used quite successfully in an SSC dipole with a mass flow of M=50 g/sec only.

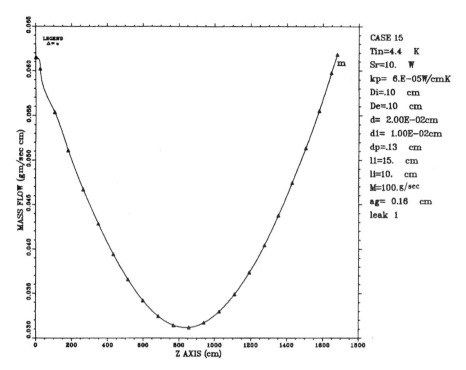

Figure 6.   Transverse He mass flow *m*

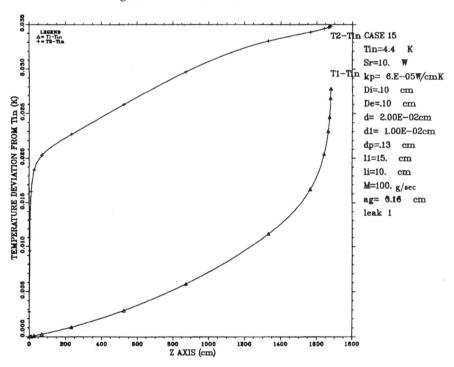

Figure 7.   Bypass temperatures T1–Tin, T2–Tin.

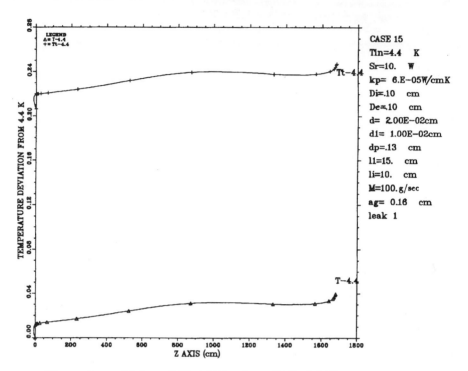

Figure 8. Main coil T–4.4 K, trim coil Tt–4.4 K temperatures.

PRACTICAL ASPECTS OF BUILDING TRANSVERSE COOLING

There are two main steps involved in building the transverse cooling scheme into the magnets. First, yoke laminations must be periodically substituted by laminations with channels. A building scheme where it would be impossible to insert incorrect laminations could be devised to make the installation foolproof. Second, the channels in the collars must be blocked. This can be done by inserting several blank collars in the center of the collar packs. The collaring press would then have to accommodate these blank collars. The present tooling is such that pressure is applied on the collars at the channels, therefore it is not possible to insert blank laminations and indium blocks were used. With new tooling, however, it may be possible to have collars with no channels in the center of the packs.

5 cm DESIGN AND FURTHER WORK

A major source of helium leaks comes from gaps between yoke laminations, whose size depends on the lamination packing fraction. The 1/16" laminations have a packing fraction of 99.4%. If the yoke lamination thickness is changed to 1/4", there will be one quarter the leakage, provided the lamination gap width remains the same. This value could very well increase considerably with thicker laminations as they might not nest as well as the 1/16" laminations. Another source of leaks which was already addressed occurs between the collars and the yokes, particularly at the poles and at the keys where the collars are not in contact with the yoke. Finally leaks from the bypasses to the busbar passages must also be considered.

The effect of blockages in the coil cooling passage on the mass flow and consequently the temperature distribution are currently investigated. Blockage is mostly a problem for the present narrow coil cooling gap and when the coil walls are not smooth because of the fiber glass–epoxy bonding. Should the gap become wider and the epoxy–impregnated fiber glass be replaced by polyimide, blockages would be unlikely to occur.

Will this method still be relevant for the new 5 cm design? With the disappearance of the trim coil and a possibly wider coil cooling passage gap the simple longitudinal cooling by diffusion might be adequate. Previous calculations showed that a gap of 5 mm may allow $S_r = 8$ W. The relevance of the transverse cooling method will be assessed for the new magnet design as soon as the new design parameters become available. In any event the following remains to be determined: the confirmed heating conductivity of Kapton, the maximum expected synchrotron radiation heating during the lifetime of the SSC, the maximum allowable increase of the magnet coil temperatures, and the maximum pressure drop available.

THE CRYOGENIC SYSTEM FOR THE SSC MAGNET TEST LABORATORY[1]

R. Byrns, B. Heer, P. Limon, M. McAshan, J. Rasson and
P. VanderArend

Superconducting Super Collider Laboratory[*]
Accelerator Division
2250 Beckleymeade Avenue
Dallas, TX 75237

INTRODUCTION

The fabrication of a large series of magnet development models, 70 pre-production models (beginning in 1993), as well as many thousands of production magnets requires the early construction of a Magnet Test Laboratory (MTL) to provide the testing capability for design, quality control and acceptance testing. The exact testing protocol is still in planning, although outlines of the magnet test program are sufficiently well understood to permit development of the performance and fabrication specifications for the MTL.

The requirement for production testing begins with the imminent need to test prototype and pre-production dipoles. The magnet production rate will rise to a peak level of 2,500 per year and all of the magnet production is to be completed at the end of fiscal 1997. The total dipole number in the two rings of the collider is about 9000 and there is one quadrupole for every five dipoles. Testing in the MTL will be performed on between 10 and 20 percent of the magnets with the larger percentage being tested early in the production period.

In addition to the production magnets of the collider, there are many other kinds of superconducting magnets in the SSC complex that will require prototyping and production testing. These include dipoles and quadrupoles for the High-Energy Booster Ring, vertical and combined dipoles and special-purpose quadrupoles for the IRs. The SSC program also has a need to test such components as recoolers and completed spool pieces either as part of the development of such components, or as part of the quality control of their manufacture.

---

[*]Operated by the Universities Research Association, Inc., for the U.S. Department of Energy under Contract No. DE-AC02-89ER40486.

MAGNET TEST PROGRAM

The accelerator magnet assembly consists of a cold mass (beam tube, superconducting coils, stainless steel collars, and iron yoke) installed in a cryostat (0.6 m dia. × 15 m long). A cross section of the magnet is shown in Figure 1. The cold mass contains passages for supercritical liquid helium flow that maintains the superconducting coils near 4 K. The weight of each cold mass, roughly eight tons, must be supported by posts that extend into the warm outside world, but limit the leakage of heat into the cold regions.

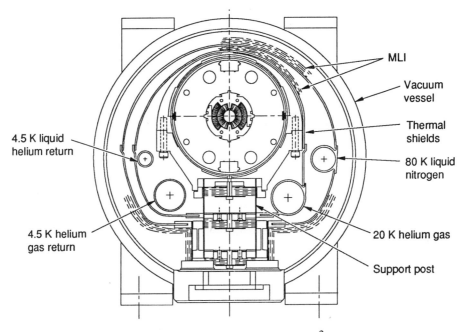

Figure 1. Magnet Cross Section.[2]

Table 1 lists a preliminary sequence for production cold testing of dipoles. A single magnet is a component and is not capable of stand-alone operation. In order to carry out the testing program, the magnet is installed on a test stand. The stand completes the electrical, mechanical, and cryogenic systems of the magnet and provides the operating environment needed for the testing. The stand also contains instrumentation necessary to verify the operating conditions and to determine the behavior of the magnet under test. Although locations and instrumentation for warm tests must be provided, the cold testing sequence is the critical path in the operation of MTL and determines its schedules.

The total cold testing cycle for a single magnet takes six days, therefore the throughput per test stand is five magnets per month. Six stands are needed to satisfy the basic system requirement of 30 per month. Under these conditions, a single additional stand is needed for quadrupoles, and together with three stands for R&D testing, the MTL needs to have 10 test stands in total.

Table 1. Preliminary Magnet Test Sequence 3-Shift Operation

Activity	Sequence Duration, hrs	Total
Prepare Stand, Place, and Align Magnet	2	2
Weld Beam Pipe	2	4
Wiring Connection and Checkout	16	20
Leak Check Beam Pipe		
Install Warm Bore		
Connect Leads		
Weld Small Cryostat Piping		
Air HiPot		
Weld Remaining Cryostat Piping	1	21
Leak Check	12	33
Pump and Purge		
Wiring Check	1	34
Install Shields	2	36
Install Vacuum Bellows	12	48
Pump Cryostat		
Leak Check		
Helium HiPot		
Allowance for Delay	8	56
Cooldown	8	64
Instrument Check & Evaluation		
Quench Protection System Setup		
Magnet Testing	48	112
Quench Testing		
Magnetic Field Measurement		
Allowance for Delay	8	120
Warmup	12	132
Remove Vacuum Bellows and Shields	3	135
Cut Piping and Single Phase Bellows	4	139
Wiring Inspection and Disconnection	1	140
Disconnect Beam Pipe	1	141
Remove Magnet	1	142
Allowance for Delay	2	144

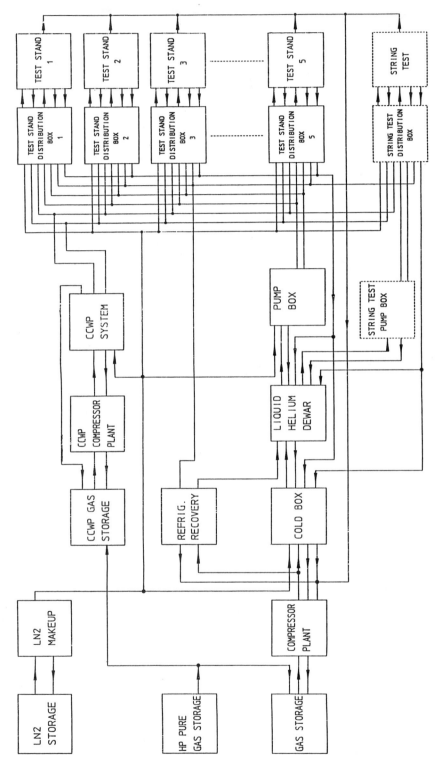

Figure 2. Block Diagram of MTL Refrigeration System.

The following types of tests are expected to be conducted at the SSC Magnet Test Laboratory:

(1) Cryogenic Tests
    (a)   Vacuum and mechanical integrity of the cryogenic system throughout the operating range.
    (b)   Magnet cooldown and warmup rates; temperature differences.
    (c)   Pressure drop in cryostat as a function of mass flow.
    (d)   Heat loads at 4 K, 20 K, and 80 K and simulation of synchrotron radiation loads.
    (e)   Half-cell tests (five dipoles, one quadrupole, one spool piece).

(2) Magnet Tests
    (a)   Dipole quench characteristics.
    (b)   Dipole field quality at operating temperatures.
    (c)   Magnet electrical integrity tests, including insulation tests (short circuits), power leads, and sensing elements.
    (d)   Magnet vibration during operation (around 5 Hz).
    (e)   Other magnets including special dipoles, quadrupoles, and correction elements.

(3) Tests of Other Equipment
    (a)   Spool pieces and recoolers.
    (b)   Liquid helium circulating pumps.

DESCRIPTION OF THE MAGNET TEST LABORATORY

Figure 2 shows a block diagram of the MTL refrigeration system. Following is a list of equipment for the MTL.

(a) Helium gas storage system with a storage capacity of 3,500 std. m^3 at 18 bar, 20°C.

(b) Helium refrigerator/liquefier system with a capacity of 2000 W at 4.5 K and 20 g/s of liquid helium. The staged compressors are oil-flooded rotary screw compressors with full oil removal. Two parallel sets are selected for efficient turn-down and redundant reliability. The cold box thermal cycle is dual pressure which has the potential for:
    1.   Turbine isolation from quench pressure excursions.
    2.   Contributes to better efficiency with a better match to staged compressors.
    3.   Minimizes the compressor sizing for the first stage.

(c) Liquid helium dewar with a volumetric capacity of 40,000 liquid liters and a gas ullage of 4 m^3. The dewar also serves as a quench tank capable of allowing the system to handle a rapid succession of quenches occurring at a rate of 20 quenches per 8 hours.

(d) Liquid helium pump system which consists of two circulation pumps with total flow rate of 500 g/s at 0.5 bar differential pressure and a booster pump providing 200 g/s at 3.0 bar differential pressure. The system also includes precooler, after-cooler and surge vessels.

(e) Test stand distribution system with individual subcooler for each test stand. The distribution system is suitable for five test stands and a future expansion capability of another five test stands. Each subcooler is capable of providing 100 g/s at 3.6 K or 50 g/s at 2.5 K.

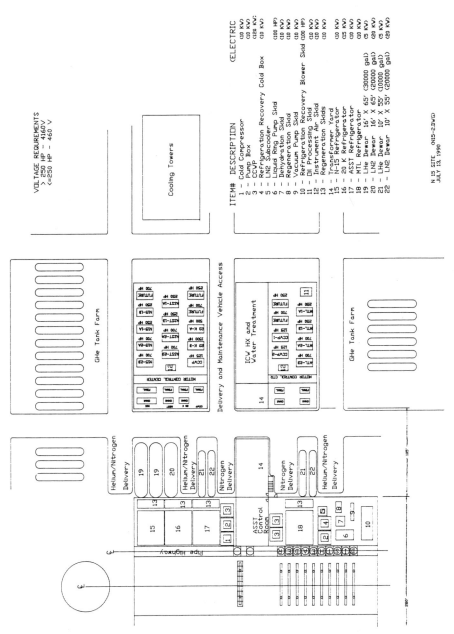

Figure 3. MTL Floor Plan and Equipment Layout.

(f) Cleanup, cooldown, warmup, and purification system (CCWP) capable of providing 110 g/s of pure helium mass flow rate at any temperature between 80 K and 300 K. This system consists of gas storage, compressors and CCWP modules. Table 2 shows the requirement of the CCWP functions.

Table 2. CCWP Function Requirements.

Case	Mass flow [g/s]	Supply temperature [K]
cleanup	15	300
cooldown	80	77–300
warmup	50	300
purification	10	300

(g) Refrigeration recovery module consisting of warm or cold vacuum pumps and a heat exchanger system. This system is capable of maintaining the low pressure in the subcoolers and also recovers refrigeration from the subcoolers' sub-atmospheric streams.
(h) Liquid nitrogen tank and distribution system suitable for providing the magnets shield and other modules with 77 K LN supply.
(i) Interconnecting vacuum insulated transfer lines.
(j) Control system consisting of instrumentation, data acquisition and process control hardware and software.

Figure 3 shows the proposed equipment layout and floorplan for the MTL and also the ASST (Accelerator System String Test).together with ten test stands for magnets and other test components.

REFERENCES

1. RFP-SSC-90A-01107, Superconducting Supercollider, Request for Proposal. MTL Cryogenics System. Section 2, Technical Specification.

2. *Supercollider 1*, Proceedings of the International Industrialization of Superconducting Super Collider, New Orleans, 1989.

A DYNAMIC MODEL FOR HELIUM CORE HEAT EXCHANGERS

W. E. Schiesser

Lehigh University, Bethlehem, PA 18015
and
Superconducting Super Collider Laboratory,* Dallas, TX 75237

H. J. Shih

Superconducting Super Collider Laboratory,* Dallas, TX 75237

D. G. Hartzog, D. M. Herron, D. Nahmias and W. G. Stuber

Air Products and Chemicals, Inc., Allentown, PA 18015

A. C. Hindmarsh

Lawrence Livermore National Laboratory, Livermore, CA 94550

ABSTRACT

To meet the helium (He) requirements of the superconducting supercollider (SSC), the cryogenic plants must be able to respond to time-varying loads. Thus the design and simulation of the cryogenic plants requires dynamic models of their principal components, and in particular, the core heat exchangers. In this paper, we detail the derivation and computer implementation of a model for core heat exchangers consisting of three partial differential equations (PDEs) for each fluid stream (the continuity, energy and momentum balances for the He), and one PDE for each parting sheet (the energy balance for the parting sheet metal); the PDEs have time and axial position along the exchanger as independent variables. The computer code can accommodate any number of fluid streams and parting sheets in an adiabatic group. Features of the code include: rigorous or approximate thermodynamic properties for He, upwind and downwind approximation of the PDE spatial derivatives, and sparse matrix time integration. The outputs from the code include the time-dependent axial profiles of the fluid He mass flux, density, pressure, temperature, internal energy and enthalpy. The code is written in transportable Fortran 77, and can therefore be executed on essentially any computer.

* Operated by the Universities Research Association, Inc., for the U.S. Department of Energy under Contract No. DE-AC02-89ER40486.

Introduction

The conceptual design of the SSC calls for ten liquid He refrigeration plants to be located at one-tenth intervals around the 53-mile ring (Ref. 1). A basic feature of the operation of the refrigeration plants is related to the unsteady state or dynamic operation of the SSC; typically it will go through transient periods of operation, such as cool down, operation as a particle accelerator, then limited shutdown for maintenance and preparation for the next experiments. Also, unexpected disturbances will occur, such as magnet quenches, that must be accommodated. Because of the low temperatures required for superconducting magnets, several heat shields will be used which require coolant streams at different temperatures. These streams interact through a series of heat exchangers in the refrigeration plants to achieve desired coolant temperatures. The heat exchangers are therefore a major component of the refrigeration plants, and the design of the heat exchangers is a central consideration in determining the operating performance of the refrigeration plants. The dynamic nature of the operation of the refrigeration plants to accommodate normal and unexpected transient conditions within the SSC therefore requires that the modelling and computer simulation of the heat exchangers involve the solution of unsteady-state (time-dependent) partial differential equations.

Core (or plate-fin) heat exchangers are used for the SSC refrigeration plants, mainly for their high heat transfer efficiency and relatively low cost. The heat transfer section of the prototypical core heat exchangers consists of parallel flat plates of metal (called parting sheets), with corrugated metal connecting the plates (called fins). He flows past the fins parallel to the parting sheets, cocurrently or countercurrently. Figure 1 illustrates the stacking of parting sheets and fins and a parallel flow pattern in a core heat exchanger. The passage arrangement for the prototypical core heat exchangers is periodic, i.e., an adiabatic group consisting of a small number of passages is repeated many times, and the adiabatic group has a mirror symmetry about the middle passage. Figure 2 illustrates the passage arrangement within an adiabatic group consisting of four fluid streams. It also illustrates the indexing convention in our model for streams and parting sheets, i.e., the index increments toward the symmetry passage. Our model for the prototypical core heat exchangers is focused on the dynamic behavior of the fluid streams and parting sheets in an adiabatic group.

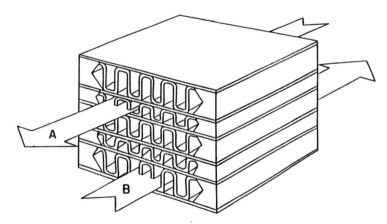

Figure 1. Stacking pattern of parting sheets and fins in a prototypical core heat exchanger. The He streams are flowing parallel to the parting sheets.

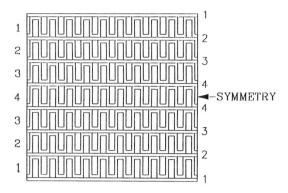

Figure 2. Passage arrangement within an adiabatic group consisting of four streams. Stream 4 is the symmetry stream.

In Section 2 we derive the model PDEs from the conservation principles of mass, energy and momentum. In Section 3 we detail the numerical method for solving the model PDEs, the computation of various quantities, other than the dependent variables, that appear in the model PDEs and the implementation of initial and boundary conditions. In Section 4 we present the simulation results for a particular core heat exchanger in a 4K refrigeration plant. In Section 5 we discuss an approximate first-order model for speeding up the dynamic simulation. Lastly, in Section 6 we give some conclusions.

Mathematical Model

The dynamic model consists of a system of partial differential equations (PDEs) which are derived from the basic conservation principles of mass, energy and momentum applied to the fluid He streams and that of energy applied to the parting sheets. The approach in deriving these equations is to consider a section of length δz of an exchanger in the axial direction in which the fluid He flows. The conservation equations are written for this section, and then δz taken to the limit of zero to arrive at the final PDEs. Since the equations are dynamic (unsteady state) and distributive, time (t) and axial position (z) are the independent variables. The dependent variables will be defined as each PDE is described.

Application of the mass, energy and momentum balances to one He stream gives

$$\frac{\partial \rho}{\partial t} = -\frac{\partial w}{\partial z} , \tag{1}$$

$$\frac{\partial(\rho u)}{\partial t} = -\frac{\partial(wh)}{\partial z} + \frac{Q}{S} , \tag{2}$$

and

$$\rho \frac{dv}{dt} = -\frac{\partial P}{\partial z} + F_{ext} , \tag{3}$$

respectively, where ρ, w, u, h, v, and P are the density, mass flux, internal energy, enthalpy, velocity and pressure of the stream respectively, Q is the heat transfer rate to the stream per unit length of the core, S the flow cross-sectional area of the stream, and F_{ext} the force per unit volume acting on the stream in addition to the fluid pressure. Equation (1) gives the time derivative of the He density ρ. Substituting Eq. (1) in Eq. (2), we obtain the time derivative of the He internal energy u

$$\frac{\partial u}{\partial t} = \frac{1}{\rho} \left\{ (u - h)\frac{\partial w}{\partial z} - w\frac{\partial h}{\partial z} + \frac{Q}{S} \right\} . \tag{4}$$

Equation (3) is the momentum balance in Lagrangian form, whereas Eqs. (1) and (4) are in Eulerian form. To have the momentum balance in Eulerian form, we apply the transformation between Lagrangian and Eulerian coordinates

$$\frac{dv}{dt} = \frac{\partial v}{\partial t} + v \frac{\partial v}{\partial z}$$

to Eq. (3) and obtain, using the definition $w = \rho v$ and Eq. (1), the time derivative of the He mass flux w

$$\frac{\partial w}{\partial t} = -\frac{\partial P}{\partial z} - \frac{\partial}{\partial z}\left(\frac{w^2}{\rho}\right) + F_{ext} . \tag{5}$$

In our model F_{ext} includes friction and gravitation, and is thus given by

$$F_{ext} = -\frac{f}{2D}\frac{|w|w}{\rho} - \rho g , \tag{6}$$

where f is the friction factor, D the hydraulic diameter and g the gravitational acceleration. Substituting Eq. (6) for F_{ext} in Eq. (5) then gives

$$\frac{\partial w}{\partial t} = -\frac{\partial P}{\partial z} - \frac{\partial}{\partial z}\left(\frac{w^2}{\rho}\right) - \frac{f}{2D}\frac{|w|w}{\rho} - \rho g . \tag{7}$$

The friction factor f is calculated from the Blasius formula for smooth pipes (Ref. 2), i.e.,

$$f = \frac{0.316}{R_e^{0.25}}, \qquad R_e = \frac{Dw}{\mu} , \tag{8}$$

where μ is the He viscosity. We apply a correction to f to account for the effect of the rough corrugated flow passages.

Applying the energy balance to one parting sheet for longitudinal heat conduction, we obtain the time derivative of the parting sheet temperature T_m

$$\frac{\partial T_m}{\partial t} = \frac{1}{\rho_m c_m}\left\{\frac{\partial}{\partial z}\left(k_m \frac{\partial T_m}{\partial z}\right) + \frac{Q_m}{S_m}\right\} , \tag{9}$$

where ρ_m, c_m, k_m, and S_m are the density, specific heat, thermal conductivity and cross-sectional area of the parting sheet, respectively, and Q_m is the heat transfer rate per unit length to the parting sheet. The need for determining the temporal dependence of the parting sheet temperature is obvious since streams of different temperatures exchange heat through parting sheets.

Equations (1), (4), (7), and (9) are then the PDEs in our dynamic model for core heat exchangers, and the He density ρ, internal energy u, mass flux w, and the parting sheet temperature T_m are the four dependent variables. Of course, the model PDEs must be applied to each fluid stream and parting sheet in the adiabatic group of the core, and the indexing convention, illustrated in Fig. 2, is used to distinguish them.

Computer Implementation

To obtain the numerical solution for our core heat exchanger model, we apply the Method of Lines to the model PDEs. That is, we divide the entire core in the axial direction into N intervals of equal length, evaluate the spatial derivatives at the center of each interval by finite difference approximations and integrate the temporal derivatives at the center of each interval by an ordinary differential equation integrator. Specifically we use the two-point biased and three-point centered approximations implemented in the Fortran code DSS/2 (Ref. 3) to evaluate the first-order and second-order spatial derivatives, respectively, and the sparse matrix integrator implemented in the Fortran code LSODES (Ref. 4) to integrate the temporal derivatives. It is essential in the Method of Lines that the temporal derivatives can be determined at any instant (including the initial moment) to allow their integration, i.e., the right-hand sides of the model PDEs must be computed. Thus we need to not only evaluate the spatial derivatives, but also compute all quantities, other than the four dependent variables, that appear in the right-hand sides of the equations. These quantities are h, P, μ, Q, Q_m, D, S, S_m, ρ_m, c_m, and k_m. In the following discussion, the calculation of these quantities is detailed along with the specification of the initial and boundary conditions required to solve the model PDEs. Note that a simplified fin structure, shown in Figure 3, is adopted in our model to facilitate the analysis of heat exchange with the fins. As indicated in Fig. 3, h_f is the fin height, b_f the fin thickness, and n_f the number of fins per unit width of the core.

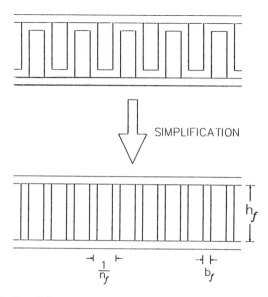

Figure 3. Simplification of the fin structure adopted in our model.

Geometrical Quantities

The flow cross-sectional area S, hydraulic diameter D, and parting sheet cross-sectional area S_m are geometrical quantities. In terms of the fin parameters b_f, h_f and n_f, the flow cross-sectional area S is given by

$$S = w_c n_f \left(\frac{1}{n_f} - b_f \right) h_f , \tag{10}$$

where w_c is the width of the core. The hydraulic diameter D is defined to be four times the ratio of the flow cross-sectional area S to the flow cross-sectional perimeter C which, in terms of the fin parameters, is given by

$$C = 2 \left(\frac{1}{n_f} - b_f + h_f \right) w_c n_f . \tag{11}$$

In general different streams in a core heat exchanger have different fin parameters and thus different flow cross-sectional areas S and hydraulic diameters D. The parting sheet cross-sectional area S_m is given by

$$S_m = b_m w_c ,\qquad (12)$$

where b_m is the parting sheet thickness. S_m is constant for each parting sheet.

Other Thermodynamic Properties

Given the He density ρ and internal energy u from Eqs. (1) and (4), we use a He thermodynamic model (Ref. 5) to obtain the He enthalpy h, pressure P and temperature T. The He temperature T, as we will see below, is required to calculate the heat transfer rates Q and Q_m. The He viscosity μ and the parting sheet density ρ_m and specific heat c_m are essentially constant in the range of the He thermodynamic properties under which the core heat exchangers are operated; suitable values have been used for each of them in the model. The parting sheet thermal conductivity k_m, however, varies significantly with temperature. This temperature dependence is taken into account in the model. Because of this, k_m is inside the second derivative with respect to z in Eq. (9).

Heat Transfer Rates

Figure 4 indicates the various heat fluxes that contribute to the cooldown and warmup of the jth stream and parting sheet: $q_{1,j}$ is the heat flux from parting sheet j to stream j; $q_{2,j}$ is the heat flux from parting sheet $j+1$ to stream j; $q_{3,j}$ is the heat flux from parting sheet j to stream $j-1$; $q_{4,j}$ is the heat flux from the fins in stream j to stream j; $q_{5,j}$ is the heat flux from parting sheet j to the fins in stream j; and $q_{6,j}$ is the heat flux from the fins in stream $j-1$ to parting sheet j. If $A_{1,j}$, $A_{2,j}$, $A_{3,j}$, $A_{4,j}$, $A_{5,j}$, and $A_{6,j}$ denote the corresponding heat transfer areas per unit length for these heat fluxes, the heat transfer rate per unit length to stream j, Q_j, and the heat transfer rate per unit length to parting sheet j, $Q_{m,j}$, are given by the following equations:

$$Q_j = q_{1,j} A_{1,j} + q_{2,j} A_{2,j} + q_{4,j} A_{4,j} \qquad (13)$$

and

$$Q_{m,j} = -q_{1,j} A_{1,j} - q_{3,j} A_{3,j} - q_{5,j} A_{5,j} + q_{6,j} A_{6,j} . \qquad (14)$$

Note that the above heat fluxes are directional, i.e., when their values are positive, the directions of heat flow are indicated by Fig. 4 and when their values are negative, the directions are reversed. With Fig. 4 in mind, the following equations can be written for the various heat fluxes:

$$
\begin{aligned}
q_{1,j} &= h_j(T_{m,j} - T_j) & j &= 1, 2, \ldots, M , & (15)\\
q_{2,j} &= h_j(T_{m,j+1} - T_j) & j &= 1, 2, \ldots, M-1 , & (16)\\
q_{3,j} &= h_{j-1}(T_{m,j} - T_{j-1}) & j &= 2, 3, \ldots, M , & (17)\\
q_{4,j} &= h_j \Delta T_j & j &= 1, 2, \ldots, M , & (18)\\
q_{5,j} &= -k_{m,j}\left(\frac{dT_{f,j}}{dx}\right)_{x\,=\,0} & j &= 1, 2, \ldots, M , & (19)
\end{aligned}
$$

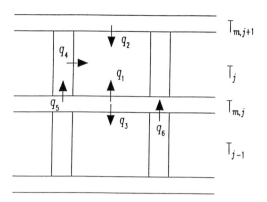

Figure 4. Various heat fluxes that contribute to the cooldown and warmup of the jth stream and parting sheet.

and

$$q_{6,j} = -k_{m,j} \left(\frac{dT_{f,j-1}}{dx} \right)_{x = h_{f,j-1}} \qquad j = 2, 3, \ldots, M \,, \tag{20}$$

where M is the number of different streams in the adiabatic group, h_j the heat transfer coefficient of stream j, ΔT_j the average temperature difference between stream j and the fins within stream j, and $T_{f,j}(x)$ the temperature of the fins between parting sheets j and $j+1$ at position x along the fin height. In the next section we derive $T_{f,j}(x)$ and define ΔT_j. Periodic arrangement of the adiabatic group in the core and mirror symmetry inside the adiabatic group allow us to determine $q_{3,j}$ and $q_{6,j}$ for $j = 1$ and $q_{2,j}$ for $j = M$ respectively,

$$q_{2,j=M} = q_{1,j=M} \tag{21}$$

and

$$q_{3,j=1} = q_{1,j=1}, \qquad q_{6,j=1} = -q_{5,j=1} \,. \tag{22}$$

The associated heat transfer areas (again, referring to Fig. 4) are given by the following equations:

$$A_{1,j} = w_c n_{f,j} \left(\frac{1}{n_{f,j}} - b_{f,j} \right) \qquad j = 1, 2, \ldots, M \,, \tag{23}$$

$$A_{2,j} = A_{1,j} \qquad j = 1, 2, \ldots, M \,, \tag{24}$$

$$A_{3,j} = w_c n_{f,j-1} \left(\frac{1}{n_{f,j-1}} - b_{f,j-1} \right) \qquad j = 2, 3, \ldots, M \,, \tag{25}$$

$$A_{4,j} = 2 w_c n_{f,j} h_{f,j} \qquad j = 1, 2, \ldots, M \,, \tag{26}$$

$$A_{5,j} = w_c n_{f,j} b_{f,j} \qquad j = 1, 2, \ldots, M \,, \tag{27}$$

and

$$A_{6,j} = w_c n_{f,j-1} b_{f,j-1} \qquad j = 2, 3, \ldots, M \ . \tag{28}$$

Again, the periodic arrangement of the adiabatic group gives $A_{3,j}$ and $A_{6,j}$ for $j = 1$,

$$A_{3,j=1} = A_{1,j=1}, \qquad A_{6,j=1} = A_{5,j=1} \ . \tag{29}$$

Since we are given in the heat exchanger specifications the total heat transfer rates per degree for each stream U_j (not the heat transfer coefficients h_j), the following equation is used to deduce h_j,

$$h_j = \frac{U_j}{C_j L n_j} \ , \tag{30}$$

where C_j is given by Eq. (11) for stream j, L the length of of the core, and n_j the total number of stream j in the core.

Fin Temperature Distribution

Assuming that the fins between two parting sheets are in steady state at any instant, application of an energy balance to the fins gives

$$\frac{d^2 T_f}{dx^2} - \frac{2h}{b_f k_m}(T_f - T) = 0 \ . \tag{31}$$

In deriving Eq. (31) we have assumed k_m does not vary with x. The general solution of Eq. (31) is

$$T_f = T + C_1 \sinh(\alpha x) + C_2 \cosh(\alpha x) \ , \tag{32}$$

where $\alpha = \sqrt{2h/b_f k_m}$. Equation (32) should be applied to the fins in each stream. Using the boundary conditions $T_f(x = 0) = T_{m,j}$ and $T_f(x = h_{f,j}) = T_{m,j+1}$ for the fins in stream j, we determine $C_{1,j}$ and $C_{2,j}$ as follows,

$$C_{1,j} = \frac{(T_{m,j+1} - T_j) - (T_{m,j} - T_j)\cosh(\alpha h_{f,j})}{\sinh(\alpha h_{f,j})} \qquad j = 1, 2, \ldots, M-1 \tag{33}$$

and

$$C_{2,j} = T_{m,j} - T_j \qquad\qquad j = 1, 2, \ldots, M \ . \tag{34}$$

The mirror symmetry inside the adiabatic group gives $C_{1,j}$ for $j = M$,

$$C_{1,j=M} = \left.\frac{(T_{m,j} - T_j) - (T_{m,j} - T_j)\cosh(\alpha h_{f,j})}{\sinh(\alpha h_{f,j})}\right|_{j = M} \tag{35}$$

The fin-to-stream heat flux q_4 is determined from the following expression,

$$q_4 = \frac{2 w_c n_f \left(\int_0^{h_f} h(T_f - T)dx\right)}{2 w_c n_f h_f} \ , \tag{36}$$

i.e., the ratio of the total heat transfer rate per unit length of the core to the total heat transfer area per unit length. Since $q_4 = h\Delta T$ by definition, we have

$$\Delta T = \frac{1}{h_f}\int_0^{h_f}(T_f - T)dx \ , \tag{37}$$

which is the average of the temperature difference $T_f - T$ over the fin height h_f. It is straightforward to obtain the following expressions from Eq. (32):

$$\Delta T = \frac{1}{\alpha h_f} \{C_1 \cosh(\alpha h_f) + C_2 \sinh(\alpha h_f) - C_1\} \,, \tag{38}$$

$$\left(\frac{dT_f}{dx}\right)_{x=0} = C_1 \alpha \,, \tag{39}$$

and

$$\left(\frac{dT_f}{dx}\right)_{x=h_f} = C_1 \alpha \cosh(\alpha h_f) + C_2 \alpha \sinh(\alpha h_f) \,. \tag{40}$$

Equations (38), (39) and (40) are used in Eqs. (18), (19) and (20) to compute q_4, q_5 and q_6, respectively.

Initial and Boundary Conditions

For boundary conditions, we have chosen to specify the inlet temperature and pressure and the outlet pressure for the He streams. Also we use the insulated (Neumann) boundary condition at both ends of the parting sheets, i.e.,

$$\left(\frac{\partial T_m}{\partial z}\right)_{z=0} = \left(\frac{\partial T_m}{\partial z}\right)_{z=L} = 0 \,,$$

to specify that the heat leak from the ends of the parting sheets is negligible. The inlet mass flux is treated as another time-dependent variable whose temporal behavior is governed by the momentum balance equation, Eq. (7). The spatial derivatives $\partial w/\partial z$, $\partial h/\partial z$, and $\partial(w^2/\rho)/\partial z$ are approximated by two-point upwind finite differences, and thus the inlet conditions are built into the approximations. The spatial derivative $\partial P/\partial z$ is approximated by two-point downwind finite differences, and thus the outlet pressure is built into the approximations. In applying the momentum balance to compute the inlet mass flux, the spatial derivative $\partial(w^2/\rho)/\partial z$ at the inlet is assumed to be the same as that at the first grid point (the center of the first spatial interval) downstream. If the adiabatic group has M different streams and the length of the core is divided into N intervals, applying the Method of Lines to the four model PDEs then requires integration of $4MN + M$ temporal ODEs.

To provide a consistent initial condition for the dynamic model, a steady state calculation is included. This is done by setting the temporal derivatives in the model PDEs to zero, approximating the spatial derivatives by the same finite difference methods as used in the dynamic model, and solving the resulting algebraic equations with the IMSL subroutine DNEQNF (Ref. 6). If uniform mass flux is assumed, i.e., the temporal derivative in the continuity equation is automatically set to zero, only three temporal derivatives, i.e., $\partial u/\partial t$, $\partial w/\partial t$ and $\partial T_m/\partial t$, are required to be zero. For M different streams in the adiabatic group and N axial intervals in the core, we then need to solve $3MN + M$ nonlinear algebraic equations for $3MN + M$ unknowns which are ρ, u and T_m at the center of each spatial interval and the steady state mass flux or the outlet pressure for each stream.

A scaling factor s_f is used to adjust the friction factor f to achieve realistic pressure drops in the core heat exchangers, since the Blasius formula for friction factor, Eq. (8), which is valid for smooth pipes, gives unrealistically small pressure drops. Typical values of the scaling factor are about 10 from our simulation, which are well below 40, the upper bound for realistic designs (Ref. 7). Because the steady state mass flux, friction scaling factor and outlet pressure are interrelated, our model allows that any two of these three variables can be specified in the steady state calculation and the third will be determined by the IMSL nonlinear equation solver. The scaling factor used or determined in the steady state calculation is assumed to remain the same in the dynamic simulation.

Simulation Results

To check the code for our static and dynamic core heat exchanger models, we have simulated heat exchanger R2 in a SSC 4K prototype refrigeration plant (Ref. 8). The passage arrangement of heat exchanger R2 is the adiabatic group BAB repeated 18 times where A is the high pressure He stream flowing down and B the low pressure He return stream flowing up. Listed below are the parameters for R2 required as input to the model:

Parameter	Stream A	Stream B	Unit
b_f	0.010	0.008	in
h_f	0.281	0.380	in
n_f	20.2	14.7	in^{-1}
U	96321	96319	BTU/hr-°F
$L = 10$ ft			
$w_c = 16.75$ in			
$b_m = 0.032$ in			

The following data on temperature, pressure and mass flux are used in the initial steady state calculation:

Parameter	Stream A	Stream B	Unit
P_{in}	264.00	11.17	psia
T_{in}	−315.69	−417.63	°F
P_{out}	263.96	10.94	psia
w	−4361	1437	lb/hr-ft^2

where the subscripts *in* and *out* refer to the inlet and outlet streams, respectively. The friction scaling factor, as determined from the IMSL nonlinear equation solver, is 8.685 for stream A and 9.741 for stream B.

We simulated the unsteady state resulting from increasing the inlet temperature of stream A by 10°F. The simulation results for 20 sec are shown in Figures 5, 6 and 7, i.e., the axial profiles of mass flux, temperature and pressure, respectively. We see from Figure 5 that the mass flux of each stream varies from one uniform value to another as time passes, indicating that starting with one steady state we approach another; Figure 6 indicates that the temperatures of both streams increase at the inlet of stream A but do not change at the inlet of stream B, as expected. It is interesting to note in Figure 7 that the pressure of stream A decreases and then rises in the direction of the flow. This results from our inclusion of the gravitational force in the momentum balance equation.

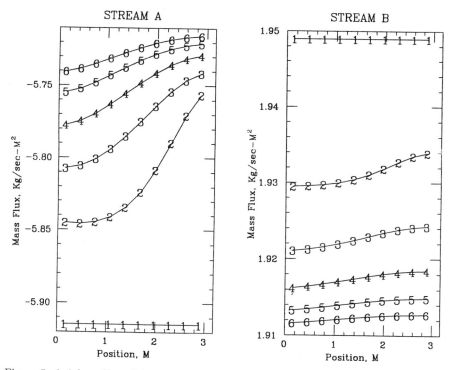

Figure 5. Axial profiles of the stream mass flux. Each profile is 4 seconds apart. The curve index increases with time.

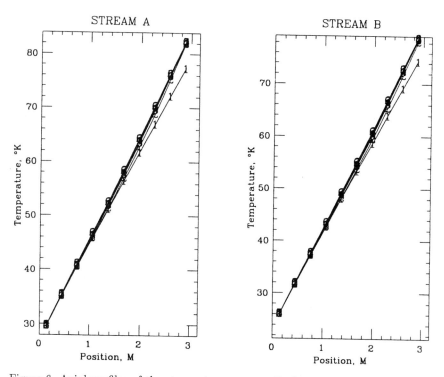

Figure 6. Axial profiles of the stream temperature. Each profile is 4 seconds apart. The curve index increases with time.

We found that it was important to restrict the order of the algorithm in the ODE integrator LSODES. Specifically, we have found that by restricting the maximum order of the algorithm of LSODES to two (instead of the usual upper limit of five) the computer time is greatly reduced (by two orders of magnitude). This suggests that the greatly increased efficiency of the ODE integration by LSODES is due to the existence of nearly imaginary eigenvalues of the ODE system. For the BDF (Gear) methods on which LSODES is based, the order three and above algorithms have poor stability near the imaginary axis (Ref. 9), and will therefore take small steps if the ODE system has nearly imaginary eigenvalues; computing the temporal eigenvalues of our model ODE system indicated this is the case.

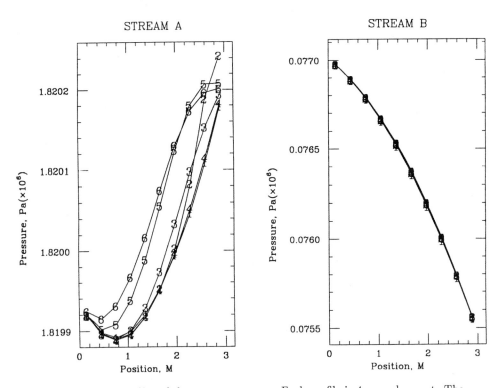

Figure 7. Axial profiles of the stream pressure. Each profile is 4 seconds apart. The curve index increases with time.

Reduced Model

Using the simulation results from our rigorous core heat exchanger model, we developed an approximate low-order model representation (Ref. 10) which could be used when the execution times of the rigorous dynamic model are excessive, e.g., when several heat exchangers are included in a complete refrigeration plant simulation. The low-order model relates the following inputs and outputs:

Inputs:	Inlet temperature T_{in}
	Inlet pressure P_{in}
	Outlet pressure P_{out}
Outputs:	Inlet mass flux w_{in}
	Outlet temperature T_{out}
	Outlet mass flux w_{out}

The procedure generally involves estimating the time constants of the inlet mass flux, outlet temperature and outlet mass flux responses for each stream from rigorous static and dynamic simulations, then approximating the transient response of these three state variables with single exponentials (first-order model) for a series of step inputs. The details are illustrated by the following example calculation for heat exchanger R2.

Let X denote one of the three output variables w_{in}, T_{out} and w_{out}. From the steady state calculation of our rigorous model, we obtain X_i and X_f, the initial and final values of X, for a given (arbitrary) step input. Assuming the transient $X(t)$ is governed by a single exponential with a time constant t_c, we calculate the value of X at $t = t_c$ according to

$$X_c = X_i + 0.362(X_f - X_i).$$

From the rigorous dynamic simulation we obtain two consecutive times t_s and t_{s+1} ($\Delta t = t_{s+1} - t_s$ should be reasonably small) such that $X_s < X_c < X_{s+1}$ or $X_{s+1} < X_c < X_s$, where $X = X_s$ at $t = t_s$ and $X = X_{s+1}$ at $t = t_{s+1}$. The time constant t_c is then obtained by linear interpolation, i.e.,

$$t_c = t_s + \frac{(X_c - X_s)}{(X_{s+1} - X_s)}(t_{s+1} - t_s).$$

Listed below are the time constants (in seconds) determined this way for heat exchanger R2 for a step increase in the inlet temperature of stream A of 10°F:

	w_{in}	T_{out}	w_{out}
Stream A	0.9	22.4	12.0
Stream B	7.2	1.7	9.1

To compute the response of the first-order model to a general time-varying input, we first approximate the input by a series of step functions, then we obtain the final steady state for each step change from the steady state solution of our rigorous model. With the time constants determined above, the transient response for each time step, denoted with a superscript (s), is computed according to:

$$X^{(s)} = \left(X_i^{(s)} - X_f^{(s)}\right) \exp\left(-\frac{(t - t^{(s)})}{t_c}\right) + X_f^{(s)},$$

where $t^{(s)}$ is the time at the beginning of step (s). The above formula is just the solution of the following linear ODE:

$$\frac{dX^{(s)}}{dt} = -\frac{1}{t_c}(X^{(s)} - X_f^{(s)}), \qquad X^{(s)}(t^{(s)}) = X_i^{(s)}.$$

Thus each of the three outputs is described in the reduced model by a linear ODE. Figures 8, 9 and 10 show the transient responses of w_{in}, T_{out} and w_{out} from the first-order model for a series of step changes in the inlet temperature of stream A: $+5°F$ at $t = 0$ sec, $+5°F$ at $t = 5$ sec, and $-10°F$ at $t = 15$ sec. For comparison, the transient responses from the rigorous model subject to the same input are also shown. In general this reduced model works rather well considering it reduces a large number of nonlinear ODEs $(4MN + M)$ to a small number of linear ODEs $(3M)$. It has the major advantage that it is always correct at steady state (by "correct," we mean that it agrees with the rigorous model at steady state).

In principle, this model reduction technique can be applied to any of the core heat exchangers, i.e., it appears to be a general methodology, although further testing is needed to establish its general utility. Also, the use of the method presupposes that the steady state calculations can be done fast enough to compute the series of X_f required along the transient. This in turn assumes the iterative solution of the nonlinear algebraic equations by subroutine DNEQNF will converge rapidly. This convergence will be enhanced by the fact that for relatively small steps, X_f will not be too far from X_i, and therefore, X_i serves as a reasonably good initial guess for DNEQNF. However, we cannot guarantee in advance that the model reduction technique, or for that matter, the rigorous dynamic simulation code, will produce a solution. Much depends on the success of the steady state calculation with DNEQNF, and the solution of large systems of nonlinear algebraic equations is a notoriously difficult computational problem. The choice of the initial estimate of the solution is critical in this process.

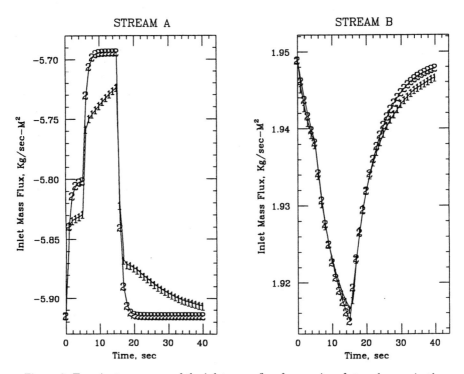

Figure 8. Transient responses of the inlet mass flux for a series of step changes in the inlet temperature of stream A: $+5°F$ at $t = 0$ sec, $+5°F$ at $t = 5$ sec, and $-10°F$ at $t = 15$ sec. Curve 1 is from the rigorous model and curve 2 from the reduced model.

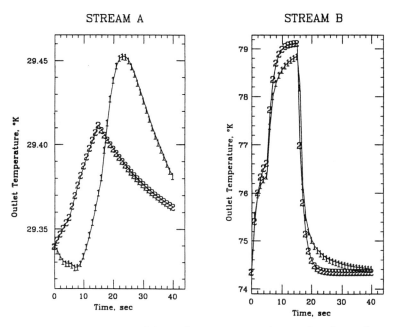

Figure 9. Transient responses of the outlet temperature for a series of step changes in the inlet temperature of stream A: $+5°F$ at $t = 0$ sec, $+5°F$ at $t = 5$ sec, and $-10°F$ at $t = 15$ sec. Curve 1 is from the rigorous model and curve 2 from the reduced model.

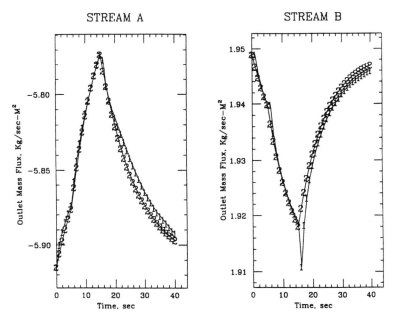

Figure 10. Transient responses of the outlet mass flux for a series of step changes in the inlet temperature of stream A: $+5°F$ at $t = 0$ sec, $+5°F$ at $t = 5$ sec, and $-10°F$ at $t = 15$ sec. Curve 1 is from the rigorous model and curve 2 from the reduced model.

Conclusions

A detailed model for He core heat exchangers has been described in this paper. The associated Fortran 77 implementation has a user-selected flag for the following options:

(1) A dynamic model consisting of a system of partial differential equations expressing the conservation of mass, momentun and energy for the He and the conservation of energy for the parting sheets. Time and position along the exchanger are the independent variables of this model. The time integration is performed by a sparse matrix implementation of the backward differentiation formulas; the spatial derivatives in the PDEs are approximated by upwind and downwind finite differences.

(2) A steady state model obtained by setting the temporal (time) derivatives of the dynamic model, (1), to zero. The resulting system of nonlinear algebraic equations is solved by Newton iteration. In particular, this steady state model provides a consistent initial condition for the dynamic model, (1).

(3) An approximate, low-order model, for the dynamic model, (1), which can be used in simulation and control studies.

Special features of (1) and (2) include: (a) a choice of rigorous and ideal gas He thermodynamic properties (the latter can often be used with good accuracy to save computer time), (b) detailed analysis of the thermal circuits which leads to six heat flux terms in the He energy balance, and (c) a modified Blasius pressure drop correlation with scaling computed by the code to account for the complex flow patterns within the exchanger.

The output from the dynamic model, in particular, gives: (a) the He state variables, e.g., temperature, pressure, density, enthalpy and internal energy, (b) the He mass flux, and (c) the parting sheet temperature as a function of time and position along the exchanger. Additionally, the six heat fluxes are available as a function of time and position which provide insight into the point-to-point operating characteristics within the exchanger. These variables can be calculated for any number of fluid streams and parting sheets within an adiabatic group; therefore, we anticipate the model and code can be used to study the steady state and dynamic performance of any of the heat exchangers in a SSC refrigeration plant (Ref. 8).

References

1. M. S. McAshan, "Refrigeration Plants for the SSC," SSC Central Design Group Report SSC-129, May 1987.

2. R. B. Bird et. al., *Transport Phenomena*, John Wiley & Sons, New York (1960).

3. J. C. Pirkle, Jr. and W. E. Schiesser, "DSS/2: A Transportable Fortran 77 Program for Ordinary and One, Two, and Three-Dimensional Partial Differential Equations," Proceedings of the 1987 Summer Computer Simulation Conference, Montreal, July 1987.

4. A. C. Hindmarsh, "ODEPACK, A Systematized Collection of ODE Solvers," in *Scientific Computing*, R. S. Stepleman et. al. (eds.), North-Holland, Amsterdam (1983).

5. "Helium Thermodynamic System," Report by Air Products and Chemicals, Inc. to the SSC Central Design Group, March 1988.

6. IMSL MATH/LIBRARY—*Fortran Subroutines for Mathematical Applications*, Version 1.1, User's Manual, Vol. 2, January 1989.

7. D. M. Herron, Air Products and Chemicals, Inc., private communication.

8. "SSC Site-Specific Conceptual Design Report," Vol. 1, Superconducting Super Collider Laboratory, Dallas, Texas, p. 447, December 1989.

9. C. W. Gear, *Numerical Initial Value Problems in Ordinary Differential Equations*, Prentice-Hall, Englewood Cliffs (1971).

10. D. G. Hartzog, Air Products and Chemicals, Inc., private communication.

SIMULATION OF THE SSC REFRIGERATION SYSTEM USING THE ASPEN/SP

PROCESS SIMULATOR

Joseph Rasson

Superconducting Super Collider Laboratory[*]
2250 Beckleymeade Avenue
Dallas, TX 75237

Jay Dweck

V. P. of Technology
Simulation Sciences Inc.
6000 E. Evans Avenue
Denver, CO 80222

INTRODUCTION

The SSC Magnet must be maintained at a superconducting temperature of 4 K. The proposed refrigeration cooling processes consist of fairly simple closed cycles which take advantage of the Joule-Thompson effect via a series of expansions and compressions of helium gas which has been precooled by liquid nitrogen. The processes currently under consideration consist of three cycles, the 20 K shield cooling, the 4.0 K helium refrigerator and the helium liquefier. The process units which are to be employed are compressors, turbines, expanders, mixers, flashes, two stream heat exchangers and multiple stream heat exchangers. The cycles are to be operated at or near steady state.

Due to the large number of competing cooling sector designs to be considered and the high capital and operating costs of the proposed processes, the SSC Laboratory requires a software tool for the validation and optimization of the individual designs and for the performance of cost-benefit analyses among competing designs. Since these processes are steady state flow processes involving primarily standard unit operations, a decision was made to investigate the application of a commercial process simulator to the task.

---

[*]Operated by the Universities Research Association, Inc., for the U.S. Department of Energy under Contract No. DE-AC02-89ER40486.

Several months of internal evaluations by the SSC Laboratory revealed that while the overall structure and calculation approach of a number of the commercial simulators were appropriate for this task, all were lacking essential capabilities in the areas of thermodynamic property calculations for cryogenic systems and modeling of complex, multiple stream heat exchangers. An acceptable thermodynamic model was provided and a series of simple, but three software vendors. Based on the results of the benchmark tests, the ASPEN/SP process simulator was selected for future modeling work.

ENHANCEMENT OF THE ASPEN/SP PROCESS SIMULATOR

While the ASPEN/SP process simulator has the basic flowsheeting capabilities and unit operation models required for this task, several enhancements were needed to facilitate development of the SSC process models. These enhancements can be classified into three categories: thermodynamic models, two-phase flash algorithm, and unit operation models. Each is discussed below.

Thermodynamic Models

Unlike most chemical processes, the SSC cooling sector involves the flows of only pure components. Essentially pure helium is the fluid used in the cycles, while pure nitrogen is used external to the cycle to precool the helium to the maximum temperature at which Joule-Thompson cooling can take place. While the lack of mixtures should simplify the physical property calculations, the peculiarities of helium actually lead to greater physical property problems than for even fairly complex chemical systems.

Nitrogen does not exhibit any unusual behavior and, as such, is fairly easily modeled via conventional equations of state. For the purposes of this work, a highly accurate 32-term variation of the BWR equation of state developed by[1] Jacobsen is employed. This equation is valid from 63.15 to 1900 K and pressures up to 1000 bar. The uncertainties in the calculated temperatures are under 0.5% while the enthalpy is within 3.0 joules/k-mole across the entire range. Calculation times are fairly long owing to the difficult nature of the root finding procedure for this equation.

Helium exhibits phase, enthalpy and transport behavior unlike any other substance. Liquid helium forms two distinct phases, the normal fluid and the superfluid. The specific heat decreases with temperature to about 2.5 K, then increases dramatically to the lambda point (2.172 K), then decreases dramatically. The thermal conductivity of the normal fluid actually decreases with decreasing temperature, which is similar to the behavior of a gas. On the other hand, the thermal conductivity of the superfluid is so high, that bubble formation cannot take place during boiling. The superfluid also displays a behavior termed superfluidity, where it acts as if it had zero viscosity.

To provide for the simulation of the proposed cycles, a physical property model capable of accurately calculating the liquid and vapor phase fugacities, enthalpies, entropies and densities of helium at temperatures from 0.8 K to 500 K and pressures up to 3,000,000 N/sqm is required. Since the majority of the proposed designs operate above the lambda temperature of helium (2.172 K), the first model which was implemented is the well-known[2] McCarty model, distributed by the National Bureau of Standards. This model was designed as a standalone program for the calculation of vapor and liquid helium properties.

While the model can accept various input specifications, the only specifications which are useful for ASPEN/SP are temperature and pressure.

When the temperature and pressure are specified, the model determines the phase of the helium and returns a list of properties, including density, enthalpy, entropy and heat capacity. In ASPEN/SP, however, the flash algorithm is responsible for determining the phase of the system. In order to make this determination, the flash algorithm requires the fugacity and enthalpy of both the liquid and vapor phases at the specified condition. If the system does not exist in one of the states at the condition, extrapolated hypothetical values must be returned by the physical property models. Since the McCarty model is not designed to return data for hypothetical phases, modifications had to be made.

The equation of state has multiple roots, two of which represent either the actual or extrapolated liquid and vapor roots. The root finding algorithm of the original McCarty model, which employs a Newton-Raphson approach, is not robust enough to locate the extrapolated root in the general case. The cause of this problem was isolated to the division of the temperature-pressure plane into four regions, each with different coefficients, to enhance the accuracy of the model. The function is not smooth across the boundaries, limiting the effectiveness of the Newton-Raphson algorithm when the boundaries are crossed. This problem was solved by modifying the Newton-Raphson algorithm to use the analytical derivatives to calculate an approximate next value of the iteration variable, then recalculating the derivative numerically via finite difference between the new point and the original point. This derivative then is used by the Newton-Raphson technique to determine the actual new value. This approach while only slightly slower does eliminate the robustness problems.

For temperatures below the lambda point, a proprietary model developed by Air Products was implemented. This model also appears to perform reliably above the lambda point, although errant phase determinations did result when applying this model near the critical point. This problem was not observed with the ASPEN/SP implementation of the McCarty model. The two models are implemented in such a fashion that the user can select either model by setting a single input parameter.

Two-Phase Flash Algorithm

The two phase flash algorithm contained in ASPEN/SP is extremely robust and efficient for a wide variety of chemical systems over broad ranges of temperature and pressure. However, in order to accommodate the temperature range encountered in the proposed cooling sector designs and the temperamental nature of the McCarty thermodynamic model, two minor modifications were required. First, the lower temperature bound was changed from 50 K to 2 1 K. (For the later work, the bound was lowered to 0.5 K.) Second, functions were developed to aid in the generation of initial temperature estimates for specified pressure - enthalpy flashes.

The initial estimate functions are necessary because of the highly irregular nature of the enthalpy of the hypothetical vapor root solution to the McCarty equation. In certain circumstances, the enthalpy of the hypothetical vapor actually will decrease with increasing temperature. This can cause the erroneous calculation of a high temperature vapor instead of a low temperature liquid at a particular pressure and enthalpy if a poor initial temperature estimate is selected. The initial estimate functions are based on regressed fits of the saturated liquid and vapor enthalpies of helium as a function of pressure.

Unit Operation Models

In addition to the two-phase flash algorithm enhancement described above, three unit operation enhancements were required to simulate the proposed designs. These three enhancements involve the development of user subroutines to facilitate off-design calculations with the turbine, robustness improvements for the two stream heat exchanger and the development of a multiple stream heat exchanger model. Each of these enhancements is discussed below.

When a turbine is operated at other than the design pressures and flow rate, the efficiency is different from the design efficiency. This variation in efficiency is described via a turbine curve. The user subroutines function by calculating the inlet nozzle diameter for the turbine at the design condition, assuming choke flow. The efficiency in off-design conditions then is computed as a fraction of the design efficiency via a look-up of a turbine curve at the desired condition, given the inlet nozzle diameter. A subroutine which calculates the speed of sound in helium was developed to facilitate this calculation.

The two stream heat exchanger algorithm needed minor enhancements to handle certain types of phase transitions in the helium and to facilitate heat leak calculations. For chemical systems involving mixtures, the system must pass through a two-phase region between the vapor and liquid states. For pure components, however, a supercritical fluid which is at a pressure above the critical pressure, and is cooled at constant pressure can undergo an immediate transition to a saturated liquid without passing through the two-phase region. This behavior is observed in several of the heat exchangers and necessitated a modification to the two stream heat exchanger phase transition logic. The exceedingly low temperatures and odd geometries of the heat exchanger necessitate a provision for heat leak calculations. A capability to specify either a positive or negative heat leak as a fraction of the total heat transferred was added to the two stream heat exchanger model.

A significant amount of effort was expended in the design of a multiple stream heat exchanger algorithm. Multiple stream heat exchangers are quite complex, involving the splitting of each feed stream into a number of flows which are exchanged in a countercurrent fashion against the opposite feed streams in a series of channels. Each hot stream flow contacts two cold stream flows simultaneously and visa versa. A certain fraction of the heat exchanger areas is devoted to each type of exchange.

Two algorithms were designed to model these exchangers. The first considers the area distribution inside the exchanger and simultaneously calculates the various types of exchanges, combining the outlet flows to achieve the final temperatures. The second is a highly simplified algorithm, referred to as the lumped-approach, which considers the exchanger to be a series of two stream heat exchangers. At the cold side, the coldest feed stream is heated against the combined hot streams, until the stream reaches the temperature of the next coldest feed stream. At this point, the two cold streams together are exchanged against the combined hot streams. The hot streams are treated analogously. That is, the hottest stream is cooled against the combined cold streams until the stream reaches the temperature of the next hottest stream. At this point, the two hot streams together are exchanged against the combined cold streams.

For the purposes of this work, the simplified, lumped-approach algorithm was adopted. This approach imposes a number of limitations upon the model, including the uniformity of outlet temperatures on each side, the specification of a single heat transfer coefficient for all phase regimes, imprecise determination of pinch-point conditions, and limited applicability to off-design and rating situations. Independent specification of pressures and physical property methodology for individual streams are allowed and have been implemented.

The algorithm which has been implemented has a great deal of flexibility, allowing the specification of either the hot or cold stream outlet temperature, the exchanger duty, the minimum approach temperature or the overall exchanger area. The algorithm begins by calculating the overall heat duty and temperature at which each of the streams on either side begin to participate in the heat exchange. The first three types of specifications then are straightforward, involving no iterative calculations. The overall heat duty is the primary iteration variable for the minimum approach and area specifications. The heat duty is a superior iteration variable to the temperature due to the linear nature of the convergence and the elimination of convergence difficulties associated with phase transitions of pure components.

The multiple stream heat exchanger algorithm has performed very well, converging reliably and efficiently for all of the cases which have been tried. All of the specification options have been exercised and shown to produce consistent results.

EXAMPLE APPLICATION

The simulation of a sample design is discussed here. Figure 1 is a simulation flowsheet of the combined 20 K shield and 4 K refrigerator for the sample design. This flowsheet contains ten two stream heat exchangers and assorted compressors, turbines and heaters. The nitrogen precooling has been neglected and the multiple stream heat exchangers have been represented as groups of two stream heat exchangers.

The special cryogen physical property option set has been employed. The simulation is performed in rating mode, with the areas of each of the heat exchangers being specified. Since the flowsheet forms a series of complex loops, tear streams have been specified and estimated flow rates and phase conditions specified. The tear streams are to be converged simultaneously.

Representative simulation results are contained in Figures 2 and 3. Figure 2 is a plot of the temperature profile in heat exchanger EXR5, which subcools a high pressure liquid stream against a low pressure vapor stream. The pronounced curvature of the hot stream cooling curve results from the unusual temperature variation of the helium heat capacity in the near-critical region. Figure 3 is a plot of the temperature difference profile in exchanger EXR4. The peak in this curve results from the curvature of the hot stream cooling curve. Figure 4 is an excerpt of the stream report for the simulation which contains the hot outlet stream from exchanger EXRS. The tear streams were converged in nine overall flowsheet iterations in approximately four hours on a 16 MhZ, 386 PC.

CONCLUSIONS

Our work has demonstrated that reliable, predictive simulation models of the cooling sector of the SSC can be developed with ASPEN/SP. Verification and preliminary

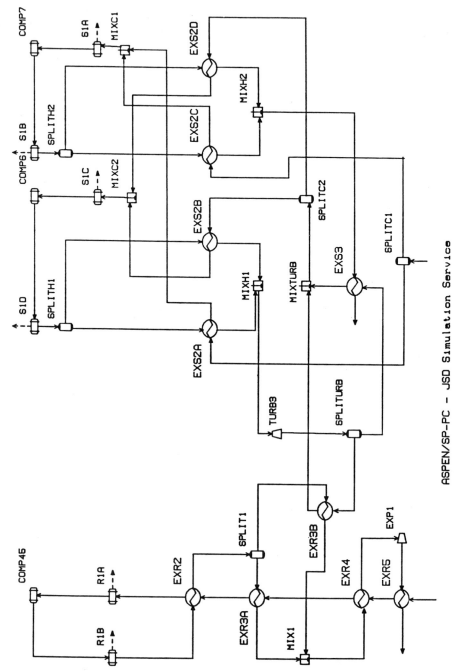

ASPEN/SP-PC - JSD Simulation Service

Figure 1. Simulation flowsheet of combined 20K shield and 4K refrigerator.

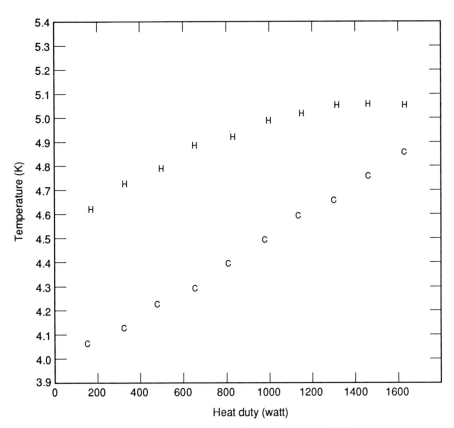

Figure 2. Plot of temperature profile in heat exchanger EXR4.

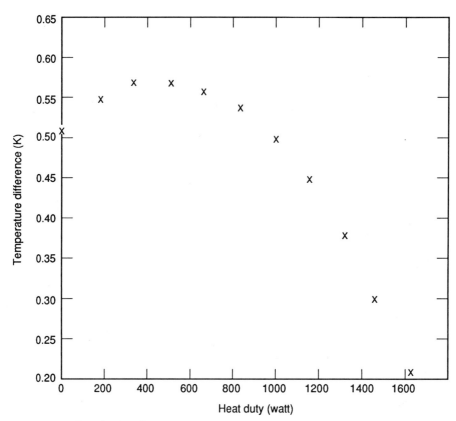

Figure 3. Plot of Temperature Difference Profile in Heat Exchanger EXR4.

```
ASPEN/SP Run On  9/ 4/87 by JSD Simulation
ASPEN/SP Version  1.5 Released By JSD, Inc., Denver, Colorado on JUNE 30, 1985
                   SIMPLIFIED CRYOGENIC REFRIGERATION CYCLE
                   STREAM SECTION

256        236       172       128
           FLOW DETAILS          256       236       172        128
========================================

COMPONENT FLOWS <KMOL/SEC>
...............................

NITROGEN                        0.0       0.0       0.0        0.0
HELIUM                          0.0617    0.0329    0.0621     0.0621
                                .........  .........  .........  .........
TOTAL                           0.0617    0.0329    0.0621     0.0621

PHASE SPLITS
...............................

VAPOR   FRACTION <MOLE  BASIS>  1.0000    1.0000    0.0        1.0000
LIQUID  FRACTION <MOLE  BASIS>  MISSING   MISSING   1.0000     0.0
SOLID   FRACTION <MOLE  BASIS>  MISSING   MISSING   0.0        0.0

INTENSIVE PROPERTIES
...............................

TEMPERATURE <K>                 16.7322   16.7322   4.5034     4.8731
PRESSURE <ATM>                  2.1000    2.1000    2.0000     0.7900
MOLECULAR WEIGHT                4.0030    4.0030    4.0030     4.0030
ENTHALPY <J/KMOL>              -.58562+07 -.58562+07 -.62110+07 -.61100+07
ENTROPY <J/KMOL-K>            -.66360+05 -.66360+05 -.11112+06 -.85263+05
DENSITY <KMOL/CUM>             1.5445     1.5445    31.1068    2.3222
```

Figure 4. Excerpt of the stream report for the simulation that contains hot water outlet
stream from EXR4.

optimization of individual designs also has proven feasible via the simulation models. Finally, the simulation models have proven useful for the performance of rudimentary cost-benefit analyses of competing process designs. The ability to perform meaningful off-design and optimization studies is somewhat limited due primarily to the inadequacies of the multiple stream heat exchanger and compressor-turbine unit operation models.

RECOMMENDATIONS FOR FUTURE WORK

While the modeling work has been successful and has served most of the desired purposes, a few additional enhancements would prove quite useful for process optimization and studying the performance at off-design conditions. The first of these enhancements is a simplified equation of state for helium, which executes faster than the McCarty model with no significant loss of accuracy. Air Products has developed a proprietary implementation of such a model, although a thorough study of the accuracy and robustness of the model has not yet been performed. A second desirable enhancement is a compressor - turbine model which is capable of performing fairly sophisticated rating calculations, including the prediction of choke flow conditions and off-design efficiencies. Such a model is essential for off-design si3ulations. A third recommended enhancement is the addition of an axial heat conduction calculation to the two stream heat exchanger. Axial conduction is significant due to the geometry of the cryogenic exchangers employed. A final recommended enhancement is a more rigorous multiple stream heat exchanger model, which considers the area distribution inside the exchanger. Such a model would be very useful in optimizing heat exchanger design and in predicting off-design performance.

REFERENCES

1. Jacobsen, R. T., Stewart, R. B., McCarty, R. D. and Hanley, H. J. M., Thermophysical Properties of Nitrogen from the Fusion Line to 3500 R [1944 K] for Pressures to 150,000 PSIA [10342 × 10^5 N/sqm], Nat. Bur. Stand. (U.S.), Tech. Note 648 (Dec 1973)

2. McCarty, R. D., Thermodynamic Properties of Helium 4 from 2 to 1500 K at Pressures to 10^8 Pa, J. Phys Chem. Ref. Data 2, No. 4, 923-1042 (1973)

6. Detectors I

CAD TOOLS FOR DETECTOR DESIGN

John Womersley

Florida State University, Tallahassee, FL 32306-3016

Nicholas DiGiacomo and Kendrick Killian

Martin-Marietta Astronautics Group, Denver, CO 80201

ABSTRACT

Detailed detector design has traditionally been divided between engineering optimization for structural integrity and subsequent physicist evaluation. The availability of CAD systems for engineering design enables the tasks to be integrated ·by providing tools for particle simulation within the CAD system. We believe this will speed up detector design and avoid problems due to the late discovery of shortcomings in the detector. This could occur because of the slowness of traditional verification techniques (such as detailed simulation with GEANT). One such new particle simulation tool is described. It is being used with the I-DEAS CAD package for SSC detector design at Martin-Marietta Astronautics and is to be released through the SSC Laboratory.

INTRODUCTION

The large, elaborate detectors to be constructed for experiments at future hadron colliders will need careful modelling and simulation of many important features of the design. In the past decade, the design of large detectors systems has usually proceeded something like the schematic shown in Figure 1.

Firstly engineers have designed the structural elements of the system, following general guidelines from physicists. The goal is to have a system which will support itself, satisfy the laboratory safety committee, and be able to be assembled without collapsing. The guiding information from the physicists would usually include such admonitions as "don't put too much material in front of the calorimeter", or "make this part as thin as possible", but detailed analysis of the engineering design has had to wait until blueprints were produced by the engineers. At this point, the physicists would make back-of-the-envelope calculations, and if things looked

DETECTOR DESIGN IN THE 1980'S

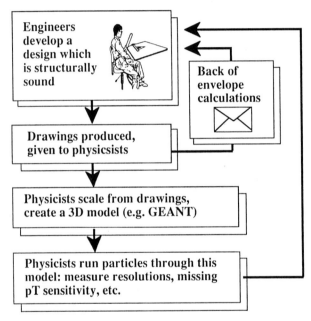

FIGURE 1

Shematic illustration of the traditional process of detector design.

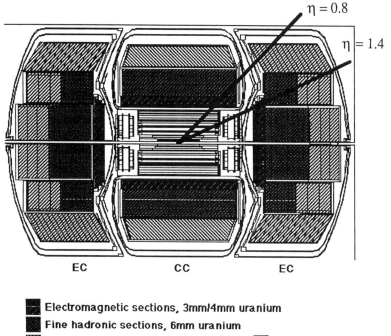

η = 0.8

η = 1.4

EC   CC   EC

■ Electromagnetic sections, 3mm/4mm uranium
■ Fine hadronic sections, 6mm uranium
▨ Coarse hadronic sections, 47mm steel   ▧ 47mm copper

FIGURE 2

Cross section of the D-Zero liquid argon calorimeter for the Fermilab Tevatron collider, showing the cryostat vessels and the large amount of uninstrumented material between pseudorapidities of 0.8 and 1.4.

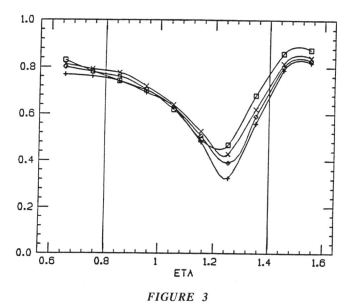

FIGURE 3

Fraction of incident energy recovered in live calorimetry, as a function of pseudorapidity of the jet axis, for the D-Zero calorimeter. QCD jet events with transverse momentum between 120 and 160 GeV/c.

reasonably satisfactory, would proceed to model the preferred detector design using (for example) a simulation program such as GEANT. The geometrical details would have to be entered again by hand, often by taking a ruler and scaling from the engineering blueprints, then typing these numbers into a computer. Particles would then be run through the GEANT model of the detector, and its potential physics capability assessed, with the possibility of design changes being required if the capability is less than expected.

The disadvantages with this approach stem from its manifest slowness. It can take months or even years from the appearance of a realistic engineering design to the point where it has been explored by physicists. This makes it impractical for the tight schedules required of SSC and LHC detectors. It is also prone to leave experiments locked into a non-optimal geometry, simply because it was discovered too late. An example of the latter effect is the region between the central and endcap calorimeter cryostats in the D-Zero detector at the Fermilab tevatron collider. As may be seen in Fig. 2, there is a great deal of uninstrumented material in the form of cryostat walls and supports between pseudorapidities of 0.8 and 1.4   This has the effect that the average energy recovered in "live" calorimetry for jets which point into this region is only about half what is recovered for jets pointing toward clean regions of the detector (Fig. 3). To avoid ruining the missing p_T sensitivity of the detector, extra instrumentation has had to be added in the form of massless gaps and scintillators. The detector now performs satisfactorily, but the need to add the extra elements should have been avoided through better design.

NEW TECHNOLOGY

In the 1990's we can transform the process of design, by taking advantage of the fact that cheap and readily available CAD/3D modeller systems are becoming commonplace. These systems contain many useful engineering tools: structural analysis, fluid low and so forth. They are a great help in the design of detectors and much use is already being made of them. An example is the 3D model of the EMPACT detector study created using the I-DEAS 3D modeller from SDRC. (Fig. 4). To see how far things have progressed, one need only compare this computer-generated, 1990 view with the laboriously hand-drawn picture of D-Zero circa 1985 (Fig. 5).

THE EXISTING TOOL

In the calorimeter design work at Martin-Marietta undertaken as part of the Generic SSC R&D program, and subsequently for the EMPACT proposal, we have made a first step toward adding a **particle tracking tool** to the CAD package. The aim is to allow engineers to begin to optimize the design for physics at the same time as it is optimized structurally --- the key to elimination of the design bottleneck.

At present the tool incorporates ray tracing within the I-DEAS geometry, and an algorithm for estimating the effect of uninstrumented material and leakage on calorimeter resolution. This algorithm has been described elsewhere [1] and seems to be reasonably accurate for our purposes.

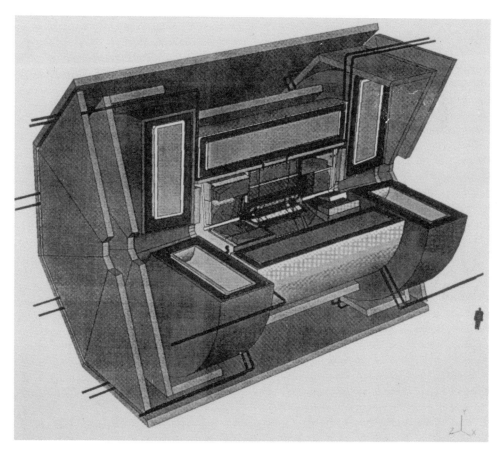

FIGURE 4

3D model of the EMPACT SSC detector study.

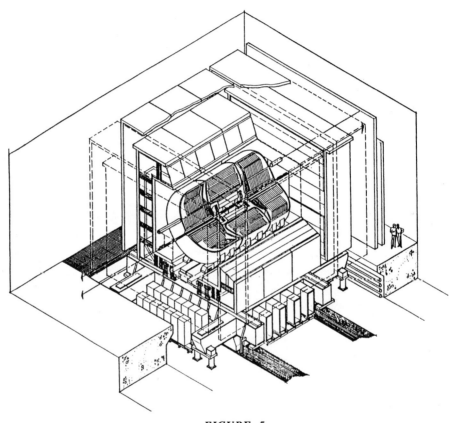

FIGURE 5

3D picture of the D-Zero detector.

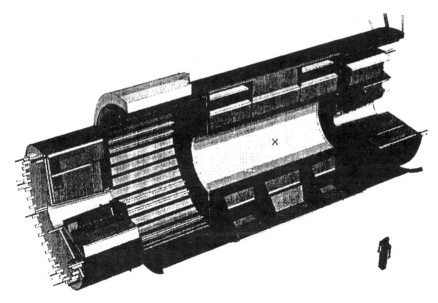

FIGURE 6

Hermetic liquid argon calorimeter SSC R&D project.

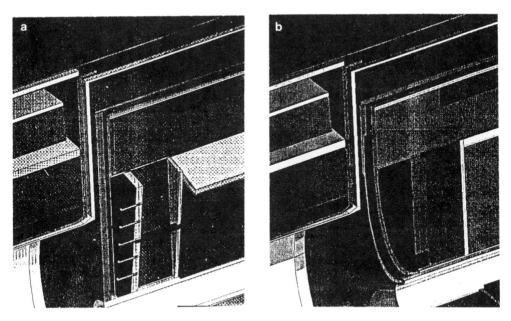

FIGURE 7

Two possible configurations of endcap cryostat head walls for the calorimeter shown in Fig. 6. (a) stayed flat heads, and (b) elliptical heads.

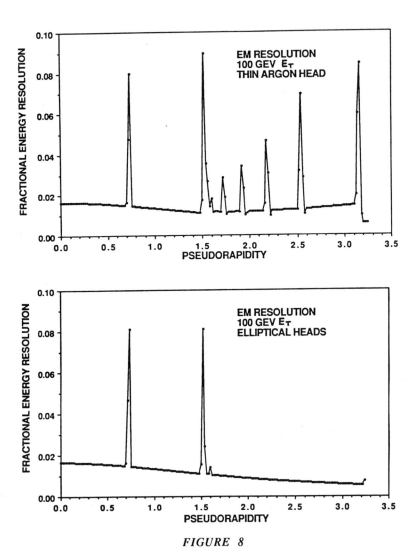

FIGURE 8

Estimated calorimeter resolution for single electrons at a transverse energy of 100 GeV, as a function of pseudorapidity, for the two calorimeter configurations shown in Fig. 7.

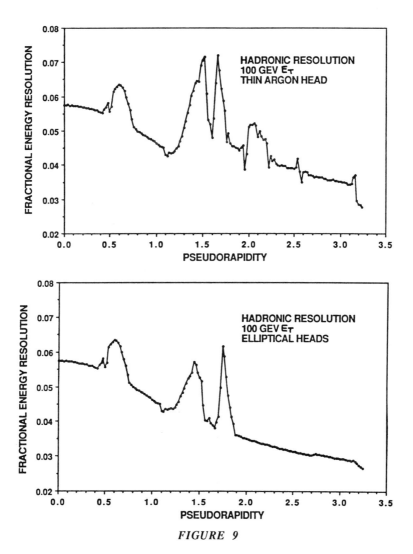

FIGURE 9

Estimated calorimeter resolution for single charged pions at a transverse energy of 100 GeV, as a function of pseudorapidity, for the two calorimeter configurations shown in Fig. 7.

The program runs on a VAX workstation. Licenses for the use of I-DEAS for research purposes at universities are currently avialble for under $10,000 per year. I-DEAS is also the Fermilab standard CAD system.

An example of the usefulness of this approach is in the comparison of difdferent possible endcap cryostat head wall configurations for the liquid argon calorimeter in the generic R&D study. The calorimeter is shown in Fig. 6, and the possible endcap configurations in Fig 7 a and b. By tracing rays through the 3D model of the detector and estimating the effect of the dead material on energy resolution, plots such as those shown in Figs 8 and 9 are obtained for electrons and hadrons. As may be seen, the favored configuration in this case is the elliptical heads (Fig 7 b) which give better calorimeter resolution, and this is what has been followed in the development of this calorimeter project.

FURTHER DEVELOPMENT

We propose to release the existing code to the high energy physics community through the SSC Laboratory, and to continue to develop this tool with funding from the SSCL and Martin-Marietta. Future extensions and developments include adding transverse shapes to the simulated showers, fluctuations, the capability to take input from event generator programs such as ISAJET or PYTHIA, and the addition of readout segmentation schemes to the calorimeter models.

We are also investigating the possibility that we may be able to create a relatively quick-and-dirty I-DEAS to GEANT geometry interpreter based upon this system. This would avoid the need to re-create a 3D model in GEANT when one has already been made using the CAD system, and would coincide with the future plans of the GEANT group at CERN[2].

CONCLUSION

Even if you do not wish to use the actual code when it is released through the SSCL, we hope that you will agree with, and follow, the philosophy that it embodies, namely that the integration of engineering design work with physics optimization will make better decetors, and enable them to be designed in less time. This must be the direction in which we move over the next decade, because SSC and LHC detectors are simply too large and expensive for any less care to be taken with them.

This work has been supported in part by the US Department of Energy.

REFERENCES

[1]   W.J. Womersley, FERMILAB-CONF-89/155-E, to be published in Nuclear Instruments and Methods A (in press).

[2]   R. Brun, in proceedings of the Workshop on Physics and Detector Simulation for SSC Experiments, Dallas, January 1990.

A PROJECTIVE GEOMETRY LEAD FIBER

SCINTILLATOR DETECTOR

H. Paar, D. Thomas, M. Sivertz, B. Ong, D. Acosta, T. Taylor
and B. Shreiner

Physics Department, B-019
University of California, San Diego
La Jolla, CA 92093

BACKGROUND

The Superconducting Super Collider (SSC), presently under construction near Dallas, Texas requires highly sophisticated particle detectors. The energy and particle flux at the SSC are more than an order of magnitude higher than the highest machine located at the Fermi National Accelerator near Chicago. An important element of particle detectors for the SSC is the calorimeter. It measures a particle's energy by sampling its energy deposit in heavy material, such as (depleted) uranium or lead.

The sampling medium must be interspersed with heavy absorber material. In the case of scintillating plastic, two methods are under consideration: plates and fibers. In the case of plates, a 'sandwich' of scintillator plates and uranium plates is constructed. In the use of fibers (still in the proto-type stage), 1 mm. diameter cylindrical scintillating fibers are inserted into grooves that are machined into lead layers. The layers are stacked and epoxied together to form the required geometrical shape of the detector.

Lead and scintillating plastic sampling can meet the physics require-ments of the detector. This has been shown in an R & D program which is underway at the University of California at San Diego (UCSD), High Energy Physics Group. This R & D is funded by the Department of Energy, High Energy Physics and SSC Divisions.

PRESENT PROTOTYPE CONSTRUCTION AT UCSD

Presently, the second of two 56 cm. projective modules is under con-struction at UCSD. The first module was completed last year and has been taken to CERN and operated in a testbeam. Each module consists of nine mini-modules assembled in a honeycomb steel holder. Each minimodule consists of 15 lead layers glued together with scintillating fibers inserted into grooves. There are 196 fibers in each minimodule. The fibers from all nine minimodules converge on the other side of the holder to a hexagonal plexi-glas 'cookie.' The 'cookie' butts up against a hexagonal lightpipe, at the other end of which is connected a photomultiplier tube. The entire back end of the module is enclosed in mu metal shielding.

A typical minimodule is a 56 cm. long tapering tower, 3.22 cm. per side on the readout end, converging down to 2.74 cm. per side on the front end. 16 layers of lead 2 mm. thick are glued together, then milled, to produce this shape. In the first module, grooves are machined into the lead on each side, 1 mm. in diameter and 2 mm. apart. Fibers are placed into the grooves prior to assembly for gluing. Assembly takes approximately two hours, and the minimodule is ready for further machining after overnight curing of the epoxy.

To accommodate the holder, the outer layer of fibers on the readout end of each minimodule is machined away. Each holder has holes drilled into it for alignment pins. The holder wall thickness is 2 mm., the same as one lead layer. The holder also provides a foundation for the plexiglas 'cookie,' lightguide, photomultiplier tube and shielding.

The second 56 cm. module under construction will utilize a newly developed rolling fixture to emboss the grooves into the lead layers instead of machining them. In the rolling fixture, the 2 mm. thick sheets are drawn through two aluminum rollers. This technique has reduced both the cost of grooving and the handling time by a factor of 100.

Another development has been a gluing fixture designed to press the layers into the desired shape. The gluing fixture consists of several pieces of angle iron separated by aluminum plates. Adjustment screws pass through the plates and push against alignment bars which press the newly glued layered stack to the proper dimensions.

Future developments include the construction of four 2 m. long projective modules, utilizing both the rolling and gluing fixtures. Also, an outside company is looking at the mass production of modules, specifically the commercial availability of fiber/lead sheets.

CONCLUSION

The SSC needs a calorimeter. The group at UCSD has been working on a lead/fiber concept that has been shown at other conferences to be a viable candidate. One 56 cm. prototype module has been constructed and tested at CERN, another one is under construction, and four 2 m. long modules will be constructed, utilizing cost and time-saving techniques such as rolling instead of machining. Further, the mass production of full scale modules by outside vendors is under construction.

A HIGH RESOLUTION BARIUM FLUORIDE CALORIMETER[1]

Ren-yuan Zhu[2]

Lauritsen Laboratory
California Institute of Technology
Pasadena, CA 91125

ABSTRACT

A high speed and highly radiation-resistant barium fluoride crystal array is under development by the Caltech group in collaboration with the Brookhaven National Lab, the Shanghai Institute of Ceramics in China, the KEK in Japan and the AccSys Technology Inc. in California. The barium fluoride array will serve as a prototype for a high precision electromagnetic calorimeter at the SSC, which has a unique capability in detecting the Higgs particles in the mass region between 80 to 180 GeV. The research and development program involves 3 aspects: (1) production of barium fluoride crystals with very large size (800 cm^3) and high radiation resistance (10 MRads); (2) development of a fast readout system with an optical and electrical suppression on the slow component; and (3) development of an effective precision calibration system based on radiative capture of a pulsed proton beam from an Radio Frequency Quadrupole accelerator on a calcium fluoride target.

INTRODUCTION

This report present a design study of a very high precision electromagnetic calorimeter composed of large size barium fluoride (BaF$_2$) crystals for the Superconducting Super Collider (SSC) [1]. The speed (gating in a single beam croosing), granularity (typically $\Delta\eta \times \Delta\phi = 0.04 \times 0.04$ per crystal), radiation hardness ($\geq 10^7$ Rads), high scintillation light output and the large acceptance ($|\eta| \leq 3.8$) of BaF$_2$ calorimeter will provide an excellent precision detector for the SSC to probe new physics in the high intensity SSC environment [2].

The BaF$_2$ electromagnetic calorimeter will have energy resolution of $1.3\%/\sqrt{E} + 0.5\%$, comparable with what obtained from the L3 BGO electromagnetic calorimeter [3]. The physics motivation for a precision EM calorimeter is to make very precise energy measurement for photons and electrons. The unique capability of this detector is shown, for example, in the ability to detect Higgs particles in a mass region of 80 to 180 GeV [4].

It is well known that there exists a gap between the upper limit for Higgs detection at LEP Phase II (80 GeV) [5] and the lower limit for Higgs detection at the SSC using the four lepton final state (180 GeV) [4]. The precision EM detector will cover this gap by measuring the Higgs in its $\gamma\gamma$ and ZZ* decay channel.

---

[1] Work supported by U.S. Department of Energy Contracts No. DE-AC03-81-ER40050.
[2] Representing the team of H. Newman (Caltech), H. Ma, C. Woody (BNL), Z.Y. Wei, Z.W. Yin (SIC, China), T. Matsuda, F. Takasaki (KEK, Japan) and R. Hamm (AccSys, California).

The high resolution calorimeter also will be used to search for new massive gauge bosons through their decays into e^+e^-. This will allow us to explore the energy regime where high energy symmetries, such as superstring-inspired E6, break down to form the gauge structure observed at low energies.

PHYSICS OF PRECISION PHOTONS AND ELECTRONS

The history of High Energy Physics over the last 20 years has shown that precision measurements of leptons have been a key factor in most important discoveries in the field. Examples are the discovery of the J/ψ, Υ, $W^\pm$, and Z^0. The most important discoveries at the SSC could result from the detection of entirely new phenomena at the 1 TeV scale, outside of the Standard Model or its simplest extensions. These discoveries could appear as new multilepton final-states, or as new configurations of leptons and jets. The ability to distinguish and precisely measure electrons and photons, as well as muons, at the SSC will allow isolate the new physics signals from the backgrounds of the *"Standard"* physics, such as the inclusive production of Z's and W's together with jets.

Should the Standard Model survive in the SSC era, then the highest mass resolution, and good acceptance will be required to detect the Higgs, through its distinctive decays into multileptons or photons.

$H^0 \to \gamma\gamma$ Search for $M_{H^0} \in (80, 150)$ GeV

In this section we demonstrate the ability of the BaF$_2$ EM detector in searching the Higgs by using its $\gamma\gamma$ decay mode. The study was done by using a Monte Carlo program PYTHIA version 5.3 [6], and has been cross-checked with ISAJET version 6.27 [7].

Figure 1 shows a lego plot of transverse energies in the $\eta - \phi$ plane for a 150 GeV Higgs which decays into two photons together with associated QCD jets. The distinctive feature of isolated photons in the Higgs decay can be used to reject the QCD background.

The main background in the $H^0 \to \gamma\gamma$ search is direct photon production: $q\bar{q} \to \gamma\gamma$ (59 pb) and $gg \to \gamma\gamma$ via a box diagram (493 pb)[9]. This so-called *"irreducible background"* has to

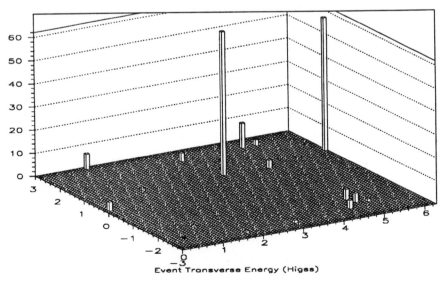

Event Transverse Energy (Higgs)

Figure 1. Transverse energy distribution of a $H^0 \to \gamma\gamma$ event in the $\eta - \phi$ plane.

be reduced by event selection cuts. Figure 2 shows the distributions of photon rapidity (η_γ), photon transverse energy (E_T^γ), rapidity of the 2 photon system ($\eta_{\gamma\gamma}$) and $\cos\theta^*$, where θ^* is the polar angle of photons in the $\gamma\gamma$ rest frame, for $H^0 \to \gamma\gamma$, $q\bar{q} \to \gamma\gamma$ and $gg \to \gamma\gamma$. The cuts used in event selection are:

- $|\eta_\gamma| < 2.8$;

- $E_T^\gamma > 20$ GeV;

- $|\eta_{\gamma\gamma}| < 3$ ($\Rightarrow$ to reduce $qq \to \gamma\gamma$).

- $|\cos\theta^*_\gamma| < 0.8$ ($\Rightarrow$ to reduce $gg \to \gamma\gamma$).

The background cross sections after event selection cuts are reduced to 91 pb for $gg \to \gamma\gamma$ and 14 pb for $q\bar{q} \to \gamma\gamma$. Since the Higgs width in this mass region is very narrow ($\sim$10 MeV) a very high resolution electromagnetic calorimeter, such as the BaF$_2$, is needed to detect the 0.3 pb signal [8] from $H^0 \to \gamma\gamma$.

There are copious π^0's, and thus photons, produced at the SSC. The QCD two jet cross section is 1.2 mb for $\hat{P}_T > 20$ GeV and $|\eta| < 2.8$. There are also processes producing one real photon and a QCD jet, such as $qg \to q\gamma$ (119 nb) and $qq \to g\gamma$ (6 nb), which will fake a two-isolated-photon final state with a higher probability. An isolation cut is used to reject photons produced from fluctuations in jet fragmentation. The cut requires that the sum of the transverse energy in a cone of radius R=0.6 (R=$\sqrt{\Delta\eta^2 + \Delta\phi^2}$), excluding the transverse energy of the photon itself, is less than 10% of the transverse energy of the photon plus 5 GeV.

By using this isolation cut, the rate of one isolated photon from a jet is reduced by a factor of 10^{-4}. The rate of two isolated photons from two jets is thus reduced by a factor

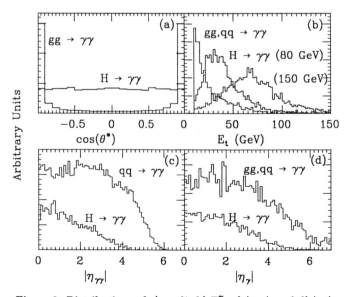

Figure 2. Distributions of a) $\cos\theta^*_\gamma$, b) E_T^γ, c) $|\eta_{\gamma\gamma}|$; and d) $|\eta_\gamma|$.

of 10^{-8}, assuming that the two jets are statistically independent. Figure 3a) shows the signal/background ratio of 5, 9, 15, 27 standard deviations for Higgs mass of 80, 100, 120 and 150 GeV, respectively, with BaF$_2$ calorimeter. For comparison, the same signal is shown in 3b) for a conventional electromagnetic calorimeter with energy resolution of $15\%/\sqrt{E} + 1\%$. The Higgs peaks after background subtraction is shown in figure 3 c) and d), respectively. The detector with a conventional resolution will certainly not be able to do this physics.

$H^0 \rightarrow ZZ \rightarrow e^+e^-e^+e^-$ Search for $M_{H^0} \in (140, 800)$ GeV

The *"gold plated"* signature for the heavy Higgs decay mode $H^0 \rightarrow ZZ \rightarrow e^+e^-e^+e^-$ is four isolated electrons. All of the QCD background, including contributions from charm, bottom and even top decays (computed with top mass = 140 GeV), is rejected with the isolation cut described in previous section (with a reduced radius R=0.3). The only remaining background is pp $\rightarrow$ ZZ $\rightarrow e^+e^-e^+e^-$ which has a cross section of 24 femtobarns (fb) for two real Z^0.

When the Higgs mass is below the $2M_{Z^0}$ threshold (180 GeV) the process becomes $H^0 \rightarrow ZZ^* \rightarrow e^+e^-e^+e^-$ where one Z^0 is off mass shell. The background in this mass region is pp $\rightarrow$ ZZ$^* \rightarrow e^+e^-e^+e^-$ and pp $\rightarrow$ Zt$\bar{t} \rightarrow e^+e^-e^+e^-$X. However, these calculations are yet to be implemented in the PYTHIA or ISAJET program. In principal, one expect that by measuring this decay mode one might be able to search for the Higgs down to 140 GeV.

Figure 4 shows the 150 GeV Higgs signal computed for the proposed BaF$_2$ EM calorimeter in a period of one SSC year. For comparison, the same signal is shown for a conventional electromagnetic calorimeter with a conventional energy resolution of $15\%/\sqrt{E} + 1\%$.

Z$'$ Search

The BaF$_2$ EM detector has excellent potential to probe physics beyond the Standard Model. We will search for the Z$'$ in its two electron decay mode, i.e. by looking for two isolated electrons in the final state. The major backgrounds in the Z$'$ search are from the Drell-Yan process and from heavy quark decays. Figure 5 shows the Z$' \rightarrow e^+e^-$ signal, on top of the combined Drell-Yan and QCD backgrounds for $M_{Z'} = 150$ GeV, computed for a period of one SSC year with BaF$_2$ resolution. The result expected for a calorimeter with conventional resolution is shown for comparison.

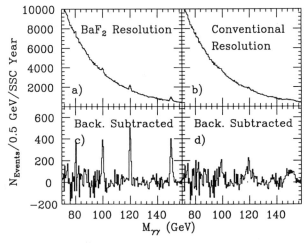

Figure 3. $H^0 \rightarrow \gamma\gamma$ Detection for electromagnetic calorimeters with a) and c) BaF$_2$ EM resolution, and b) and d) conventional resolution.

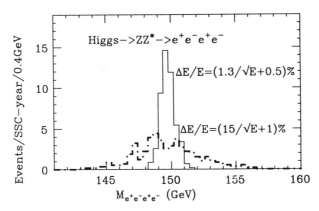

Figure 4. $H^0 \rightarrow ZZ^* \rightarrow e^+e^-e^+e^-$ detection for electromagnetic calorimeters with BaF$_2$ resolution and conventional resolution in one SSC year.

BaF$_2$ E-M CALORIMETER

The proposed barium fluoride electromagnetic calorimeter consists of 26,000 large size (50 cm long, 24 radiation lengths) crystals. The calorimeter has three parts:

- A central barrel calorimeter has an inner radius of 75 cm and an outer radius of 140 cm. It covers the rapidity of $\pm$ 1.45.

- Two endcaps, located at $z=\pm$ 150 cm, cover the rapidity of $1.45 \leq | \eta | \leq 2.8$.

- The forward and backward calorimeters, located at $z=\pm$ 10.5m, cover the rapidity range of $2.8 \leq | \eta | \leq 3.8$.

Figure 6 is a schematic showing the central BaF$_2$ calorimeter consisting of the barrel and the end caps. Also shown in the figure is a typical crystal with vacuum photodiode readout at the back. The crystals are approximately square in cross section, with sizes ranging from 2×2 cm^2 in the front for the smallest crystal, to 5.2×5.2 cm^2 in the back of the largest crystal. The typical crystal covers a $\Delta\eta \times \Delta\phi$ interval of 0.04 × 0.04. The total volume of crystals is 17.2 m^3 with 83.6 tons.

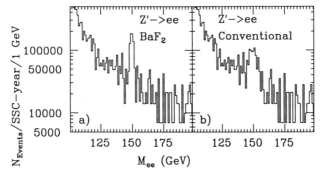

Figure 5. $Z' \rightarrow e^+e^-$ detection in one SSC year, for electromagnetic calorimeters with a) BaF$_2$ EM resolution, and b) conventional resolution.

273

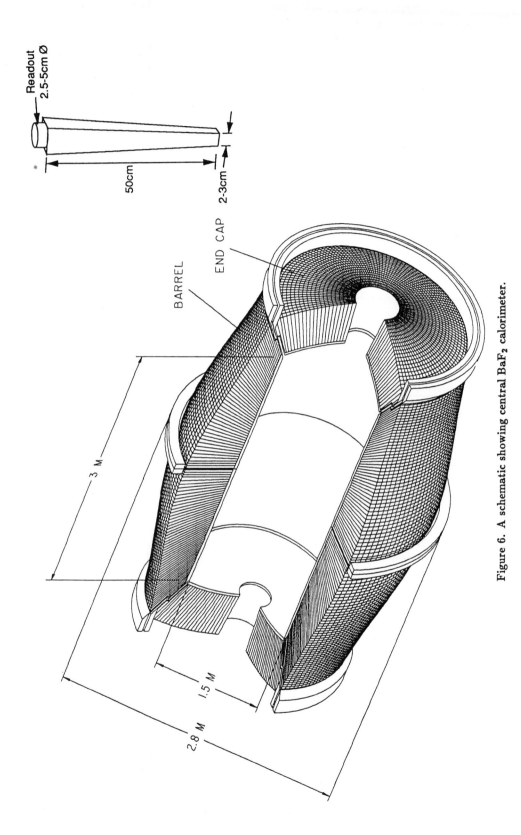

Readout
2.5-5cm Ø

50cm

2-3cm

END CAP

BARREL

3 M

1.5 M

2.8 M

Figure 6. A schematic showing central BaF$_2$ calorimeter.

The proposed BaF_2 calorimeter has following features:

- The scintillation light readout has 3 nsec peaking time, i.e. gating in a single beam crossing;

- Bipolar signal output with residual tail of less than 10^{-4} after 35 nsec;

- Radiation resistance up to at least 10^7 Rads to survive the SSC environment;

- Energy resolution of a $1.3\%/\sqrt{E} + 0.5\%$.

- π/e suppression ratio of better than 10^{-3}.

Fast Readout of BaF_2

Barium fluoride (BaF_2) is a unique high density inorganic scintillator with three emission spectra peaking at 195 nm, 220 nm and 310 nm, with decay time constants of 0.87, 0.88 and 600 nsec respectively [10]. Because of the speed of the *"short"* or *"fast"* components and the evidence that it has high radiation resistance [11], BaF_2 has gained vast interest in recent years [12].

Slow Component Suppression

Table 1 lists the properties of candidate fast scintillation materials: pure CsI, BaF_2, CeF_3 and liquid xenon.

The intrinsic intensity of the slow scintillation component (600 nsec) in BaF_2 is 5 times higher than the fast components. Within the past years an approach of producing BaF_2 crystals with the suppressed slow component was discovered by Schotanus *et al.* [13]. They found that a small amount of lanthanum added to the crystal during growth greatly suppressed the slow component without significantly effecting the fast component. A subsequent study by Woody *et al.* [14] showed that there are several dopants which produce strong slow component suppression and retain high fast component light output. However, only lanthanum still preserves the radiation hardness of the pure material up to a level beyond 10^6 rads.

Figure 7 shows the emission spectra for pure BaF_2 and BaF_2 doped with 1% of lanthanum. The peak intensity of the slow component (300nm) is reduced by a factor of about five with

Table 1. Fast Scintillators

	P. CsI	BaF_2	CeF_3	L. Xe
Density (g/cm^3)	4.51	4.88	6.16	3.05
X_{rad} (cm)	1.85	2.10	1.7	2.77
X_{int} (cm)	37.0	29.9	26.2	55
X_{rad}/X_{int}	0.051	0.068	0.065	0.050
$R_{Moliere}$ (cm)	3.5	4.4	2.6	5.6
Ref. Index	1.80	1.49	1.62	N/A
Hygroscopic	No	No	No	N/A
Lumin. (nm)	300	320	340	170
		210	310	
τ_{Decay} (nsec)	10	630	27	45
		0.9	≈ 2	
Light Output	<2000	5000	700	40,000
(photons/MeV)		1000	?	

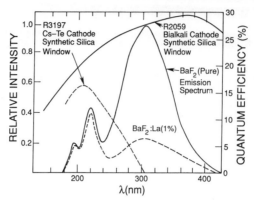

Figure 7. BaF$_2$ scintillation spectra and PMT quantum efficiencies.

little change to the fast components (195 and 220 nm). It is also known that the intensity of the fast components have no temperature dependence, while the slow component increases with decreasing temperature, at a rate of -2.4%/°C [15].

Since the fast components and the slow component of BaF$_2$ are in different wave length, they are spectroscopically separable. Figure 7 also shows the the quantum efficiencies of two photomultipliers (PMT): a special UV-sensitive, solar-blind PMT with a Cesium Telluride (Cs-Te) photocathode and a synthetic silica (quartz) window (Hamamatsu R3197) and a PMT with a bialkali photocathode and a quartz window (Hamamatsu R2059). By using the solar-blind photocathode (Cs-Te), one would able to collect only the fast scintillation light.

Figure 8 shows pictures of the scintillation light pulses recorded on an HP54111D digital scope by using bialkali photocathode (R2059), Cs-Te photocathode (Hamamatsu R3197) and Cs-Te photocathode together with a La-doped BaF$_2$ crystal. The optical suppression factors for the slow component, defined as the photoelectron number of the fast components divided

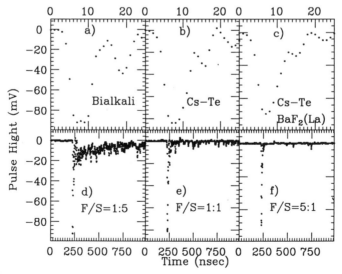

Figure 8. BaF$_2$ scintillation Light pulse observed by a) and d) a Hamamatsu R2059 PMT; b) and e) a Hamamatsu R3197 PMT; and c) and f) a Hamamatsu R3197 PMT and a La-doped BaF$_2$ crystal.

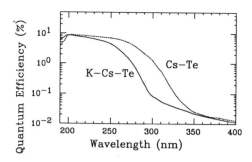

Figure 9. Quantum efficiency of a Hamamatsu K-Cs-Te photocathode comparing with a conventional Cs-Te photocathode.

by the photoelectron number of the slow component (F/S), are also shown in the figure. With lanthanum doping a fast to slow ratio of 5 is obtained. The rise time of the scintillation light pulse in the picture was completely dominated by the 2.3 nsec rise time of the PMT's.

An additional optical suppression of the slow component can be achieved by using a new photocathode (K-Cs-Te) recently developed by the Hamamatsu corporation [16]. Figure 9 shows a comparison of the quantum efficiency responses of this new solar-blind (K-Cs-Te) photocathode and a conventional solar-blind photocathode (Cs-Te). The K-Cs-Te photocathode has 10 times better suppression power in terms of Q.E.@220nm/Q.E.@300nm, comparing with the Cs-Te photocathode.

Readout Design

The principal readout components of BaF_2 are designed as follows:

- A vacuum photodiode with K-Cs-Te photocathode and Quartz window as the photosensitive device. This device will suppress the slow component of BaF_2 by a factor of 50. The device also works in a magnetic field.

- A preamplifier, mounted on the base of vacuum phototube, has 45 nsec decay time. The typical noise of this preamplifier together with vacuum photodiode is 1000 electrons.

- A fast shaper with 3 – 10 nsec peaking time. This shaper has two pole zero cancellation: one for residual slow component (~600 nsec) and the other for the preamplifier decay (45 nsec). It has two outputs: a bipolar output for the signal and a unipolar output for RFQ calibration. The residual signal of the bipolar out put after 35 nsec is less than 10^{-4}.

Figure 10 shows the input and output pulse shape simulated with a PSPICE program on an IBM-PC for the designed circuit. The input signal, which has 3 nsec rise time and a Fast/Slow ratio equals one, is an average of digitized BaF_2 scintillation pulses obtained from a solar-blind PMT with Cs-Te photocathode.

Also shown in Fig. 10a) is the output from the preamplifier. Both bipolar and unipolar output pulses are shown in Fig. 10b). This very fast readout system will enable us to operate the detector in a single beam crossing, thus reduce the pile-up to the minimum.

Radiation Resistance

Early work done by S. Majewski and D. Anderson [11] showed that no color centers were formed in BaF_2 up to a dose of 1.3×10^7 Rads in an 800 GeV proton beam. The crystals tested were from Harshaw. Many other works [11] confirmed this early observation for irradiations from either charged particles or photons. It is also known that impurities in the crystal will cause radiation damage. An absorption band around 205 nm was identified as originating from Pb contamination [17], and this was correlated with the susceptibility to radiation damage.

We have also tested the radiation resistance of BaF_2 crystals. Our preliminary study shows that the BaF_2 is one of the highest radiation resistant crystals known. The radiation damage caused by γ-ray or neutron irradiation is recoverable by annealing the crystal at 500°C for three hours. This indicates that neutrons, as well as photons, do not cause permanent damage in BaF_2 crystals. We thus assume that the BaF_2 damage mechanism from neutrons is the same as that from photons. Most crystal samples, however, suffer a limited initial damage, and a saturation occur after a small amount of dosage. The result of these measurements indicates that one can obtain radiation hard crystals with limited initial irradiation (training). A further investigation on the correlations between the impurities in the crystal and the small initial damage will be carried out. It is expected that highly radiation resistant BaF_2 crystals will be produced following this investigation.

Linearity

The scintillation light linearity of BaF_2 is known to be as good as the BGO. A CERN test beam done by E. Lorenz et al. [18], showed that the linearity of the light output of BaF_2 is excellent up to 50 GeV. We expect the light output linearity of BaF_2 will remain good for at least four orders of the magnitude.

Energy Resolution

The energy resolution of a total absorption calorimeter, such as BaF_2 calorimeter, can be factorized as:

$$(\frac{\sigma}{E})^2 = (\frac{a_0}{\sqrt{E}})^2 + (\frac{a_1}{E})^2 + c^2$$

The contributions of each term to the resolution are:

- a_0 is the contribution from the photoelectron statistics.

- a_1 is the contribution from electrical noise, sum over a few Moliere radii.

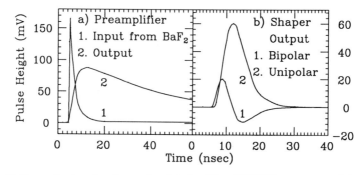

Figure 10. BaF_2 readout pulse shape simulated with a PSPICE program: a) input and after preamplifier and b) outputs.

278

- c is a constant term, $c^2 = \sigma_G^2 + \sigma_C^2 + \sigma_n^2$:
  - σ_G (geometry effect): shower leak (front, side and rear) and inactive material between cells;
  - σ_C (intercalibration error);
  - σ_n (physics noise): fluctuation of shower and light yield etc.

Light Yield

Figure 11 shows the measured photoelectron yield per MeV as a function of the integration time obtained from a R3197 PMT for six BaF$_2$ crystals. The data points in the figure were fitted to a sum of contributions from two fast components (F) and from the slow component (S) with an exponential decay time constant τ: y = F + S [(1 - exp(-t/τ)]. The numerical results of the fit are also shown in the figure.

Table 2 lists the fit results, photoelectron yield, the ratios of the fast components to the total light yield and the FWHM energy resolutions measured by using a ^{137}Cs γ-ray source for 6 different crystals. In summary, around 80 fast photoelectrons/MeV and F/S $\sim$ 1 were observed. The crystal doped with 1% of lanthanum has a F/S $\sim$4. This 80,000/GeV photoelectron yield is more than enough to provide 1.3%/$\sqrt{E}$ resolution.

Electrical Noise

As described in section 3.1.3 the expected noise of a single readout channel is $\sim$1000 electrons. The contribution of this term to the total energy resolution is thus negligible. In the construction of the BaF$_2$ calorimeter we will impose electrical isolation between channels so that the correlated noise will be reduced to a minimum.

Summary of Resolution

A GEANT simulation was carried out to calculate the intrinsic resolution of the BaF$_2$ calorimeter. The simulation was calculated for an array made by 121 (11 × 11) BaF$_2$ crystals

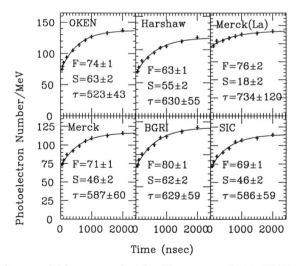

Figure 11. Photoelectron yields measured with a Hamamatsu R3197 PMT as a function of the integration time for six BaF$_2$ crystals.

Table 2. Result of the Light Yield Measurement with a Hamamatsu R3197 PMT

Crystal	OKEN	Harshaw	Merck(La)	Merck	BGRI	SIC
Dimension(cm)	$\phi1"\times1"$	$2\times2\times5$	$\phi1"\times1"$	$\phi1"\times1"$	$\phi1"\times1"$	$1.7\times1.7\times1.9$
P.e/MeV(55ns)	79±2	66±2	76±2	74±2	82±2	72±2
P.e/MeV(2μs)	137±3	118±2	94±2	116±2	142±3	114±2
S.G./L.G.(%)	58	56	80	64	58	63
$\Delta E/E(\%)$(55ns)	54±3	64±3	57±3	54±3	48±3	53±3
(FWHM)(2μs)	31±1	35±1	47±1	34±1	31±1	35±1
Fast P.e./MeV	74±1	63±1	76±2	71±1	80±1	69±1
Slow P.e./MeV	63±2	55±2	18±2	46±2	62±2	46±2
Fast/Total(%)	54	53	81	61	56	60
τ_{slow}(ns)	523±43	630±55	734±120	587±60	629±59	586±59

with proposed size. Effects included in the simulation are: 250 μm carbon fiber wall between crystals, shower leakage because of summing limited number of crystals and 30% radiation length of dead materials at the front of the BaF$_2$ array, which is the sum of the materials in the central tracking detector (20% r.l.) and the support structure.

Table 3 lists the result of the BaF$_2$ resolution, including intrinsic and other contributions. With an assumption of 0.4% precision calibration, it is clear that the BaF$_2$ calorimeter will be able to provide the design goal of $1.3\%/\sqrt{E} + 0.5\%$ resolution for electrons and photons.

Calibration

Precise, frequent calibration *in situ* is vital in maintaining the high resolution of the precision. As shown in table 3, the dominant contribution to the resolution of BaF$_2$ calorimeter is the uncertainty of the calibration for high energy electrons and photons. The L3 experiment has extensive experiences in maintaining the high precision of BGO calorimeter [3].

Our primary calibration will use a novel technique based on a Radio Frequency Quadrupole accelerator (RFQ) [19]. This technique has been developed by Caltech and AccSys in past five years for calibrating the L3 BGO electromagnetic calorimeter. The first calibration RFQ will be installed in L3 soon. The detailed description of the RFQ calibration is given in the R&D section of this report. This technique will provide a relative calibration with a precision of 0.4% in a few minutes.

Complementary to the RFQ calibration, we will also use the high flux of minimum ionizing particles (MIP) produced by the collider as a calibration souce. A MIP passing through a BaF$_2$

Table 3. Energy Resolution (%)

E (GeV)	5	10	100	500
Photoelectrons	0.2	0.1	0.04	0.02
Noise	0.8	0.40	0.04	0.008
Calibration	0.40	0.40	0.40	0.40
GEANT	0.67	0.60	0.42	0.36
Total	1.1	0.83	0.58	0.54

crystal longitudinally would deposit 0.33 GeV energy in the crystal which would be readout by the crystal with a few percent resolution. With a statistics of a few hundreds tracks, the peak position or the calibration point can be determined to 0.4%. The multiplicity of charged hadrons at SSC is large enough to provide a calibration within 12 hours during SSC running.

The daily calibrations with these techniques will be confirmed by the measurement and invariant mass reconstruction of inclusive Z^0, Υ and J production, where the vector bosons decay into e^+e^-. With a p_T cut of 10 GeV on the vector boson, a typical crystal in the central calorimeter will see 20 e^+ or e^- from Z^0decay, and 200 from Υ decay [20] for each week of running.

Mechanical Structure

Because of the density of the crystals, their brittleness, and the constraint of having to minimize the dead material in the calorimeter to maintain the best resolution, we have chosen to follow the safe, classical design concept of the L3 BGO calorimeter. The structural material is confined to thin cell walls around the crystals, and to a cylindrical inner tube which carries the weight and which is attached to a conical funnel which transmits the load to roller-bearing pads. This results in the best solid angle coverage and energy resolution. Each crystal is located in a cavity composed of thin walls made by Carbon fiber composite. The Carbon fiber cavities are bonded together, to form a rigid honeycomb with wall thickness of 250 μm.

The carbon fiber cavities in the barrel are bonded to an inner cylindrical sandwich with shell acrylic foam core and thin carbon fiber composite skins. As in L3, the mechanical stability of the assembly is increased by using a spring loaded fixture at the back face of each crystal to push the crystals against the inner shell. The inner shell thus takes the weight of the crystals and transmits it to the end-flanges.

Mechanical tests on the material resistance of large BaF$_2$ crystals (i.e. their ultimate compressive strength, bending strength, crack propagation, etc.) have begun, along with cutting and polishing tests to improve these properties. These tests, which will be completed as part of the R&D program presented in the next section, will be used to measure the maximum load that may be taken by the crystal material under compression, as part of the mechanical structure.

The thickness of the cellular structure, and the tolerances which allow for clearance and for light-diffusing reflector (for light collection uniformity), will be reduced to the limit of technical feasibility. Our initial studies show that it should be possible to achieve the same cavity wall thickness (250 μm) and assembly tolerances (100-300μm) between the crystals and the cavity walls.

R&D PROGRAM

BaF$_2$ Crystal Production — R&D at SIC and Caltech

Production of 26,000 large pieces over a period of few years will require a monthly yield of up to 1000 pieces. In terms of crystal weight, the required yield is less than the current capacity of Merck (BDH) for CsI(Tl) production. The world's production capacity of CsI is 30 metric tons per year with 220 furnaces. The commercial price of CsI is now \$1.6 – \$2/cm^3.

The Shanghai Institute of Ceramics (SIC) has supplied BGO crystals for the L3 with a production rate up to 400 finished crystals per month. Accumulated with 6 years of experience on research, development, and production of BGO crystals, collaborated with LAPP (Annecy, France), CERN, and with Caltech, SIC has been working on BaF$_2$ scintillator crystal research since October, 1989. In 1989, SIC grew small pure and doped BaF$_2$ crystals for tests at Caltech, followed by exploration of making large size crystals.

In a collaboration with the Beijing Glass Research Institute (BGRI), SIC has delivered a ϕ40 mm × 220 mm crystal to Caltech. SIC has also invested a new large furnace facility. Adding contribution in other equipments, raw materials and labor, thereby, considered as in Western terms, SIC has invested an equivalent value in excess of \$250,000 in 1989.

The research and development program at SIC will extend to three years (from 1 January 1990 to 31 December 1992) and will focus on the following three aspects:

1. Development of a reliable method of growing large size, highly transparent and radiation resistant crystals.

2. Systematic study of precision BaF_2 machining, including cutting, lapping and polishing.

3. Systematic study of rare-earth doped BaF_2 crystals, especially lanthanum-doped BaF_2 crystals.

The manpower devoted to this research and development program will be totaled 13.0 FTE's in 1990, 16.6 FTE's in 1991, and 19.0 FTE's in 1992 (1 FTE equals to 40 hours per week). Crystals produced at SIC will be delivered to Caltech. The transmittance, the yield and the radiation resistance will be test at Caltech.

Light Sensitive Readout — R&D at KEK, BNL and Caltech

In this project, a high speed, high dynamic range, solar blind phototube, with maximum sensitivity in the 200 nm range will be developed by Hamamatsu, and will be tested at KEK and Caltech. The analog readout system is under design at BNL. With help of the Instrument division of BNL, fast preamplifier and shaper will be developed and tested. The final design will be produced in a quantity of 100 which will be installed in a test matrix of 49 crystals at Caltech.

Multistage microcomputer-based readout systems capable of the necessary speed will also be developed for the BaF_2 array calibration. The digital, higher levels of the readout will be adapted from the VME and Motorola 68020-based, stand-alone systems developed and now used at Caltech for the RFQ calibration project [21].

Precision Calibration — R&D at AccSys and Caltech

Over the last 5 years, the Caltech group has developed and tested a novel calibration technique [19] based upon the radiative capture of a pulsed proton beam from an RFQ accelerator in a lithium target, $^7Li(p,\gamma)^8Be$. The resultant flux of 17.6 MeV photons can be used to calibrate the thousands of BGO crystals in the L3 electromagnetic calorimeter at once, with an absolute accuracy of better than 1% in 1–2 hours. When installed in the L3 experiment at LEP this system will help maintain the high resolution of the electromagnetic calorimeter during running. An experimental test of a 4 × 5 BGO crystal array at the RFQ facility at AccSys Technology, Inc. in Pleasanton, California, was carried out in November, 1987 [19]. An absolute calibration precision of 0.8% was achieved in the test.

Following the 1987 test, we realized the possibility of a new technique, appropriate for the calibration of calorimeters at the SSC or other accelerators in the TeV range. By using the radiative capture of protons from a pulsed beam in a fluoride target, $^{19}F(p,\alpha)^{16}O^*$ [22], and the subsequent decay of the excited oxygen nucleus, $^{16}O^*$, hundreds to thousands of 6 MeV photons could be produced per millisteradian per pulse. These *"equivalent high energy photons"* would serve as a calibration source for electromagnetic calorimeters.

A milestone beam test was carried out in September, 1988, at AccSys under a Phase I grant from the DoE Small Business Innovative Research (SBIR) program, where 4 BaF_2 counters

Table 4. Measured and Calculated Equivalent Photon Energy

Proton Energy (MeV)	2.0	2.5	3.0	3.5	3.85	4.0
EPE_{meas} (GeV)	1.5	13	22	30	37	42
EPE_{cal} (GeV)	2.6	14	24	33	38	42

were set up together with a 7×7 L3 BGO crystal array as a reference counter [19]. The test demonstrated that this technique functions as a clean *"pulse generator"* of scintillation light with energy up to few GeV, originating several cm inside the crystals. The clean, narrow Gaussian-distributed light pulses produced with this technique were shown to provide relative calibrations with a precision of 0.4% or better in a few minutes.

There are much stronger fluorine resonances between 2.0 and 4.0 MeV [22] beyond the 1.92 MeV beam energy at AccSys. By using a 3.85 MeV RFQ and a CaF_2 target, which would have no neutron production as a by-product below 4.05 MeV [22], an equivalent photon energy (EPE) per calorimeter element of 30 GeV/(0.1 μCoulomb) or more is expected [19]. Table 4 lists equivalent photon energies, defined as the sum of photon energies in 1.6 msrad from 0.1 μCoulomb beam charge, measured with a CaF_2 target bombarded with a proton beam from a Van de Graaff at Kellogg Lab at Caltech. The expected equivalent photon energies calculated with an integration of the resonances are also listed in the table. It is clear that with a 3.85 MeV proton beam up to 40 GeV/1.6 msrad/0.1 μCoulomb is achievable.

During the AccSys tests we typically used a beam pulse of 0.10–0.15 μCoulomb with pulse length of a few μsec. In order to adapt this technique for future hadron colliders, such as the SSC or LHC, however, the pulse length of several μsec may not match the experimental readout electronics. Recent computer simulation studies at Saclay have shown that it should be possible to compress 4 MeV beam pulses of 0.1 μCoulomb into 100 nsec time or less, by using multiturn injection into — and single turn extraction from — a small storage ring [23]. This development will allow this technique to be used in calibrating calorimeters at the SSC and LHC. A new pulse time compressor at the output of their RFQ system will be developed by AccSys under a Phase II grand from the DoE SBIR program [23].

Caltech group will develop a precise measurement of total beam charge on the target for each pulse, to provide a direct normalization which can be cross-checked against the normalization to a standard. The design of this target has already been partially worked out. It will employ a tantalum collimator, a precise inductive current pickup and an integrator.

After the BaF_2 array is constructed, we will carry out a series of test at the AccSys RFQ test facility, with a typical beam intensity of 0.1 μCoulomb on a CaF_2 target to simulate 30 GeV pulses. These tests will span a long time period (up to several weeks), and will aim at proving the long term systematic stability as well.

ACKNOWLEDGEMENTS

I am grateful to Drs. F. Takasaki and D. Scigocki and to Merck, Inc. who provided BaF_2 crystal samples for this test. Mr. M. Chen and D. Kirby at Caltech carried out the Physics Monte Carlo simulations described in this report. Dr. Z.Y. Wei tested all BaF_2 samples at Caltech. Dr. H. Ma carried out the analog readout design for the BaF_2 readout at BNL. Mr. M. LeBeau, M. Rennich and R. Barber provided many valuable informations about the mechanical structure design for the BaF_2 calorimeter. Many useful discussions with Drs. A. Gurtu, M. Chen, J. Gunion, R. Hamm, H. Newman, T. Sjostrand, S. Ting, S .Suzuki, C. Woody, Z.W. Ying and B. Zhou are greatly appreciated. The assistance of S. Sondergaard, J. Hanson, and many members of the Caltech technical staff is also acknowledged.

References

[1] M.D.G. Gilchriese, in *"Proceedings of the 1984 Summer Study on the Design and Utilization of the Superconducting Super Collider"*, edited by R. Donaldson and J.G. Morfin, Snowmass, Co.(1984) 607; H.H. Williams, in *"Proceedings of the 1986 Summer Study on the Physics of the Superconducting Super Collider"*, edited by R. Donaldson and J. Marx, Snowmass, Co.(1986) 327; SSC Central Design Group, *"Report of the Task Force on Detector Research Development for the Superconducting Super Collider"*, SSC-SR-1021 (1986).

[2] D. E. Groom, in *"Proceedings of the 1988 Summer Study on the High Energy Physics in 1990's"*, Snowmass, Co. (1988).

[3] L3 Collaboration, *"L3 Technical Proposal"*, May 1983; and J. Bakken et al., Nucl. Instr. and Meth. **A254** (1987) 535 and **A228** (1985) 294.

[4] J. Gunion et al., *"The Higgs Hunter's Guide"*, UCD Preprint, 89-4 (1989).

[5] CERN Green Book, *"ECFA Workshop on LEP 200"*, Aachen, September, 1986.

[6] H. Benson and T. Sjostrand, *"A Manual to the Lund Monte Carlo for Hadronic Processes"*, PYTHIA version 5.3, November, 1989.

[7] F. Paige and S.D. Protopopescu, *"ISAJET 6.27 A Monte Carlo Event Generator for P-P and $\bar{P}$-P Reactions"*, April, 1990.

[8] The cross-section of $H^0 \rightarrow \gamma\gamma$ is scaled up by a factor of two according to J. Stirling, a talk given in a recent *"ECFA-LHC meeting"* at CERN. An argument was given that one should use a running b quark mass (in the spirit of perturbative QCD) in evaluating the $H^0 \rightarrow b\bar{b}$ coupling. Since the b quark mass is smaller than the scale of the Higgs mass, the width for this channel is reduced by almost a factor of two. Also other quark channels are reduced by the same argument. Since the effective width into $\gamma\gamma$ is unchanged, the actual branching ratio thus is increased. According to T. Sjostrand, this increase will be implemented into the PYTHIA program in near future. See also [4].

[9] D. Dicus and S. Willenbrock, Phys. Rev. **D37** (1988) 1801.

[10] M. Lavel et al., Nucl. Instr. and Meth. **A206** (1983) 169; and P. Schotanus et al., Nucl. Instr. and Meth. **A259** (1987) 586; and IEEE–NS **34** (1987) 76.

[11] S. Majewski and D. Anderson et al., Nucl. Instr. and Meth. **A241** (1985) 76; A. J. Caffrey et al., IEEE Trans. Nucl. Sci. **NS-33** (1986) 230; and S. Majewski et al., Nucl. Instr. and Meth. **A260** (1987) 373.

[12] D.F. Anderson et al., Nucl. Instr. and Meth. **A228** (1984) 33.

[13] P. Schotanus et al., IEEE–NS **34** (1987) 272; and Nucl. Instr. and Meth. **A281** (1989) 162.

[14] C.L. Woody et al., IEEE–NS **36** (1989) 536.

[15] P. Schotanus et al., Nucl. Instr. and Meth. **A238** (1985) 564; and Kobayashi et al., Nucl. Instr. and Meth. **A270** (1988) 106.

[16] S. Suzuki, private communication.

[17] P. Schotanus et al., Nucl. Instr. and Meth. **A272** (1987) 917 and Technical Univ. at Delft Preprint **88-1**.

[18] E. Lorenz et al., Nucl. Instr. and Meth. **A249** (1986) 235; and J. Giehl et al., Nucl. Instr. and Meth. **A263** (1988) 392.

[19] R.Y. Zhu et al., *"Supper Collider I"*, Plenum Press, edited by McAshan, (1989) 587; H. Ma et al., Nucl. Instr. and Meth. **A274** (1989) 113; and Nucl. Instr. and Meth. **A281** (1989) 469.

[20] H. Bergstroem *et al.*, Penn State Preprint PSU/TH/63, March, 1990.

[21] R.Y. Zhu, "A portable BGO readout system for RFQ test", talk given in L3 collaboration meeting at Geneva, May 1987.

[22] F. Ajzenberg-Selove, *Nucl. Phys.* **A475** (1987) 1; and H. B. Willard *et al.*, *Phys. Rev.* **85** (1952) 849.

[23] R. Hamm, *"SSC Electromagnetic Calorimeter Calibration Source"*, Phase II proposal to the DoE SBIR program, December, 1988.

FAST LIQUID SCINTILLATORS BASED ON ORGANIC

DYE-POLYMER CONJUGATES

Guilford Jones II* and Elise G. Megehee

Chemistry Department
Boston University
Boston, MA 02215

ABSTRACT

A new family of organic photopolymers is proposed for use in liquid scintillators to be deployed in SSC radiation detectors. The polymers are based on simple derivatives of polymethacrylic acid (PMAA) which are conjugated to a primary chromophore such as p-terphenyl (PTP). Electronic excitation energy resulting from particle bombardment of solvent molecules is harvested by the photopolymers through energy transfer. Excitation is subsequently entrained along polymer backbones via PTP groups and deposited in traps which consist of a second fluorophore which is also polymer bound as a minority species. The energy transfer concept is also applied to fluorophores used in combination (co-bound) with hydrophobic domains of the polyelectrolyte PMAA in water. Primary and secondary fluorophores can be freely substituted in order to select wavelengths of light appropriate for visible detectors (500-750 nm). The photopolymers are to take advantage of the following unique features: (1) the selection of organic polymers which adopt highly ordered, helical conformations for efficient longitudinal energy transfer to trap sites, (2) large frequency shift capabilities associated with energy transfer from PTP moieties to co-bound dye, utilizing the upper (second electronic state) of the dye, (3) the photochemical stability (radiation hardness) associated with the very fast removal of excitation energy through non-radiative decay of upper electronic states.

INTRODUCTION

The development of particle tracking detectors for the SSC provides an array of challenges for the materials research community. Materials must be resistant to ionizing radiation yet responsive to this same radiation on the nanosecond timescale.[1] Fabrication requirements are also high, since large arrays of elements with dimensions in the tens of meters with associated readout electronics will be used. This work focuses a portion of the development effort regarding generic detector materials on a new class of organic polymers which are chemically modified (conjugated) with dyes for the efficient detection of dye emission in the red. This project focuses on the study of polymers incorporated as solutes in scintillation liquids, although the new materials also could be used in other forms (e.g., composite plastics, gels).

An important model for the design of a scintillation system of exceptional quantum efficiency and radiation resistance is found in the mechanism of natural photosynthesis. The chloroplasts of green plants are able to harvest photons and localize excitation energy in "reaction centers" with extraordinary speed and efficiency.[2] Electronic excitation migrates or is entrained over dozens of chlorophyll molecules and accessory pigments before photochemical reaction at "special pair" sites, all within about 5 psec following photoexcitation.[3] This system operates with virtually no quantum losses due to photodegradation, in part as a result of the assembly of reactants within a polymer matrix of considerable complexity (thylakoid membrane protein). This matrix insures the close proximity, arrangement and ordering of groups; i.e., all events are essentually intramolecular and highly efficient.

We propose to design a series of new scintillators based on simple organic polymer structures which will mimic in a rudimentary way the efficient energy harvesting capabilities of the natural system. Polymer chains will be modified with primary chromophores (majority groups) which are photoactive in the uv and also lightly "doped" secondary chromophores which are photoactive in the visible. The latter minority groups will be selectable (in fact, replaceable) since they will be electrostatically rather than covalently bound to polymer chains. The secondary chromophores may be selected from the class of laser dyes which emit with high efficiency in the red.

Several additional design features will be deployed in the new scintillators. (1) Primary and secondary chromophores will be chosen so that individual steps of excitation energy for the harvesting polymers will be exceptionally efficient. The strategy will match secondary dye fluorophores having significant absorptions in the uv, involving excitation to upper electronic levels (normally the second singlet excited state, S_2) with a primary fluorophore which is a strong uv emitter (large radiative decay constant) at a matched uv wavelength. (2) So that energy is entrained along polymer chains with exceptional efficiency, polymers will be investigated which display vectorial properties which favor "one-dimensional" energy migration. This scaffolding feature of conformation for selected polymers (indeed, a common property of protein subunits) insures that energy is not deposited at non-emissive trap sites which are found in the coils of more conventional polymers (e.g., formation of excimers).[4] Instead, electronic excitation is required to "flow" along an axis until a minority dye residue is excited. (3) A third consideration involves the host medium. It has been known for some time that the photostability of organic scintillators (laser dyes) is much reduced by plastic of crystalline hosts vs dissolution in a flowing liquid.[5] Therefore, advantage is to be taken of the <u>rapid, efficient removal of excess energy that is characteristic of organic liquids</u> (especially low dielectrics). A circulating liquid provides opportunities for restoration or filtering of partially degraded media. Also possible is the design of subtly different (exchangable) detection media, in principle for the discrimination of particle type or energy, not requiring the physical reconfiguration of apparatus.

The proposed energy deposition mechanism is shown in Scheme 1. Following particle beam excitation of a solvent fluorophore, energy is transferred to primary chromophores which are themselves polymer co-bound with laser dye. Energy is transferred in a series of "hops" among primary fluorophores which constitute the "antenna" of the polymer system before it is trapped irreversibly at minority sites at which laser dye is bound. The upper dye electronic excited state releases much of its excess energy as heat to the medium with such speed (relaxation in psec time)[6] that photochemical degradation is not competitive. The lower S_1 state of the dye fluoresces brightly (nsec time scale) in a routine fashion. The system takes advantage of the very high rate of energy transfer via the Foerster dipole mechanism which occurs when fluorophores are bound together.[7] Emission of primary fluorophores and reabsorption by secondary dye fluorophores are not required. Also, without co-binding on a polymer chain, the long wavelength dye species would have to beused in a highly concentrated form to insure energy collection, and a host of problems associated with dye aggregation and inner filter effects would interfere.[8]

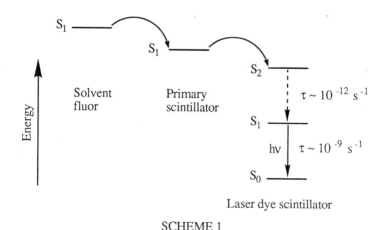

Laser dye scintillator

SCHEME 1

BACKGROUND

The deployment of organic molecules as scintillators in liquids or in plastics has been known for several decades.[9] The advantages that organic materials offer include fast response times (<200 psec for initiating events and nsec time scale fluorescence) and the potential for economical use of large detection volumes (up to kiloliter scale). Although much is known about the mechanisms of scintillation for organic molecules, very significant improvement if likely to be achieved by further survey of newer classes of fluorescent compounds. For example, the class of organic dyes, usually referred to as "laser dyes",[10] has not been investigated significantly in terms of scintillation applications. These structures, which include well known classes of dyes such as coumarins, xanthenes (rhodamines) and cyanines are very efficient emitters in normal photoluminescence, and are commonly used in the amplifying medium of dye lasers.[11] Found among the types of laser dyes are the polyphenyls and oxazoles (POP and derivatives) which have found wide use in scintillation counting. Particularly attractive is the future development of organic dyes as emitters or wavelength shifters in scintillation applications for which detection at longer wavelengths in the visible is desireable. The generation of silican photodetectors which operate at >600 nm will require new materials which are responsive and efficient fluorophores for the visible.

Energy migration via "hopping" or exciton transport among like chromophores in solutions of films of conventional organic polymers has been studied for some years.[12] Recent advances include the demonstration of "antenna effects" in which itinerant energy

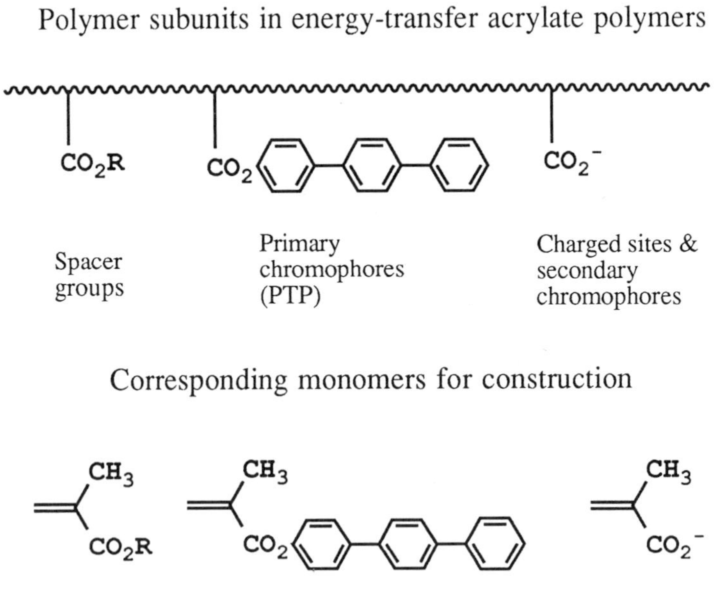

Polymer subunits in energy-transfer acrylate polymers

| Spacer groups | Primary chromophores (PTP) | Charged sites & secondary chromophores |

Corresponding monomers for construction

SCHEME 2

flows from primary chromophores to remote minority sites in vinyl polymers bearing aromatic side chains.[13] These studies have not yet included investigation of laser dye energy acceptors which offer unusually large frequency shift capabilities. Nor have these studies deployed vectorial polymers which appear to offer an advantage in reducing the probability of energy trapping at non-emissive polymer sites.

PROPOSED SYSTEM: POLYACRYLATE CONJUGATES OF P-TERPHENYL AND LASER DYES

The proposed system consisting of polymer and associated (conjugated) chromophores is shown in Scheme 2. p-Terphenyl (PTP) is the primary fluorophore selected for study on the basis of its high fluorescence quantum efficiency (0.98) and high rate of radiative decay (fluorescence lifetime, $\tau_f = 1.0$ nsec).[14] PTP thus displays an exceptionally high oscillator strength for a chromophore active in the uv and has been used for wavelength shifting in scintillation films of conventional organic polymers has been reported.[15] In addition, is has been shown that excitation of PTP to its fluorescent state by ionizing radiation is highly efficient on fast electron bombardment of the dilute hydrocarbon solutions of PTP.[16]

The model of polymer-assisted energy migration will employ PTP in energy transfer to laser dyes which fluoresce in the 450-750 nm range. Candidate energy acceptor dyes (Scheme 3, counter-ions not shown) are well suited, not only due to their efficient absorption at long wavelengths but also due to their significant absorption in the uv; that is, excitation of

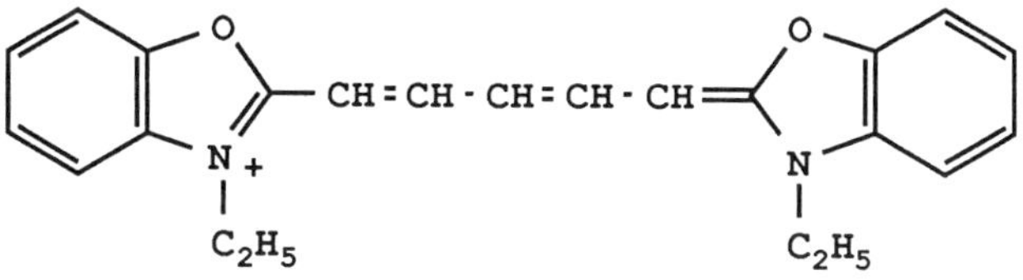

1

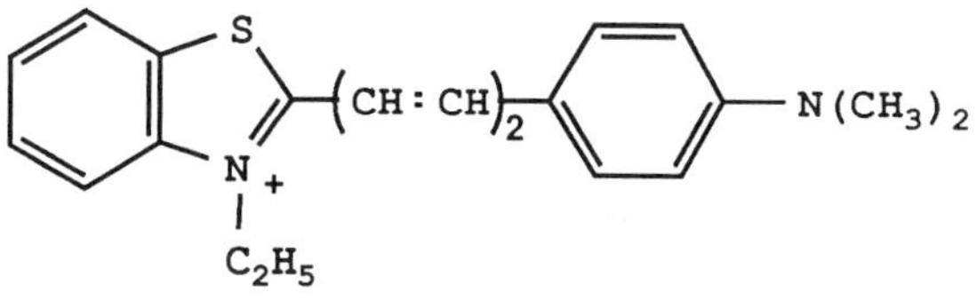

2

SCHEME 3

dye fluorescence is a good spectral match for the emission of donor PTP. The transition for
DODC(1) and a similar band for Styryl 7 (2) are associated with excitation to an upper state
of the chromophore (S$_2$, Scheme 1). Energy which reaches these laser dye structures from
PTP groups will be rapidly degraded, populating the lower fluorescent dye level (S$_1$) for
efficient fluorescence in the red. The polymer main chain is based on the familiar esters of
polymethacrylic acid. However, an important innovation is the structural integrity and
vectorial properties which have been discovered in a small subset of the acrylate polymers.
Okamoto, et al.,[17] have shown that substitution of the R in -COOR monomer acrylate
groups (Scheme 2) with hindered (bulky) substituents results in enforced regularity in the
preferred arrangement of groups as monomer units are successively added in the process of
polymerization. This regularity results in the development of helical structures (Scheme 4).
The specific orientations shown can be independently achieved through the use of chiral
(optically active) polymerization catalysts.

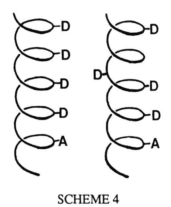

SCHEME 4

A MODEL SYSTEM: DEMONSTRATION OF ENERGY TRANSFER BETWEEN
FLUOROPHORES COBOUND WITHIN THE "HYPERCOILS" OF
POLYMETHACRYLIC ACID

Polymethacrylic acid (PMAA) in water provides a good model system for studying
energy in a polymer domain. PMAA is a polyelectrolyte with ionizable side chain
substituents linked to a hydrocarbon backbone which at pH<4 provides hydrophobic
microdomains in aqueous solution. PMAA is known to exist in three distinct conformational
forms at different pH regimes depending on the degree of ionization[18] (Scheme 5). At pH
lower than 4, the unionized PMAA attains a globular form with the methyl groups projecting
toward the interior of the globule due to hydrophobic interaction. The carboxylic acid groups

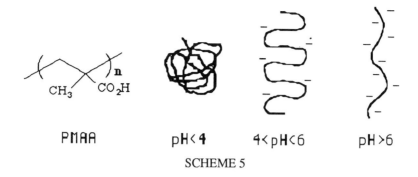

PMAA   pH<4   4<pH<6   pH>6

SCHEME 5

project outward and interact with the water phase. The interior of the PMAA globule has been found to be rather non-polar as measured by a pyrene emission probe.[19] Polycyclic aromatic hydrocarbons such as phenanthrene, anthracene and pyrene derivatives have been solublized by unionized PMAA (MW~10^5) aqueous solutions.[20] At higher pH(>7), the carboxylic acid groups are almost completely ionized, and the PMAA chain is stretched out due to repulsions of the anionic groups. The hydrocarbon solubilizing power of the polymer is abruptly lost under such conditions. In the intermediate pH range of 4 to 6 the partially ionized and partially open polymer attains a coiled conformation.[21]

Significant enhancement of solubility of coumarin and rhodamine laser dyes in water has been observed in the presence of high concentrations of PMAA (P/D > 500) at low pH (<5).[22] For some dyes a significant blue-shift of the emission maximum and enhancement of fluorescence quantum yield were observed. Highly polarized fluorescence (P~0.3) of the PMAA solubilized dyes indicates the prescence of dye molecules in the hydrophobic microdomain of the PMAA hypercoil.

In the present study PTP and cationic laser dyes have been incorporated into the hydrophobic microdomains of low molecular weight poly(methacrylic acid) in water. Shifts in the absorption and emission maximum as well as highly polarized fluorescence relative to water alone will be used to determine if the dyes are in the PMAA hypercoil. The laser dyes used were chosen so that their S_2 absorption overlapped the PTP emission. Fluorescence depolarization will be used to identify energy transfer between PTP and the laser dyes.

EXPERIMENTAL METHODS

<u>Materials</u> The structures of the dyes employed in the present study are shown in part in Scheme 3. PTP was obtained from Aldrich. DODCI (<u>1</u>), Styryl 7 (<u>2</u>), Rhodamine 6G Perchlorate (R6G) and Coumarin 102 (C102) were laser grade dyes obtained from Eastman Kodak. All dyes were used as received after checking purity by thin layer chromatography.

Poly(methacrylic acid) used in the present study was prepared as previously described[22] (Molecular weight ~15,000 and 400,000).

Dyes were employed at a concentration of 10^{-5} M (DODC was 10^{-6} M) for the measurement of absorption and emission spectra. Aqueous PMAA solutions were prepared by diluting 10^{-3} (10^{-4}) M stock dye solution in ethanol to make 10^{-5} (10^{-6}) M solution. Solutions for energy transfer measurement were 1×10^{-5} M in PTP, 1×10^{-5} M in laser dye and 0.01 M in PMAA (monomer concentration). Comparison solutions of dye in aqueous PMAA and in water alone were prepared. The PMAA and dye concentrations were kept constant. Since PTP is not soluable in water, a mixed 1:1 ethanol: water reference solvent was used. The pH of the solutions were measured using a pH meter and adjusted to pH 3.6 to 3.8 using dilute HCl (0.1M).

Absorption spectra were recorded on a Perkin-Elmer 552 spectrophotometer. An SLM Instruments Model 48000S multi-frequency phase fluorometer is employed for measurement of fluorescence emission and excitation spectra. Strong evidence for energy migration among chromophores can be obtained through the observation of fluorescence depolarization which is also measured with the 48000S instrument.[12] Fluorescence polarization values (P) were measured by placing polarizers in the paths of the excitation and emission beams and measuring the emission intensities with different combinations of the plane of excitation and emission polarization. For energy transfer studies polarization was measured at the emission maximum of the laser dye (acceptor) and the excitation maximum of PTP (donor).

RESULTS AND DISCUSSION

When the dyes and p-terphenyl were dissolved in aqueous PMAA solutions at pH<4 and P/D>500, all exhibited shifts in the absorption and emission maxima and in most cases increased polarization(P) of emission relative to water alone (see Table 1). This behavior is indicative of the dyes residing inside the PMAA hypercoil.[22] C102 exhibits a strong solvatochromic blue shift of the fluorescence maxima on going from water to aqueous PMAA. Such solvatochromic shifts of emission maxima for coumarin dyes is well documented and gives a measure of the polarity of the solvent.[23] The blue shift reflects migration of C102 from a polar aqueous environment to a non-polar hydrophobic domain of the polymer.

By contrast R6G, PTP, DODC and Styryl 7 all show red shifts in their absorption and fluorescence maxima in aqueous PMAA at high P/D and low pH. Styryl 7 shows a strong solvatochromic red shift of the absorption maximum while R6G, PTP and DODC show small shifts. Further Styryl 7 shows ~100 fold enhancement in it fluorescence intensity on going from water to aqueous PMAA.

Table 1. Absorption and Emission Maxima for p-Terphenyl
and Dyes in Water and Aqueous PMAA, 0.01M
(monomer concentration) and pH 3.8.

System	λ_a(nm)	λ_f(nm)
PTP/EtOH:H$_2$O (1:1)	280	339
PTP/PMAA	278	342
C102/H$_2$0	396	495
C102/PMAA	399	458
DODC/H$_2$0	576	600
DODC/PMAA	586	607
R6G/H$_2$0	527	558
R6G/PMAA	537	564
Styryl 7/H$_2$0	522	701
Styryl 7/PMAA	563	702

PTP, R6G, C102 and DODC all show large increases in the polarization of their fluorescence in the PMAA hypercoil environment. Upon excitation into their S2 band there appears to be a depolarization of the fluoresence relative to excitation into the S1 band. Styryl 7 shows a large polarization in water itself with no increase upon entering the PMAA hypercoil suggesting that it is already a very rigid molecule. For R6G (5μM) and PTP (10μM) this polarization of emission is enhanced by going to PMAA with molecular weight of 400,000 (0.01M monomer concemtration). Figure 1 shows the excitation spectra of R6G in PMAA (MW~400,000). Upon addition of PTP there in an enhancement of the 280 nm excitation of the R6G emission which is interpreted as resulting from PTP absorbing 280 nm light and undergoing energy transfer to R6G. R6G in PMAA (MW~400,000) displays polarization of P = -0.151 (EX=280 nm, EM=560 nm), while addition of PTP causes this polarization to fall perceptively (P = -0.045 EX=280 nm, EM=560 nm). This loss of polarization is interpreted in terms of the proposed energy transfer mechanism.

ACKNOWLEDGEMENT

This work is supported through the SSC Generic R and D Detector Program under Department of Energy contract DE-AC02-89 ER40509.

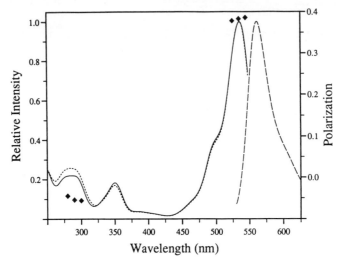

Figure 1. Excitation (solid line) and emission spectra (long dashed line) of R6G (5×10^{-6} M) in PMAA (MW~400,000, 0.01 M monomer concentration) and excitation spectrum in the presence of 1×10^{-5} M.PTP (short dashed line). Polarization data (right-hand y-axis) for the PTP/R6G/PMAA case are shown as diamond symbols.

REFERENCES

1 M. G. D. Gilchriese, Ed., "Radiation Effects at the SSC," SSC Central Design Group, Berkeley, 1988.

2 S. M. Danks, E. H. Evans, and P. A. Whittaker, "Photosynthetic Systems," John Wiley and Sons, New York, 1983.

3 S. G. Boxer, Biochim. Biophys. Acta, 726:265 (1983).

4 M. Sisido, Macromol. Chem. Suppl., 14:131 (1985).

5 a) I. P. Kaminow, L. W. Stulz, E. A. Chandross, and C. A. Pryde, Appl. Optics, 11:1563 (1972).

  b) G. Neumann and S. Reich, Appl. Phys. Lett. 25:119 (1974).

6 N. J. Turro, "Modern Molecular Photochemistry," The Benjamin Cummings Publ. Co., Inc., Menlo Park, 1978, chp 1.

7 L. Stryer, Ann. Rev. Biochem., 47:819 (1978).

8 K. H. Drexhage, in "Dye Lasers," F. P. Schaefer, Ed., Springer-Verlag, New York, 1977.

9 F. D. Brooks in "Detectors in Nuclear Science," D. A. Bromley, Ed. (Nucl. Instrum. Methods, 162:447 (1979)).

10 M. Maeda, "Laser Dyes," Academic Press, New York, 1984.

11 G. Jones, II, "Photochemistry of Laser Dyes," in "Dye Lasers and Applications," F. Duarte, Ed., Academic Press, 1990, chp 7.

12 J. Guillet, "Polymer Photophysics and Photochemistry," Cambridge University Press, Cambridge, 1985, chp 9.

13 X. Ren and J. E. Guillet, Macromolucles, 18:2012 (1985).

14 J. B. Birks, "Photophysics of Aromatic Molecules," Wiley-Interscience, 1970, New York, chp 4.

15 W. Viehmann and R. L. Frost, Nucl. Instrum. Methods, 167:405 (1979).

16 H. T. Choi, F. Hirayama and S. Lipsky, J. Phys. Chem., 88:4246 (1984).

17 Y. Okamoto, K. Suzuki, K. Ohta, K. Hatada, and H. Yuki, J. Am. Chem. Soc. 101:4763 (1979).

18 A. Katchalsky and H. Eisenberg, J. Polym. Sci.6:145 (1951).

19 T. Chen and J. K. Thomas, J. Polym. Sci. Part A-1, 17:1103 (1979).

20 G. Barone, V. Crescenzi, A. M. Liquori and F. Qudrifoglio, J. Phys. Chem. 71:2341 (1967).

21 A. F. Olea and J. K. Thomas, Macromolecules, 22:1165 (1989).

22 G. Jones, II and M. A. Rahman, Proc. Int. Conf. Lasers, 1989, No. HN 9.

23 G. Jones II, W. R. Jackson, S. Kanoktanoporn, and A. M. Halpern, Optics Comm. 33:315 (1980).

A DETECTOR FOR BOTTOM PHYSICS AT THE SSC

P. Skubic

for the BCD Collaboration
Physics and Astronomy Department
University of Oklahoma, Norman, OK, 73019

ABSTRACT

A detector concept optimized to study B-physics at the SSC
will be described. An overview of the detector design,
including Monte Carlo simulations, and designs for individual
detector sub-systems will be presented. The status and future
plans of detector R&D programs associated with the sub-systems
will be described including: a solid state microvertex detector
using pixel and double-sided silicon microstrip detectors; a
straw tube tracking system with fast TDC; VLSI readout
electronics and data acquisition system. Areas where there is
close collaboration with industry will be emphasized.

INTRODUCTION

The SSC is expected to produce 1.8×10^{11} bottom particles
in 1000 hours of running at a luminosity of 10^{32} cm^{-2}t^{-1}. We
describe here a design for a detector optimized to take
advantage of this large rate which will be capable of
collecting data at the above luminosity and which can
reconstruct a large fraction of the bottom particle decays.
This large data sample of B decays can be used to study CP
violation as has been described in detail elsewhere.[1] An
overview and description of the central tracking subsystems
proposed for such a detector by the Bottom Collider Detector
(BCD) collaboration in an Expression of Interest (EOI)
submitted to the SSC Laboratory[2] will be given below. The
vertex detector, which is the most important single subsystem,
is emphasized.

OVERVIEW OF BCD DETECTOR

Fig. 1(a) shows the entire BCD detector system in an
isometric view. Fig. 1(b) shows plan and section views with

each subsystem labeled. The detector consists of a central
spectrometer with a large dipole magnet. The silicon vertex
detector occupies the region in radius between 1.3 cm and ~10
cm. A straw tube tracking system surrounds the vertex detector
and occupies a volume of 180 x 180 cm^2 in cross section and 6 m
along the beams. A particle ID system consisting of TOF
counters and TRD's will be used in triggering, followed by a
RICH detector. An electromagnetic calorimeter and muon
detectors enclose the central detector. Fig. 1 also shows a
forward arm spectrometer on one side for measuring particle
decays in the region of pseudorapidity between 3.5 and 5.5.

VERTEX DETECTOR

    The silicon vertex detector consists of three barrel layers
of silicon detectors oriented parallel to the beam line,
interleaved with "disk" silicon detectors which are
perpendicular to the beam. The disk detectors are needed
because the B decay angle distribution according to Monte Carlo
simulation is expected to peak in the forward and backward
direction at small angles with respect to the beam line as
shown in Fig. 2. We presently envision the inner layer region
of barrel and disk detectors to consist of pixel devices which
are described in another talk at this conference.[3] The outer
region will then consist of double-sided silicon microstrip
detectors.[4] In the present "straw man" design, this detector
is held together by the active elements which are glued to one
another to form modules and then the whole detector is
supported by a ~10 mil thick beryllium channel as shown in Fig.
3. Fig.4 shows a Monte Carlo simulation of a 40 TeV event in
the vertex detector.

    Tests of this mechanical design are being carried out at
Fermilab by Hans Jostlein and Carl Lindenmeyer[5]. A brass model
of the silicon detector geometry was assembled and the heat
load of the electronics was simulated with resistors. The model
was cooled by blowing air through the assembly along its axis
(the disk detectors are offset along the beam line to form a
helical path for the gas) and temperature and position
measurements were made. The maximum observed temperature rises
were about 10° C for a 76 g/s flow rate and a heat load of 0.1
Watts/cm^2. The total pressure needed to drive the cooling air
was 12 inches of water.

    The proposed method of assembly in this design uses a
modified 3-axis coordinate measuring machine (remotely
controlled by a computer) which handles the silicon planes with
a vacuum holder. The alignment of each plane with respect to
the vacuum holder is checked by moving each plane below a
position measuring microscope under computer control and
reading out multiple targets. The planes are then moved to
their appropriate glueing location on a hexagonal mandrel which
is mounted on a precision turn table. Experiments were done
with a UV curable polymer glue made by Norland Products Co.
Tiny glass prisms are used in the corners to provide a very
thin bead to avoid shrinkage on curing.

    An active R&D program is underway to develop the detectors
and VLSI readout electronics necessary for this vertex
detector. Double-sided silicon microstrip detectors with 25

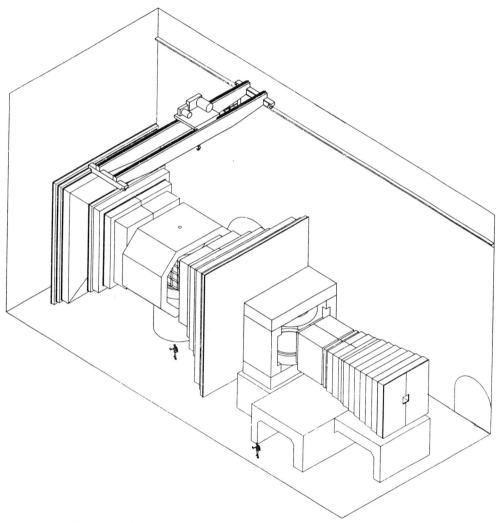

Fig. 1(a). Isometric view of the BCD detector.

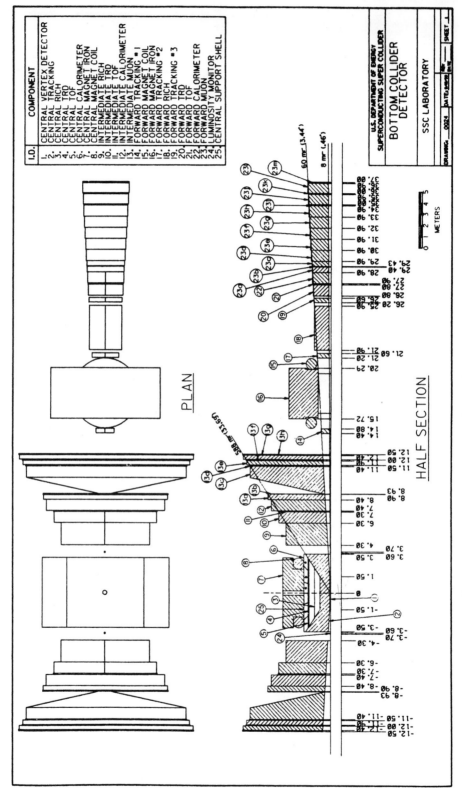

Fig. 1(b). Plan and section views of the BCD detector.

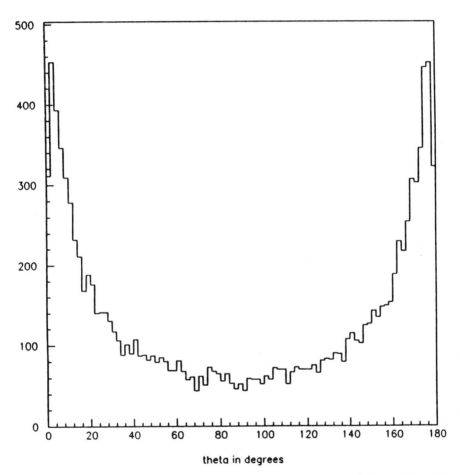

Fig.2.   Distribution in polar angle for particles from B-
         decay.

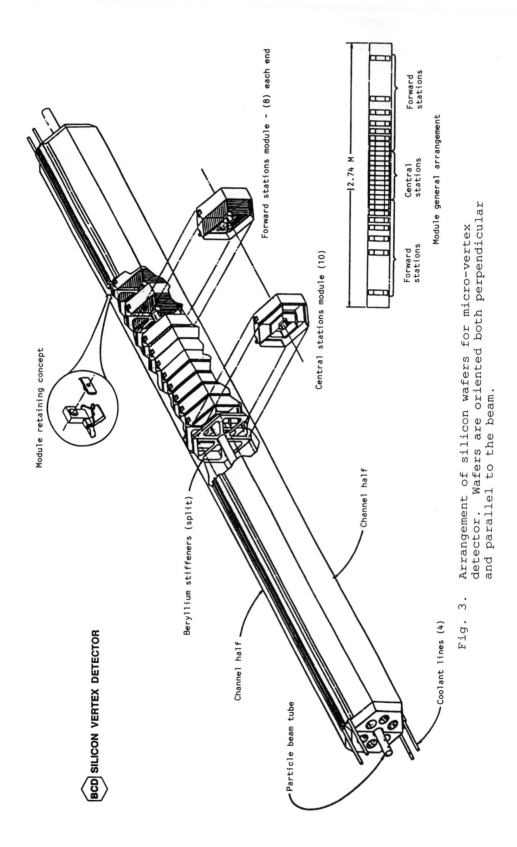

BCD SILICON VERTEX DETECTOR

Module retaining concept

Forward stations module - (8) each end

Central stations module (10)

Channel half

Beryllium stiffeners (split)

Channel half

Particle beam tube

Coolant lines (4)

12.74 M

Forward
stations

Central
stations

Forward
stations

Module general arrangement

Fig. 3. Arrangement of silicon wafers for micro-vertex detector. Wafers are oriented both perpendicular and parallel to the beam.

304

micron pitch are being developed by Micron Semiconductor, Inc.
in collaboration with the University of Oklahoma. Beam tests
with single and double sided detectors using existing readout
chips (SVX-D and CAMEX) will be done in the MTEST beam at
Fermilab during the 1990-91 fixed target run. An important
issue is whether radiation resistant detectors and readout
electronics can be developed to allow operation at small
radius. Radiation tests are in progress in collaboration with
Oak Ridge National Laboratory (ORNL) to investigate the
radiation hardness of existing detectors and readout chips.
Design of a VLSI readout circuit for a Fermilab version of the
experiment is in progress by engineers at Fermilab and ORNL.
To perform the SSC experiment, this design will have to be
upgraded and fabricated with a radiation-hard process. ORNL is
investigating the capabilities of rad-hard processes at Harris
Semiconductor.

The pixel detector development is being done by a
collaboration of the following institutions: UC-Berkeley/Space
Sciences Lab., UC-Davis, Hughes Aircraft Co., Iowa State U.,
Lawrence Berkeley Lab., U. of Oklahoma, U. of Pennsylvania,
Princeton U., SLAC, and Yale U. The work by Hughes Aircraft
on the development of "hybrid" pixel detectors, which consist
of PIN diode arrays, made by Micron Semiconductor, indium-bump
bonded to Hughes developed readout chips, has been described in
previous publications.[6] The major goal of the work at Hughes
during the next year is to develop a hybrid prototype device
which has sparse readout (which selectively reads out only
those pixels containing interesting data along with their
coordinates), and time stamping of the hit pixels with a 50 ns
resolution. A design has been developed by Hughes and LBL for
a 64 by 32 pixel array which has a pixel area of 50 microns by
150 microns containing 23 transistors and six capacitors. Each
pixel is designed to dissipate about 20 micro-watts of power
and have about 200 electrons rms noise.

STRAW TUBE TRACKING DETECTOR

The silicon vertex detector is designed to reconstruct
bottom particle decay vertices. The design is optimized to
reduce the errors on measuring the three dimensional decay path
of the primary particles. This detector is followed by a straw
tube tracking system which will measure the momentum of the
charged decay products. The straw tubes will be about 5 mm in
diameter and up to 2 m long, and will be operated as drift
chambers with a designed spatial resolution of ~60 microns per
tube. Each straw is a two-ply laminate of an inner aluminized
polycarbonate film about 14 microns thick surrounded by a layer
of 12.5 micron thick Mylar. The anode wire is 20 micron
diameter gold-plated tungsten. A likely chamber gas is
CF_4/Isobutane which has a saturated drift velocity of about 100
microns/ns. An R&D program has begun to develop large-scale
straw tube systems for BCD.[7]

The electronics front-end system is based on a design
developed by the SSC Front-End Electronic Subsystem effort.[8] It
uses a bipolar preamplifier, shaping amplifier with 5-10 ns
measuring time, and a discriminator followed by a CMOS time-to-
voltage converter, analog-storage array, and readout ADC.[9]

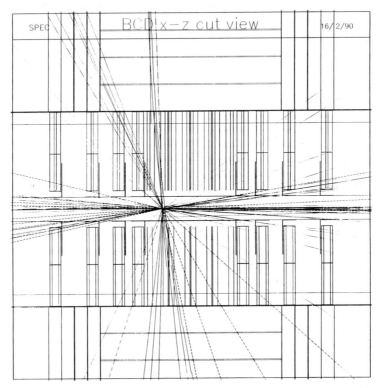

Fig. 4. Monte Carlo simulation of a 40-TeV B event in the
vertex detector. The scale along the beam is
compressed three times relative to the transverse
scale.

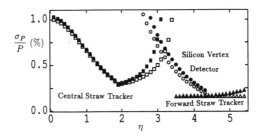

Fig. 5. The momentum resolution versus pseudorapidity of
the tracking system for particles with transverse
momentum (to the beams) of 2.5 GeV/c and momentum
vector perpendicular to the magnetic field. The
contribution of multiple scattering to the
resolution is also shown, and is dominant.

Monte Carlo simulation was done to estimate the momentum resolution of the straw tube tracking system. The straw tracking planes were formed in superlayers of 8 individual planes, all straws in each superlayer having the same spatial orientation (x,y,u, or v). Fig. 5 shows the results from the simulation where the momentum resolution versus pseudorapidity of the tracking system is shown for particles with transverse momentum of 2.5 GeV/c and momentum vector perpendicular to the magnetic field. The spatial resolution of the straw-tube chambers is taken to be 50 microns and that of the silicon vertex detector is 5 microns. Multiple scattering dominates the resolution in all parts of the detector based on these input assumptions.

TRIGGER AND DATA AQUISITION

A 50 kHz rate of bottom production at a luminosity of 10^{32} $cm^{-2}t^{-1}$ is 0.5% of the total interaction rate of 10 MHz. Therefore the trigger must be relatively loose if we are to collect a large number of events for studies of CP violation. A single level of prompt trigger is proposed which provides a modest event rate reduction in 1.5-2 μs. A large scale computer ranch then implements software triggers based on full-resolution data from several sub-systems of the detector.[10] Fig. 6 shows the data acquisition architecture. We estimate that 1000 events/s will be archived, each consisting of 0.5 MBytes. Four triggers have been investigated: single electron, single muon, a high-luminosity J/ψ trigger, and a secondary-vertex trigger.

Hughes Aircraft has proposed use of a recently developed "3-Dimensional" computer[11] to combine primary and secondary-vertex position estimates with other types of detector data in order to generate a level 1 or level 2 trigger. This computer exploits wafer-to-wafer interconnection to achieve high speed at low power. Wafers of 128 x 128 processors give a combined performance of 390 Mflops for 32-bit words. Up to 15 such wafers can be interconnected vertically in a volume less than 330 cm^3 and with a power dissipation less than 100 watts. Fig. 7 shows a schematic representation of this 3-D approach. The silicon detector information can be used to form a secondary vertex trigger which takes advantage of the straight line tracks in the non-bend view. The vertexing algorithm can also take advantage of the small transverse size of the beams. Silicon tracking and fast vertex algorithms are under development.

CONCLUSION

A detector design for intermediate p_t physics at the SSC has been presented. The present status of an R&D program to develop a silicon vertex detector capable of reconstructing bottom particle decays has been described. This program to develop double-sided microstrip detectors, pixel detectors, and associated VLSI readout chips is a collaboration between industry (Hughes Aircraft and Micron Semiconductor), national laboratories, and universities. Similar vigorous R&D efforts associated with the BCD collaboration to develop straw tube tracking detectors, particle ID detectors, and data acquisition electronics are also in progress.

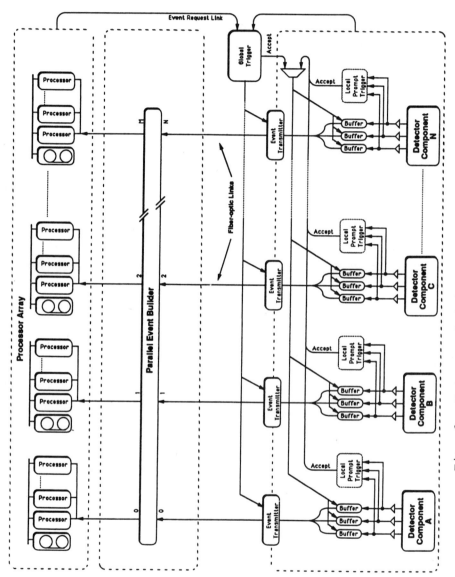

Fig. 6. Data-acquisition-system architecture.

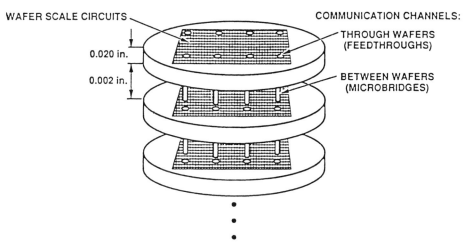

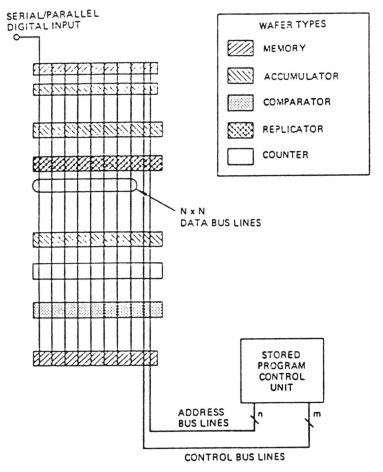

Fig. 7.   Schematic representation of the concept and
          architecture of the 3-D Computer. Only five
          modular wafer types are necessary to efficiently
          execute a wide variety of applications.

309

ACKNOWLEDGEMENTS

The author would like to thank all members of the BCD collaboration for help in the preparation of this report. This work is supported in part by DOE contract number DE-AS05-80ER10629.

REFERENCES

1. P. Krawczyk, D. London, R.D. Peccei, and H. Steger, "Predictions of the CKM Model for CP Asymmetries in B Decay", Nucl Phys. B307, 19 (1988).

2. H. Castro, et al.,"Expression of Interest for A Bottom Collider Detector at the SSC", (May 25, 1990).

3. J.G. Jernigan, et al., "Performance Measurements of Hybrid PIN Diode Arrays", talk (VI-B-5) presented at IISSC (March 1990).

4. G. R. Kalbfleisch, P. L. Skubic, M. A. Lambrecht, C. D. Wilburn, "Charge Correlation Measurements from a Double-Sided Solid State Ministripe Detector", IEEE Trans. Nucl. Sci., NS36, 272 (1988).

5. H. Mulderink, N. Michels, and H. Jostlein, "Mechanical and Thermal behavior of a Prototype Support Structure for a Large Silicon Vertex Detector (BCD)", Fermilab TM-1616 (August 23, 1989); H. Jostlein and J. Miller, "Heat Resistance and Air-Pressure Drop in a Model of the BCD Silicon Vertex Detector", Fermilab TM (Jan. 1990); C. Lindenmeyer, "Proposed Method of Assembly for the BCD Silicon Strip Vertex Detector Modules", Fermilab TM-1627 (Oct. 16, 1989).

6. S. Shapiro, et al., "Silicon PIN Diode Array Hybrids for Charged Particle Detection", Nucl. Instr. Meth. A275, 580 (1989); J. Garrett Jernigan, et al., "Performance Measurements of Hybrid PIN Diode Arrays", SLAC-PUB-5211 (May 1990).

7. C. Lu, et al., "Proposal to the SSC Laboratory for R&D of a Straw-Tube Tracking Subsystem", Princeton U. preprint DOE/ER/3072-56 (Sept.30, 1989).

8. H.H. Williams et al., "SSC Subsystem Proposal for Front-End Electronics", (Sept. 1990).

9. L. Callewaert, et al., "Front End and Signal Processing Electronics for Detectors at High Luminosity", IEEE Trans. Nucl. Sci. NS-36, (1989).

10. E. Barsotti et al., "Digital Triggers & Data Acquisition Using New Microplex & Data Compaction IC's", Proceedings of the Workshop on High Sensitivity Physics at Fermilab (Nov 11-14, 1987) J. Slaughter, N. Lockyer, M. Schmidt, editors, p. 369.

11. M.J. Little et al., "The 3-D computer", Proc. IEEE International Conference on Wafer Scale Integration (San Francisco, January, 1989).

SUPERCONDUCTING AIR CORE TOROIDS FOR PRECISION

MUON MEASUREMENTS AT THE SSC

S.R. Gottesman,[2] P. Bonanos,[9] T. Brown,[2] A. Carroll,[1] I.H. Chiang,[1] A. Favale,[2] J.S. Frank,[1] J. Friedman,[5] E. Hafen,[5] J. Haggerty,[1] P. Haridas,[5] L.W. Jones,[7] H.W. Kendall,[5] K. Lau,[3] L. Littenberg,[1] J.N. Luton,[8] M. Marx,[10] R. McNeil,[4] W. Morse,[1] J. Mueller,[2] L. Osborne,[5] I. Pless,[5] B.G. Pope,[6] J.A. Pusateri,[2] L. Rosenson,[5] P. Spampinato,[2] R.C. Strand,[1] E. Schneid,[2] R. Verdier,[5] and R. Weinstein[3]

1. Brookhaven National Laboratory
2. Grumman Corporation
3. University of Houston
4. Louisiana State University
5. Massachusetts Institute of Technology
6. Michigan State University
7. University of Michigan
8. Oak Ridge National Laboratory
9. Princeton Plasma Physics Laboratory
10. State University of New York at Stony Brook

ABSTRACT

A superconducting air core toroid design is currently being evaluated for use as a high-precision muon spectrometer (EMPACT) at the SSC. The design offers the possibility of measuring transverse momentum P_T with uniform resolution over a wide range of rapidity η. The overall weight of the detector is substantially reduced by the elimination of an iron yoke (for flux return). As currently conceived, all muon trajectories are measured outside the toroid in a nonmagnetic environment, thus eliminating the requirement for access to the magnet interior. Initial studies indicate that super-conducting air core toroids capable of performing high-precision muon momentum measurements can be constructed in this configuration using a conservative design.

INTRODUCTION

Superconducting air core toroids are currently being developed for use as a precision muon spectrometer at the SSC.[1] The development of these toroids offers substantial advantage over conventional magnet designs for detection and characterization of muons by (a) reducing the amount of multiple scattering a muon experiences while traversing the magnet, and (b) making optimal use of the magnetic field geometry which is always perpendicular to the particle trajectory. Other advantages stem from the substantial reduction in weight that is achievable with superconducting air core toroids compared to iron magnets (approximately an order of magnitude) and the fact that particle tracking can now be optimized in a nonmagnetic environment.

The baseline detector concept used for this study is the EMPACT detector[2] shown in Fig. 1, in which all major detector components are identified. A quadrant of the muon spectrometer system is shown in Fig. 2. The muon spectrometer system consists of a single central toroid of inner radius 4m and outer radius 8m, with two end toroids each having an inner radius of 1m (dewar corner) and outer radius of 9m. The maximum magnetic field strength B_{max} is 1.5T in the central toroid and 4.5T in

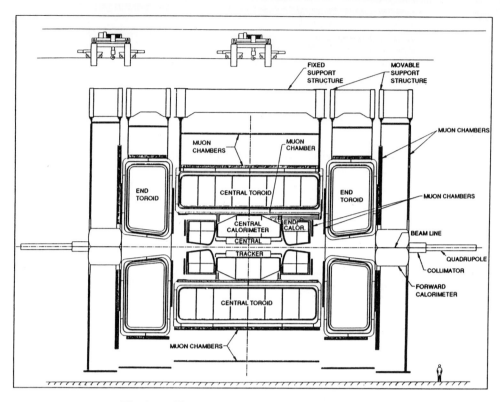

Fig. 1.   Elevation view of the EMPACT detector.

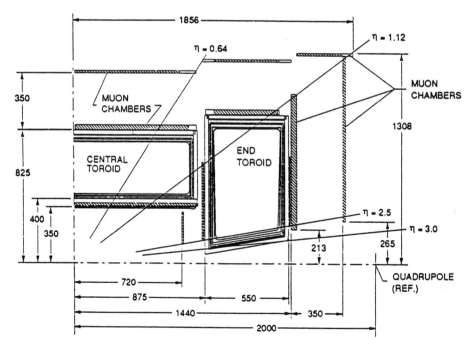

Fig. 2.   One quadrant of the EMPACT muon spectrometer system showing key
dimensions (cm) of toroids and muon detectors.

the end toroids. This spectrometer has a magnetic field integral (B·dl) of 3.6Tm at $\eta=0$ ($\theta=90$ deg.) which increases to a maximum of 18.6Tm at $\eta=2.5$ ($\theta=9.4$ deg.). Additional design specification detail may be found in reference 3.

Although large superconducting air core toroids are new in the high energy physics arena, they have been developed and successfully applied in the plasma fusion energy program. The feasibility of using large superconducting air core toroids for high energy physics applications has been verified in an earlier study[4] commissioned by the SSC Lab which concluded that they are indeed feasible, and furthermore, could be built with a conservative design, despite the fact that they would be the largest superconducting air core toroids ever constructed. However, the toroid configuration studied in reference 4 was not as thin as the one considered here, and essential FEM stress and stability analysis must still be performed. Preliminary results from our current analysis effort are encouraging.

Initially two design configurations were considered. For the first configuration, muon chambers would be positioned within the magnetic field volume and hence would require access to the magnet interior. This design was rejected after considering the consequences of introducing penetrations through the cold support structure. In addition there were concerns regarding the operation of high-precision muon chambers in a strong magnetic field. A second configuration (Figs. 1 & 2) places all muon trajectory measurements outside the magnet (a nonmagnetic environment) and requires no access to the interior, considerably simplifying the toroid design. It is this second configuration which has been selected for further development, and which now forms the baseline concept.

MOMENTUM MEASUREMENT OF MUONS

Muon momentum is determined by measuring the angular deflection a muon experiences while traversing a known magnetic field. The precision with which one determines momentum is thus a function of how accurately one measures the muon's trajectory; any distortion in this measurement will contribute an error to the momentum estimate.

There are essentially two sources of error contributing to the momentum measurement. First, there is the error introduced by multiple coulomb scattering as the muon traverses the structural material of the magnet. This contribution can be parameterized as

$$\left(\frac{\Delta p}{p}\right)_{ms} = \alpha \tag{eq. 1}$$

where Δp is the error in the momentum measurement and p is the magnitude of the muon momentum. This quantity varies as the square root of the amount of material (radiation lengths) traversed and inversely as the strength of the magnetic field integral $\int B \cdot dl$.

Since α is a function of the amount and type of material traversed by a muon in passing through the spectrometer, sufficient care must be exercised in the material selection if multiple scattering is to be kept to a minimum. As an example, 8.9 cm of aluminum equals one radiation length of material, whereas for copper it is only 1.43 cm. Therefore a 1.43 cm block of copper will have the same scattering effect on muons as 8.9 cm of aluminum. Since multiple scattering is to be kept to a minimum, aluminum stabilized NbTi superconductor is preferable to one stabilized with copper. The same type of reasoning has led to the selection of aluminum instead of steel as the preferred material for the toroid structure.

In addition to the multiple scattering contribution to the measurement error, one must include the degradation in momentum resolution caused by the finite spatial precision with which one measures a muon's trajectory. This measurement error may be parameterized to first order as

$$\left(\frac{\Delta p}{p}\right)_{sp} = \beta p \tag{eq. 2}$$

where β is a parameter incorporating the details of the trajectory measurement.

Since the multiple scattering (ms) and spatial (sp) contributions are statistically independent they are to be added in quadrature:

$$\left(\frac{\Delta p}{p}\right)^2 = \left(\frac{\Delta p}{p}\right)_{ms}^2 + \left(\frac{\Delta p}{p}\right)_{sp}^2 \tag{eq. 3}$$

or,

$$\left(\frac{\Delta p}{p}\right)^2 = \alpha^2 + \left(\beta p\right)^2. \qquad \text{(eq. 4)}$$

PERFORMANCE

As a means of establishing a procedure for the quantitative comparison between different toroid designs, we have chosen as a figure-of-merit the effective mass resolution in reconstructing Z^0 particles from Higgs decay for the following decay mode,

$$H \rightarrow Z^0 Z^0 \rightarrow \mu^+ \mu^- \mu^+ \mu^-. \qquad \text{(eq. 5)}$$

The ideal spectrometer would have negligible multiple scattering with excellent track measurement capability (i.e. negligible α and β terms in eq. 4) such that it would be possible to reconstruct Z^0's close to the natural width. In order to gain insight as to what acceptable values for α and β are, we have simulated 5,000 Higgs decays for two different Higgs masses (M_H=400 and 800 GeV/c^2) using the ISAJET event generator. Then for each event the muon momentum vectors were varied by an amount $\Delta \vec{p}$ with probability commensurate for a Gaussian distribution of width $\sigma = (\Delta p/p)$ as given by eq. 4. Using these modified momentum vectors, the two Z^0 masses were reconstructed for each Higgs decay and the dimuon effective mass distribution histogrammed. Repeating this process for different α, β pairs yielded a set of mass distributions, each one of which was subsequently fitted with a Gaussian function.

The width $\Delta M_{\mu\mu}$ of each Gaussian distribution represents the reconstructed mass resolution of the Z^0 particle for different values of α and β. The results of this analysis are presented in Fig. 3 which shows the reconstructed mass resolution in four different rapidity bites as a function of β. Since the primary objective of this study was to understand the instrumental effects of multiple scattering and spatial resolution for reconstructing the Z^0 mass, all Z^0's were generated with zero natural width. The curves of Fig. 3 thus include only instrumental effects. However, a horizontal dashed line representing the Z^0 natural width has been superimposed on the figure for convenience of comparing the instrumental resolution to the Z^0 natural width. From this figure it is to be observed that a toroid characterized by a multiple scattering coefficient of α=4% would be incapable of measuring momentum with sufficient precision to reconstruct the Z^0 close to its natural width; however an α=2% toroid would be capable of accomplishing this goal.

These considerations have led to the baseline configuration described in reference 3, with aluminum being the preferred material. Using this baseline design, rays were projected through the spectrometer system for different angles (θ) with respect to the beamline, and for each ray the multiple scattering term α calculated. The results are plotted in Fig. 4, where it can be seen that the multiple scattering term is $\alpha \approx 2.5\%$ for muons with trajectories perpendicular to the beam axis, and decreases to $\alpha \approx 1.0\%$ for forward trajectories of θ=10 degrees. The spiked feature between 30 and 40 degrees represents the transition region between the central and end toroids where the ray penetrates four walls instead of two.

A momentum resolution function[5] ($\Delta p/p$) has been calculated for this design under the assumption that the spatial measurement error will be smaller than 200 μm for each detector plane, with an effective multilayer resolution of 50 μm for each set of measurements. This function is shown in Fig. 5 for different muon momenta. The curve denoted by boxes ($\square$) corresponds to the average resolution for muons from an 800 GeV/c^2 Higgs decay as in eq. 5.

Finally we show in Fig. 6 the reconstructed Z^0 mass distribution for a Higgs mass of 400 GeV/c^2. In (a) the Z^0 is shown at its natural width, as it would be seen with perfect resolution. The second plot (b) shows what can be expected from a toroid having the baseline momentum resolution of Fig. 5. In (c) the Z^0 is reconstructed from a calorimeter measurement of $Z^0 \rightarrow e^+e^-$ with resolution $\Delta E/E = .15/\sqrt{E} + 0.01$. In (d) are seen the results for an iron toroid with typical $\Delta p/p$=10% momentum resolution.

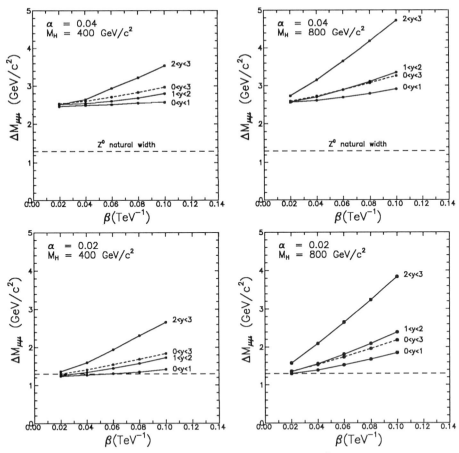

Fig. 3   Effective mass resolution $\Delta M_{\mu\mu}$ of reconstructed Z^0's from Higgs decays. The different curves in each plot are for different bites in rapidity (y).

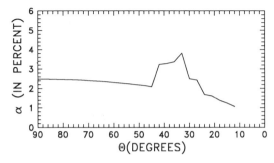

Fig. 4.   Multiple scattering parameter α versus θ, the angle of the muon trajectory with respect to the beam.

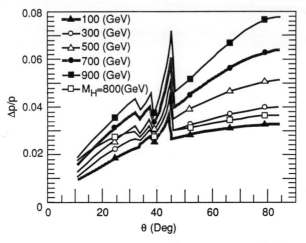

Fig. 5.   Momentum resolution function $\Delta p/p$ as a function of θ, the angle of the muon trajectory with respect to the beam, for different muon momenta. The curve marked ($\square$) is the average momentum resolution for muons resulting from 800 GeV/c^2 Higgs decay.

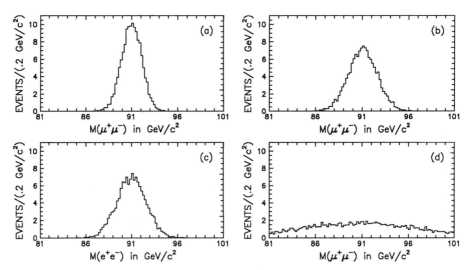

Fig. 6.   Reconstructed Z^0's from 400 GeV/c^2 Higgs decay for (a) perfect momentum resolution, (b) superconducting air core toroid, (c) calorimeter with $\Delta E/E = .15/\sqrt{E} + .01$ resolution, and (d) iron toroid, $\alpha = 10\%$.

CONCLUSIONS

Superconducting air core toroids show great promise for use as high precision muon spectrometers at the SSC. Among their advantages are (a) more accurate momentum measurement resulting from reduced multiple scattering, (b) maximum use of magnetic field due to particle trajectories always being perpendicular to the magnetic field, (c) substantial reduction in weight (up to an order of magnitude) compared to iron magnets, and (d) the opportunity to optimize tracking in a completely nonmagnetic environment.

A baseline conceptual design is currently being developed and assessed for its physical performance. By using aluminum wherever possible in the toroid structure and superconducting coil design, multiple scattering effects have been kept to a minimum, rendering the capability of reconstructing Z^0's from Higgs decays close to their natural width.

REFERENCES

1. T. Fields et al., "Proposal for Development of Superconducting Air Core Toroids as a Precision Muon Spectrometer for SSC Experiments," October, 1989.

2. M. Marx, "EMPACT (Electrons, Muons, Partons with Air Core Toroids) — An Alternative Approach to a High P_T SSC Experiment," SSC-219, May 1989. Also, "EMPACT — Electrons Muons Partons with Air Core Toroids — An Expression of Interest for an Experiment to be Performed at the Superconducting Super Collider," submitted to SSCL May 25, 1990.

3. J.A. Pusateri et al., "The SSC Superconducting Air Core Toroid Design Development," these proceedings.

4. P. Bonanos and J. Luton, "Toroidal Detector Scoping Study," SSCL-N-697 (1990).

5. L. Rosenson, "Momentum Resolution and Track Reconstruction Quality for the EMPACT Detector," EMPACT Note 224 (1990).

R AND D PROGRAMME IN SWITZERLAND FOR FUTURE DETECTORS

H. Rykaczewski[1], D. Bernat[2], M. Bourquin[3], H. Brändle[4],
R. Castella[5], G. Faber[1], W. Farr[6], H. Hofer[1], M. Jongmanns[1],
W. LeCroy[6], P.A. Munch[2], E. Rösch[5], W.G. Rüegg[4], and
S. Stueflotten[7]

1. Institut für Hochenergiephysik, Eidgenössische
   Technische Hochschule, Zürich, Switzerland
2. National Elektro, Geneva, Switzerland
3. Université de Genève, Département de Physique Nucléaire
   et Corpusculaire, Geneva, Switzerland
4. Asea Brown Boveri Ltd. Research Center, Baden-Dättwil
   Switzerland
5. Groupe DIXI, Le Locle, Switzerland
6. LeCROY S.A., Meyrin/Geneva, Switzerland
7. EB Norsk Kabel, Asker, Norway

ABSTRACT

Initiated by the growing demand for intensive interactions between physics and industry in order to develop new products and technologies needed for future high energy physics detectors, a large research and development programme has been elaborated by Swiss universities (ETH Zürich and University of Geneva) and leading edge Swiss industry. In this programme all costs are shared equally by both partners, science and industry. To maximize the use of existing industrial infrastructure, physicists and engineers are delegated from their home institutes to work directly in the participating firms and thus have full access to all facilities.

Based on the recent experience collectively gained in the construction and operation of the large L3 Detector at CERN (Geneva), and the strong general interest to actively participate in the eventual L3+1 Detector at LHC or the L* Detector at SSC, substantial Swiss efforts shall focus on the conception, design, construction, installation and operation of a high precision muon spectrometer.

The need for an integrated research and development programme has been identified and will be carried out in the following areas in close collaboration with industry as mentioned: alignment systems (Groupe DIXI), precision electronics (LeCROY S.A.) and parallel optical data transfer (Asea Brown Boveri Ltd., National Elektro, EB Norsk Kabel). The entire programme deals with a broad variety of high technology developments. It is expected to lead to a large number of applications far beyond the limited market of high energy physics experiments.

EXPERIMENTATION IN PARTICLE PHYSICS

Discoveries of new particles and experimental observations of new phenomena in high energy physics within the last 30 years (ν_e, ν_μ, J, τ, T, W and Z) [1 - 6] have been made by high precision experimentation on leptonic final states. Many of these discoveries came as surprises and were not predicted by the various current theories at the time the experiments were designed. Although not predicted and needed by theory, the experimental observations of these new particles have had an extremely important impact on our picture of the subatomic world. All new particles play an essential role in present fundamental understanding of elementary particle physics and give us a high level of confidence in the correctness of the Standard Model.

In general, the decay rate of heavy particles into leptonic final states is much suppressed compared with the hadronic decay channel, but the background in the leptonic signal is also smaller. Thus, in spite of the rare occurrence of such leptons, one usually can clearly resolve the signal, provided the leptonic final states are identified and measured with a sufficient accuracy. Recent experience has proven that common features of all experiments discovering new particles were an accurate lepton identification and measurement and an efficient hadron rejection. These basic characteristics for the rich discovery potential of high precision lepton detectors has, in view of the new multi-TeV hadron colliders, led to conceptual a design of the L* Experiment [7,8].

DESIGN CONSIDERATIONS FOR A SSC DETECTOR

The Superconducting Super Collider SSC will provide a totally new and exciting field for elementary particle physics. Its two unique properties, the high energy of 40 TeV and the high luminosity of 10^{33} cm^{-2} s^{-1}, offer the opportunity to search for new, rare and unexpected phenomena. However, these same features must also be seen in view of the many difficult and challenging technical problems one will face in experimentation at the SSC.

The most obvious and outstanding characteristic to be considered in the design of a SSC detector is the enormous particle rate produced at highest energy and luminosity. Under these conditions, at an expected production rate of 10^{10} hadrons per second, unprecedented conditions on the various elements of the detectors will be imposed. At such high rates it will become exceedingly difficult to study the details of hadron production. As a result, we will concentrate most of our efforts in the precision study of muonic final states which will offer a reliable and promising approach to search for new particles and phenomena.

Already from the very beginning of our considerations in designing an experiment for multi-TeV hadron colliders, two major features of the detector were identified as essential: the clean and unambiguous identification of muon pairs and the precision measurement of their momenta. Optimization of these prime objectives were considered to have highest priority. These goals are achieved by the following principle sequence of measurements:

1. Measurement of the muon momentum by a track chamber installed near the interaction region with a resolution of about 50% at very high momentum.
2. Tracking the muon behavior in a fine grain sampling absorber calorimeter. Energy losses of the muons due to relativistic

Bremsstrahlung and pair production are determined with an accuracy of a few percent.
3. High precision momentum measurement of muons traversing the calorimeter system in a large, material free magnetized volume. The design goal of about 2% for the muon's momentum (at 500 GeV) resolution determine the basic parameters of the detector.

THE L* DETECTOR

The concept of a precision lepton detector follows the basic ideas of a series of experiments which have been conducted at various accelerators within the last 25 years [9-11]. Through the permanent application of new and refined technologies the performance of these experiments have been substantially improved over this period. This experience on precision instrumentation as well as a realistic assessment of new technological developments within the next few years have led to a series of possible designs of the L* Detector. Schematics of detector configurations presently under study are shown in Figures 1 and 2.

Inner Track Detector

Surrounding the interaction region is a track chamber system to measure muon momenta with a 50% accuracy. It is a very difficult problem to design and build such a device which can operate at expected is the high particle rate and the high radiation environment. Efforts on the design and construction of the chamber system are presently concentrated in the United States.

Calorimeter System

After traversing the inner track detector particles enter a fine sampling calorimeter system covering the total solid angle down to 10°. A very forward calorimeter system will cover the remaining region down to the beam line. The calorimeter system will have a minimum thickness of 12 interaction lengths. Such a configuration not only allows us to accurately measure the energy of electromagnetic and hadronic showers completely contained within the detector, but furthermore it has the important function of absorbing all particles but muons. Several approaches on the design are presently under discussion. A large research and development programme has been started in Germany, the Soviet Union and the United States.

Muon Spectrometer

The inner detectors discussed above will be completely surrounded by a system of 160 large area drift chambers (up to 6.70 m long and 3.30 m wide). These chambers will precisely determine the trajectory of the only charged particles which can pass through the calorimeter absorber, the muons. The trajectory of muons is curved since the entire experiment will be operating within a large volume magnetic field. Precision measurement of the curvature to about 20 μm accuracy allows the determination of the muon momentum to the required level of approximately 2%. The central region of the muon spectrometer (polar angles between 33° and 147°) will be installed at radial distances between 3.5 m and 9.5 m. The length of the outmost elements of the muon spectrometer is about 30 m. The forward region of the muon spectrometer, still in discussion, should maintain a muon momentum resolution of a few percent to polar angles as small as 5°. At present, institutes from Spain, Switzerland and the United States are working on the L* Muon Spectrometer.

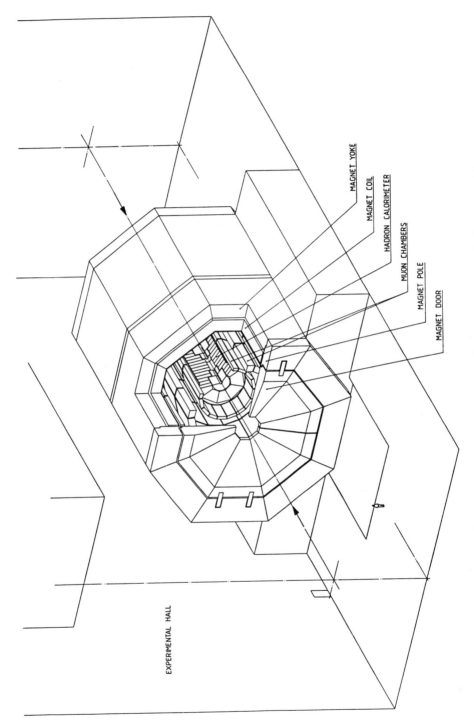

EXPERIMENTAL HALL

MAGNET YOKE

MAGNET COIL

HADRON CALORIMETER

MUON CHAMBERS

MAGNET POLE

MAGNET DOOR

Figure 1. Perspective View of a L* Detector Layout (Warm Coil)

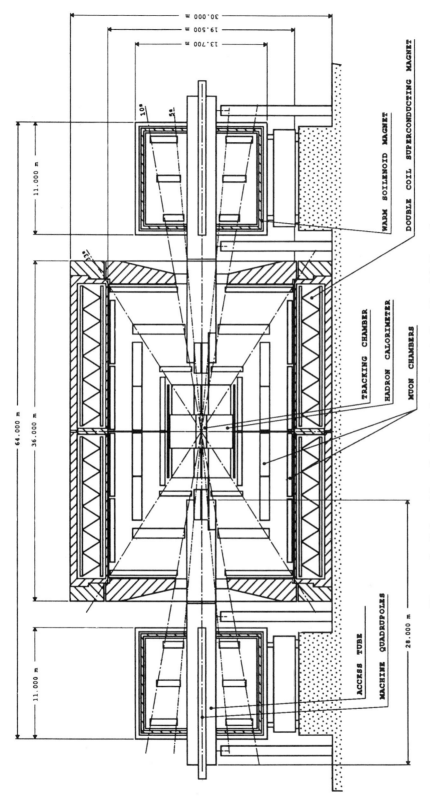

Figure 2. Schematic of a Detector Layout for the L* Experiment
Using a Superconducting Coil

323

Magnet

All detector systems of the L* Experiment will be imbedded in a magnetic field of 7.5 kGauß. Conceptual magnet designs are carried out in the Soviet Union and the United States. All options in discussion have very specific impacts on the conception of the individual detector elements.

The magnet option shown in Figure 1 operates with a warm coil. Immediately surrounding the central part of the muon spectrometer a 6,400 tons, octagonally shaped, aluminum coil operating at a current of 65 kA will be installed. The radial thickness of the coil is 1.4 m. The power consumption will be 20 MW. The coil itself will be again surrounded by an iron yoke of 2 m thickness with a weight of 32,000 tons. The two magnet poles consist of 15,000 tons of iron. They are conceived as moveable magnetic doors allowing access to the inner detectors. This magnet configuration has an overall height of 26.9 m and a length of 32.5 m. The magnetic volume is roughly 8,000 m³. Figure 2 illustrates an option using a superconducting coil in the central region and two warm solenoidal magnets in the forward directions.

RESEARCH AND DEVELOPMENT FOR MUON SPECTROMETERS

The successful implementation of this complex precision detector and its operation with the designed precision, however, strongly depends on substantial concentrated efforts on the many new challenging technical problems related to its construction and performance.

This perception has led to worldwide efforts on several research and development programmes for the entire L* Experiment. In particular, one complementary research and development programme on precision instrumentation for the study of muons in the TeV region has been defined [12] in collaboration between U.S. Institutes (Massachusetts Institute of Technology, California Institute of Technology, Carnegie Mellon University, Charles Stark Draper Laboratory, Harvard University, Johns Hopkins University, Los Alamos National Laboratory, Oak Ridge National Laboratory, Princeton University and Purdue University) and Swiss Institutes (Eidgenössische Technische Hochschule ETH Zürich and University of Geneva).

Six principal directions of the common research and development programme have been identified and can be summarized as follows:

1. Large Area Drift Chambers: The goal is to design precision muon chambers and to establish a standard for chamber mass production. The chamber itself must provide good single wire accuracy, must be able to be aligned to ±20 µm tolerance and must be optimized for high reliability. One of the basic objectives of this project is to substantially increase the number of track sampling wires in order to improve the overall track resolution.

2. Gas Mixtures: The total chamber volume of the L* Muon Spectrometer is about 2,200 m³. Safety considerations demand the use of nonflammable gas mixtures. In order to achieve the necessary high tracking accuracy, the selected gas must have a low diffusion. High rate capability and reliability are mandatory requirements. Any gas mixture considered must have a Lorentz angle small enough to allow precision operation of the chambers in the 7.5 kGauß magnetic field.

3. Support Structure: An essential condition for the precision measurements to be performed by the drift chambers is the accurate and stable mechanical support of the chambers. The more stringent tolerances, the extended dimensions of the chambers and the increased weight of chamber modules compared with the L3 Experiment calls for an adequate structural design. The entire system must be designed and constructed such as to maintain an overall mechanical stability of about 20 μm over distances of 9 m.

4. Alignment Systems: Imposed by the smaller mechanical tolerances which can be accepted for the individual detector elements, improved alignment systems with higher accuracy have to be developed.

5. Precision Electronics: To match the anticipated spatial resolution of the chambers, precision electronics with improved time resolution will be developed and constructed. Special attention is to be given to the radiation hardness of the electronical components.

6. Data Transfer: The number of signals to be readout from the drift chambers will be about one order of magnitude larger than in the L3 Experiment. Obviously, the handling of such an enormous number of signals in the very specific environment of a SSC experiment cannot be simply done as before. Efforts will concentrate on the development of a new readout system using fiber optical components to transmit the data from the experiment to the recording electronics located in the counting rooms.

The common research and development programme on precision instrumentation for the study of muons is organized such that the projects on Large Area Drift Chambers, Gas Mixtures and Support Structure are under the responsibility of the participating institutes from the United States. The other three complementary fields of research and development, Alignment Systems, Precision Electronics and Data Transfer, will be carried out in Switzerland.

A CONCEPT OF RESEARCH AND DEVELOPMENT IN SWITZERLAND - A SYNERGY BETWEEN SCIENCE AND INDUSTRY

The part of the research and development programme for future detectors to be conducted in Switzerland will be done by joint efforts between Swiss universities and specialized industry. These collaborative efforts are organized within the frame of a programme sponsored by the Swiss Commission for the Promotion of Applied Scientific Research, KWF (Kommission zur Förderung der Wissenschaftlichen Forschung). Financial support through this governmental commission imperatively demands a contribution to the funding at a level of at least 50% by the participating firms.

In view of the new type of challenges introduced by the tight technical specifications of the anticipated instrumentation, research and development will no longer be completely conducted within the premises of the universities, as traditionally done, but dominantly at the industrial centers. This new approach of close interaction with industry will incorporate various important advantages. For example, already existing modern infrastructure which is suited for high precision industrial production can be used in an efficient and cost effective way. Collaborative research and development carried out inside an existing industrial infrastructure will facilitate the implementation of high technology manufacturing procedures in the very early stage of designing the final product. Thus, attention can focus on the important aspect of developing suitable methods applicable to mass production of precision instruments.

The motivation to participate and support the research and development programme for the L* Experiment in the frame set by the regulations of KWF are manifold. The scientific community is searching collaboration with leading edge industry to develop products and manufacturing technologies suitable for well defined problems within their specific field of research. In view of the large scale of the next generation of high energy physics experiments this early stage industrial involvement will become more important than ever before, since technical specifications, mass quantities and sizes have reached new dimensions.

Industry's interest in such a research and development programme is certainly the potential market created by the anticipated experiments to be built at new accelerator facilities, either for those built at the SSC in Texas or for those of the Large Hadron Collider LHC presently under discussion at CERN. The size of these projects have now reached a level to become attractive for large, long term investments. Further than this aspect and of even much more importance are the foreseen technological spin-offs leading to products and markets applicable to a community far beyond the limited market of the new high energy physics experiments. Experience has shown that close collaboration with universities in high technology areas strengthens the competence and competitiveness of the industrial partners and has often led to a substantial time lead against competition. Finally, one of the main motivations for industrial engagement in this type a research and development programme is the exploration of new fields of activities as well as the possibility to get acquainted with scientific experts and engineers, who, within the development of the industrial interest, may join the firms at a later stage.

This research and development programme elaborated between Swiss universities and Swiss industry will be carried out until the end of 1993. The total financial volume is about 20 million Swiss Francs.

THE SWISS R AND D PROGRAMME FOR MUON SPECTROMETERS

Alignment Systems

A critical component in the ultimate performance of any large precision system, such as the L* Muon Spectrometer, is the alignment of all system elements. Permanent control and eventual corrections to movements and deformations induced by external influences, like temperature changes or mechanical loads, are essential. This monitoring and rectifying system, of course, becomes of greater relevance the smaller the mechanical tolerances are defined. For large, high precision mechanical structures even the selection of special materials will not prevent system movements to be of similar magnitude than the tolerable accuracy. In order to measure and compensate these effects, the entire system must be instrumented with monitoring equipment with intrinsic accuracies of about one order of magnitude better than the system deformation.

The requirements as described above for the L* Muon Spectrometer are practically identical to those imposed for the performance of the large precision machine tools of the next generation. Mechanical tolerances of a few microns will have to be maintained over distances of several meters. External influences to the tools must be detected and corrected for, just as in the case of high precision instrumentation for high energy physics experiments. The similar complexity of the alignment requirements of both mechanical systems can be seen in Figures 3, 4 and 5. The recognition of the same basic technical problems has led to the collaboration with the Groupe DIXI, located in Le Locle, a Swiss manufacturer of large (up to 6 m * 6 m * 2 m) precision machine tools.

DIXI's industrial applications goes far beyond the production of precision machine tools. DIXI's involvements include automotive industry, industrial refrigeration, aerospace and helicopter industry, watch production, nuclear industry, research and development etc. on an international market. The industrial infrastructure at DIXI available for the R and D programme is particularly well adapted for our purposes. It includes large, highly accurate tools, temperature stabilized facilities, computer controlled manufacturing, mass production capacity, quality control systems etc.

The research and development project will focus on topics relevant for precision instrumentation like alignment systems and distance measurements. Both fields of application, high energy physics experiments and large precision machine tools shall profit from this activity. The following areas of activity have been identified:

1. Linearity Measurements: Industrial machine frames are designed such as to obtain good stability and minimal deformation. After installation of precision machines and structures their alignment is verified and regularly repeated. Continuous alignment verification of relevant structural elements and eventual corrections by servo mechanisms compensate movements and thus increase the overall accuracy.
   The next generation of high precision mechanical structures to be used either for drift chambers or machine tools demands continuous alignment with an accuracy of about 5 μm for distances as long as 20 m. In our programme we shall explore new alignment methods using multiple point alignment along the alignment axis. Work in this area will concentrate on improvements of electro optical sensor components as well as refinements of existing methods [13].
2. Distance Measurements: Accurate length measurements done with interferometer systems seem to be ideal for many industrial applications. However, these devices are rarely used in industrial applications because of several severe technical restrictions and their relatively high investment cost. These systems generally can only be used to measure distances without any obstacles in the way of the optical center line. Furthermore, present precision systems are sensitive to changes in temperature, humidity and pressure. In the frame of this project we want to develop a low cost interferometer system allowing the simultaneous length determination along three independent axes. External influences deteriorating the measurement precision shall be substantially reduced compared with present commercial products.
   A first detailed study has led to the conception of a novel interferometer design which allows to increase the precision by one order of magnitude at substantially reduced cost compared with commercial systems. The resolution of the system is expected to be better than 2 μm over distances of at least 20 m. The principle of the concept shall be proven, the performance under standard operating conditions shall be explored and the feasibility of a wide spread application

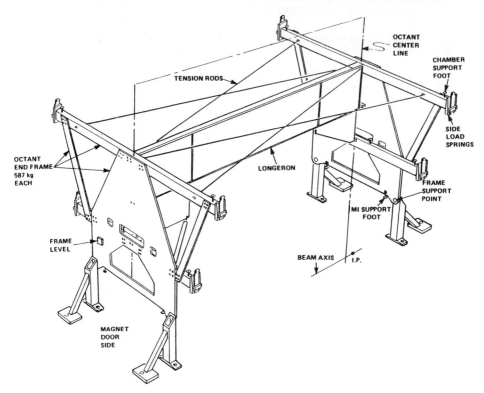

Figure 3. Mechanical Structure of L3 Muon Spectrometer

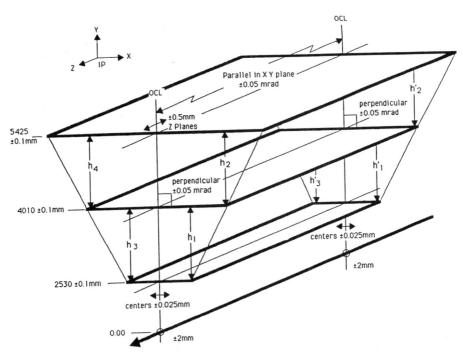

Figure 4. Dimensions and Tolerances of
Mechanical Support for L3 Muon Spectrometer

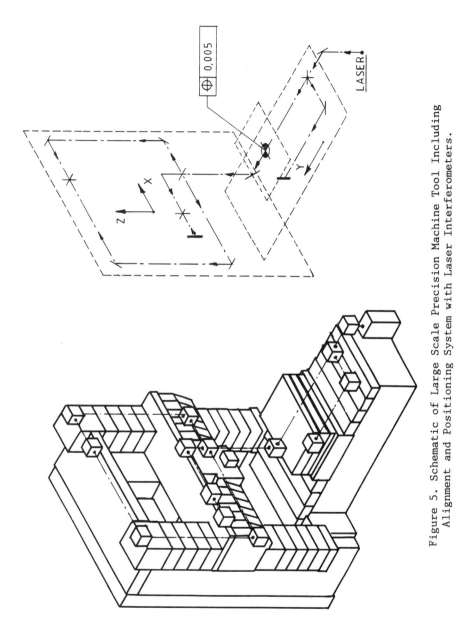

Figure 5. Schematic of Large Scale Precision Machine Tool Including Alignment and Positioning System with Laser Interferometers.

329

shall be analyzed in the research and development programme. Application for a patent on the basic concept of this novel interferometer system has been submitted.

Critical components in the foreseen research and development programme are lasers, optical splitters, optical fibers, intermediate targets (i.e. lenses, slits, cross-hairs, ...) and charged coupled devices [14]. Principal directions of research is the integration of existing techniques into a global measurement system. Areas of research include the optimization of splitting the laser light in order to create coherent light beams. The problem of efficient coupling of this light into optical fibers shall be investigated. Suitable intermediate targets to be integrated in the system are to be selected. Optical information will be readout by charged coupled sensors. A system shall be developed to analyze the sensor data and offer the user comprehensive digital display.

In addition to the activities briefly described here, other projects will address the problems of planarity measurements, angular measurements in magnetic fields and alignment verification systems.

Precision Electronics and Data Transfer

The readout of information from the detectors to the attached electrical and electronical devices is one of the major general problems of any experiment using very large arrays of detectors. The close interplay of the electronic devices used to generate and encode information with the elements needed to transmit this information implies to consider the research and development of all these aspects together. For this project wide spread industrial participation has been centrally organized and coordinated by LeCROY S.A. (Precision Electronics) and the Asea Brown Boveri Ltd. Research Center (Data Transfer).

In order to grasp the complexity of a possible readout system for the L* Muon Spectrometer one can extrapolate from our recent experience with the L3 Experiment. About 28,000 signal readout wires for the momentum measuring chambers (p chambers) of the L3 Muon Spectrometer have

Readout System

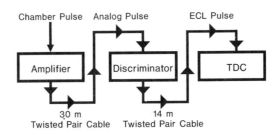

Figure 6. Schematic of Signal Readout Used for the L3 Muon Spectrometer

been connected to a total of 14,000 amplifiers via 82 Ohm decoupling resistors. The amplifiers convert incoming currents to voltages with a conversion factor of 25 mV/μA. The differential output (typically 200 mV for a muon) is connected via a 30 m long twisted pair cable to discriminators set to 20 mV threshold. The logical "time over threshold" signal is conducted through a 14 m twisted pair cable to a 500 Mc Time Digitizer (LeCroy LRS 1879 FASTBUS TDC) which continuously records until the common stop from the beam crossing arrives. The TDC covers a range of 1,100 ns with 2.2 ns least hit accuracy. The system is stable to 0.2 ns as checked by the calibration system which electronically induces pulses onto the chamber wires. The organization of the basic electronical components and cables is sketched in Figure 6.

In a similar fashion, about 8,000 readout wires of the chambers measuring the muon's coordinate along the beam line (z chambers) are processed by time recording channels which are also read out by FASTBUS TDCs. Parallel outputs without time processing are used to form fast road triggers.

Another major subsystem where data are transferred from numerous sensors located inside the experimental area to the data-taking electronics installed in the counting rooms is the Monitor System of the L3 Muon Spectrometer. For the purpose of monitoring, calibrating and controlling the entire spectrometer a total of a about 20,000 signals are recorded. The signals, which are generated at various locations inside the detector and which have a wide range of different electrical characteristics, are transferred by a total of about 1,500 multiwire cables to the recording electronic devices.

For the proper, stable and safe operation of all the Muon Chambers a precision high voltage system operating up to 4.5 kV has been developed to allow us to remotely control some 12,000 high voltage channels. A total of 65 km of multiwire cable (up to 24 conductors) and about 48,000 high voltage contacts are installed between the high voltage control panels inside the counting rooms and the chambers.

Evidently, the proper installation of such an enormous quantity of signal lines is a challenging technical and organizational problem. Special attention had been given to the routing of the all cables which had to be installed in the limited space between the outer part of the L3 Support Tube and the doors of the L3 Magnet. In order to completely understand and resolve the difficult problems related to the installation of this large quantity of cables in the very limited installation time of only several weeks, we have conducted a full-scale, partial preinstallation of the cables six months before starting installation in the underground experimental area.

A summary of the wires, cables and connectors installed for the Muon Spectrometer of the L3 Experiment can be seen in the second column of Table I. We have estimated the number of these elements needed for the L* Muon Spectrometer assuming the same transmission schemes as for the L3 Experiment. This extrapolation is also listed in Table I.

It is very clear that such a proportional increase of cables and connectors using present technologies cannot be tolerated:

1. Cable volume will become too big and will introduce an unacceptable dead space deteriorating the hermeticity and quality of the detector.

TABLE I
DATA TRANSFER FOR MUON SPECTROMETERS

	L3 Experiment	L* Experiment
Number of wires	200,000	1,500,000
Number of cables	8,000	50,000
Number of connectors	25,000	150,000
Wire length	8,000 km	80,000 km
Cable length	160 km	4,000 km

2. Forces introduced by large quantities of heavy cables attached to the position sensitive Muon Chambers need to be carefully considered and compensated in the structural design of the spectrometer. These forces therefore must be kept to their absolute minimum.
3. The large number of individual connectors leads to a complex system which needs a careful organization in order not to confuse. Even a low failure rate may, because of the large number of individual components, result in a considerable and unacceptable number of errors.
4. The installation time of the cable systems will, since parallel installation is restricted due to limited space, be much too long to be integrated with any realistic detector installation schedule.

Initiated by the recent experience gained in the construction and installation of the L3 Muon Spectrometer, we want to explore the possibilities of developing new data transfer systems. In a first step we shall concentrate on the development of a new readout scheme justified by the fact that the central part of the L* Muon Spectrometer alone will consist of about 205,000 channels to be read out. The addition of the L* Forward Muon Spectrometer will increase this number of readout channels to about 300,000 - 400,000.

First discussions have led to two basic approaches as sketched in Figures 7 and 8. In both cases we intend to combine the amplifier and discriminator into one low power unit attached on the ends of the chambers. Transfer of information of the chamber pulses to the TDC will be done by an optical fiber system. The two readout schemes differ in the way the optical signal is produced.

In the suggestion outlined in Figure 7 light is produced by an electro-optical conversion unit which will be directly plugged into the discriminator's output. Inside this conversion unit the electrical signals are used to drive fast, low power light diodes. The diodes themselves are coupled to the individual fibers of the multi-bundle fiber optical cable which transmit the signal to the recording electronics. For economical reasons the electro-optical conversion unit will function as an efficient active connector. An interesting aspect of this connector design is that all connections will be purely electrical, not optical. Light guided through the fibers will reach the opto-electrical conversion unit which will transform the optical signal again into an electrical one. Also this opto-electrical conversion unit will be plugged directly into the TDC and therefore also acts as an active connector.

In the second version [15] (Figure 8) the fibers themselves will be coupled to a laser light source. This laser will inject light into fibers such that light is constantly transmitted throughout the whole transfer path, from discriminator output to TDC input. The amplifier/discriminator unit itself is directly connected with a fast switching device which, whenever a chamber pulse exceeds the threshold, will "switch off" the light for a given light fiber. The exact timing when a fiber has been switched off will be recorded by a TDC. Also in this approach there will be a conversion unit transforming the light signal into an electrical signal just before the input of the TDC.

In both approaches it must be guaranteed that the sensitive timing information of the chamber pulses will not be influenced by any of the components of the fiber optical system. The readout chain must operate and preserve rise times in the nanosecond region.

Proposed Readout - System A

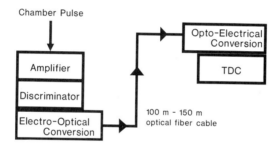

Figure 7. Proposed Readout System for L* Muon Spectrometer - Version A

Proposed Readout - System B

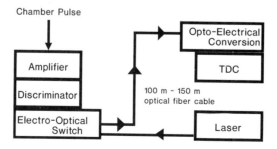

Figure 8. Proposed Readout System for L* Muon Spectrometer - Version B

The advantages of fiber optical readout systems as described can be summarized as follows:

1. It allows to develop a new type of low power discriminator serving either as LED driver or a switch. Therefore the heat dissipation of the discriminator will be so low that it can be integrated with the amplifier and mounted on the chambers. Consequently, there is no need for transmission of analog signals over long distances.
2. Attenuation of signals being transmitted over even several hundred meters is negligible. This was not the case for the L3 Experiment where the attenuation of the ECL signal between the discriminator and the TDC was important over a distance of only 14 m. In order to guarantee the proper operation of the TDC, the decrease of the ECL signal had to be kept below 50%, which was ensured by appropriate cable of short length. Evidently, at distances of about 100 meters or more, signals will be attenuated much too much when using cables similar to those of the L3 Experiment.
3. Fiber optical systems are immune to electromagnetic interference.
4. Since optical fibers can be more densely packed than copper cables, more individual signals can be transmitted within a cable of a given diameter.
5. Higher packing of signal lines in a cable allows to use multipin connectors with a larger number of contacts and thus reduces the number of individual connectors.
6. The cable weight per signal line is drastically lower for fiber optical cables than for copper cables. As a consequence forces to be compensated by the mechanical structure will be considerably reduced.

In addition to the many technical improvements of using an optical fiber readout system, a rough estimation on the installation time for the cables of the L* Muon Spectrometer shows that the conventional approach as realized in the L3 Experiment would be impossible.

Under very optimistic assumptions the distance between the L* Muon Chambers and the readout electronics may be as short as only 100 m. Conventional copper cables with acceptable diameter carry typically 16 signals. Thus about 13,000 cables of 100 m length need to be installed for the 205,000 signal wires of the central L* Muon Spectrometer. In favorable conditions ("easy" cable routing, no obstacles, easy access, no interference with other installations) an experienced technician may be able to install about 30 m of cable per hour. Assuming these good conditions, the laying of the signal cables will need at least 43,000 manhours or about 23 manyears. Obviously, it will be extremely complicated to integrate such a large and time consuming activity into tight detector installation schedules. This problem will definitely be of less importance when using more densely packed fiber optical cables with up to 96 signal lines per cable, thus reducing the number of cables and their installation time by a factor of six.

The objective of this research and development project is the construction of a working model of the proposed L* Muon Spectrometer Readout System (see Figures 7 and 8). It shall demonstrate its feasibility and its operation in the specific environment of the L* Experiment. One of the main considerations in the choice of any material is its resistance against radiation which, at multi-TeV colliders, reach a level of up to several hundred Mrad per year [16,17].

In the frame of developing the L* Muon Spectrometer Readout System we shall concentrate on the following specific system components:

1. Amplifier: Based on the experience gained in constructing the L3 Muon Spectrometer, improvements of the existing amplifiers shall be pursued. Since the L* Muon Spectrometer will contain a large volume, non flammable gases must be used for safety reasons. Unfortunately, these gases usually have a relative low gas gain. Therefore, one of our goals is to achieve an increase of gain by a factor of five, resulting in a gain of about 100 mV differential output per 1 μA of signal input. The electronic noise must be kept well below 2 mV. Power consumption of the amplifiers must be significantly less than 50 mW per channel in order to ensure that the heat dissipation of the electronics mounted on the chambers does not affect the chambers' mechanics.

2. Discriminator: Development of a low power discriminator is needed. The discriminator shall be combined with the amplifier as one single unit. This economical design will avoid large quantities of cables and connectors. However, this approach must be carefully studied, tested and improved to ensure that interference or feedback to the amplifiers does not cause any noise.
   For precise time measurements the discriminator threshold shall be remotely controlled and adjusted in a range between 10 mV and 500 mV collectively for large groups of wires. Normalized small pulses will be sent to the next components of the readout chain. The amplifier/discriminator unit must prove to be insensitive to temperature variations as well as it must demonstrate stable and reliable operation in a radiative area. High quality standards and outmost reliability towards the amplifier/discriminator unit are determined by the fact that repairs or replacements are normally possible only once a year.

3. Electro-Optical Conversion: This device will act as an active connector. Electrical signals from the discriminator's output will drive fast lightdiodes. The LEDs selected for this unit must prove high reliability, stable performance and must have rise times in the nanosecond range. Time jitters of the conversion unit must be kept below of a fraction of a nanosecond.

4. Electro-Optical Switch: Development and construction of an electronical device performing as a fast switch is necessary. This device switches the light from the laser source by using the signals coming from the discriminator. The muon chamber signals exceeding the discriminator threshold will "switch off" the light emitted from the laser, which else is transmitted to the higher levels of the readout system. The entire switching mechanism must be constructed in such a way that switching times are stable and kept below one nanosecond under all environmental conditions. Channel to channel fluctuations are to be keep below about 0.1 nanoseconds.
   Materials used in this fast, active switch will be subjected to significant radiation doses with potential effects on the elements. Since the fast switch will use opto-electronic parts as, for example, optical fibers, electro-optical crystals, and glasses, attention will be given to proper doping of the used materials in order to increase the resistance against short and long effect radiation damages.

Because of the large number of muon chamber channels the fast switch must integrate a large number of individual signals (about 100 to 200 signals per switch). This number of signals shall match with all other components of the readout system.

5. Connectors: Great attention must be given to the construction of high quality connectors, which must ensure good and stable contacts. The connectors therefore need a mechanical construction with tight tolerances. Furthermore, they must handle many signals at once, perhaps up to 200 contacts. The connectors shall incorporate a polarization pin, thus avoiding false wiring, as well as an easy to use, but solid and reliable locking mechanism. Since the entire Muon Spectrometer Readout System requires several hundred thousand contacts, care must be taken in designing and defining all steps involved in the actual mounting of any connector. Methods have to be developed to optimize a fast, reproducible and efficient way of splicing fibers and mounting connectors in workshops as well as in situ.

The need for an efficient connector mounting system and technique is evident when considering a given installation time of six months in situ for connecting 400,000 contacts. This precision operation can only be completed in the given time provided the complete mounting of one contact does not exceed an average time of only 9 seconds.

6. Optical Fiber Cables: Technologies developed in modern telecommunication systems shall be the basis of the research and development programme for the optical fiber cables. Despite the recent success in using optical fiber cables in many fields, the particular environment and use of the cables for the L* Muon Spectrometer require extensive improvements and considerable efforts in order to fulfill the following conditions:

a) High packing of about 100, perhaps even 200, signal lines per cable of maximum diameter of 15 mm.

b) Radiation insensitive fibers which can tolerate both long term low radiation exposure as well as short radiation shocks. Fiber material must be carefully selected such that substantial variation of radiation along the cable routing (6 orders of magnitude) will not influence the optical transmission.

c) Temperature variations along the cable routing can be as high as 40° C. Thermal effects on the performance of the fibers must be excluded in a range between 15° C and 55° C.

d) Aging effects caused by the specific environments must be excluded. Small changes of attenuation along the cable routing may be tolerated as long as the calibration of the entire readout system can compensate this effect.

e) The outer sheath of the cable must fulfill all present safety requirements. In addition to the standard mechanical requirements, they must especially be made of flame retardant and halogen free materials. The outer sheath must furthermore be resistant against high levels of radiation over a period as long as 15 years. The mechanical construction of the cable shall be laid out in such a way that on one side rough treatments (scratching, pulling, squeezing, stepping onto, etc) during installation will not cause any damages, but also on the other side that the cable still is flexible enough for small bending radii.

7. Laser: The light to be used as transmitting medium between the output of the discriminator unit and the input of the digitizing electronics is channeled into the readout system by

a high power, high frequency laser system. Lasers, located in the permanently accessible areas of the control rooms, must be designed and constructed to have sufficient power to serve the order of 100 to 200 readout channels. The laser signal shall also be modulated such as to have an overlaid structure of spikes. These spikes must have rise times in the sub-nanosecond region. The research and development programme of the laser system shall include aspects of low cost mass production as well as the integration into fast and simple monitor and control systems of the entire readout chain.

8. Opto-Electrical Conversion: These units, in principle, can be constructed by using the same basic concepts as already applied in advanced, modern telecommunications systems. Research and development activities will concentrate on ensuring the reliability of large quantities of transducers as well as they shall concentrate on questions related to instabilities and time jitters introduced by the transducers. Also for the transducers, as well as for any element of the readout system, compatibility and efficient installation is of major concern.

Depending on the configuration of the readout system, the opto-electrical conversion must be designed to match either of the two proposed readout schemes, which use either LEDs or a fast electro-optical switch.

9. Time Digitizers: Each channel of the entire readout system must have a resolution better than one nanosecond. The channels themselves shall be combined in a high packing TDC unit of eventually 128 channels or more.

Time over threshold is to be recorded in order to enable the measurement of the pulse width. Experience has shown that best chamber resolution is obtained at thresholds of about 10% of the average pulse height. At this low level small perturbations, like reflections, can severely and intermittently affect the pulse width measurement. Because of this effect, a width measurement done at a pulse height level of about five times the threshold level is absolutely essential. Sophisticated analysis of data recorded by the L3 Experiment has proven that accurate corrections can be made by using the measured pulse width, but not with the pulse heights.

Additional efforts on the improved performance of time digitizers will concentrate on higher accuracies of the common stop signal. Presently the accuracies are limited by the systematic uncertainties of the occurrence of the stop signal within the clock cycle. This uncertainty is typically 1.2 ns corresponding to a systematic displacement of the track of up to 60 μm in one chamber cell. This effect can cause up to 120 μm errors in the sagitta measurement, far above any acceptable level.

OUTLOOK

This research and development programme to be carried out in Switzerland in close collaboration between universities and industry concentrates on attractive high technology progression. The identified fields of the programme offer a large variety of interesting aspects such that both partners, universities and industry, will benefit. The work focuses on the specific problems related to the L* Muon Spectrometer and addresses many general technical challenges such that widespread applications are predictable.

REFERENCES

1. G. Danby et al., Phys. Rev. Lett. 9, 36 (1962).
2. J.J. Aubert et al., Phys. Rev. Lett. 33, 1404 (1974).
3. M.L. Perl et al., Phys. Rev. Lett. 35, 1489 (1975).
4. S.W. Herb et al., Phys. Rev. Lett. 39, 252 (1977).
5. G. Arnison et al., Phys. Lett. 122B, 103 (1983) and M. Banner et al., Phys. Lett. 122B, 476 (1983).
6. G. Arnison et al., Phys. Lett. 126B, 398 (1983) and P. Bagnaia et al., Phys. Lett 129B, 130 (1983).
7. H. Rykaczewski, Proceedings of the XXI Rencontre de Moriond, 1986, Les Arcs, France, "Strong Interactions and Gauge Theories", 387 (1986).
8. U. Becker et al., Nuclear Instruments and Methods in Physics Research A253, 15 (1986).
9. U. Becker et al., Nuclear Instruments and Methods 128, 593 (1975).
10. The MARK-J Collaboration: D.P. Barber et al. Phys. Reports 67 (7), 337 (1980).
11. The L3 Collaboration: B. Adeva et al., Nuclear Instruments and Methods in Physics Research A289, 35 (1990).
12. G. Gratta et al., "A Research and Development Program to Continue the Development of Precision Instrumentation for the Study of Muons in the TeV Region", submitted to the Superconducting Super Collider Laboratory, October 1989.
13. P. Duinker et al., Nuclear Instruments and Methods in Physics Research A273, 814 (1988).
14. R. Fabbretti et al., Nuclear Instruments and Methods in Physics Research A280, 13 (1989).
15. H. Brändle, W.G. Rüegg, L. Schultheis, "Stecker für ein Optisches Kabel", Application for Patent submitted to Bundesamt für Geistiges Eigentum, Bern, Switzerland, Ref. No.: 3362/89-7, September 14, 1989.
16. SSC Central Design Group, Radiation Effects at the SSC, M.D.G. Gilchriese, Ed., SSC-SR-1035, 74 (1988).
17. H. Schönbacher, Proceedings of the ECFA Study Week on Instrumentation for High-Luminosity Hadron Colliders - Vol. 1, 1989, Barcelona, Spain, CERN 89-10, 129 (1989) and references therein.

7. Materials and Magnets I

NBTI SUPERCONDUCTORS WITH ARTIFICIAL PINNING STRUCTURES

L. R. Motowidlo, P. Valaris, H. C. Kanithi, M. S. Walker,
and B. A. Zeitlin

IGC Advanced Superconductors Inc.
1875 Thomaston Ave.
Waterbury, CT, 06704

ABSTRACT

Multifilament NbTi superconductors have been fabricated with artificial pinning centers (APC) by present state of the art wire processing. Copper and niobium are each tried as the pinning material. The APC superconductor was designed such that the flux lattice spacing can be comparable to the spacing of the pinning center. The pinning strength of the APC superconductor was found to depend on the pinning material, and its volume percent relative to the NbTi. Preliminary critical current densities of 2893 A/mm^2 at 5 Tesla and 9517 A/mm^2 at 2 Tesla in experimental materials containing one micron filaments have been obtained.

INTRODUCTION

Type II superconductors such as NbTi are characterized by two critical fields, H_{c1} and H_{c2}. Between these two critical fields, referred to as the mixed state, the superconductive material is penetrated by a triangular array of magnetic flux lines of equal spacing, the flux line lattice (FLL). The flux lines themselves are screened from the superconductor by circular super currents leaving most of the superconducting material free of field. The spacing between the fluxoids is uniquely determined by the applied magnetic field. When a current is passed through a type II superconductor in the mixed state, the flux lattice experiences a Lorentz force,[1] which tries to move the flux lines at right angles to both the current and magnetic field. If the fluxoids move, energy is dissipated and an increase in temperature may ensue which can drive the superconductor normal. Fortunately, this motion of flux lines can be impeded or stopped by introducing metallurgical defects, such as voids, grain boundaries, alpha-Ti precipitates, etc. into the material. These normal state defects interact with the fluxoids to effectively pin the FLL. Recent developmental work on optimized NbTi has recognized that very fine alpha-Ti precipitates play a major role in pinning the FLL.[2]

Improvements in the critical current density have been achieved thus far, by combining several precipitation heat treatments with large degrees of cold work. Critical current densities as high as 3680 A/mm^2 at 5T and 4.2K in the optimized material have been achieived .[3] Despite these significant achievements in improving the J_c and recognizing the role of the alpha-Ti as effective flux pinners, improvements beyond the 4000 A/mm^2 mark are uncertain by the conventional method presently utilized to get optimum properties. Further advances in the J_C may be realized if the pinning centers can be produced artificially by using appropriate composite geometry at the beginning of conductor fabrication[4,5] with a microstructure which can thus be created both regular and independent of the heat treatment procedures.

In the present approach,[6] shown schematically in Figure 1, the superconducting NbTi rod is jacketed by a normal metal, M, which becomes the pinning matrix. When the hexagonal lattice of flux lines intersect the NbTi/M composite filaments perpendicularly, the fluxoids distort around the NbTi element as they thread their way in the pinning material. In this paper we present preliminary results which compare the pinning strength of APC superconductors designed with varying volume percent of the pinning material as well as the effect of the type of material utilized for the pinning center.

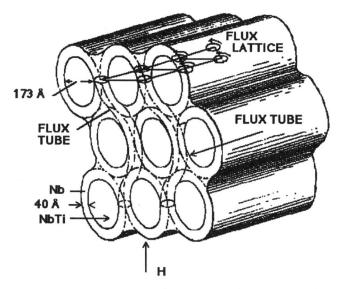

FIGURE 1. Fluxoids of the FLL thread continuously through the Nb artificial pinning center matrix.

EXPERIMENTAL METHODS

NbTi rods jacketed with 12, 25, and 50 percent niobium were placed within 15.56 cm O.D. copper billets. Each billet was extruded and drawn to an outer diameter of 0.155 cm. The copper was then stripped off, leaving 0.142 cm diameter monofilaments (subfilaments) surrounded by Nb. NbTi rods jacketed with 20 and 33 percent copper were similarly prepared. Restack billets each containing 61 of these 0.142 cm rods were assembled in copper cans of 1.75 cm O.D.. These were drawn to 0.142 cm diameter multifilament wires. This process was repeated with a second restack assembly and cold draw to 0.179 cm. A third restack billet was assembled with 37 elements and processed to 0.179 cm O.D.. A fourth and final restack was then assembled with 37 elements and reduced to a final wire of 0.609 mm. A schematic of the process is shown in Figure 2.

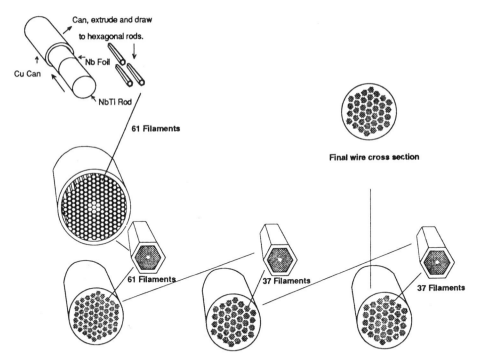

FIGURE 2. A schematic of the manufacturing process.

The resulting superconducting wires all contain 83,509 filaments, each made up of 61 subfilaments. Each subfilament is a superconducting NbTi core with an approximate 200 Angstrom diameter embedded in a surrounding pinning matrix of niobium or copper. These dimensions correspond to a wire diameter of 0.609mm. Figure 3 is an overall cross section of the final multifilament configuration. Figure 4 is a closeup of the third restack assembly and Figure 5 shows further magnification of the second restack and of each filament. Each of the filaments in Figure 5 is composed of 61 subfilaments. Samples were processed to various final sizes having NbTi/M dimensions in the range of interest for pinning. No precipitation heat treatments typical of conventional NbTi conductors were given to these APC conductors. Samples of approximately 200 cm long, were cut for critical current measurements in external magnetic fields from 1 through 7 Tesla. A standard four point probe on a helically wound sample at 4.2K was utilized. Voltage taps were spaced 75 cm apart. The critical current was defined at occuring at a resistivity unset of 10^{-12} Ohm-cm for I_c results.

RESULTS AND DISCUSION

Critical current densities, J_c, as a function of applied magnetic field, B, are plotted in Figure 6 for the various APC conductors at those subfilament sizes giving peak J_c's(B=5T) for each pinning material. The current densities shown in Figure 6 have all been calculated based on the sum of the areas of NbTi and normal pinning material in the matrix of the subfilaments, treating them together as a superconducting phase. The critical current density thus defined for the 25Nb is 2893 A/mm^2 at 5 Tesla and 9517 A/mm^2 at 2 Tesla.

These results demonstrate clearly that the fabrication of conductors with artificially introduced flux pinning centers is feasible by mechanical metallurgy. Furthermore, the results were obtained without heat treatment procedures normally required in a conventionally processed NbTi superconductor. This is both a potential cost saving and a control against alloying of the finely divided constituents of the

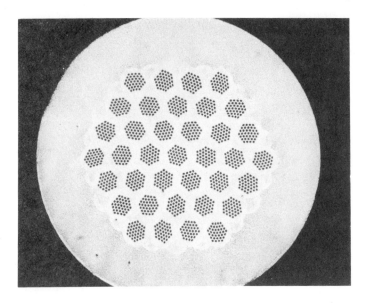

FIGURE 3. An overall cross-section of the 25 percent niobium APC superconductor. The total number of subfilaments is 5,094,049. Magnification 50X .

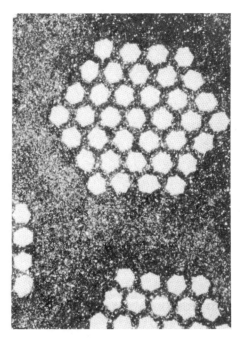

FIGURE 4. A closeup of the third restack bundle. Magnification 400X .

FIGURE 5. Further magnification showing the second restack bundle. Each filament is composed of 61 subfilaments. 1000X.

composites. These results were obtained with final filament diameters on the order of a micron. Each filament is composed of the initial stack of subfilaments dimensioned such that the magnetic field dependent flux lattice spacing is on the order of the subfilament diameter. The results are also believed to be related to the proximity effect since coherence length in the composite are also comparable to the subfilament dimension and spacing shown. It is particularly interesting that the peak properties occur at increased subfilament spacing as the percentage of normal pinning material is increased. For example, 50 Nb has a peak in the J_c at a larger subfilament diameter than in the case of 25Nb or 20Cu.

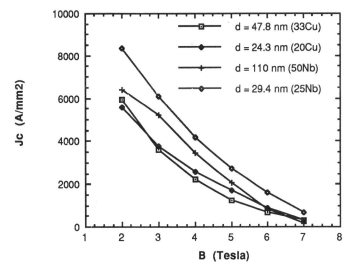

FIGURE 6. The critical current density versus applied magnetic field for various Nb and Cu pinning centers and subfilament diameters (d).

In Figure 7 we compare the bulk pinning force of the APC design and conventionally optimized NbTI superconductors. If we compare the niobium APC designs we observe that the pinning force for 25Nb is the largest. Niobium is shown to be a stronger pinning material than copper. If the pinning mechanism is related to the scattering electrons within the pinning center, then the greater pinning force observed for niobium compared to copper may reflect its greater normal state resistivity. Work is underway both theoretically and experimentally to understand how in fact the flux lines interact with the structure and to what extent the proximity effect modifies this interaction in the APC superconductors. It has been shown both theoretically and experimentally that when two metals, a superconductor and normal metal are formed contiguously, the superconducting properties, such as the H_{c2} and T_c are profoundly altered.[7,8] In particular, the dimensions for the artificial flux pinning in our samples are in the regime were the proximity effect would be important. In Figure 8, we show a Kramer's type plot for three APC designs. The effect of the niobium thickness on the H_{c2} is illustrated for 12Nb, 25Nb, and 50Nb. It is observed that 12Nb has a larger apparent H_{c2} than for the case of 50Nb. Finally in Figure 9, we compare typical current densities of Superconducting Super Collider (SSC) strand with the results obtained for the APC 25Nb. We observe that the J_c of both cross-over at the magnetic field of 5T. At low fields the APC superconductor shows significant improvement over the conventional wire. At high fields the J_c falls off rapidly possibly due to proximity effects. Further improvements in the J_c at mid and high field range may be realized if

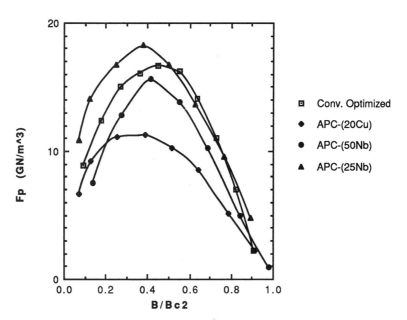

FIGURE 7. The bulk pinning force versus the normalized magnetic field for conventional NbTi and APC superconductors of Figure 6.

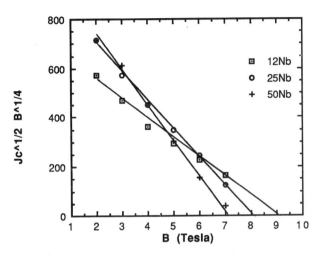

FIGURE 8. A Kramer's plot of the APC superconductors with 12Nb, 25Nb, and 50Nb.

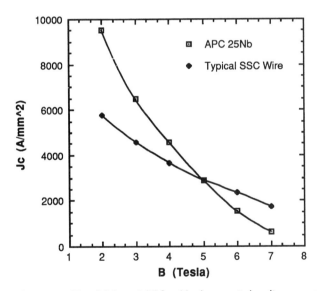

FIGURE 9. The SSC and APC critical current density versus the applied magnetic field. Wire diameters are 0.647mm and 0.609mm respectively.

a better pinning material is found. Because the APC approach does not add any further complication of metallurgical changes such as might occur from heat treatment procedures, it is well suited to study the pinning mechanism in NbTi and may further enhance our understanding of the conventionally optimized NbTi superconductors.

CONCLUSIONS

At low fields (1-3 Tesla) the APC superconductor significantly exceeds the best J_c of conventionally optimized NbTi and is roughly comparable to the best conventional NbTi superconductor at 5 Tesla. The APC superconductor is fabricated by conventional wire processing and does not require intermediate anneals as currently required in present state of the art NbTi superconductors. We have also shown that the thickness of the barrier as well as the pinning material influence the pinning strength of the APC superconductors. In the present study we have found Nb to be a stronger pinner than Cu. Furthermore 25% Nb shows the best J_c results compared to 12 and 50% Nb designs. IGC Advanced Superconductors Inc. is continuing theoretical and experimental work necessary for further understanding of H_{c2} as well as J_c in the APC superconductors.

ACKNOWLEDGEMENTS

The authors would like to acknowledge the partial financial assistance of the Lawrence Berkeley Laboratory.

REFERENCES

1. J. Friedel, P. G. DeGennes, J. Matricon   Appl. Phys. Lett., Vol.2, No.6, pp.119, (1963).

2. P. J. Lee, D. C. Larbalestier   J. of Mat. Sci., 23 , pp.3951, (1988).

3. Li Chengren and D. C. Larbalestier   IEEE MAG 23, No.2, March (1987).

4. B. A. Zeitlin, M. S. Walker, L. R. Motowidlo   United States Patent, No. 4803,310, Feb.7, (1989).   Work initiated in 1983.

5. G. L. Dorofejev, E. Yu. Klimenko, S. V. Frolov, E. V. Nikulenkov, E. I. Plaskin, N. I. Salunin, and V. Ya. Filkin   Proc., 9th Int'l Conf. Magnet Technology, p. 564, Boston, Mass., (1985).

6. L. R. Motowidlo, H. C. Kanithi, and B. A. Zeitlin   CEC/ICMC, Los Angelos, CA, Aug., (1989).

7. N. R. Werthamer, Phys. Rew. , Vol. 132, No.6 , Dec.15, (1963).

8. P. G. DeGennes, Rev. Mod. Phys., p.225, Jan., (1964).

RECENT DEVELOPMENT OF SSC CABLE IN FURUKAWA

M. Ikeda, H. Ii, S. Meguro and K. Matsumoto[*]

Superconducting Products Department and Yokohama
R&D Laboratories[*], The Furukawa Electric Co., Ltd.
2-4-3, Okano, Nishi-ku, Yokohama, 220 Japan

ABSTRACT

After the First International Industrial Symposium on the Super Collider, the following developmental work on NbTi superconducting Rutherford cable for the Superconducting Super Collider (SSC) was done at Furukawa Electric. (1) Improvement in the critical current density of the SSC wire, based on volume production: The manufacturing process enabling us to achieve the critical current density of more than 3,000 A/mm^2 at 5 T is being established. (2) Development of large keystone angled cable to be used for modified SSC dipole magnets: Critical current degradation due to cabling is nearly proportional to cross-sectional area reduction of strands. A 5-cm bore, 1-m long dipole magnet having no wedge inside the coils was fabricated by using the large keystone angled cable and successfully tested by KEK. (3) Magnetization measurements on the SSC wire samples: The long term magnetization decay imposing a serious problem in the accelerator design is possibly associated with flux creep phenomenon occurring in a proximity-coupled matrix between NbTi filaments.

INTRODUCTION

Furukawa Electric presented their activities in development of the SSC cable at the First International Industrial Symposium on the Super Collider[1]. Thereafter, some new developments were made. This paper describes progress in the following three areas:

Many superconductor manufacturers have succeeded in producing the SSC wire. However, their critical current densities are still in the range of 2,800 to 2,900 A/mm^2 at 5 T, which can meet the SSC specification but is not sufficiently high enough. When considering volume production of the SSC wire and better performance of the SSC dipole magnets, it is necessary to achieve a critical current density of more than 3,000 A/mm^2 at 5 T. Although it may be possible to achieve this in a laboratory scale production, there are many restrictions in volume production. Therefore, extensive work was planned and is now in progress, based on volume production.

In the current design of the SSC dipole magnet, there are wedges inside the coils. In order to eliminate the wedges which degrade the mechanical stability and the winding efficiency, Rutherford cables having a large keystone angle of up to 4.8 degrees have been trial-fabricated. Some results were reported elsewhere[1,2]. Cables having keystone angles larger than the SSC specification were wound into a modified

SSC dipole magnet and successfully tested by KEK.[3,4] In this paper, better under-standing of the critical current degradation due to cabling is discussed.

Recently, long term decay of the sextrupole field has been observed in Tevatron, HERA and SSC dipole magnets. This serious problem has been considered in the SSC design. Goldfarb et al found that the source of the problem might be flux creep in the proximity-coupled matrix.[5] In order to make it clear, time depend-ence of magnetization was measured on three kinds of the SSC wire.

IMPROVEMENT IN CRITICAL CURRENT DENSITY

Volume Production Process and Sample Preparation

Much effort to improve the critical current density of NbTi has been made. The world record of the highest critical current density, 3,830 A/mm^2 at 5 T, was achieved by Furukawa Electric, although the wire does not meet the SSC specifica-tions.[6] Generally, there are the following preferable conditions for realizing high criti-cal current density: Highly homogenized NbTi billets or bars, long heat treatment time, many heat treatment steps, heavy cold reduction and so on. However, these conditions can be adopted only for prototype scale production. From the standpoint of cost, superconductor manufacturers face many restrictions. For example, commercially available, highly homogeneous NbTi billets or bars should be used, heat treatment should be reasonably long with a reasonable number of steps, and cold reduction rate should be limited by available extruders and available drawing benches.

Considering these circumstances, extensive experiments for the improvement in the critical current density are planned, aiming at the establishment of a volume production process for the SSC wire. Samples were prepared according to the follow-ing conditions; use of highly homogeneous billet made by ourselves, 90 mm in hot-extruded bar diameter, three variations in the intermediate heat treatment, three variations in the intermediate cold reduction rate, five variations in the final cold reduction rate. The samples were finally twisted and annealed to meet the existing SSC specification.

The structure of the samples is the same as the SSC outer wire; 0.648 mm in wire diameter, 6 microns in NbTi filament diameter and 1.8 in copper-to-NbTi ratio. It is noted that optimization of processes should be made for the real wire and that the outer wire is not preferable for achieving high critical current density to the inner wire because of its higher copper-to-NbTi ratio.

Test Results

The critical densities were measured and calculated according to the SSC specification. Although the experiments are continued, some results which have been obtained so far are shown here. Two curves shown in Fig. 1 correspond to two series which have the same intermediate cold reduction rate, but different heat treatment steps. Their total heat treatment times are nearly the same. A critical current density of more than 3,000 A/mm^2 at 5 T is obtained at an optimum cold reduction rate. This process is expected to be applicable to volume production of the SSC wire.

Other data is plotted in Fig. 2 regarding the critical current density ratio of 5 T to 8 T. Each point corresponds to the individual processes. The ratios are not constant but scattered. The existing SSC specification is based on critical current densities of 2,750 A/mm^2 at 5 T and 1,100A/mm^2 at 8 T. It is shown in Fig. 2 that even if the critical current density at 5 T exceeds 2,750A/mm^2, the points over the dashed line cannot exceed 1,100A/mm^2 at 8 T. Most discussion has been focussed on the critical current density at 5 T, but the critical current density should be opti-mized at 7 T for the SSC inner wire and at 5.6 T for the outer, according to the existing SSC specifications.

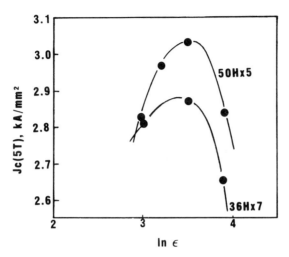

Fig. 1. Critical current density at 5 T, Jc(5T), versus final cold
reduction rate, ε. Intermediate heat treatments are 50 hrs. x 5
steps and 36 hrs. x 7 steps.

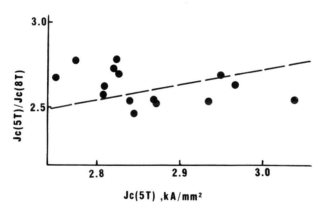

Fig. 2. Relationship between critical current density at 5 T, Jc(5T),
and critical current density ratio of 5 T to 8 T,
Jc(5T)/Jc(8T). The dashed line corresponds to critical current
densities of 2,750 A/mm^2 at 5 T and 1,100 A/mm^2 at 8 T.

LARGE KEYSTONE ANGLED CABLE

Trial-Fabrication

Using the SSC inner strands (0.808mm in wire diameter), large keystone angled cables were trial-fabricated.[2] The cables have the same number of strands (23) and the same cable width (9.3mm) as the SSC inner layer cable. The packing factors are 90 and 95%. The keystone angles are 2.4, 3.0, 3.6, 4.2 and 4.8 degrees, while 1.6 degrees in the SSC inner layer cable.

In the cross-sectional views of the cables having larger keystone angles, the strands near the wide edge are less deformed. Therefore, the strand lay is not as good for the 4.8 degree cable. The other cables are mechanically sound and exhibit no irregularities, such as excess twist.

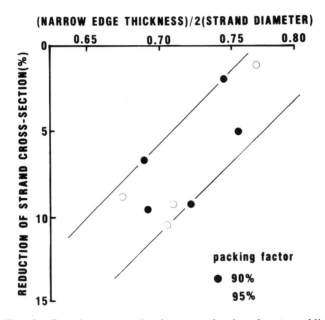

Fig. 3. Strand cross-sectional area reduction due to cabling.

Critical Current Degradation due to Cabling

The most important problem with large keystone angled cables is critical current degradation due to cabling. In the trial-fabricated cables, heavy cross-sectional area reduction was found, especially at the narrow edge of the cable, as shown in Fig. 3. The value of the narrow edge thickness over 2 times of the strand diameter indicates how heavily the strand are deformed.

Fig. 4 shows the relationship between the strand cross-sectional area reduction and the critical current degradation. It is clear in comparison with the dashed line in Fig. 4 that the critical current degradation is nearly proportional to the stand cross-sectional area reduction and that the critical current degradation due to filament sausaging and filament breakage is not so much.

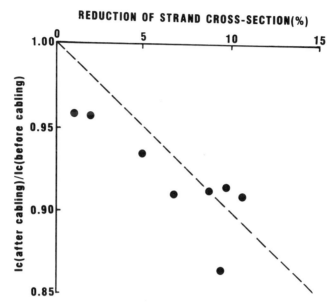

Fig. 4. Relationship between critical current degradation due to cabling and strand cross-sectional area reduction. On the dashed
line, the critical current degradation equals to the strand
cross-sectional area reduction.

Modified SSC Dipole Magnet

Based on the above results, a 5-cm bore 1-m long dipole magnet was fabricated and successfully tested by KEK.[3,4] The dipole coils have an ideal arch structure because they have no wedges. The large keystone angled cable enables us to realize such a structure. The strands have quite similar parameters to the SSC wires on strand diameters, filament diameters, copper-to-NbTi rations, etc. The cables have the same number of strands and the keystone angles are 3.1 degrees for the inner layer cable and 1.7 degrees for the outer.

The dipole magnet was tested and produced the designed field after several quenches. After the coil was tightly clamped, the dipole magnet was tested once again. The quench current reached the designed current at the second quench. In two series of the excitation tests, the quench current reached 99% of the short sample critical current after training. These test results imply a possibility that the large keystone angled cable can successfully modify the SSC dipole magnet.

MAGNETIZATION OF SSC WIRE

Samples

Magnetization measurements were made on three kinds of the SSC wire having nearly the same critical current density. Parameters of the samples are shown in Table 1. Samples #1 and #2 are the outer wire. Sample #1 corresponding to the existing specification was fabricated by a single stacking method. On the other hand, Sample #2 corresponding to the old specification was fabricated by a double stacking method and, consequently, has a short filament spacing which is not preferable in magnetization.

Sample #3 is the inner wire recently developed. Because there exists a highly resistive Cu-0.5%Mn alloy matrix between filaments, low magnetization is expected, in spite of a shorter filament spacing.

Table 1. Parameters of SSC wire samples for magnetization measurement.

Sample	#1	#2	#3
Wire diameter (mm)	0.648	0.648	0.808
Cu/CuMn/NbTi ratio	1.8/0/1	1.8/0/1	1.0/0.5/1
Filament diameter (μm)	6.0	4.8	2.5
Filament spacing (μm)	1.1	0.6	0.45
Jc (A/mm^2) at 5 T	2,650	2,700	2,780

Magnetization Decay

Long term magnetization decay was measured in a magnetic field of 0.3 T after a cycle of 0-1-0 T using a SQUID flux meter. The magnetic field of 0.3 T corresponds to the injection field of 1 TeV of the SSC. Although the injection field has been changed to 0.66 T corresponding to 2 TeV, it is better to measure the magnetization decay at 0.3 T because of its large signal.

Measured results are shown in Fig. 5 and Fig. 6. In comparison between Samples #1 and #2, Sample #2 shows a larger magnetization decay than Sample #1. It is thought that this is because Sample #2 has a shorter filament spacing. Sample #3 shows a very small magnetization decay, as expected. It is thought that Sample #3 has a highly resistive matrix between filaments in spite of its short filament spacing. From the measured results on three samples, it is clear that the magnetization decay is associated with the filament spacing and/or the matrix resistivity between filaments. As Goldfarb et al pointed out, a possible source of the magnetization decay is associated with flux creep in the proximity-coupled matrix between filaments.

CONCLUSIONS

1. A volume production process enabling us to achieve a critical current density of more than 3,000 A/mm^2 at 5 T is being established. Further study must be made on the critical current densities at 7 T for the SSC inner wire and at 5.6 T for the outer.

2. Large keystone angled cables of up to 4.8 degrees have been trial-fabricated. A mechanism of the critical current degradation due to cabling is mainly due to the strand cross-sectional area reduction. The large keystone angled cables show a possibility that they can successfully modify the SSC dipole magnet through the elimination of wedges in the coils.

3. Long term magnetization decay of the SSC wire is associated with flux creep phenomenon occurring between filaments. It is reduced by a long filament spacing and/or a highly resistive matrix.

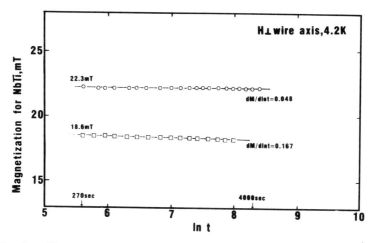

Fig. 5.　Magnetization decay of the SSC outer wire, Samples #1 (upper) and #2 (lower), having a copper matrix.

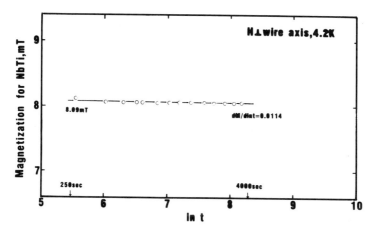

Fig. 6.　Magnetization decay of the SSC inner wire, Sample #3, having a Cu-0.5%Mn alloy matrix.

ACKNOWLEDGEMENTS

Authors are grateful for the continual encouragement of Dr. T. Isobe of Furukawa Electric and also for the valuable comments from Prof. H. Hirabayashi and Assistant Prof. T. Shintomi.

REFERENCES

1. M. Ikeda, "Development of SSC cable in Furukawa," Supercollider 1, 243 (Plenum Publishing Corporation, 1989)
2. H. Ii, Y. Nagasu, M. Ikeda, T. Shintomi and H. Hirabyashi, "Trial-fabrication of superconducting Rutherford type cable having large keystone angle," 11th International Conference on Magnet Technology, MA-01 (Tsukuba, 1989)
3. T. Shintomi, A. Terashima, H. Hirabyashi, M. Ikeda and H. Ii, "Development of large keystone angle cable for dipole magnet with ideal arch structure," International Cryogenic Materials Conference, HX-08 (Los Angeles, 1989)

4. T. Shintomi, A. Terashima and H. Hirabayashi, "Development of superconducting dipole magnet with ideal arch structure using large keystone angle cable," 11th International Conference on Magnet Technology, JG-03 (Tsukuba,1989)
5. R. B. Goldfarb and R. L. Spomer, "Magnetic characteristics and measurements of Nb-Ti wire for the Superconducting Super Collider," International Cryogenic Materials Conference, FX-02 (Los Angeles, 1989)
6. K. Matsumoto and Y. Tanaka, "High critical current density in multifilamentary NbTi superconducting wires," 6th US-Japan Workshop on High Field Superconductors (Boulder, 1989)

A NEW DEVICE FOR PRODUCTION MEASUREMENTS OF FIELD INTEGRAL

AND FIELD DIRECTION OF SC DIPOLE MAGNETS

H.Preissner, R.Bouchard, P.Lüthke, A.Makulski,
R.Meinke, and K.Nesteruk

Deutsches Elektronen-Synchrotron DESY
D-2000 Hamburg 52

ABSTRACT

The performance of all superconducting magnets for HERA is tested in the
DESY magnet test facility and their magnetic field is measured. For dipole magnets
the magnitude and the direction of the field is measured point by point along the axis
with a mole-type probe which is transported through the beam pipe. The positioning
of the probe is done via a toothed belt with an accuracy of 1 mm. The probe houses
two Hall probes perpendicular to each other, a gravitational tilt sensor and an NMR
probe. The field in the plateau is measured by NMR, the fringe field is measured by
the Hall probes and the field direction relative to gravity is obtained from the ratio of
the two Hall voltages and the tilt sensor. The field integral is determined with an
accuracy of 10^{-4} and the average field direction is measured with an accuracy of 0.2
mrad.

INTRODUCTION

The magnetic performance of the 9 m superconducting dipole magnets[1] for the
proton ring of the HERA collider[2] is tested in several measurement procedures before
installation in the tunnel. Their magnetic length and field direction is measured at
2 T. For this purpose they are equipped with a thermally isolated pipe with an inner
diameter of 43 mm and a total length of 13 m (including the path through the cold
boxes). Because of the length and the relatively small warm bore standard
measurement techniques are not well suited. For the mole-type device developed at
DESY there are essentially no limitations from the length of the magnet. Up to now
more than 200 magnets (about half of all HERA dipoles) have been measured with
the probe designed for field integral and field direction measurements. In the next
chapter the main design features of the probe are presented, in the following chapter
the system for transporting the probe through the magnet is described. A brief
overview on the control system is given in the third chapter, and in the last chapter
the methods for calculation of field values and some results of measurements are
shown. More details on the mole system are documented in a thesis[3].

DESIGN OF THE PROBE

Along the 12 m of the mole's travel through the measuring pipe a rotation of the mole by ~ 30 degrees around its axis cannot be avoided. The orientation of the various sensors in the mole, however, should not vary by more than ± 60 mrad (± 3.5 degrees). Therefore the sensors are mounted on a gravitational pendulum which is suspended between the two end pieces of the mole (see fig.1). Each end piece is equipped with a non-magnetic ball bearing (glass/plastics or copper-beryllium) for support of the pendulum and two of the mole's wheels. Furthermore the end pieces hold the electrical connectors and connections for the gas supply and for the toothed belt of the transport system. The mechanical connection between the end pieces is achieved by a titanium tube which forms the outer cover of the mole.

The pendulum is of cylindrical shape with a copper bar at the bottom for the required eccentricity. For calibration purposes the motion of the pendulum can be blocked. The wiring between the end pieces and the sensor unit on the pendulum is designed to keep additional torques on the pendulum low. For this aim thin and elastic wires are fed through the inner bore of the ball bearings.

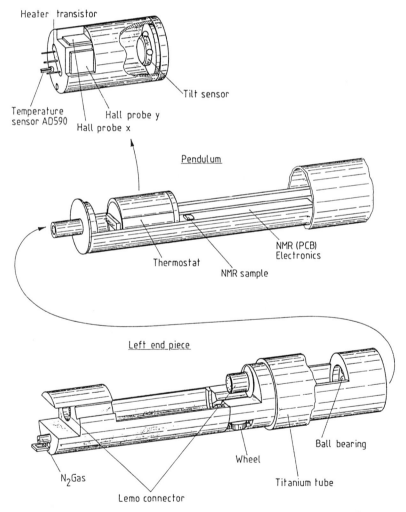

Fig. 1. Schematic view of the probe with the outer cover removed. The upper drawing shows the assembly of the sensors on the copper block inside the thermally isolated capsule.

The sensor unit contains on one side the NMR probe with rubber as resonating material and the RF regulating ciruits. The performance of rubber as probe material is excellent and, unlike a water probe, it is safe at low temperatures. The rest of the sensors is enclosed in a thermally isolated capsule with its interior stabilized to 30 ± 0.1 C. Two Hall probes, a gravitational tilt sensor and an IC temperature sensor are mounted on one copper block which is heated with up to 4 W by a transitor (fig.1). The Hall probes are parallel to the axis of the pipe, one horizontal and one vertical. They are connected in series for the supply current and are not equipped with resistors for linearization. The tilt sensor[4] is additionally enclosed in a copper case to avoid internal temperature gradients. It contains an electrolytic fluid between 3 electrodes and acts as voltage divider for the supplied AC voltage. When using a capacitve AC coupling, the stability of the sensor is improved with two parallel resistors ($\sim$ 20 kΩ) from the center electrode to the outer electrodes. A preamplifier is mounted in one of the end pieces to transform the output signal of the tilt sensor to low impedance.

TRANSPORT SYSTEM

A systematic view of the transport system for the mole device is shown in fig.2. The mole runs on four plastic wheels and it is pulled through the bore of the magnet via toothed belts attached to both ends. The front belt is driven by a step motor to the requested position. An AC motor applies a constant force to the rear belt which, by driving an angular encoder, gives a measure for the actual position of the mole. The number of encoder counts per travelled distance depends on the tension on the belt, but the calibration proved to be constant over a year within 1 mm per 10 m.

3 cables with 8 wires each plus a BNC cable for NMR control are needed to supply and to read-out the sensors in the mole. Each side of the mole connects to 2 cables. In addition nitrogen gas flow is supplied through plastics hoses to both ends of the mole to prevent too large a drop in temperature and the condensation of moisture in the warm bore. During the motion of the mole a special cable transport system picks up cable on one end and releases cable at the same rate on the other end from variable length cable loops.

Schematic lay-out of probe driving mechanism

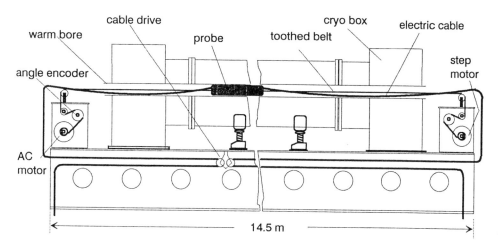

Fig. 2. Transport system for the probe on a magnet test stand.

CONTROL AND READOUT

An LSI computer is used to control the transport system, the measuring system and the current supply for the dipole magnet. Most of the functions are controlled via a GPIB interface. The AC signal from the tilt sensor is first converted to a DC voltage by a dedicated commercial hybrid circuit[4]. The reading of the sensor voltages is sequential by switching the read-out to a voltage-to-frequency converter. Switching is done by relais in a switch box which is also responsible for switching the supply voltages and currents. For safety the switch box refuses to perform any of these connections unless it recognizes the proper probe-type identifier, a resistor of a defined value built into the probe. This way probes for various purposes and with different wiring can safely be connected to the same cables.

In a standard measurement the probe takes readings at a number of predefined positions along the magnet. A position is reached after sending the corresponding number of pulses to the step motor. At each point the values of the magnet current, the temperature, the NMR, the two Hall probes, the tilt sensor and the angular encoder (for the actual position) are read out and written to a file. The tilt sensor and the vertical Hall probe are read interleaved (5 and 4 times, respectively) to reduce errors for the field direction arising from residual motion of the pendulum which is long-lasting due to eddy currents in the copper bar. The NMR is expected to lock in resonance in the plateau of the magnetic field; on failure the measurement routine pauses giving the operator the chance to retune the NMR controller and to repeat the read-out. Pausing of the program can also be caused by unexpected values of the other read-outs.

CALIBRATION AND PERFORMANCE

Standard measurements are performed with a magnet current of 2100 A yielding a plateau field of about 2 T. The field integral $\int Bdl$ is approximated by the sum over those points where readings were taken. B is the magnitude of the magnetic field as given by the NMR in the region of the plateau, and it stands for the vector sum of the two field components perpendicular to the magnet axis when the field is measured by the Hall probes:

$$B = \sqrt{B_x^2 + B_y^2} \ .$$

As the vertical Hall probe is almost parallel to the field the following approximation can be used:

$$B \simeq B_x \cdot (1 + a \cdot U_y/U_x + b \cdot (U_y/U_x)^2) \ ,$$

where U_x and U_y are the voltages of the horizontal and the vertical Hall probes, respectively (reduced by their 0-voltages). The coefficients a and b take account for different sensitivities, for non-perpendicular mounting of the Hall probes and the non-linearity of the Hall voltage U_y to a relative accuracy of 10^{-5} in the plateau. The field dependence of the Hall ratio U_y/U_x is ignored, causing a maximum error of 20 Gauss in the fringe field. Both Hall probes are not linearized by a resistor; the nonlinearity of the horizontal probe is corrected for by a hyperbola function (fig.3) to an accuracy of 2.0 Gauss between 5 and 20 kGauss:

$$B_x = c \cdot U_x \cdot (1 + d \cdot (\sqrt{U_x^2 + e^2} - e)) \ .$$

For each measurement the horizontal Hall probe is auto-calibrated by the NMR readings in the plateau in the sense that the coefficient c is redetermined and only d and e are taken from a calibration table.

The field direction ϕ is measured relative to the orientation of the pendulum by the Hall ratio U_y/U_x and to obtain its value relative to gravity the angle given by the tilt sensor is subtracted. The relation between angle and Hall ratio is field dependent

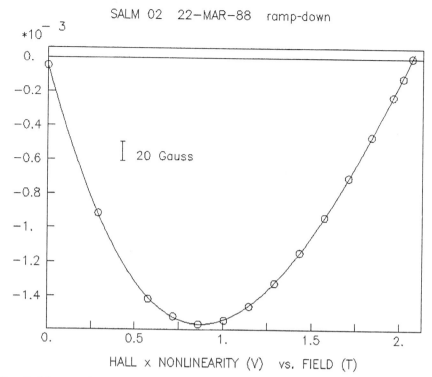

SALM 02   22-MAR-88   ramp-down

$*10^{-3}$

HALL x NONLINEARITY (V)   vs. FIELD (T)

Fig. 3. Measured non-linearity of the Hall voltage and its parametrization (solid
line) as a function of the magnetic field. The non-linearity is displayed as
the deviation from a staight line through the origin to the measured point
at 2.1 T. At 2 T the voltage is 0.1 V.

but for the average angle of the magnetic field this dependence can be ignored. When
the deflection of the pendulum is less than 3.5 degrees, then the angle to be measured
by the Hall probes is proportional to U_y/U_x within 0.1 mrad and

$$\phi = \alpha \cdot U_y/U_x - \beta \cdot U_T - \gamma \ .$$

where U_T is the read-out for the tilt sensor and α, β, γ are calibration constants. The
accuracy of the measurement of ϕ is 0.2 mrad and it is limited by the stability of the
tilt sensor and its control module.

Results from a standard measurement with read-outs at 91 positions are shown
in fig. 4. The spacing between the positions is 2 cm in the fringe field and 28 cm in
the plateau.

CONCLUSIONS

The mole device and its prototypes have been in operation with high reliability
and no requirements on maintenance since more than two years. Installation of the
probe in the transport system is relatively simple and a standard measurement
(lasting about 1 hour) needs almost no attention from the operator.

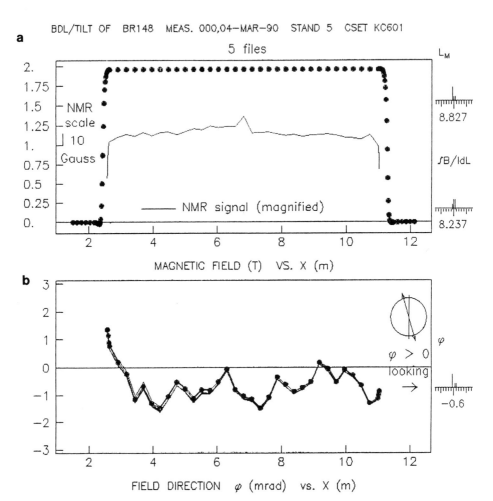

Fig. 4. Results of field measurements for a selected magnet. (a) field magnitude,
(b) field direction in the plateau. The small histograms on the right show
the reproducibility of 5 repeated measurements for the magnetic length
(1 mm/tic), for the integral of field / current (Tm/kA) and for the field
direction (0.1 mrad/tic).

REFERENCES

1. H. Kaiser, Design of Superconducting Dipole for HERA, 13th Intern. Conf.
    on High Energy Accelerators, Novosibirsk, USSR, Aug. 1986.
    S. Wolff, Superconducting HERA magnets, IEEE Trans. Mag. 24, No. 2,
    719 (1988).
    B. H. Wiik, Design and Status of the HERA Superconducting Magnets,
    World Congress on Superconductivity, Houston, Tx., Feb. 1988.
    H. R. Barton Jr., R. Bouchard, Yan-Fang Bi, H. Brück, M. Dabrowska, D.
    Darvill, Wen-Liang He, Zhuo-Min Chen, Zheng-Kuan Jiao, D. Gall, G.
    Knies, J. Krzywinski, J. Kulka, A. Ladage, R. Lange, Liang-Zhen Lin,

A. Makulski, R. Meinke, F. Müller, K. Nesteruk, J. Nogiec, H. Preissner, W. Rakoczy, P. Schmüser, M. Surala, E. Schnacke, Z. Skotniczny, Wen-Long Shi, Performance of the Superconducting Magnets for the HERA Accelerator, DESY report HERA 89-20 (1989) and contribution to 11th Int. Conf. on Magnet Technology, MT-11, Tsukuba, Japan, Aug. 1989.

2. For a recent review see :
   B. H. Wiik, HERA Status, 1989 IEEE Particle Accelerator Conf., Chicago, Ill., Mar. 1989.
3. R. Bouchard, thesis, unpublished.
4. SPECTRON sensor L-211U and TECHTRIUM MUPI-module, distributed by G+G Technics AG, CH-4411 Lupsingen, Switzerland.

AN "IN HOUSE" CABLING FACILITY*

H. Kanithi, F. Krahula and B. Zeitlin

IGC Advanced Superconductors Inc.
1875 Thomaston Avenue
Waterbury, CT 06704

ABSTRACT

While considerable effort has been expended in industry in recent years to improve the quality and current carrying capability of the strand for the Superconducting Super Collider (SSC) dipole and quadrupole magnets, much of the work on cabling equipment and techniques has been left to the National Laboratories. IGC, in cooperation with AFA, and with financial assistance from the Small Business Innovative Research program of the US DOE, has recently established an "in house" cabling facility. It is anticipated that this equipment will provide an economical method of producing both inner and outer cables for SSC magnets. The equipment, which is described in some detail, has now been installed in our facilities and the results of some of the preliminary "shake down" tests carried out on it are described in this paper.

INTRODUCTION

Flat Rutherford type superconducting cables have long been used in High Energy Physics Particle Accelerator magnets. The Superconducting Super Collider project requires approximately 25,000 km of cable for the main ring of dipole and quadrupole magnets. The 4 mm aperture magnet design calls for two types of cables[1] - a 23 strand inner grade and a 30 strand outer grade. The dimensional and electrical requirements of the cables are more stringent than for any cables made in the past. The technical specifications[1] are the result of extensive effort by Royet and coworkers at the Lawrence Berkeley Laboratory (LBL). They built an experimental machine to accomplish the initial development which led to the generation of a specification[2] for an industrial cabling machine.

It is estimated that a total of 10 to 12 cabling machines will be required to meet the production demand for the SSC over a 4-year period. Dour-Metal, a Belgian company has made the first industrial prototype cabler which has been installed in New England Electric Wire plant in Lisbon, N.H. The conceptual design and the characteristics of the machine have been reported by J. Grisel[3] et al.. After initial shake-down tests and subsequent modifications, this machine is now being routinely used to produce the cables required by the SSC magnet development program. The first industrial production results were summarized by Royet[4] and coworkers.

In this paper we describe an "in-house" cabling facility capable of making up to 30 strand cables. The cabling machine was built by AFA Industries, Garfield, N.J. to the LBL specification, some aspects of which, were modified by IGC.

---

*This work is partially supported by the US Department of Energy under a Small Business Innovation Research Contract (No. DE-AC02-86ER80387)

MACHINE FEATURES

The entire cabling facility, as it is presently set up at IGC, is shown in Figure 1. It consists of five main components:

1) a planetary rotor assembly,
2) a turks head,
3) a caterpillar drive,
4) a size measuring device,
5) a take-up unit.

1) Rotor Assembly

Driven by a 15 kW (20 h.p.) motor, the rotor assembly (Figure 2) has two bays, each carrying 15 spools. Except for the shaft and supporting frame which are made of steel, all other parts are of structural aluminum. Planetary motion of the spools is generated by mechanical chain drive connected to an independent 2.2 kW (3 h.p.) motor which is synchronized with the main drive through a micro processor. The maximum planetary ratio is 5:1 allowing the pay-off spools to make up to 5 revolutions for every revolution of the rotor in either clockwise or counter clockwise direction. The spools are of plastic and of a readily available size - 200 mm flange x 115 mm barrel x 150 mm traverse. Each can hold up to 15 kg of strand. Being light weight, they add little to the cetrifugal forces of the moving rotor. The maximum rated speed of the rotor assembly is 120 rpm which corresponds to a cabling rate of ~10 m/min. The cable pitch or lay length can be adjusted between 63 and 89 mm by means of locked change gears.

One of the critical requirements of the cabling is the precise strand tension control. In the present machine, the tension can be adjusted between 2 and 4 kg. Once set, it is required to stay within +/-0.25 kg of the nominal value from the full spool condition to the empty spool condition. The tension is also to remain constant within +/-0.1 kg from spool to spool. Two types of mechanisms are included in the strand pay-off to accomplish this stringent tension requirement. One is friction pads situated at the side of each spool flange and the other is in-line magnetic clutches. The strand from either bay travels straight until it is wound on the cylindrical drums of the magnetic clutches. There is an active wire break detection system which uses externally wired photoelectric cells and is activated by the loss of strand tension.

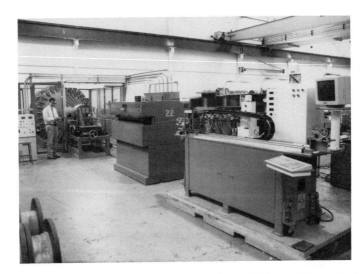

Figure 1. An over view of the cabling facility at IGC consisting of rotor assembly, turks head, caterpillar capstan, cable measuring machine and take-up unit (not in picture).

Figure 2. Rotor assembly with two bays of spools carrying 15 spools each.

Figure 3. An over all view and a close up of the turks head.

All strands pass through individual thin stainless steel tubes which support the wires as they rotate at high velocities before reaching a gathering or spider die. The strands then wrap around a core pin which is rigidly held in position by locking screws at the back of the machine. The position of the core pin relative to the turks head can be adjusted.

2) Turks Head

The cabling facility is equipped with a non-powered turks head (Figure 3) of type "Fenn 4" containing 10 cm dia. rollers. The turks head is mounted on a frame which has three degrees of freedom to facilitate accurate alignment with the core pin. It is believed that non-powered rollers are adequate to form the cable without excessive line tensions. If it deemed necessary to have powered rollers after exhaustive production trials followed by the characterization of the resulting cable, an appropriate power driven turks head may be set up in the future in place of the existing unit.

3) Caterpillar Drive

The cable pulling force to overcome all the strand tensions and the turks head rolling friction, is provided by a linear caterpillar capstan. The latter is driven by a shaft directly coupled to the main rotor drive motor. Special synthetic belts of the caterpillar can exert up to 225 kg tension while contacting 1.4 m of the cable.

4) Cable Measuring Machine

A recent version of a cable measuring machine was supplied by the SSC laboratory. It is capable of verifying cable thickness, width, and keystone angle at selected intervals at cabling speeds of up to 10 m/min. It is equipped with an IBM personal computer for recording the data.

5) Take-up Unit

The present cabling facility is equipped with a take-up unit designed to handle up to 76 cm diameter reels. The maximum capacity is ~300 kg of finished cable. The reel traverses laterally at predetermined rates so that the cable always stays in a fixed location. This type of arrangement exerts little disturbance to flat cables. For larger than 76 cm dia. reels, the present take-up machine will have to be replaced with another bigger unit. The overall length of the facility including the take-up unit, is approximately 17 m.

TEST RUN

The initial rotational tests were done using commercial bronze wire. Being stiffer than superconductor strand, the bronze wire took a considerable amount of time to form the cable. The tendency for the bronze cable was to collapse during the early set up at the slightest provocation. After several successful attempts with the bronze wire, we have loaded the machine with SSC outer grade superconducting strand and made ~100 m of 30 strand cable. Since the cable measuring machine had not been commissioned at that time, size verification was done by manual gauging. The highest speed we ran the cabler was 100 rpm. The noise level of this machine at this speed was remarkably low as compared to the Dour machine at New England Electric Wire plant.

In order to determine any degradation in critical current due to cabling, several strands were carefully unravelled, straightened and tested for I_c at 5.6 Tesla. The degradation based on strand Ic prior to cabling was calculated to be <3%. Although this value is typical for such cables. Further cabling with this facility and a full evaluation of the resulting cables are yet to be done.

SUMMARY

IGC, in cooperation the SSC Laboratory, with financial assistance from the Small Business Innovative Research program of the US DOE has recently established an "in house" cabling facility capable of producing 30 strand Rutherford type cables. All the major components of the facility were made commercially in the US. The rotor assembly offers a significantly different design than the first machine, built by Dour Metal in Belgium. Preliminary shake down tests were successfully completed. A pilot sample cable of the SSC outer grade was made, tested and found to have acceptable critical current performance. Further cabling and a full evaluation of the resulting cables are planned in the near future, to make the facility truly "production ready". When the SSC changes over to the new 5 mm bore magnet design from the 4 mm design, the present facility will be able to produce 30 strand inner cable. Modifications to the rotor assembly to be able to produce 36 strand new outer cable are not practical at this point. However, the design and construction features of this machine may be easily adapted to future 36 strand cablers which are required to support the SSC production needs.

REFERENCES

1) Material Specification titled "NbTi Superconductor Cable for SSC Dipole Magnets", Prepared by R. Scanlan, Lawrence Berkeley Laboratory, No. SSC-MAG-M-402 Rev. No. 2 dated Oct. 3, 1988.

2) Specification titled "General Superconducting Super Collider - Cabler - 30 spool", prepared by R. Wolgast, No. M686, Code AA0100, Lawrence Berkeley Laboratory, dated Aug. 13, 1986.

3) J. Grisel, J.M. Royet and R. M. Scanlan and R. Armer, "A unique Cabling Machine Designed to produce Rutherford type Superconducting Cable for the SSC project", IEEE Trans. on Magnetics, Vol. 25, No. 2, March 1989, p1608.

4) J. Royet, R. Armer, R. Hannaford and R. Scanlan, "An Industrial Cabling Machine for the SSC", Proc. of the International Industrial Symposium on the Super Collider, Feb. 8-10, New Orleans, M. Mc Ashan, Ed., Plenum Press, New York(1989) p 273.

CONCEPTUAL DESIGN FOR THE SSC HIGH ENERGY BOOSTER

T.H. Nicol, F.A. Harfoush, M.A. Harrison, J.S. Kerby
K. P. Koepke, P.M. Mantsch, T.J. Peterson, and
A.W. Riddiford

Fermilab National Accelerator Laboratory
Box 500
Batavia, IL 60510

ABSTRACT

A tremendous amount of work has been done over the course of the past several years on the design of dipole magnets for the SSC main collider. Although they dominate the total magnet cost for the project and thus deserve the tremendous research and development effort they have received, the main ring magnets represent only part of the magnet requirements for the SSC. This paper presents the work to date on the design of dipole magnets for the SSC high energy booster (HEB). A complete discussion of the design is beyond the scope of this work.[1] Rather it serves as an overview of the main aspects of the complete design.

INTRODUCTION

The high energy booster serves as the injector for the main collider ring as well as a source for test beams. The dipoles for the HEB are required to produce a peak operating field of 6.26 T which corresponds to a maximum energy of 2 Tev. The injection energy of 200 Gev defines the minimum field level of 0.626 T. The magnet design is consistent with the overall goals of the accelerator operation which require continuous ramping with a two minute cycle. The relationship of the HEB to each of the other components of the complete accelerator are shown in Figure 1.

The design requirements for the HEB magnets differ from those of the main collider in several respects. The good field aperture requirement gives a bore that is 40% larger; 70 mm compared to 50 mm for the collider dipole and the magnetic field strength is slightly less; 6.26 T as compared to 6.6 T for the collider. The HEB must also be capable of bipolar operation and continuous ramping and must be stable up to 10% above its nominal operating field. The design of the HEB dipole includes a cable specification which ensures adequate operating margin, coil and iron geometries to provide field shape, mechanical structures to contain coil stresses both from fabrication preload and Lorentz forces, and a complete cryostat system. This paper will address each of these topics in some detail with special emphasis on some of the more interesting aspects of the design. Table 1 lists as many of the design parameters as we were able to deduce from a preliminary specification of the HEB magnet system or to conclude as a result of our design work. Figure 2 illustrates a two dimensional cross section through the complete magnet assembly developed during the course of this study.

Supercollider 2, Edited by M. McAshan
Plenum Press, New York, 1990

Table 1. HEB Dipole Magnet Parameters

Number of dipoles	864
Overall dipole slot length	8.25 m
Magnetic length	7.75 m
Coil inner diameter	70 mm
Mass of conductor	363 kg
Cold Mass	6708 kg
Central Field	6.26 T
Current	5582 A
Inductance	80 mH
Stored energy	1.25 MJ
Beam tube inner diameter	56 mm
Inner layer number of turns	72
Outer layer number of turns	60
Iron Yoke inner diameter	20 cm
Iron Yoke outer diameter	42 cm
Cryostat outer diameter	72 cm
Magnet mass	7973 kg

COIL CROSS SECTION

The most significant change to the coil cross section is the introduction of an offset coil design which improves the profile of the field error, dB/B. The idea has been studied by T. Collins[2], and more recently by Ishibashi[3]. They studied the simplified problem of one constant current density area per group of conductors, neglecting the insulation wrapped around each conductor.

The conductor is assumed to have the same critical current density as the conductor used in the collider dipole (2750 A/mm^2 at 5 T and 4.2K). The primary objective in selecting the conductor for this magnet was to ensure that the magnet have a performance margin of 10% over the nominal operating field. A copper-to-superconductor ratio of 1.5:1 was selected to ensure coil stability and to facilitate quench protection.

Figure 3 shows a typical cross-section through the coil assembly. A wedge is not needed in the outer layer for the given coil inner diameter. The keystone angle is large enough to keep the conductors roughly radial. However, the inner layer of conductors needs at least one wedge to keep conductors radial. It has been found that two wedges permit solutions with smaller harmonic coefficients. Furthermore, each of the wedges allows a solution to be found which zeros out one of the coefficients. The method discussed here uses two wedges and zeros out the sextupole and decapole harmonic coefficients. The magnet performance is summarized in Table 2.

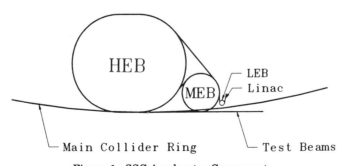

Figure 1. SSC Accelerator Components

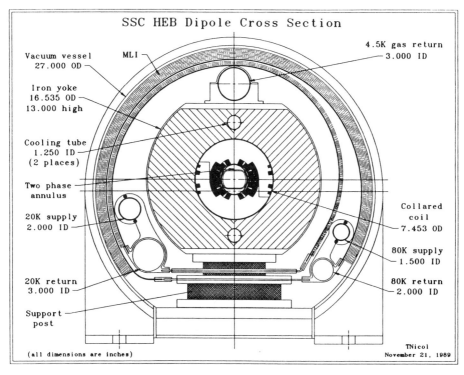

Figure 2. Complete HEB Two Dimensional Cross Section

Table 2. HEB Dipole Magnet Performance

Imax (amps)	6200
Bmax (T)	7.521
B (@Imax) central field (T)	7.021
Bmax/B central	7.14%
I operating current (amps)	5582
B (@I operating) central field (T)	6.26
Transfer function (Tesla/kA)	1.132
Number of turns	132
Inner iron radius (inches)	3.789
Coil offset (inches)	0.163
dB/Bmax (x10000)	1.006
Radius at dB/Bmax (inches)	1.05

Harmonic coefficients at 1 inch
in units of 10^{-4}:

2 pole:	10000	26 pole:	-2.18
6	0.0	30	0.42
10	0.0	34	-0.02
14	0.42	38	0.01
18	1.62	42	0.03
22	0.70		

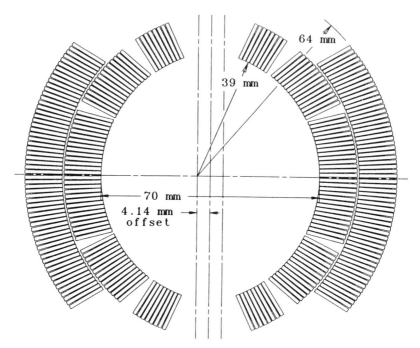

Figure 3. HEB Offset Coil Geometry

YOKE MAGNETIC DESIGN

The iron yoke, in addition to holding the coils rigidly in place acts to give a substantial field enhancement. The iron contribution to the field can be analytically computed for a finite or infinite permeability. The problem becomes analytically intractable when the permeability itself becomes a function of field, resulting in a nonlinear problem. Numerical methods, such as POISSON, have been widely used to solve such problems. Our main concern here in designing the yoke is to achieve a level of saturation sextupole less than 2 units at 1 inch and an amplification factor of less than 2%. A convenient unit of measurement of the multipole coefficients is in units of $10^{-4} B_0$, that is, one part in 10,000 of the main dipole field. Since this magnet is to have a bipolar operation, an important question is the effect of remnant field on the field quality. We shall address all these questions and explain the steps that lead to our suggested yoke design.

First we looked at the effect of varying the iron inner radius on the sextupole. It has been observed from numerical simulations that for a given inner radius the sextupole variation versus dipole field can have one of two shapes. The first and most common shape is where the sextupole magnitude increases to a maximum value and then decreases. This behavior is attributed to the saturation in the immediate vicinity of the coil causing the increase in sextupole. Saturation on either side of the coil will cause the sextupole magnitude to decrease. It is important to note that such a behavior is very dependent on the inner radius value. For a relatively large inner radius, the notion of immediate vicinity becomes less significant. This results in a reduction in the peak value until a point is reached where any increase in inner radius will lead to the second shape; that of a monotonically decreasing sextupole versus dipole field.

An increase in inner radius drives the sextupole more negative and a decrease drives it more positive. This implies that the negative contribution is due to the midplane saturation and the positive contribution to the pole saturation nearest to the coil. We have found that for an inner radius of 4 inches, the sextupole value at both low and high fields is negative.

As the inner radius increases the curve becomes a monotonically decreasing function. Based on the two types of behavior just discussed we then have two options in our yoke design. Either select a radius such that the sextupole value over a given field range does not exceed a limit value or have a monotonically decreasing value that will not reach a critical value at a desired high field point. Increasing the iron inner radius will decrease the iron saturation and therefore result in a better amplification factor. The opposite is true if we decrease the inner radius. An inner radius in the range of 3.5 to 4 inches will limit the amplification factor to less than 2% at 6.6 T fields. The effect of the outer radius on sextupole is less predictable.

By reducing the thickness of the iron we will significantly increase the sextupole component. The effect of the outer radius on the dipole field quality is measured by the amplification factor. By increasing the thickness of the iron a smaller amplification factor (stronger dipole field) is obtained up to a point beyond which any increase in radius will have little effect on the field and the outer radius acts as if it were infinite. The opposite is true if we reduce the thickness of the iron.

Another important parameter in the yoke design are the fringe fields. Looking at the midplane cross section along the horizontal axis it is possible to plot the field value in air in the immediate vicinity of the iron. For an outer radius of 8.66 inches the fringe field is less than 1 kG. We have chosen this value as our minimum outer yoke radius.

To summarize the result of our findings the inner radius of the iron should be selected in the range of 3.5 to 4 inches and the outer radius should have a minimum radius of 8.66 inches. For this range of data the amplification factor is still in an acceptable range of less than 2%.

BEAM TUBE DESIGN

The HEB dipole design incorporates a novel beam tube assembly which provides for the dissipation of the anticipated 4.2K heat loads during continuous ramping. The assembly allows for a redesigned flow loop that shortens the thermal path between the heat sink and the primary heat source. The new cooling scheme achieves more uniform coil temperatures through a continuous single phase to two phase heat exchange.

The beam tube assembly shown in Figure 4, consists of two concentric, constant thickness stainless steel tubes. The inner tube, 0.09375 inch thick, separates the bore tube vacuum from the two phase helium flow located in the annular space of the two tubes. This provides for a physical beam aperture of 1.6 inch vertical, and 2.2 inches horizontal. The two phase helium flows through an annulus 0.0625 inch wide at the mid-plane and 0.5 inch wide at the poles. The outer tube, 0.0625 inch thick, separates the single phase flow in the collared coil region from the two phase return flow. The single phase helium then flows through a 0.0625 inch gap concentric to the inside diameter of the coil. A thin (0.003 inch) copper plating is applied to the exterior of the single phase/two phase tube to moderate any azimuthal temperature distribution in the wall. Mechanically the inner beam tube is located with respect to the single phase/two phase tube by axially intermittent spacers. These spacers also integrate the tubes structurally by distributing the loads between the two. The single phase/two phase tube is located by contact with the collar packs in the pole region. This interference serves to limit the deflection of the tubes when pressurized and maintain material stresses within acceptable limits.

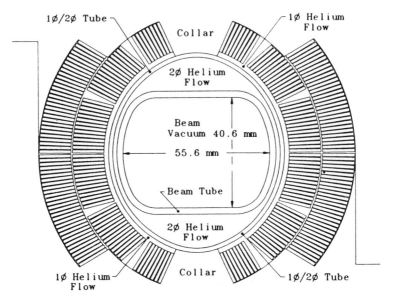

Figure 4. HEB Dipole Beam Tube Assembly

COLLAR DESIGN

The collar design of the HEB follows the collider ring magnet closely. The collars are high strength aluminium, 3.724 inches OD with front and back collars joined by semi-perfs. The locking mechanism is provided by eight keys located symmetrically, two per quadrant. Unlike other designs the HEB dipole collars have a stainless steel skin. The

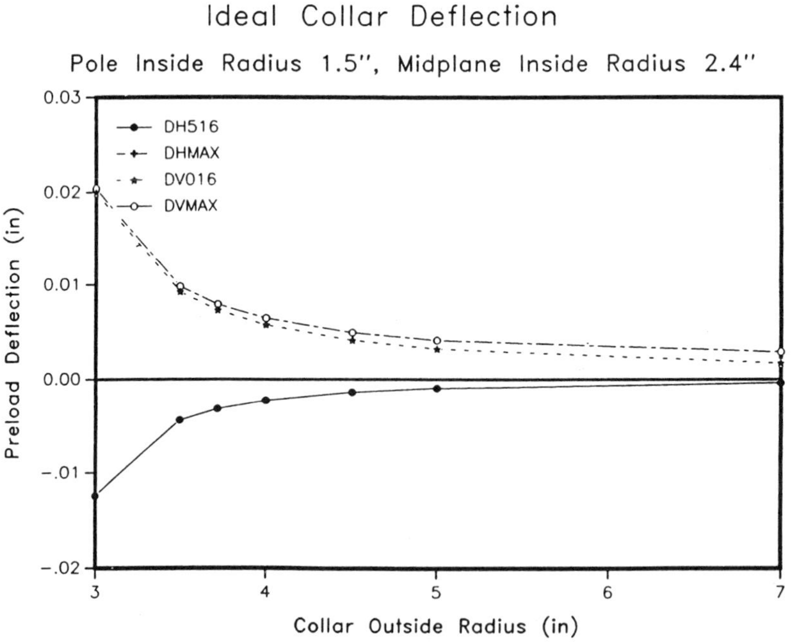

Ideal Collar Deflection
Pole Inside Radius 1.5″, Midplane Inside Radius 2.4″

Figure 5. HEB Dipole Collar Deflection vs. Collar Thickness

skin, 0.0625 inch thick, serves to limit the single phase helium volume, but is not a structural member in the assembly.

The magnetic analysis of the yoke saturation properties indicated that a yoke inner radius of 3.8 inches was optimum. This result coincides well with the structural analysis which indicates that a collar outer radius greater than 3.5 inches is needed. A collar outer radius of 3.724 inches was selected which results in 0.003 inch deflections in the horizontal parting plane and 0.008 inch in the vertical from preload forces without any yoke support.

With the collar dimensions chosen, a keying mechanism is needed which makes the actual collar approach the ideal case as closely as possible. The keys provide the locking mechanism which prevents the collars from springing apart when the collaring process is over. Furthermore, they must be sized so that the deflection of the collared coil is minimized while the maximum bending stresses remain below the elastic limit of the collar material.

Each of the collar keys are 0.2 inch wide each. The four keys nearest the horizontal midplane are 0.375 inch deep, the remaining ones are 0.25 inch deep. The keys are all stainless steel. The finite element model used for the analysis consists of front and back collars which, by symmetry, provide a complete model of the collar lamination packs. This method is identical to that used in the analysis of collider dipoles. Forces arising from 20 kpsi inner and 10 kpsi outer preloads (corresponding to a 8.8 T central field) were applied directly to the inner surface of the coil cavity in the collars. Keys were created of various size, shape, and number to simulate schemes of potential interest.

The first collar design used a single key per quadrant. The collar deflection was limited to 0.012 inch in the vertical plane. However a local maximum von Mises stress of 96 kpsi was predicted in the collar at the midpoint of the keyway, 37% larger than the elastic limit of the collar material. To distribute the stresses more evenly, a second key was added to each quadrant, both keys being 0.25 inch deep. The vertical deflection is reduced slightly, to 0.011 inch, and the maximum stress is reduced to 78 kpsi at the midpoint of the upper key. The midplane key is shielded by the action of the upper key. Increasing the depth of the midplane key to 0.375 inch reduced the vertical deflection to 0.010 inch and reduced the maximum stress to 70 kspi by distributing the locking action of the keys more uniformly. This was the design chosen.

YOKE MECHANICAL DESIGN

The HEB dipole is a cold iron magnet with a vertically split, dry iron yoke. Cold iron has several advantages over its warm iron counterparts. First, it provides the maximum field enhancement due to the close proximity of the iron to the collared coil assembly. Second, there are no forces imposed on the suspension system resulting from the coil and iron centerlines not being coincident. The disadvantage of cold iron lies principally with the fact that it represents a significant load on the refrigeration system during cooldown.

There are two differences between the HEB dipole yoke and the collider dipole yoke. First, the HEB dipole uses a vertically split yoke. This ensures that a parting plane gap represents no perturbation to the magnetic field. Contact between the yoke and the collar is ensured in the horizontal plane. Mechanically this increases the cold mass resistance to the Lorentz forces by increasing the structural stiffness. This assembly style has recently been tested on collider dipole short magnet models and is also used in the sucessfully prototyped LHC dipole magnets.

Second is in use of dry iron. In this system the single phase helium flow is contained in two pipes 1.25 inches in diameter which pass through the yoke laminations. The advantages of the dry iron concept are that the total helium volume of the magnet is reduced and that the cold mass outer skin is no longer required to satisfy pressure vessel codes. This latter feature allows the yoke laminations to be cropped at the top and bottom thus reducing the yoke mass and providing space within the cryostat for the cryogenic piping. Concerns about the dry iron concept include ensuring thermal contact between the single phase pipe and the laminations at all times, and the quench characteristics of the magnet with reduced helium volume leading to significant pressure rises. The quench behavior is expected to be similar to the warm iron Tevatron dipoles where all the helium volume is contained in the collared coil region as in this magnet. Analysis of these effects is still in progress.

COLD MASS SUSPENSION SYSTEM

The suspension system in a superconducting magnet performs two functions; it resists structural loads imposed on the cold mass assembly during shipping, handling, seismic disturbances, and quenches, and it insulates the cold mass from heat radiated and conducted from the environment. The suspension system for the HEB dipole follows that of the collider dipoles. The major components of the suspension system are shown in Figure 6. The magnet assembly is supported vertically and laterally at two places along its length. To accommodate axial shrinkage during cooldown, the magnet assembly is free to slide on one support while the other serves as the anchor position. To distribute any imposed axial load to the supports, a tie bar is used to connect the supports. The structural design criteria are listed in Table 3.

Each post assembly consists of inner and outer composite tubes connected by an intermediate stainless steel transition tube. Stainless steel and aluminium discs and rings serve to join the tubes and act as tie points to other cryostat components. A cross section through a typical re-entrant support post is shown in Figure 7. Calculations and tests have shown that the bending loads resulting from the axial and lateral loads produce the highest stresses in the post assembly. In order to satisfy the load cases a computer program was written to calculate the tube thicknesses required to satisfy the structural requirements from a given set of input criteria.[4] The overall height and diameter of the support post and the ratios of the various thermal path lengths are determined in large part by the cryostat configuration and the conductive heat load constraints.

The support post design which satisfies the structural requirements results in a fiberglass composite outer tube 8 inches in diameter with a wall thickness of 0.14 inch. The inner tube is a 6 inch diameter graphite composite with a wall thickness of 0.15 inch. The maximum stresses are 20 kpsi and 30 kpsi respectively. The estimated heat load from this design to the 4.2K system is 0.12 W per post assembly. The 20K and 80K heat loads are 0.82 and 6.2W respectively.

CRYOSTAT DESIGN

The cryostat for the HEB dipole also has its origin in the SSC main ring dipole. The cryostat consists of a vacuum vessel, 80K and 20K shields, cold assembly, multilayer insulation (MLI), and suspension assembly. Unlike its collider counterpart, which will have a synchrotron radiation heat load to 4.2K of 0.15 W/m, the AC heating due to ramping of the magnet to full field and back will result in a heat load of approximately 2.0 W/m. This higher heat load accounts for most of the differences in the piping schemes between the two cryostat systems particularly in the two phase cooling around the beam tube.

Table 3. HEB Dipole Suspension System Structural Design Criteria

Shipping and Handling Loads	vertical 2.0 G
	lateral 1.0 G
	axial 1.5 G
Seismic load guidelines	Nuclear Regulatory Guide 1.61
	1.61 vertical and horizontal
	spectra scaled by 0.3
Maximum axial quench load	15000 lb

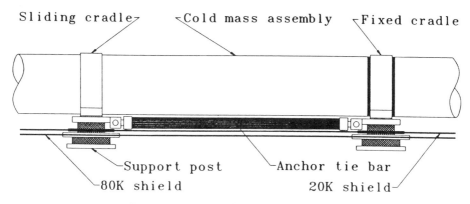

Figure 6. HEB Dipole Suspension System

The vacuum vessel is a 27 inch diameter carbon steel pipe. It is fabricated from three lengths of pipe joined at stiffening rings located at the two support post locations. The stiffening rings are required to transmit internally generated loads from the support posts to ground.

The two shells radially inward from the inner surface of the vacuum vessel are thermal radiation shields. They serve as heat sinks to minimize radiative heat transfer to the collared coil assembly. The 80K shield intercepts heat radiating from the 300K surface of the vacuum vessel. The 20K shield intercepts heat radiating from the inner surface of the 80K shield. Both shields also intercept heat conducted through the support system The shields are aluminum shells welded to their respective cooling tubes, and both shields are covered with MLI in order to minimize radiative heat transfer to their surfaces. The 80K shield is covered with 3/4 inch of reflective mylar and a nylon spacer material. Approximately 3/8 inch cover the 20K shield. A cross section of the complete cryostat system is shown in Figure 2.

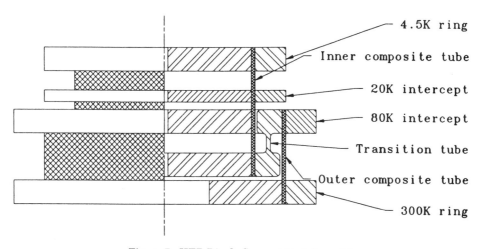

Figure 7. HEB Dipole Support Post Assembly

CONCLUDING REMARKS

As mentioned in the introduction, a complete description of the design work to date on the high energy booster is beyond the scope of this paper. We have tried to present an accounting of the areas in which we have made significant progress and which represent interesting aspects of the design. This work is clearly not the last word in the HEB design. Much R&D needs to be done, particularly in those areas of the magnet which represent new or unproven technology, e.g. the offset coil geometry, the annular two phase flow inside the beam pipe, indirect iron cooling, etc. Also the details of the lattice need to be addressed so that the physical parameters of the dipole can be finalized. However, we feel that this work serves as a good starting point from which a final design can begin.

REFERENCES

1.   SSC Laboratory, "Site-Specific Conceptual Design Report," November 22, 1989.

2.   T.Collins, Fermilab Internal Report, TN1406.

3.   A.Ishibashi, et al., Design study on the superconducting dipole magnets with non-circular aperture coils in application to future colliders, in: "Supercollider 1," Volume 1, Plenum Press, New York.

4.   T.H. Nicol, et al, Design and analysis of the SSC dipole magnet suspension system, in: "Supercollider 1," Volume 1, Plenum Press, New York.

SELF-PROPELLED IN-TUBE SHUTTLE AND CONTROL SYSTEM FOR AUTOMATED

MEASUREMENTS OF MAGNETIC FIELD ALIGNMENT

W.N. Boroski and T.H. Nicol

Fermi National Accelerator Laboratory
Batavia, Illinois

S.V. Pidcoe

General Dynamics, Space Systems Division
San Diego, California

R.A. Zink

SSC Laboratory
Dallas, Texas

ABSTRACT

A magnetic field alignment gauge is used to measure the field angle as a function of axial position in each of the magnets for the Superconducting Super Collider (SSC). Present measurements are made by manually pushing the gauge through the magnet bore tube and stopping at intervals to record field measurements. Gauge location is controlled through graduation marks and alignment pins on the push rods. Field measurements are recorded on a logging multimeter with tape output. Described is a computerized control system being developed to replace the manual procedure for field alignment measurements. The automated system employs a pneumatic walking device to move the measurement gauge through the bore tube. Movement of the device, called the Self-Propelled In-Tube Shuttle (SPITS), is accomplished through an integral, gas driven, double-acting cylinder. The motion of the SPITS is transferred to the bore tube by means of a pair of controlled, retractable support feet. Control of the SPITS is accomplished through an RS-422 interface from an IBM-compatible computer to a series of solenoid-actuated air valves. Direction of SPITS travel is determined by the air-valve sequence, and is managed through the control software. Precise axial position of the gauge within the magnet is returned to the control system through an optically-encoded digital position transducer attached to the shuttle. Discussed is the performance of the transport device and control system during preliminary testing of the first prototype shuttle.

INTRODUCTION

An electrical gauge to measure magnetic field angle has been developed at Fermilab[1]. The gauge is used to measure the field angle as a function of axial position along the length of SSC cold iron superconducting magnets. At present, measurements are made by inserting the gauge into a magnet beam tube and pushing it through the length of the bore with long aluminum rods. Measurements of the field direction are made and recorded at 7.62 cm. intervals. This method is

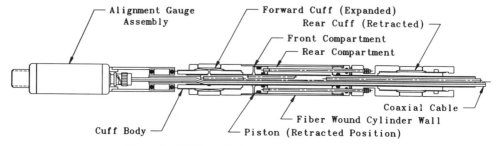

Figure 1. Self Propelled In-Tube Shuttle (SPITS).

slow and labor-intensive; therefore, a development program was initiated to automate the measurements. The result has been a unique integration of mechanics and software that will permit the automated acquisition of magnetic field data at precise locations inside the 4 cm. bore of an SSC dipole magnet beam tube.

SELF-PROPELLED IN-TUBE SHUTTLE

Integral to the automated system is a transport device called the Self-Propelled In-Tube Shuttle (SPITS). The device is shown in Figure 1. The SPITS is capable of pushing or pulling the field alignment gauge through the magnet bore tube. The alignment gauge assembly is attached to the SPITS through a threaded connection. This connection will allow other measurement devices to be readily adaptable to the SPITS. The transport device is constructed entirely of non-magnetic materials which do not interfere with the magnetic field measurements of the alignment gauge.

Internal to the transport is a pneumatic piston which controls the length of the stroke. The prototype SPITS assembly has a maximum stroke of 7.62 cm. Separate compartments on each side of the piston control the direction of transport travel. Four internal ports with pneumatic passages connect the piston compartments and inflatable cuffs to the supply gas. The present SPITS design can accommodate gas pressures to 620 kPa.

The transport device contains two independent inflatable cuffs. One cuff assembly is mounted near the front of the device; the second at the rear. Figure 2 illustrates a prototype cuff assembly and is representative of the assemblies within the SPITS device. The neoprene membrane of each cuff is bonded to three glass-epoxy composite pieces separated at 120° angles. The composite pieces are contoured to fit the inner diameter of the beam tube, and contact the beam tube when the membrane is pressurized.

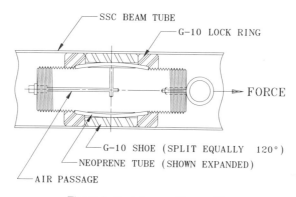

Figure 2. Prototype cuff assembly.

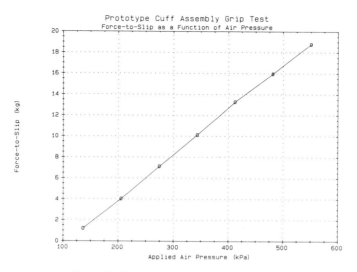

Figure 3. Force-to-slip verses cuff air pressure.

Tests were conducted on the prototype cuff assembly to determine the holding power of the cuff design. A commercial tensile-testing machine was used to apply an axial force in the direction shown in Figure 2. Data was taken with the cuff pressurized to several pressures between 138 kPa and 552 kPa. The force which caused the assembly to slip was measured for each pressure with the force-to-slip measurement repeated three times. Figure 3 shows the average force-to-slip as a function of internal air pressure.

SPITS CONTROL SYSTEM

The supply gas used to control the movement of the transport device is routed to the device through a series of four miniature solenoid-controlled air valves. Each valve uses two solenoids which control the flow of the inlet and exhaust gas, respectively, and provide substantial control over the pressurization and exhaust timing of each component. Needle valves throttle the gas flow through each exhaust port, and provide smooth forward and reverse movement of the transport. Fluid movement of the transport through the magnet bore tube is critical given the sensitive nature of the field alignment gauge.

The solenoids are energized through a series of solid state relays. The relays are part of a modular programmable controller system connected through a RS-422 serial interface to an IBM-compatible host computer. The control system is shown in block diagram in Figure 4. Software was

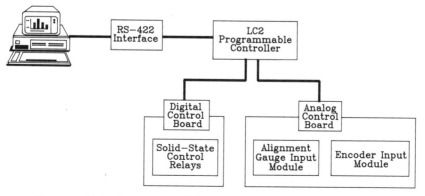

Figure 4. Block diagram of the SPITS computerized control system.

developed to control the timing sequence that produces forward and reverse movement. When fully implemented, the control portion of the software will be downloaded to the local controller. The local controller will then process the entire control sequence upon a single command from the host computer. This capability frees the host computer to perform data manipulation between measurement points.

The control sequence for one stroke of forward travel is outlined in Table 1. Reverse travel is accomplished by reversing the order of the control sequence. Figure 1 should be used to reference transport components.

Table 1. Control sequence to advance the SPITS transport.

STEP	ACTION	RESULT
1	Pressurize rear cuff	Rear cuff expands to contact the beam tube, prohibiting backward movement.
2	Exhaust front cuff	Front cuff retracts from the beam tube wall.
3	Pressurize front compartment	Pressurized gas enters the compartment before the piston, which should cause the transport to extend axially. However, because the exhaust valve controlling the compartment behind the piston is closed, the device cannot move.
4	Exhaust rear compartment	The exhaust valve on the rear piston compartment opens, allowing the gas to escape. As the gas escapes from the rear compartment, gas is allowed to enter the front compartment, causing the transport to extend axially. The exhaust gas passes through a needle valve that is throttled to slow the rate of exhaust. The result is a gentle expansion of the transport device.
5	Pressurize front cuff	Front cuff expands and contacts the beam tube, stopping forward movement.
6	Exhaust rear cuff	Rear cuff retracts, releasing the grip on the beam tube.
7	Pressurize rear compartment	Pressurized gas enters the rear compartment, which should cause the transport to contract axially. However, pressure still exists in the front compartment, so the device is held stationary.
8	Exhaust front compartment	The exhaust valve on the front compartment opens, allowing the gas to escape. As the gas escapes from the front compartment, gas is allowed to enter the rear compartment, causing the transport to contract axially. As the device is held against the beam tube by the front cuff, the rear of the device is brought forward. Again, the rate of exhaust is throttled to provide gentle movement of the device.

By design, one sequence of events moves the transport a distance of approximately 7.62 cm. An optically-encoded digital position transducer is used to maintain exact positioning. Location of the transport is determined by attaching the cable from the transducer to the rear of the transport. As the transport moves through the beam tube, the cable rotates a spring loaded shaft coupled to a rotary digital encoder. Digital pulses from the transducer at the rate of 246 per cm. of transport travel are output to the control computer. The operation of the positioning system is still under development. The objective is to have the computer determine, by means of a feedback loop, when a desired location has been reached. A command will then be sent to the solenoid valves to pressurize the foremost cuff. The rapid pressurization will stop the transport at the desired position. With the front cuff pressurized, the remaining valve sequence will be completed to ready the transport for the next stroke. Some adjustment of the pressurization timing will be necessary during start-up to achieve the desired degree of positioning accuracy.

INSTRUMENTATION CABLE / AIR LINE SUPPLY SYSTEM

A compact utility system has been designed to manage the air lines and instrumentation cables for the SPITS as the device moves along the 16.3 meter long beam tube. The take-up system is shown in Figure 5. The compact system will fit on top of a 60.5 cm. by 91 cm. tool and die table, providing elevation control and system portability.

Constant tension must be maintained on the supply lines during the entire range of SPITS travel. Tension is maintained by connecting the output reel to a constant-speed gear motor through an adjustable particle clutch, set at 0.68 kg of tension. A series of reels with different diameters achieves a step-down ratio of 125:1 so that the connecting end of the supply lines moves less than 14 cm. throughout the entire 16.3 meters of SPITS travel. Air lines and electronic cables from the control system are easily connected to the main supply lines through this interface. Winding guides will be used to lay the cables and hoses down on the reel with uniform spacing.

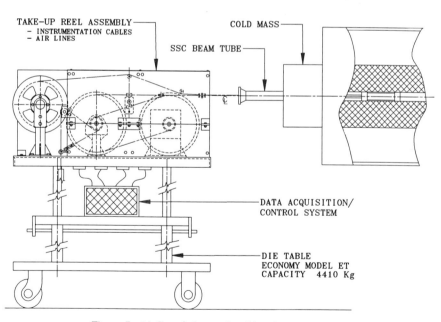

Figure 5. Air line / electronic cable take-up system.

PRELIMINARY TEST RESULTS

The first prototype SPITS device was cycled along a 1.3 meter section of SSC beam tube to measure transport performance over time. The test program called for the device to travel the equivalent of 50 SSC beam tube lengths; the actual distance travelled during testing was 838 linear meters. Fiducials placed on the body of the transport were used to measure axial and rotational creep during travel.

For these measurements, axial creep is defined as the actual change in axial displacement of the device over time. By design, there should be zero displacement due to axial creep; the device should return to its starting location. Displacement was measured as the change in axial position of a circumferential alignment mark with reference to the beam tube end from beginning to end of each run, and was recorded in cm. per meter of travel. Rotational creep is defined as the amount of azimuthal shift of the transport during travel, and was measured by the angular change in axial alignment marks from beginning to end of each run. Rotational creep was recorded in degrees per meter of travel. All measurements were made using a flexible machinist's scale with a resolution of 0.5 mm.

Measurements were taken over a 7-day period in which 11 separate measurement runs were conducted. The transport device was re-aligned with the fiducials before each run. The average time required to travel one beam tube length was 56 minutes. This is a significant reduction in time compared to that presently required to move the alignment gauge through a beam tube. Furthermore, the speed of the transport is programmable through the control software, and can be optimized to meet specific measurement needs.

Figure 6 presents the amount of rotational creep measured for each run. A slight downward trend in the amount of rotation can be seen as the test proceeded. The maximum amount of rotation for a single run was 0.82 degrees per meter of travel. Given this degree of change, the total rotation over a beam tube length of 16.3 meters would be 13.37 degrees. This is unacceptable as the maximum amount of rotation tolerable for the field alignment gauge is +/- 5 degrees. Development of a gauge-to-transport connection that will allow free rotation of the gauge with respect to the transport is under consideration. Possible solutions include frictionless mechanical bearings or spherical air bearings. Adding ballast weight at the gauge bottom with non-magnetic shims would then maintain stability of the gauge during transport travel.

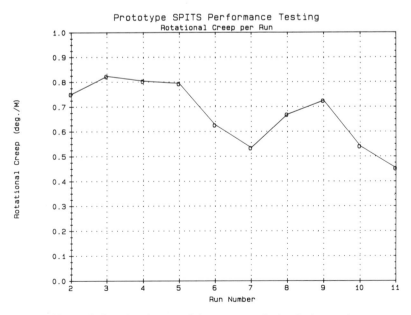

Figure 6. Rotational creep of the transport device during testing.

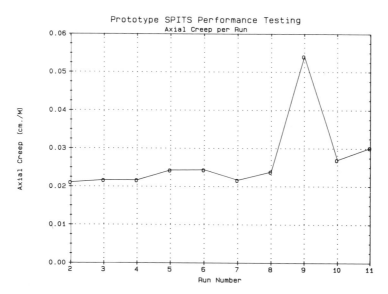

Figure 7. Axial creep of the transport device during testing.

Figure 7 presents the results of the axial creep measurements. The large amount of creep in run 9 is most likely caused by measurement error, and should be disregarded. The degree of transport creep in the remaining data points is relatively constant, although a slight upward trend is observed near the end of the testing. As final positioning of the transport will be accomplished through a digitally-encoded positioning feedback loop, the small amount of creep measured is not of concern.

SUMMARY

A transport device for automated magnetic field measurements has been designed and tested. At present, the device is also being considered for use as a laser target transport for magnet alignment measurements. Axial location of the SPITS is achievable through the use of a digital positioning transducer. Preliminary test results indicate a problem with angular rotation; however, several methods to resolve this problem are being considered. The transport system is still under development. Results of initial tests warrant continued R&D to address the shortcomings of the system. Early data indicate that the transport will significantly reduce the time required to make field angle measurements. The transport system can be operated by one person.

The work as presented was performed at Fermi National Accelerator Laboratory which is operated by Universities Research Association Inc., under contract with the U.S. Department of Energy.

ACKNOWLEDGEMENTS

The authors' express their sincere gratitude to W. McCaw for his work in the fabrication of the SPITS prototype, to D. Arnold for his assistance in illustrating this work, and to A. Lipski for his insights into addressing the rotational problems inherent to the device.

Further appreciation is given to J.D. Gonczy for his efforts in support of this work.

REFERENCE

1.   M. Kuchnir and Ed. E. Schmidt, IEEE Trans. Magn. <u>24</u>, 950 (1988).

CORRECTION OF MAGNETIZATION SEXTUPOLE IN ONE-METER LONG

DIPOLE MAGNETS USING PASSIVE SUPERCONDUCTOR

M. A. Green, R. F. Althaus, P. J. Barale, R. W. Benjegerdes,
W. S. Gilbert, M. I. Green and R. M. Scanlan

Lawrence Berkeley Laboratory
Berkeley, CA 94720

ABSTRACT

The generation of higher multipoles due to the magnetization of the superconductor in the dipoles of the SSC is a problem during injection of the beam into the machine. The use of passive superconductor was proposed some years ago to correct the magnetization sextupole in the dipole magnet. This paper presents the LBL test results in which the magnetization sextupole was greatly reduced in two one-meter long dipole magnets by the use of passive superconductor mounted on the magnet bore tube. The magnetization sextupole was reduced a factor of five on one magnet and a factor of eight on the other magnet using this technique. Magnetization decapole was also reduced by the passive superconductor. The passive superconductor method of correction also reduced the temperature dependence of the magnetization multipoles. In addition, the drift in the magnetization sextupole due to flux creep was also reduced. Passive superconductor correction appears to be a promising method of correcting out the effects of superconductor magnetization in SSC dipoles and quadrupoles.

BACKGROUND

The effect of superconductor magnetization on the quality of the field generated within a superconducting dipole was observed as early as 1970.[1] The effect of superconductor magnetization on field quality has been modeled using complex current doublet theory.[2] The mathematical model has been successfully applied to accelerator dipole and quadrupole magnets.[3]

The computer model has been used to calculate several methods of passive correction for accelerator dipoles. These methods include: passive superconductor,[4,5] ferromagnetic material,[4,6] and oriented permanent magnet material.[4] This paper describes the Lawrence Berkeley Laboratory (LBL) passive superconductor test program. The use of passive superconductors to correct the magnetization sextupole in a dipole magnet is not new. The concept was first described by Brown and Fisk in 1984.[7] Fermilab reported a test of the concept in 1986.[8]

Passive superconductor correction of magnetization multipoles has the following potential advantages for SSC dipole magnets: 1) Passive superconductor correctors are unpowered straight pieces of superconductor mounted within the magnet bore. The pieces are not connected together. 2) Passive superconductor correction corrects the magnetization multipole when the field is falling as well as rising. 3) Passive superconductor correction corrects the field over a wide range of temperatures. 4) It is hoped that passive superconductor correction will eliminate slow changes of magnetization sextupole due to flux creep.[9,10]

This paper describes the passive superconductor correctors which were built and tested in two Lawrence Berkeley Laboratory dipole magnets. The magnets include the LBL D-15-C2 four centimeter bore dipole, and the LBL D-16-B1 five centimeter bore dipole. The D-16-B1 dipole has close-in iron which starts to saturate when the magnet central induction reaches about 3 T.

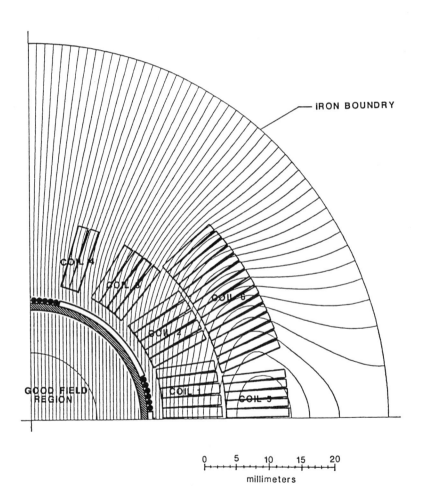

Figure 1. The D-15-C2 Dipole Coil with its Passive Corrector Installed (Magnetic Flux Lines are Shown).

PASSIVE CORRECTION OF THE D-15-C2 DIPOLE MAGNET

The LBL D-15-C2 dipole is a dipole magnet with NC-9 conductor cross-section (see Figure 1). The inner coil conductor has 5 micron filaments and a copper-to-superconductor ratio of 1.48. The outer coil conductor has 6 micron diameter filaments and a copper-to-superconductor ratio of 1.77. The superconductor used to correct the magnet has 24 micron filaments with a copper-to-superconductor ratio of 1.36.

The corrector for the D-15-C2 dipole is also shown in Figure 1. The corrector is symmetric about the midplane and the poles. Twelve 0.808 mm diameter conductors are located about the midplane and ten 0.808 mm diameter conductors are located at each pole. Forty-four 26-inch long pieces of corrector conductor are mounted on the surface of a 1.362 inch outside diameter stainless steel tube. The amount of superconductor in the corrector represents about 1.6 percent of the superconductor in the D-15-C2 magnet.

Figure 2 shows the measured magnetization sextupole of LBL D-15-C2 dipole as a function of current (the dipole transfer function is 10.3 gauss per ampere) without the passive corrector. Figure 3 shows the measured magnetization sextupole for the LBL D-15-C2 dipole, a function of current with the passive corrector installed. The passive superconductor reduced the magnitude of the magnetization sextupole by a factor of five.

The magnetization sextupole was overcorrected by the passive superconductor corrector. Some of the reasons for the overcorrection of the magnetization sextupole are as follows: 1) The filament diameter of the corrector was 24 microns instead of the 23 microns when the corrector was designed. 2) The average radius of the corrector mounted on the stainless tube is about 0.1 mm smaller than the radius used to design the corrector. 3) The largest effect is caused by an over estimate of the low field critical current density in the magnet conductor. The low field J_c was overestimated by 15 percent. The errors given above account for over 80 percent of the overcorrection observed in the LBL test of dipole D-15-C2.

The decay of the magnetization sextupole was reduced from +1.3 units per decade to about +0.2 units per decade. If the magnetization decay were completely compensated by the superconductor corrector, the decay would have been about -0.25 units per decade. It can be concluded that only 70 percent of the flux creep decay was eliminated by the passive corrector.

PASSIVE CORRECTION OF THE D-16-B1 DIPOLE MAGNET

The D-16-B1 dipole magnet built by LBL has a 5 centimeter bore with the iron against the outer coil (see Figure 4). At low fields, the dipole transfer function is about 13.0 gauss per ampere. The dipole inner coil has 6 micron filaments and a copper-to-superconductor ratio of 1.20. The outer coil has 6 micron filaments and a copper-to-superconductor ratio of 1.66. The same material was used for the D-16-B1 dipole corrector as was used for the D-15-C2 dipole corrector.

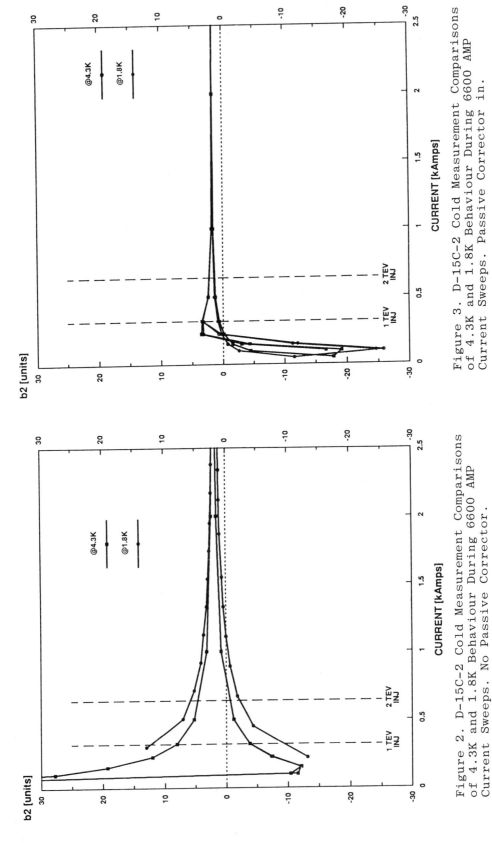

Figure 2. D-15C-2 Cold Measurement Comparisons
of 4.3K and 1.8K Behaviour During 6600 AMP
Current Sweeps. No Passive Corrector.

Figure 3. D-15C-2 Cold Measurement Comparisons
of 4.3K and 1.8K Behaviour During 6600 AMP
Current Sweeps. Passive Corrector in.

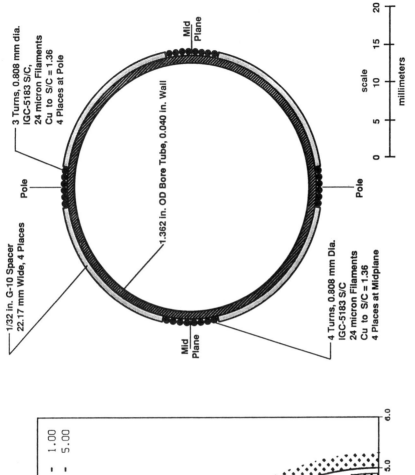

3 Turns, 0.808 mm dia.
IGC-5183 S/C,
24 micron Filaments
Cu to S/C = 1.36
4 Places at Pole

Pole

1/32 in. G-10 Spacer
22.17 mm Wide, 4 Places

1.362 In. OD Bore Tube, 0.040 In. Wall

Mid Plane

Pole

Mid Plane

4 Turns, 0.808 mm Dla.
IGC-5183 S/C
24 micron Filaments
Cu to S/C = 1.36
4 Places at Midplane

scale

0   5   10   15   20

millimeters

Figure 5. New Passive Superconductor Correction for LBL Dipole D-16B-1.

RADIUS RO (CM) = 1.00
FE RADIUS (CM) = 5.00

X DIMENSION IN RO UNITS

Y DIMENSION IN RO UNITS

Figure 4. LBL Magnet D-16B-1 Superconducting Magnet Cross Section.

393

The D-16-B1 corrector is shown in Figure 5. The corrector consists of twenty-eight 26-inch long pieces of corrector conductor mounted on the surface of the 1.362 outside diameter tube. Like the D-15-C2 corrector, the D-16-B1 corrector is symmetric about the midplane and the poles. There are six corrector wires at each pole and there are eight correctors on each side at the midplane. The superconductor in the corrector represents about 0.6 percent of the superconductor in the D-16-B1 magnet.

Figure 6 demonstrates the extent of correction by the D-16-B1 dipole correctors at the center of the magnet. The curve with correction is negative, which indicates that the corrector overcorrected the magnetization sextupole. The magnitude of the magnetization sextupole was reduced by a factor of seven to eight.

The passive corrector reduced the decay of the magnetization sextupole by about 60 percent. Measurements at 1.8K yielded decay rates which are virtually the same as the decay rates at 4.3K. (The decay rate without the corrector is about +1.0 units per decade; the decay rate with the corrector is about +0.3 units per decade.)

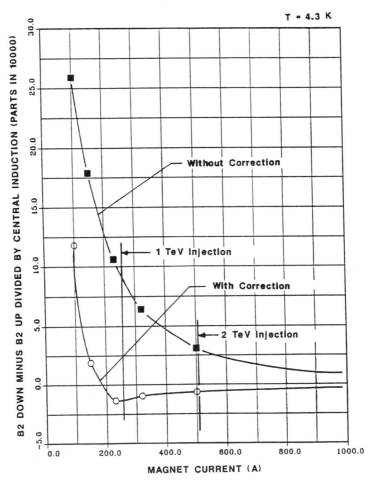

Figure 6. LBL Dipole D-16B-1 Central Field with and Without Passive Correction.

SUMMARY

The passive corrector experiments on the LBL dipoles demonstrated that correction of magnetization sextupole can be done with pieces of passive superconductor. The magnetization sextupole is corrected by the passive superconductor at a temperature of 1.8K as well as 4.3K. A reduction of the magnetization sextupole was achieved when the field was decreasing as well as when the field was increasing. The decay of the magnetization sextupole was not reduced to the same extent that the magnetization sextupole (at the start of the decay) was reduced.

Passive superconductor correction is a viable way of reducing the magnetization sextupole in the SSC dipole magnet. The superconductor in the passive corrector can be less than one percent of the superconductor in the dipole depending on the radial location of the passive correctors with respect to the dipole magnet coil.

ACKNOWLEDGEMENTS

The authors wish to acknowledge conversations that were had with A. H. Ghosh of Brookhaven National Laboratory concerning the measured magnetization of the superconducting cable in the test magnets. We also acknowledge useful conversations we have had with C. E. Taylor of the Lawrence Berkeley Laboratory.

This work is supported by the Office of Energy Research, High Energy Physics Division, United States Department of Energy under Contract No. DE-AC03-76SF00098.

REFERENCES

1.   M. A. Green, IEEE Transactions on Nuclear Science NS-18(3), p. 664 (1971).
2.   M. A. Green, "Residual Fields in Superconducting Magnets," Proceedings of the MT-4 Conference at Brookhaven National Laboratory, p. 339 (1972).
3.   M. A. Green, "Fields Generated Within the SSC Magnets Due to Persistent Current in the Superconductor," Proceedings of the Ann Arbor Workshop on SSC Issues, LBL-17249, December 1983.
4.   M. A. Green, IEEE Transactions on Magnetics MAG-23(2), p. 506 (1987).
5.   M. A. Green, "Passive Superconductor a Viable Method of Controlling Magnetization Multipoles in the SSC Dipole," IISSC Supercollider 1, Plenum Press, New York, p. 351 (1989).
6.   E. W. Collings et al., "Design of Multifilamentary Strands for SSC Dipole Magnets," Proceedings from the 1990 IISSC Conference, Miami (1990).
7.   H. C. Brown and H. E. Fisk, the idea of passive superconductor correction presented at the Snowmass SSC Workshop in 1984.
8.   H. E. Fisk, et al., Report on the Fermilab Passive Superconductor Test at the ICFA Workshop, May 1986.
9.   D. A. Herrup, et al., IEEE Transactions on Magnetics MAG-25(2), p. 1643 (1989).
10.   W. S. Gilbert, et al., IEEE Transactions on Magnetics MAG-25, p. 1459 (1989).

ELECTRICAL INSULATION REQUIREMENTS AND

TEST PROCEDURES FOR SSC DIPOLE MAGNETS

G.F. Sintchak, J.G. Cottingham, and G.L.Ganetis

Brookhaven National Laboratory[*]

ABSTRACT

The development of the basic requirements for the turn-to-turn, coil-to-coil, and coil-to-ground insulation for SSC dipoles is discussed. The insulation method is also described along with test procedures for verification of insulation integrity.

Electrical tests are performed throughout the magnet assembly and fabrication process to verify that coil integrity and insulation quality of the various components and sub-assemblies are within nominal limits. These tests are also required to certify each dipole for SSC acceptance before it is installed in the cryostat and leaves the factory for final installation. The following series of tests, which are conducted at room temperature, are listed below.

- Resistance
- Inductance and "Q"
- Hypot
- Impulse
- Ratiometer

Resistance Tests

Resistance measurements are performed using a one ampere (usually) precision constant current power supply and measuring the resultant voltage drop across the element under test. This is analogous to a four-wire ohmmeter. This test is easy to perform and is uncomplicated. The maximum output voltage required from the power supply is less than ten volts.

The main coils are connected in series as per the final wiring configuration, and, with the one ampere current flowing in the coils, the voltage drop across each coil is measured and recorded. The DC resistance test will usually indicate a turn-to-turn short. However, the coil resistance will vary somewhat with changes in room

---

[*]Work performed under contract with the U.S. Department of Energy.

(or coil) temperature. Therefore, difference voltages and temperatures are compared with previous readings. Also, the resistance of the two inner (or outer) coils should track each other closely. A voltage drop change (or difference) of 80 – 90 mVolts is usually an indication of a shorted turn.

Refer to Figure 1 for typical voltage drop DC measurements of SSC dipole coils with one ampere current flowing in the coils. Please note that the voltage drop per turn for inner and outer coils is different because of a slight difference in the superconductor cable composition.

This resistance test of the main coils is repeated frequently throughout the assembly process, and in particular, before and after the collaring operation, impulse testing, iron yoke installation, shell welding, Helium leak test, and for final testing.

Inductance and "Q" Test

The inductance and "Q" measurement provides another low voltage test on individual coils that will check for turn-to-turn shorts. This test is particularly sensitive to "soft" shorts. Q is the quality factor of the coil and is defined as the ratio of inductive reactance divided by the effective resistance of the coil. The effective resistance includes the DC resistance of the coil plus all the other resistive and eddy current losses due to the core (if any) material and manner in which the coil is wound. The inductance and "Q" measurement is done using two test frequencies. 1 kHz is used for individual open coils in an air medium, and 120 Hz is used when the dipole coils are in their final wired configuration within the yoked iron core. See Figure 2 for typical values of inductance and "Q" for various coil configurations. Turn-to-turn shorts will show a reduction of 10 – 20 % in the value of "Q".

- INNER COIL HAS 16 TURNS.

- OUTER COIL HAS 20 TURNS.

- TOTAL WINDING = (2 x 16) + (2 x 20) = 72 TURNS.

- INNER COIL VOLTAGE DROP IS APPROX. 82 mV/TURN
  82 mV x 16 TURNS - <u>1.312 VOLTS</u>

- OUTER COIL VOLTAGE DROP IS APPROX. 95 mV/TURN
  95 mV x 20 TURNS = <u>1.900 VOLTS</u>

- TOTAL MAGNET WINDING = (2 x 1.312) + 2(1.90) = 6.424 VOLTS

- CABLE VOLTAGE DROP IS 60–75 μV/INCH
  (DEPENDS ON CABLE COMPOSITION & TEMPERATURE)

Figure 1. Typical DC Measurements
SSC Dipole Coils

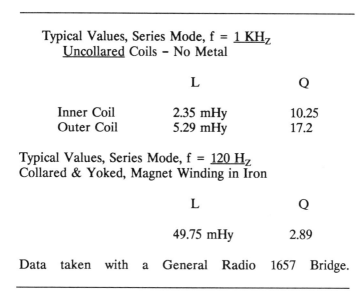

Typical Values, Series Mode, f = <u>1 KH$_Z$</u>
<u>Uncollared</u> Coils – No Metal

	L	Q
Inner Coil	2.35 mHy	10.25
Outer Coil	5.29 mHy	17.2

Typical Values, Series Mode, f = <u>120 H$_Z$</u>
Collared & Yoked, Magnet Winding in Iron

L	Q
49.75 mHy	2.89

Data taken with a General Radio 1657 Bridge.

Figure 2. Inductance & "Q" Measurements
SSC Dipole Magnet Coils

Hypot Tests

There are several insulation tests done to insure there are no shorts or excessive leakage currents between various components and sub–assemblies within the magnet dipole assembly. The high voltage, or hypot, leakage tests are done at a voltage level that exceeds what the magnet may experience during operation. In general, this test voltage is determined by doubling the expected voltage, and adding 1000 volts. The maximum test voltage required is 5000 volts, and the short circuit current should be limited to 2 mA to avoid damaging any magnet components should a flashover occur.

Simple low precision ohmmeter tests are done before making any high voltage test to be sure that the resistance between the components under test is greater than 20 megohms. The insulation hypot test requirements are as follows:

- Main coils to all other components and ground at 5 kV.
- Main lower coils to upper coils (midplane) at 3 kV.
- Trim coils to all other components and ground at 5 kV.
- Quench protection heaters to other components and ground at 5 kV.
- Ground is defined as the collar/yoke/shell and beam tube.
- Leakage current should be less than 50 μA after one minute of the applied test voltage.

The most common insulation used in magnet construction is Kapton film in various forms. Figure 3 is a list of the many types of Kapton that are used at BNL for magnet insulation.

BNL Code	Width (in.)	Thk. (in.)	Adhesive Type	Adhesive Thk.(in)	Vendor Name	Part No.
Kapton NA-1	3/8	.005	None	–	E.I. DuPont Co., Inc.[1]	500BH
Kapton NA-2	1.0	.001	None	–	E.I. DuPont Co., Inc.	100H
Kapton A-1	1/2	.005	Silicone	.0005	CHR/Carlson[2]	K104
Kapton A-2	1/2	.001	Silicone	.0005	CHR/Carlson	K105
Kapton A-3	3/4	.001	Silicone	.0005	CHR/Carlson	K105
Kapton A-4	3/8	.001	Polyester	.001	Sheldahl[3]	T320
Kapton A-5	1.0	.001	Acrylic	.0015	CHR/Carlson	K102
Kapton A-6	1/2	.001	Silicone[5]	.0015	CHR/Carlson	K250
Kapton A-7	1.0	.001	Silicone	.0015	DW/Canis[4]	304-1-1
Kapton A-8	1/2	.002	Silicone[5]	.0015	CHR/Carlson	K350
Kapton A-9	3.0	.002	Silicone[5]	.0015	CHR/Carlson	K350
Kapton A-10	3/4	.002	Silicone	.0015	DW/Canis	304-2.75
Kapton A-11	3/4	.005	Silicone	.0015	DW/Canis	304-5
Kapton A-12	1.0	.005	Silicone	.0015	DW/Canis	304-5-1
Kapton A-13		.001	Teflon Each Side	.0001	E.I.DuPont Co., Inc.	Type 120 FN 616
Kapton A-14		.001	Teflon	.0005	E.I. DuPont Co., Inc.	Type 150 FN 019
Kapton A-15		.005	None	–	E.I. DuPont Co., Inc.	Type 500 FN 131
Kapton A-16		.003	Silicone	.0015		

NOTES:

[1] E.I. DuPont Co., Inc.
Polymer Products Dept.
Northfield, MN

[2] Mfd. by: CHR Industries
New Haven, CT

[3] Mfd. by: Sheldahl Inc.
P.O. Box 170

[4] Mfd. by: Dewal Industries
Saundestown, RI

[5] Fire Retardant

Figure 3. Kapton Film Material Description.

Impulse Test

The impulse test is a high voltage test that checks the turn–to–turn voltage hold–off insulation integrity. This simulates the conditions that may occur during a quench. The coil winding insulation is stressed by discharging a capacitor that delivers a 2 kV pulse to produce (approximately) a 50 volt per turn voltage drop. See Figure 4 for a simplified diagram of the High Voltage Impulse Generator that is used for the impulse test. The resulting damped oscillation is stored on a digital storage oscilloscope and photographed or plotted. Any change in the waveform will indicate a turn–to–turn breakdown of insulation. This test is usually done before and after the collaring operation, and should not be repeated any more than necessary to avoid stressing the insulation. As mentioned before, a resistance test of the main coils is made before and after this test.

Figure 5 shows two typical waveforms from insulation breakdown during the impulse test. A partial short was made in an old outer coil from magnet DD–016 using a metal probe to pierce the insulation between adjacent turns 9/10 near the return end of the coil. The insulation breakdown voltage spike can be clearly seen occurring at the peak voltage points in the waveform. Near the end of the waveform, a sustained arc is seen, and in photograph 2, it completely dissipates the remaining stored energy.

The impulse test can also show up a more subtle type of coil defect. Figure 6 shows three photographs (30–32) of impulse tests done on magnet DD018.[1] In addition to the impulse voltage waveform (upper trace) the impulse current (lower trace) was measured using a Pearson Electronics, model 110, current monitor. Photograph 31 is a superposition of the impulse waveforms obtained from the upper and lower inner coil blocks. These waveforms don't lie on top of each other. The current crest in the lower inner coil is clearly earlier in time and higher in amplitude than that in the upper inner coil. This indicates a lower inductance in the lower inner coil and thus a dielectric failure. The inductance of the lower inner coil was computed to be 39% lower than the upper inner coil inductance and clearly indicates a defect. This method of quarter coil comparison testing is now under consideration for use in future magnet production.

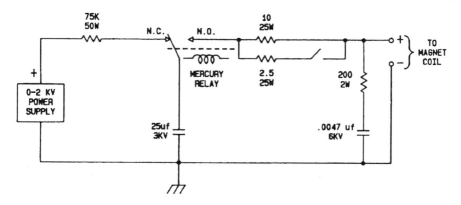

Figure 4. High Voltage Impulse Generator

[1]J.G. Cottingham and G. Sintchak; Turn–To–Turn Fault Location In Magnet DD0018, SSC Technical Note No. 87 (SSC–N–681); Nov 9, 1989.

High Voltage Impulse Test
Insulation Breakdown
Probe Induced Short

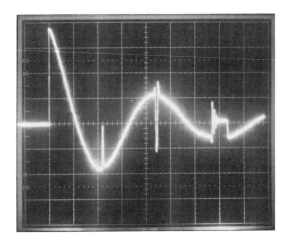

Photograph 1

500 V/Div
0.5 mS/Div

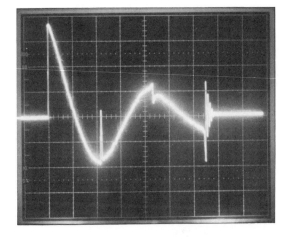

Photograph 2

Figure 5. Magnet DD016, Lower Outer Coil, LLN-024, 3-31-89

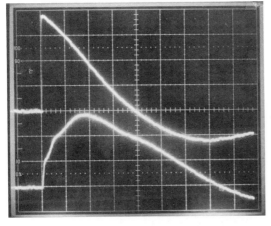

#30 Lower Inner Coil

Note: Higher current peak
occurs sooner in time.

Indicates lower inductance

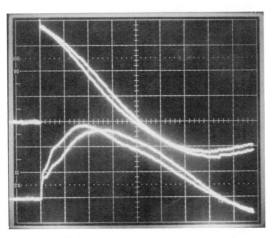

#31 Upper/Lower Coils
Overlay

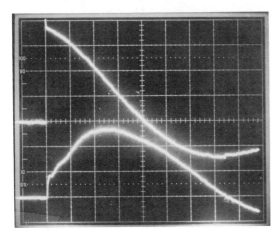

#32 Upper Inner Coil

All photographs:
500 V /Div
50 A /Div
0.1 mS/Div

Magnet DD-018   11/6/89

Figure 6

Ratiometer Test

The ratiometer test checks the turns ratio of the coils in a dipole magnet. By comparing turn ratios, it is possible to check if a coil has a shorted turn. The advantage of this method is that it is insensitive to temperature and may be used in a liquid helium environment. The test is performed with coils in iron (yoked) at 60 Hz with approximately one ampere AC current flowing in the magnet. The coils are connected in series as per the final configuration to form an autotransformer as shown in Figure 7. An AC RMS voltmeter is used to measure the voltage developed across the four individual coils using the voltage taps.

The inner upper and lower coils should develop 0.1885 of the total excitation voltage measured. The outer upper and lower coils should develop 0.3115 of the total excitation voltage. If the computed ratio is different by more than +/- 0.007 from the given ratio, it indicates there may be a shorted turn in the coil in question. Because the magnet coils are not closely coupled as in a transformer, the inner and outer ratio numbers were determined experimentally using a known good magnet at room temperature and at liquid helium temperature.

In conclusion, a cross check or different method of testing is very helpful when interpreting test results. Proper safety procedures should be followed when making any electrical tests and the equipment should be properly grounded.

BNL MAGNET RATIOMETER

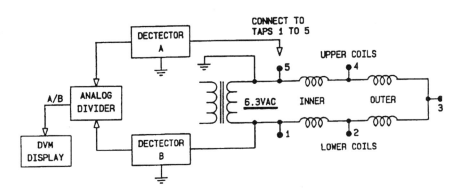

Figure 7

MULTIPLE COIL PULSED MAGNETIC RESONANCE METHOD FOR MEASURING

COLD SSC DIPOLE MAGNET FIELD QUALITY

W.G. Clark, J.M. Moore, and W.H. Wong

Department of Physics and Solid State Science Center
University of California at Los Angeles
Los Angeles, CA 90024-1547

ABSTRACT

The operating principles and system architecture for a method to measure the magnetic field multipole expansion coefficients are described in the context of the needs of SSC dipole magnets. The operation of an 8-coil prototype system is discussed. Several of the most important technological issues that influence the design are identified and the basis of their resolution is explained. The new features of a 32-coil system presently under construction are described, along with estimates of its requirements for measurement time and data storage capacity.

INTRODUCTION

The dipole bending magnets planned for the Superconducting Supercollider (SSC) will have to meet very stringent specifications for the accelerator to operate as planned.[1] One of the challenging aspects of producing these magnets is testing them cold under manufacturing conditions. This paper gives a brief description of one approach to this problem that is currently under study with support of the DOE. It uses a method of Multiple Coil Pulsed Magnetic Resonance (MCPMR).[2]

The following topics are treated in this paper. First, a review of the two main methods of mapping dipole field profiles is described along with a definition of the parameters that are to be obtained from the measurements. Then, a brief introduction of the use of Nuclear Magnetic Resonance (NMR) to measure magnetic fields is presented. This topic is followed by a description of the MCPMR method and an outline of how it is implemented in our work. The main focus of this discussion is our prototype 8-coil system; a few comments regarding our 32-coil system that is under construction are included. Then some of the leading technological issues of the method are identified and their resolution described. The paper ends with a short presentation of the main features of the new 32-coil system and a statement of conclusions.

SSC DIPOLE FIELD QUALITY AND ITS MEASUREMENT

The field of a dipole bending magnet is usually expressed in Cartesian coordinates x, y, and z as[1]

$$B_y(x,\ y)+iB_x(x,\ y)\ =\ B_0\sum_{n=0}^{\infty}(x+iy)^n(b_n+ia_n), \qquad (1)$$

where, as shown in Fig. 1, z is the (horizontal) direction of the beam tube axis, y is the (vertical) direction of the dipole field, and x is the (horizontal) direction perpendicular to both of them. The multipole moment coefficients (MMC) that specify the magnetic field, a_n and b_n, are the quantities that have to be kept within specified limits to preserve particle orbit stability. It is their measurement and that of B_0 that is the main focus of testing for dipole magnet field quality. In this paper we will follow the usual assumption that the magnetic field variation along the z-axis is inconsequential and that the problem is a two dimensional one in the x-y plane.

Alternative expressions for the MMC in polar coordinates r and θ (Fig. 1) equivalent to Eq. 1 that will be used in this paper are:

$$B_y(r, \theta) = B_0 \sum_{n=0}^{\infty} r^n [b_n \cos(n\theta) - a_n \sin(n\theta)] \tag{2}$$

and

$$B_x(r, \theta) = B_0 \sum_{n=0}^{\infty} r^n [a_n \cos(n\theta) + b_n \sin(n\theta)] . \tag{3}$$

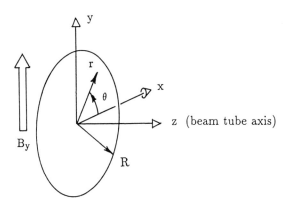

Fig. 1. Coordinates. The beam tube axis is parallel to the z-direction and the dipole field is along the (vertical) y-direction.

The main methods of measuring the MMC fall into two generic categories: rotating coils and boundary measurements. Rotating coils have been widely used for this purpose.[3] In this method relatively long coil arrays inside the beam tube are rotated about an axis parallel to the axis of the beam tube and the induced emf measured as a function of the rotation angle. From a harmonic analysis of this signal the MMC are obtained.

The other method, one example of which is the main topic of this paper, is based upon measuring the field on a boundary that encloses the region of interest. It is justified by the following reasons. Because there are no magnetic field sources inside the beam tube, each component of the field satisfies Laplace's equation. The solutions for cylindrical symmetry are given in Eqs. 2 and 3. By the standard arguments of potential theory, a measurement of the field over a closed boundary determines the field at all points within the boundary. This point is illustrated in the next section where it is shown explicitly how measuring one component of the field on the boundary is used to obtain the MMC, and therefore the magnetic field at all points inside the boundary. Note that only one of the components in the x-y plane needs to be determined; either Eq. 2 or Eq. 3 can be used to find the MMC.

IMPLEMENTATIONS OF THE BOUNDARY MEASUREMENT METHOD

We are aware of three rather different approaches to using the boundary measurement method: the MCPMR method described here,[2] the Hall sensor array proposed by B. Berkes,[4] and the NMR transmission line array method proposed by P. Starewicz and D.F. Hillenbrand.[5] In all of the cases B_y (or equally well B_x with the Hall sensor array) is measured on a circle of radius R (the boundary). In the case of the Hall array and the MCPMR methods, the measurements are made at a point along the beam tube and integrated measurements for the entire magnet are obtained by summing the individual ones. The transmission line array method would give the integrated values directly. Once B_y is determined on the boundary, the usual steps of Fourier series (Eq. 2) inversion are then used to obtain the MMC, with the results:

$$b_n = \frac{1}{\pi B_0 R^n} \int_0^{2\pi} B_y(R, \theta) cos(n\theta) d\theta \tag{4}$$

and

$$a_n = \frac{-1}{\pi B_0 R^n} \int_0^{2\pi} B_y(R, \theta) sin(n\theta) d\theta . \tag{5}$$

Equations 4 and 5 imply a measurement of B_y that is continuous on R. In principle it can be mapped out by an NMR probe or a Hall probe that moves continuously on the boundary. For various reasons, this approach appears impractical for SSC dipole magnets. The favored alternative is to use a large number (M) of discrete, non-rotating field sensors placed at the angles $\theta_m = 2\pi m/M$; $m = 0, 1, 2, ..., M-1$ on the boundary circle. In this case, the inversion formulas for the MMC become:

$$b'_n = \frac{2}{M B_0 R^n} \sum_m B_y(R, \theta_m) cos(n\theta_m) \tag{6}$$

and

$$a'_n = \frac{-2}{M B_0 R^n} \sum_m B_y(R, \theta_m) sin(n\theta_m) , \tag{7}$$

where the primes indicate that the coefficients obtained in this analysis are from measurements at discrete points on the boundary rather than continuously over it. Although the consequences of this approximation have been evaluated exactly, they are not discussed here, except to state that if the number of measurement points is sufficiently large, the error associated with using a discrete set of measurement points on the boundary is negligible.

APPLICATION OF MAGNETIC RESONANCE: 8-COIL PROTOTYPE

In this section we describe the MCPMR method developed at UCLA to measure the MMC. Equations 6 and 7 show that the MMC can be obtained by measuring B_y at R, θ_m. In turn B_y is found by measuring the Larmor frequency ω_L of the NMR precession signal detected in each coil and using the well known magnetic resonance condition[6]

$$\omega_L = \gamma \sqrt{B_x^2 + B_y^2 + B_z^2} \simeq \gamma B_0 \left(1 + \frac{B_y - B_0}{B_y} \right), \tag{8}$$

where γ is the known nuclear gyromagnetic ratio. In Eq. 8 the magnitude of the magnetic field has been expanded to lowest order in the small deviations (B_x, B_z, and $B_y - B_0$) from the dipole field. The higher order terms in the expansion are negligible for SSC dipoles. Equation 8 shows that the y-component of the field is obtained from a measurement of its magnitude. All of our work to date has used this approximation. If a situation is encountered where the approximation is not valid, an alternative analysis developed by Halbach[7] can be used to obtain the MMC from the same measurements.

The overall architecture with which the boundary measurement is accomplished in our 8-coil prototype system is shown in Fig. 2. The 32-coil system being developed is similar. Eight

NMR coils are placed on a circle in the x-y plane inside the beam tube for cold measurements. The probe is cooled to the temperature of the beam tube with a small amount of helium exchange gas in the tube. The NMR free induction decay signal for each coil is generated simultaneously by the single rf pulse transmitter operating at a reference frequency ω_R provided by the cw signal generator that is close to ω_L. Each coil has a transmission line that carries its signal to its own receiving circuit (triangle) at room temperature, where the signal is amplified and detected as discussed in more detail below. By coupling the common transmitter line to each of the coils through crossed diodes the power is fed to the coils during the rf pulse, but the transmitter is effectively decoupled from them for subsequently receiving the NMR. The detected signal is processed into digital form in a separate circuit and fed over a parallel digital bus to a microcomputer for further processing. The current system uses an IBM AT-compatible computer and ASYST software for system control, data acquisition, data analysis, and graphic presentation.

One additional step that is important is to orient the probe relative to the dipole component of the field. This task is accomplished with a Hall effect sensor that is rigidly attached to the multiple coil probe. It is oriented to provide a very sensitive null in the Hall voltage when the probe is correctly aligned with the dipole field. Alternatively, it can be used to measure the direction of the dipole field relative to the probe so that the transformation of the MMC corresponding to the rotation can be calculated in the computer.

The circuitry of an individual channel and its associated signal processing are indicated in Fig. 3. At the start of a measurement, the computer triggers an rf pulse from the transmitter at the frequency ω_R of the rf generator. The NMR free induction decay signal of the form $S(t)\cos\omega_L t$ is picked up in the sample coil and fed to the rf amplifier. The envelope function $S(t)$ decays in a characteristic time T_2^*. This decay time is related to the steady state NMR absorption signal line width δH by $\delta H = 1/\gamma T_2^*$. After passing through the mixer and a low pass filter, the lower bandpass signal $S(t)\cos(|\omega_L-\omega_R|t)$, shown in the upper waveform of Fig. 3, is sent to the comparator. The comparator converts its input to TTL signals that are high when the input is positive and low when it is negative (central waveform of Fig. 3). An "enable pulse" that is set to cover a part of the signal where it is large compared to noise selects an integral number of cycles (N) of the signal. In the digital processing circuit, N is counted and the duration of the enable pulse is timed by recording the number of ticks (n) on a clock with period τ ($\tau = 0.25$ μs in the 8-coil system). The two numbers N and n, which are sufficient to calculate the beat frequency $|\omega_L-\omega_R|$ for each coil, are sent to the computer over an optically isolated parallel bus from the digital processing circuit. In the computer B_y is

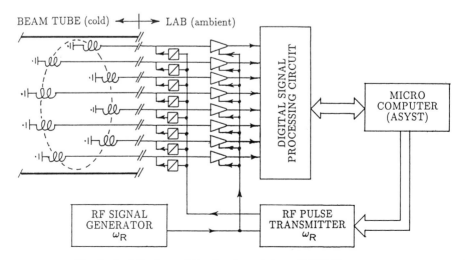

Fig. 2. Architecture of the 8-coil prototype MCPMR system.

calculated with the formula (see Eq. 8)

$$B_y \simeq |B| = \frac{\omega_R \pm \frac{2\pi N}{n\tau}}{\gamma} \ . \tag{9}$$

Then it computes the MMC with Eqs. 6-7, displays the results on the screen, and plots a contour map of the field distribution. A single data acquisition and transfer takes about 50 ms for our 8-coil prototype system, and the calculated results appear on the screen in a few seconds.

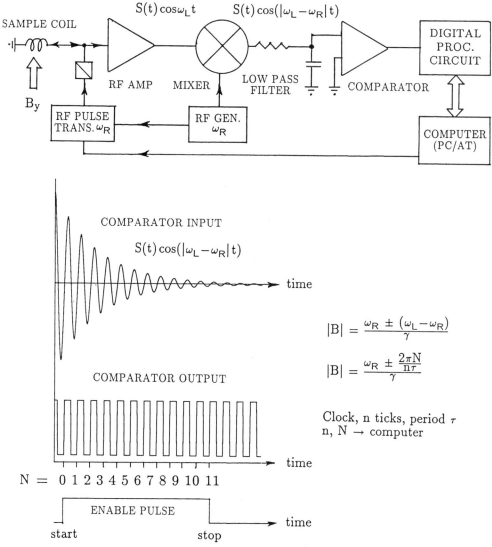

Fig. 3. Signal processing in one receiving channel of the MCPMR system. The quantity to be recorded is the beat frequency of the waveform at the input of the comparator. Its zero crossings are converted to TTL signals by the comparator. Their number (N) and duration $n\tau$ during the enable pulse is determined in the digital processing circuit and sent over the bus to the computer. The magnetic field at each coil is computed according to the formulas on the right.

SOME MAJOR TECHNOLOGICAL ISSUES AND THEIR RESOLUTION

In this section we discuss some of the factors that have an important impact on the design of an MCPMR system. The overall goals are to measure the MMC of cold dipole magnets reliably, accurately, and quickly.

One of the most critical issues is the accuracy with which B_y is measured on the boundary. Here, we discuss in very rough terms the major factors that bear on the accuracy of this measurement. Computer simulations we have carried out indicate that the individual field measurements in each coil should be accurate to about 4 parts in 10^6 to obtain an accuracy on the order of 0.01 units in the MMC using a single rf pulse, where the MMC themselves are expressed in the usual units of 10^{-4} cm^{-n}. As long as the NMR signal of the kind shown on Fig. 3 has a reasonable size relative to noise (not shown on Fig. 3), the measurements of n and N, and therefore the field, can be obtained with good accuracy. The situation is seen most easily in terms of the steady state absorption line width, δH. When the signal-to-noise ratio is large, the field can be reliably determined to approximately 10% of δH. Under these conditions, the NMR linewidth of the signal from each coil should not exceed 40 ppm of the applied field. Thus, one wants both a good signal-to-noise ratio and a narrow line width.

For the situation that applies to SSC dipole testing, δH has two important contributions: the intrinsic ("homogeneous" in the language of magnetic resonance) line width of the material and the "inhomogeneous" width caused by the inhomogeneity of the external field over the sample. A measure of the total observed line width is the square root of the sum of squares of the two contributions.

The need to reduce the inhomogeneous line width conflicts with the need for a good signal-to-noise ratio for the following reasons. Since the MMC are not zero, the field is inhomogeneous on the boundary where it is measured. It then follows that the sample volume should be small to have a small inhomogeneous δH. But as the volume is lowered to reduce δH, the signal-to-noise ratio is also reduced. It is therefore necessary to strike a compromise in the size (and shape) of the sample volume in order to optimize the MMC measurement. The compromise chosen is, of course, also affected by the inhomogeneous line width of the sample.

There are several factors that are important for optimizing the signal-to-noise side of the compromise. The first is that since the nuclear magnetization responsible for the signal is inversely proportional to the temperature, it is a great advantage to have the sample in the cold bore of the magnet. Two other factors that help are to use nuclei that have a large γ, a large nuclear spin I (as long as the symmetry of the local environment is high enough to insure that there are no problems with the nuclear quadrupole moment), and have a high concentration in the sample. The latter item means that the working nucleus should have a high natural abundance, constitute most of the nuclei in the sample, and be in a condensed phase (solid or liquid).

Since the homogeneous NMR linewidth of a liquid is normally much smaller than that of a solid, a material in the former phase is the first choice for the field mapping application described here. But at the temperature of liquid He, there is only one known substance that is liquid, He itself. Fortunately, 100% enriched ^{3}He, the only stable He isotope with a magnetic moment, is readily available. It also has a large value for γ, and therefore a strong signal in the liquid phase. It is one of the choices for our system, and is especially important for measurements at the injection field of 0.66 T, where the inhomogeneous contribution to the magnetic field is relatively small. One complication is that at a pressure of 1 atmosphere it liquefies at 3.19 K, which is below the projected operating temperature of the SSC dipole magnets. Our solution to this problem has been to operate it at a pressure of several atmospheres, so that its density is high enough to give a good signal.

At the higher fields where the inhomogeneous contribution is larger, it is possible to use a solid material as long as its homogeneous line width is not too large. In this respect we have found that the ^{27}Al resonance in powdered Al metal is useful. In our future work we will combine both materials in the same sample cell to obtain two important benefits: a shortened value of T_1 for the ^{3}He and the ability to measure two different values of the magnetic field with a probe tuned to a single frequency. These advantages are discussed in more detail below.

With these choices, we use sample coils that are on the order of 1-2 mm in diameter and 2-4 mm long.

In practice, the condition on δH described above can be relaxed substantially through statistical averaging of the measurement. It is expected theoretically, confirmed in computer simulations, and observed in practice that the precision of the MMC measurements improves as the inverse of the square root of the number of averages. As a result of this improvement, we normally do 16 repetitions for each overall determination of the MMC.

There is another major statistical averaging that is inherent in using the MCPMR method to determine the MMC. Each measurement of the MMC with the MCPMR method is an average over a section of the magnet that is the length of the NMR coil; i.e., on the order of a few mm. Since it is the *integrated* MMC that are important for orbit stability, one expects to sample the MMC at many points as the probe is moved along the magnet and to reap an additional advantage of further statistical averaging. Another consequence of the local nature of the MCPMR measurement is that it can be used as a diagnostic tool to detect flaws in magnet construction that are themselves local. Because the measurement is a quick "snapshot" of the MMC, it can also be used to measure changes in them that occur on short time scales. This feature also makes the MMC measurement relatively insensitive to small variations in the magnet current because a change in it varies the field at all points on the boundary by about the same amount, whereas the calculated MMC depend on the differences in the field on the boundary. Such behavoir has been demonstrated with the 8-coil system, and was an important factor in the success in making measurements on a short dipole at LBL.

The next important design issue is the magnetic fields at which the measurements are to be made. To obtain good sensitivity, the NMR coils are part of a tuned circuit, with the practical consequence that each probe is limited to operation at one or a small number of frequencies, and hence magnetic fields. In response to this circumstance, our work has concentrated on measurements at three values of the dipole field: the injection field of 0.66 T, the storage field of 6.7 T, and an intermediate value of about 2.3 T. The design strategy we use in this regard is to use a composite sample of Al metal powder and pressurized ^{3}He in a probe tuned to 75 MHz for measurements at 6.7 T and 2.3 T respectively, and the same materials in a probe tuned to 21.4 MHz that uses the ^{3}He signal at 0.66 T. These composite samples have an important advantage that is discussed as part of the next topic.

Another technological issue is the time required to make the measurements. It is an especially important one in a production situation where many measurements must be made, particularly if a large number of measurements is to be made along the length of each magnet. The following discussion of this point is based upon experience obtained with our 8-coil prototype system.

For each measurement, the probe must be moved to its position on the axis of the beam tube and oriented relative to the dipole field. Our experience is that these operations can be performed by hand in about 30 seconds. This time could presumably be reduced if they are automated and placed under computer control. The time to record a single measurement and store it in the computer is about 50 ms. A 10 MHz AT-compatible computer takes about 5 s to compute the field at each coil and display the MMC on the screen for the 16 repetitions that are averaged.

The step that has the potential to take the longest time is the spin-lattice relaxation time T_1 of the nuclei that are used. This quantity is the time required for the nuclear magnetization to build up significantly towards its thermal equilibrium. It represents a practical minimum waiting time for a single measurement or between repeated ones that are to be averaged. For the ^{27}Al signal in Al metal, it is (1.85 s)/T, where T is the temperature in K. Thus, for this signal the waiting time is less than one second as long as T>1.85 K. If it is averaged 16 times, only about 15 s or less is required for the projected temperatures at which cold measurements will be made.

The corresponding time for the ^{3}He signal can be much longer than 1 s, which would be a serious problem. The intrinsic T_1 for pressurized ^{3}He is so long that it is not accurately known, but it is certainly long compared to 30 min. Instead of the intrinsic relaxation rate, one invariably observes an extrinsic rate that is caused by nuclear spin relaxation at the boundary of the container. It depends on the surface-to-volume ratio of the container and the material and condition of the surface with which the ^{3}He is in contact. In the tests of our 8-coil system in a dipole magnet at LBL we found that T_1 for the coils was on the order of 10-20 s. The time required for 16 repetitions was therefore about 4 min, which is viable, but uncomfortably long.

Subsequently, we have developed a composite sample design that lowers T_1 for the ^{3}He signal to 0.7-2 s, depending on the field and temperature. It uses loosely packed Al powder that has the interstices filled with pressurized ^{3}He. This arrangement has the two advantages of a T_1 reduction for the pressurized ^{3}He and the option to measure two values of the field with a probe operating at a single frequency.

The final technological issue we discuss here is the limitation that although the MCPMR system is quite tolerant of field inhomogeneities in relation to other NMR methods that have been used to map fields, it also requires a field that is not too inhomogeneous. As a practical matter, we found in tests on a short dipole magnet charged to about 0.45 T at LBL that it was difficult to get measurements closer than about 20 cm from the end of the magnet. In this region the field is too inhomogeneous for viable measurements using the MCPMR method, so that a supplementary method is required to obtain the MMC at the ends of a dipole.

NEW 32-COIL SYSTEM

We currently have under design and construction a 32-coil MCPMR system whose purpose is to measure the MMC of SSC dipoles. It differs from our 8-coil prototype system in several important respects. The first is that it has a much larger number of coils; it will measure B_y at 32 points on the circular boundary. Because the number of MMC that can be obtained is limited by the number of measurement points on the boundary, the 8-coil prototype system was limited to 7 MMC (b_0, b_1, b_2, b_3, a_1, a_2, and a_3) for a single orientation of the probe relative to the dipole field. The published target values for the MMC[1] list terms up to b_8 and a_8. By increasing the number of coils up to 32, coefficients up to a_{15} and b_{15} can, in principle, be measured and the aliasing of higher harmonics of the field down to low order coefficients should be greatly reduced.

If it becomes important to measure even more coefficients, two strategies can be followed with the MCPMR method. The first is to add more coils and receiving channels at the price of greater complexity (there is a practical limit that is not far from 32 coils). An alternative strategy, which has worked with the 8-coil prototype system, is to replicate the effect of a larger number of coils by making measurements at different probe orientations. For example, with four orientations of the probe, one can obtain "pseudo-32-coil" performance with an 8-coil probe. The price of this option is the correspondingly longer time required to reorient the probe (about 10 s) and to make the measuements.

Another important difference in the new system is that it is designed to operate at several fields (and frequencies). The target values were discussed above. The only part of the system that is frequency specific is the tuned probe. All of the rf and digital electronics works over the entire range that is needed for SSC testing. Therefore, additional field value measurements require only a new probe that is tuned to the requisite frequency.

The third major difference is that the rf cables must be much longer than for our 8-coil prototype system to reach the entire length of the dipole. This requirement is being met with long, miniaturized transmission lines for the rf signals and by matching the impedance of the probe circuitry to the characteristic impedance of the transmission line.

We end this section with our estimate of the time and data storage requirements in using the 32-coil system to investigate one cold SSC dipole magnet. They are based upon experience using the 8-coil prototype in tests of a short dipole magnet at Lawrence Berkeley Laboratory in March, 1989. The following conditions are assumed for these projections: (1) The magnet is cold and the MCPMR probe is in its first measurement position, (2) at each position, 16 repetitions of the measurement will be done for statistical averaging, (3) measurements are made at a single field value, (4) measurements will be made at 60 points along the magnet, (5) just one orientation of the probe will be used, and (6) the procedure has become routine for those making the measurements. Under these circumstances, we estimate 15-30 seconds will be required to record the data and another 30 seconds to move the probe to the next measurement position, for a total of about 1 hour to measure 1 dipole at 1 value of the field, and about three hours for a complete set of three fields. This estimate does not include the time needed to insert the probe into the magnet, since the procedures for doing so have not been established, nor does it include time to ramp the field. On the basis of our experience with the 8-coil system, we estimate a probe change will require 15 minutes when it is routine. This time estimate also does not include breakdowns of any kind.

On the basis of the above measurement scenario, the data storage requirements are modest for such a large project. The raw data generated for one measurement with one coil uses about 3 bytes of storage. With 32 coils, 16 repetitions per coil, 60 measurement points, and coverage of three magnetic fields, the total is about 300 kB of storage for the raw data on a single dipole. The results computed from the raw data would require a negligible addition. Thus, a single 300 MB hard disk could store the raw data on 1,000 dipole magnets and a single optical disk could store the inventory of measurements on all of the dipole magnets planned for production.

CONCLUSIONS

The major design features of a multiple coil pulsed magnetic resonance system for measuring the magnetic multipole expansion coefficients of SSC dipole magnets has been described. Its operation in terms of an 8-coil prototype system was discussed. Several of the technological issues that influence the design were identified and their resolution was explained. A brief description of the new features of a 32-coil system presently under construction was given, along with estimates of its requirements for measurement time and data storage capacity.

This system shows strong promise for development and production testing of cold SSC dipole magnets. Its development is nearly complete and the technology associated with it will be ready for transfer to commercial realization in the near future.

The work reported in this paper was supported by DOE Contract DE-AC02-87ER40350 and an equipment grant from KEK (Japan's National Laboratory for High Energy Physics). One of us (WGC) thanks Prof. D. Stork for having brought the need for cold dipole MMC measurements to his attention and Prof. M. Tigner's encouragement to propose its development.

References

1. Site-Specific Conceptual Design of the Superconducting Super Collider, Final Draft, December 20, 1989.
2. W.G. Clark, T.W. Hijmans, and W.H. Wong, J. Appl. Phys. 63, 4185 (1988).
3. G. Ganetis, J. Herrera, R. Hogui, J. Skaritka, P. Wanderer, and E. Willen, Proc. 1987 IEEE Particle Accelerator Conf: Accelerator Engineering and Technology (Cat. No. 87CH2387-9), Washington, DC, USA, 16-19 March 1987 (New York, NY, USA, IEEE 1987), vol. 3, p. 1393-5. M.I. Green, P.J. Barale, W.V. Hassenzahl, D.H. Nelson, J.W. O'Neill, R.V. Schafer, and C.E. Taylor, IEEE Trans. Magn. (USA) 24, 954 (1988).
4. B. Berkes, "Cold Hall-Generator Array for Magnetic Field Multipole Measurements of Cold SSC Magnets," Workshop on Measurements of Cold SSC Magnets, 2-3 March 1989, Berkeley, CA, USA (unpublished).
5. P. Starewicz and D.F. Hillenbrand, paper III-H-18, this conference.
6. C.P. Slichter, "Principles of Magnetic Resonance" (Springer-Verlag Berlin Heidelberg, 1978) is an excellent reference for all aspects of magnetic resonance discussed in this paper.

7. K. Halbach, "Calculation of Harmonic Coefficients from Measurement of Strongly Inhomogeneous 2D fields with Probes that can Measure Only the Absolute Value of B," Workshop on Measurements of Cold SSC Magnets, 2-3 March 1989, Berkeley, CA, USA (unpublished).

8. Systems and Controls

OVERVIEW OF FERMI NATIONAL ACCELERATOR LAB CONTROL SYSTEM*

Peter W. Lucas

Fermilab
P. O. Box 500
Batavia, IL 60510

ABSTRACT

Various facets of the control of the Fermilab accelerators, in particular the Tevatron, are presented. Since Fermilab contains a superconducting machine and a sophisticated injection complex, much of the controls functionality will of necessity be the same at the SSC. The various functions required at a large laboratory are discussed; these include computer-based fire and security alarms and a cable television system, as well as computer networks connected to accelerator hardware components. A description is given of that hardware, of which much is Camac but with considerable computer backplane bus equipment also present. A large fraction of the controls hardware has access to high precision real-time clocks. Our various networks are introduced, with the physical layer being a combination of copper and more modern optic cables, with the primary intercomputer link being Token Ring. A description of the computers is presented - basically these consist of operators' consoles, host VAXs, and link driving front ends. The software effort is detailed, with emphasis on consoles and microprocessors where the majority of effort has been placed. Future plans for the system are presented briefly.

INTRODUCTION

The purpose of this paper is to detail the work of the Accelerator Controls group at Fermilab. The reason for doing so at this SSC conference is to provide a model of what controls tasks will be required at the SSCL and to indicate the level of manpower required to construct and maintain a successful system. In preparing a contribution to the recently completed SCDR we estimated the SSC's needs for controls hardware as four times those of the Fermilab system and the corresponding software requirement as 300 person-years, a level exceeding Fermilab's by about 50%.

---

*Operated by Universities Research Association, Inc. under contract with the U.S. Department of Energy

Different organizations (laboratories) have different definitions of control. Anyone interested in this work must be careful to understand what is included and what is not in the overall term. In particular the SCDR used for some subsystems the definition of controls as taken from the Fermilab organization chart, and for others an arbitrary one formulated for the 1986 SSC CDR. Both of these underestimate the total effort for a logically complete control system.

FERMILAB CONTROLS, MANPOWER AND FUNCTIONS

The current staffing level of our group is as follows:

Management	1
Engineers	9
Programmers	24
Technicians	26
Secretary	1
Total	61

Six group members are physicists, but none works in that capacity in his normal tasks.

A group of size slightly smaller than this was assembled during construction of the Tevatron, in the early 1980s. The hardware staff size is basically unchanged since that time, while on the software side there has been some growth. The areas of responsibility of this group include the Ion Source and Linac, Booster, Main Ring, Tevatron (in part), Debuncher, and Accumulator accelerators and storage rings. In addition the fixed target Switchyard is included as are all internal beam transfer lines between the above rings. Not included is the control system for running beams, once they have exited the Switchyard, to experiments; this system was designed and is maintained by a much smaller Research Division group. Also not in our area of responsibility are the controls of the Central Helium Liquifier plant, which is run by a commercial system with an RS232 connection to our network (though the control of satellite refrigerators, a major undertaking, is in our purview), and the Tevatron quench protection system and primary dipole/quadrupole bus ramp generation, both of which are handled elsewhere in the Accelerator Division. Controls for the magnet test/measurement facility are assigned elsewhere in the Laboratory.

However there are some items not directly associated with the control of accelerators, but which must exist at any large laboratory, and are assigned to our group. The first is a separate control system called FIRUS (Fire, Utility, Security) and which in the SSC SCDR was referred to as FUSE (Fire, Utility, Security, Environment). A site diagram of the Fermilab FIRUS is provided in Figure 1, together with a representative alarm display. This system is not used to assure primary safety at accelerators – hard-wired loops perform this function – but it does provide alarms and notification for a large number of safety and security related occurrences. The requirements vis a vis the control of accelerators are quite different; namely the bandwidth is by comparison modest but the necessary reliability

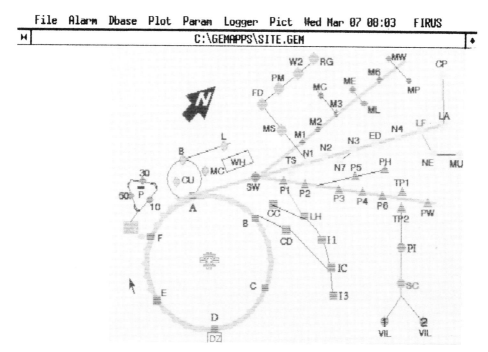

Figure 1A. Site map indicating the locations of FIRUS monitoring centers.

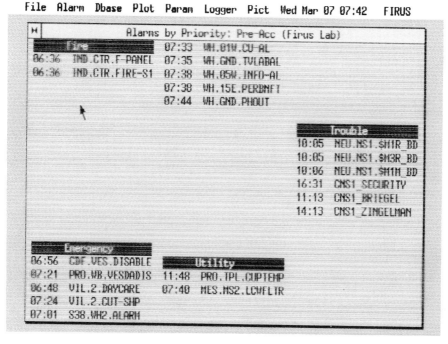

Figure 1B. Typical alarm display on a FIRUS console.

419

Table 1.   Television Equipment

Amplifiers (RF and video)	358
Cameras (all types)	437
Converters (channel)	800
Demodulators	104
Lenses (special)	268
Modulators365	
Monitors, video	329
Pan & tilt (remote control)	40
Receivers631	
Cable plant(hard line)	> 70 miles

is much greater, with many components both redundant and
battery-backed.

Also included in our responsibilities is operation of the
lab-wide CATV and CCTV systems; Table 1 summarizes the extent of
existing equipment. TV is a satisfactory technology for bringing
information from a variety of sources to the accelerator
operators, as well as for informing experimenters of accelerator
operating conditions and events. It also has a role in safety,
providing visual observations of otherwise inaccessible
locations. It is also noted that, for historical reasons,
roughly 10% of the effort of our group is devoted to the
controls and electronics of the D0 High Energy Physics
experiment.

HARDWARE OVERVIEW

Much of the remainder of this paper will refer directly or
indirectly to Figure 2, which indicates the computer nodes of
the primary controls network, ACNET (Accelerator Controls
Network), their connectivity, and the functions in the field
which are ultimately controlled.

Timing

From a hardware perspective the unifying feature of the
system is a series of digital clocks which transmit information
to essentially all control hardware. The basic timing is through
TCLK (Tevatron Clock) on which can be encoded any of 256 values
('events') attached to a 10 MHz carrier. Events carry such
information as "beginning of Main Ring proton bunch coalescing
cycle", "Tevatron start of flat top", and "720 Hz tick". There
are two other clocks which run at lower rates but are
synchronized with the RF (and thus the beam) of both the Main
Ring and the Tevatron (a third, synchronized to the Booster, is
proposed). The jitter between beam arrival and signals from
these clocks is typically of order 2 ns. They are used for
kicker magnet firings, SEM grid readout timing, and as
components of triggers for collider detectors. A separate link
closely related to timing, MDAT (Machine Data), is used to
transmit such quantities as primary bend bus current.
Transmissions on this link typically include a code to indicate

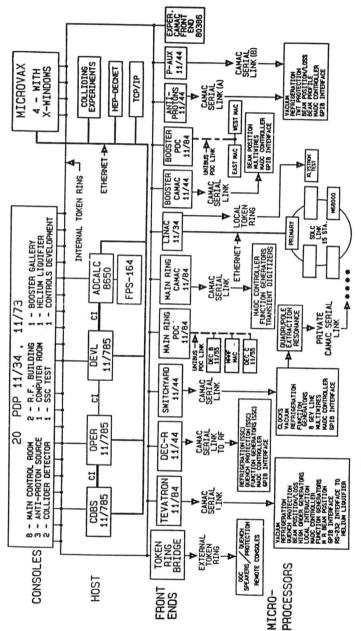

Figure 2. Tevatron control system computers.

the type of quantity being sent, followed by the value; the normal transmission rate is 720 Hz.

The ability to receive and act upon clock and MDAT signals is built into a sizable fraction of all the control hardware for the accelerators. Thus virtually all field hardware modules are designed in-house, as the ability to deal with such signals is clearly not readily available in commercial units.

Field Hardware

The modules which connect to accelerator hardware are constructed at two levels of sophistication. The simpler functions are performed in Camac (except, for historical reasons, in the Linac); such functions include digital status and control bit processing, multiplexed ADC control, ramp generation – with ramps as functions of from one to three variables, clock event recognition with pulse generation, and stepping motor control. Though referred to as less sophisticated, some of these modules contain powerful microprocessors which operate on interrupts generated by up to about 100 separate function codes. The number of installed Camac cards in our system is roughly 3150, occupying 4550 slots in 260 crates.

The more sophisticated hardware generally involves external crates (Multibus I and II and VMEbus) connected by various means to the control system. These external crates contain entire subsystems for satellite refrigeration control, beam position monitoring, quench and travelling wave tube protection, tune and chromaticity feedback correction (under development), and higher order function generation. These 'smart subsystems' have until now been connected to the rest of controls through Camac links, with gateway modules residing in Multibus or VMEbus interacting with partners in Camac crates. The communication from the controls computers to these modules is via an in-house developed protocol, GAS (see below under software). Also utilizing this protocol are some processor-based Camac modules, namely vacuum controllers and generic GPIB/RS232 interfaces. More recent subsystems, of which new quench protectors and the tune/chromaticity feedback system are the first examples, will interface directly to ACNET, each being a full fledged network node. The Linac, which has always operated with geographic local intelligent control, will now be connected to ACNET in a similar fashion.

The network contains ACNET layers for handling task-to-task communications, residing on top of IEEE 802.5 Token Ring. Token Ring was chosen over the more popular Ethernet since, for an environment where many transmissions are synchronized with beam accelerations and transfers, the token passing model provides more effective bandwidth than does the collision detecting Ethernet scheme.

Cabling

With the construction of the Tevatron there was installed over the four mile Main Ring/Tevatron length a cable consisting of nineteen channels of copper heliax. However, as the system has grown this number of channels has proven to be insufficient, and a number of other cables have been pulled. Recently

Table 2. Fermilab Main Ring and Tevatron Links

LINK NAME	LENGTH	USERS	MEDIUM	BIT RATE
DEC-T PIOX				
DEC-T PIOR	4 Mile	80 Crates	Copper	10 MBit/sec Burst
DEC-T BTR				
DEC-R PIOX				
DEC-R PIOR	1 Mile	25 Crates	Copper	10 MBit/sec Burst
DEC-R BTR				
DEC-MR PIOX				
DEC-MR PIOR	4 Miles	33 Crates	Fiber	10 MBit/sec Burst
DEC-MR BTR				
QXR PIOX	4 Miles	4 Crates	Copper	10 MBit/sec Burst
QXR PIOR				
CONSOLE PIOX	4 Miles	5 Crates	Copper	10 MBit/sec Burst
CONSOLE PIOR				
TCLK	4 Miles	~1000	Copper	10 MBit/sec CW
MDAT	4 Miles	~600	Copper	10 MBit/sec Burst
MR BEAM SYNC	4 Miles	<100	Copper	~7.5 MBit/sec CW
TEV BEAM SYNC	4 Miles	<100	Copper	~7.5 MBit/sec CW
QPM	4 Miles	25	Copper	1 MBit/sec (SDLC)
MRPSL	4 Miles	25	Fiber	10 MBit/sec Burst
MR ABORT	4 Miles	~40	Fiber	5 MBit/sec CW
TEV ABORT	4 Miles	~40	Copper	5 MBit/sec CW
FIRUS	5 Miles	~33	Copper	2.5 MBit/sec (Arcnet)
CATV	(see Table 1)			
TOKEN RING	5 Miles	90	Copper	4 MBit/sec (802.5)
(double twisted pair)				

considerable growth potential has been added by installing a
bundle of 24 multimode optic fibers over the ring circumference.
This bundle is installed in communication ducts near the
surface, not in the tunnel, so that radiation damage to the
fibers is not a concern. Table 2 indicates the signals
transferred on various Main Ring/Tevatron links. Examination of
the information in this table led us to propose a large number
of site-wide and point-to-point links for the SSC complex, with
considerable spare capacity available from inception.

COMPUTERS AND LOCAL AREA NETWORK

     The nodes and their connections as indicated in Figure 2
are explained in this section. The top row represents operators'
consoles. There are presently twenty of these, based on PDP-11s
running the RSX operating system, driving a variety of display
hardware, and networked via Token Ring. Due to the architectural
limitations of the computers, these consoles are proving
incompatible with modern software and are to be replaced. The
new machines will be VAXstations, with 8-plane color graphics,
networked via Token Ring and/or Ethernet, and running
VMS/X-Windows. The specification is that there be about fifty

such consoles at the completion of the conversion process. Many are to have two screens, the second a color X-terminal driven by client software in the station. To date we have found no terminals with performance sufficient to meet our needs.

The next row in the figure contains the host VAX computers - three 11/785s and one 8650. These are used as central file servers, as database servers, for software development, for database management, in the case of the 8650 for computationally intensive work, and for running certain application software. Included in this last category is a datalogger, logger reader, alarms distribution to consoles, and collection of accelerator data on behalf of colliding beam experiments and distribution to those experiments. It is intended in about one year to replace the three VAX-11/785s with two newer machines, as of now VAX-6410s.

The nodes of the third row are termed front ends. Each of these is the driver for one link, primarily Camac but in a few cases other technologies. These devices handle readings, both single shots and the more usual single read/repetitive reply, settings, status bit reads and control bit sets, mediation of multiple console access to given devices, and alarm condition scanning. As in the case of the consoles, the PDP-11s utilized for this task are proving to be limiting factors. At present the Tevatron front end limits certain processes, and as the number of consoles grows the potential for bottlenecks at the front ends becomes ever more serious. Thus new devices, consisting of parallel arrays of (perhaps three) 80386 microprocessors housed in Multibus II crates, are under development. The 386 was chosen primarily because it utilizes the same byte order as the PDP-11 which it is replacing; however our programmers are also finding that it provides an advantageous environment in which to work.

SOFTWARE

The Fermilab system contains over 200 integrated person-years of software effort. This work is divided among microcoded smart subsystems, front ends, networking, host VAXs, and consoles, with the last representing the largest single investment. There is provided a library of a few hundred routines, written in assembly language for the PDP-11s and in C/assembler for the VAXstations, which perform network, database, and remote file accesses, initiate and perform data acquisition, emit settings, convert to/from standard engineering units for binary quantities, and perform a wide variety of primitive graphics operations. The application programmer writes almost exclusively in Fortran, acquiring and displaying data, setting quantities, and accessing the database, through use of this extensive library. There are roughly 700 applications installed on the system. These programs range from ones to deal with very specific pieces of accelerator apparatus, to others for dealing in a non-specific manner with many pieces of equipment, to system diagnostics, and to operator and physicist general aids. The most utilized single program is a generic Parameter Page; this program was designed in conjunction with the system database and can be used to read, set, and control any scalar device. Closely allied with this is a Plot Package, which can graph any quantities, up to four, as functions either of time or of each other. Considerable work on this program over time has allowed for a wide variety of features, many related to the conditions under which plots are started and the actions taken on filling screen space. Plots can be made either in real time or as a result of reading a buffer collected at a several KHz rate and stored remotely. A program much like this one can also be used to plot as a function of time any information previously collected by the Datalogger.

The ACNET system was specified in the early 1980s, before many software tools which are now on the market were available. Thus many programs which for the SSC presumably will be purchased, were written in-house for the Tevatron. Examples are a database management system, several layers of networking and communication, code management, and the graphics library noted above. An area where much development was done by us, but where we are now trying to take advantage of modern tools, is that of microprocessors. Among the products created were GAS - a low level network package for communicating with smart subsystems, and also the name of the protocol used by this package, and OPERA - a primitive multitasking operating system for Z80s and later M68000s. Code was written primarily in the assembly languages of the various processors, and also in Pascal. However our recent developments take advantage of the far wider range of products presently available. All current work utilizes commercial operating systems, specifically MTOS and PSOS. Development tools, primarily C language compilers and object code linkers, running on both VAXs and PCs are used. In addition source level debuggers and in-circuit emulators (more for the benefit of hardware developers) are employed. Such tools have been fully integrated for the next generation of Linac geographic controller, a modernized quench protection monitor, the tune/chromaticity feedback system, and the replacement front end computers.

FUTURE DEVELOPMENTS

Modernize hardware

Examination of Figure 2 shows a few vestiges of equipment predating the construction of the Tevatron; such apparatus is used primarily for control of the RF accelerating cavities for Main Ring and Booster. Modernization will occur in these areas.

Upgrade smart subsystems

This work, insofar as it is done by Controls, is somewhat at the pleasure of the accelerator system groups, rather than initiated by ourselves. Nonetheless we are urging switchover to Token Ring communications and commercial software development tools as parts of any new efforts. At present changes are contemplated for beam position monitor, travelling wave tube protection monitor, and satellite refrigerator controllers.

Utilize workstation-based consoles

The first priority in migrating to the VAXstation-based consoles is the maintenance of the operational status quo - seeing to it that all functions now possible remain so and with changes in performance being only for the better. However once this is accomplished then we shall take advantage of the modern equipment to add functionality which does not now exist. Possibilities include construction of software-free applications through use of 'interface builders' and purchase of commercial packages for graphics and other functions.

Upgrade network

There are contemplated some changes in the ACNET network protocol, better to utilize the Token Ring bandwidth (since that protocol is our own, rather than a standard, we can alter it as we like). The major changes in this regard involve the packeting structure of network messages. Also, our Token Ring is presently running at 4 Mbit/sec; it is possible to upgrade to 16 Mbit/sec, and we shall do so when the need arises.

Make our system vendor non-specific

At present all computers (as opposed to smart subsystems) on our network are manufactured by Digital Equipment Corporation, and all run either the RSX or VMS operating system. It is generally considered disadvantageous to be restricted in this way, unable to take advantage of the offerings of other vendors. There are two serious problems to be overcome if we are to move in this direction - dealing with a different byte order and with different operating systems. When machines of opposite byte orders, or different floating point formats, are networked together it is generally essential that the network messages contain enough information that they will be decipherable by the receiving node; ACNET does not provide this at the present time. We are also presently unable to take advantage of the Unix operating system, which is growing in popularity. However the effort required in converting our networking software to any new system is estimated as ten programmer years; thus any change is something which we shall not undertake lightly.

REFERENCES

The chief sources of published information on our system are the Proceedings of the three most recent conferences on Accelerator Control Systems (the two more recent conferences also having included Controls for Large Physics Experiments). Each of these Proceedings includes several papers from our group.

Proceedings of the Second International Workshop on Accelerator Control Systems, Los Alamos, NM, October 7-10, 1985; Edited by P.N. Clout and M. Crowley-Milling; North-Holland Publishing Co.

1987 Conference, Villars, Switzerland; Proceedings to be published imminently by CERN; Edited by B. Kuiper

1989 Conference, Vancouver, BC; Proceedings to be published soon; Edited by D. Gurd and staff of TRIUMF Laboratory

A SUGGESTION FOR EXTENSION OF MAGNET CRYOGENICS INTO

ACQUISITION AND COMPUTATION ELECTRONICS AND PHOTONICS

B. E. Briley

AT&T Bell Laboratories
Naperville, Illinois

INTRODUCTION

The Superconducting Supercollider is estimated to have some 2 million liters of liquid helium and 1 million liters of liquid nitrogen associated with it (as well as some quantity of liquid argon). These cryogens are intended primarily for the purpose of inducing low-temperature superconductivity in the bending magnets (or for calorimetry duty). Data acquisition may require some 10,000, 1 Gbps links; data reduction will require up to 10 million VAX 780 equivalents; data storage needs are beyond current capabilities. The following argues that the extraordinary demands of the SSC upon electronics and photonics capabilities may justify extending the cryogens into other venues to reap benefits not widely recognized, at small incremental cost.

REVIEW OF THE EFFECTS OF LOW TEMPERATURES UPON ELECTRONICS AND PHOTONICS

Low temperatures have a variety of salutary effects upon electronics and photonics which will be enumerated categorically (low temperature as used here will be in the neighborhood of 77 K, reachable with liquid nitrogen).

ELECTRONICS

The primary source of electronics performance improvement at cryogenic temperatures (e.g., 77 K) is the increase in majority carrier mobility. Others include minority carrier freeze-out, normal conductor resistivity decrease, and thermal activation decrease.

Majority Carrier Mobility: Majority carrier mobility may increase by a factor of from two to four at 77 K [1,2]. FET devices, and particularly CMOS circuits, can benefit directly from this effect by a proportional increase in speed.

Normal Conductor Resistivity: Elemental copper experiences a 6-fold reduction in resistivity at 77 K, while aluminum resistivity drops by an order of magnitude at the same temperature [3]. (Aluminum is of particular interest because it is a commonly used interconnect conductor on semiconductor chips.) It becomes possible, therefore, to reduce the interconnect cross-section, if space is the prime concern; or enjoy the reduced RC time constants, allowing the more rapid charging and discharging of parasitic capacitance; or to choose an optimum blend of the two.

Thermal Activation Decrease: Because thermal activation energies decline by more than four orders of magnitude at 77 K, electromigration can be ignored as a limit upon the cross-section reduction discussed above [4].

Minority Carrier Freeze-out: The freeze-out of minority carriers at 77 K immobilizes bipolar transistors [5], which at first glance would appear to be a very negative result. However, examination of CMOS technology reveals a problem known as Latch-Up [6], which is caused by parasitic bipolar transistors that are a byproduct of the chip structure. Typically this problem is overcome via some isolation technique (e.g., trenches) which consumes chip area. If the parasitic bipolars are neutralized by cooling to 77 K, isolation is no longer necessary, and the chips can be made more dense.

Reduced Leakage: The leakage effects in semiconductors are reduced by some four orders of magnitude at 77 K. Among the more intriguing possibilities offered by this property is the effect upon Dynamic RAMs (DRAMs). DRAMs typically require refresh times of the order of milliseconds because the information storage mechanism for each bit is a leaky capacitor. At 77 K, the refresh time increases to hours and even days [7]. For certain classes of application, DRAMs (which are much more dense than SRAMs) can substitute for SRAMs (Static RAMs).

Improved Heat Conductivity: The thermal conductivity of typical semiconductors increases by about an order of magnitude at 77 K [8], allowing higher dissipation operation.

FIBER OPTICS TRANSMISSION SYSTEMS

Three elements of a fiber optics transmission system will be addressed: the light source, the light detecting device, and the preamplifier; the net effect of cooling all three will then be discussed.

It may be useful to remind that a given fiber optic system design may be operating either in a loss-limited or a dispersion-limited regime. That is, it may be limited in its span length because the loss budget is constraining, or because the fiber dispersion, in view of the spectral width and wavelength of the source and the nature of the fiber, is at the limit for the bit rate being transmitted. As presently envisioned, neither limit will be in evidence in the SSC application.

1. Light-Emitting Devices

The primary light-emitting devices of interest are the Injection Laser Diode (ILD) and the Light-Emitting Diode (LED).

The ILD: ILDs provide significant output power in a relatively narrow spectrum, with a small spot size and narrow radiation pattern, making them very suitable for efficient coupling into large or small core (e.g., single-mode) fiber. Their characteristics are very sensitive to temperature, and aging phenomena can be pronounced; consequently, they typically require sensing means to provide feedback controlling the bias current as the lasing threshold current shifts. This feedback can be either photonic (typically with a detector in the same package) or strictly electronic.

Early (homojunction) lasers utilized the same material type throughout, and confined the light and injected charge so inefficiently that massive currents were necessary to attain the requisite critical current density necessary to support stimulated emission; the dissipation associated with these large currents was such that cryogenic cooling was necessary to prevent destruction. As single and double heterojunction devices came into being, confinement improved to the point where room-temperature operation became possible. Typical output power from a communication system ILD is, however, relatively meager.

Though much improved over early devices, modern lasers have relatively limited lives. The failures that can occur with ILDs include catastrophic damage (e.g., facet damage due to excessive light intensity) and progressive damage due to growth of "dark-lines," etc. The catastrophic damage probabilities can be reduced with careful circuit design; the progressive damage tends to be thermally activated. Significant increases in power output are hampered by the requirement for room-temperature ambient operation because higher power reduces life expectancy. The FIT rate for even modest power devices, while acceptable for many applications, is still relatively high.

As ever higher bit-rates are sought, higher output power is needed for the same span length because the signal-to-noise ratio otherwise decreases (the number of photons devoted to representing each bit decreases as power remains constant and bit-rate increases; the sensitivity required of a receiver can be expressed for a given BER as the number of photons required per bit; the quantum limit is about 10 photons per bit for a BER of 10^{-9}).

Cooling to 77 K can improve the life expectancy of an ILD relative to thermally induced failure because thermal activation drops by a factor of from 10,000 to 100,000 [9]. The thermal conductivity of GaAs and variations (e.g., InGaAsP) will also improve by about an order of magnitude [10]. The FIT rate could potentially improve to rival those of passive components.

Because boiling nitrogen can transfer heat at from 10 to 20 watts per cm^2 before evolving nitrogen gas begins interfering with heat flow [11], the power output of a typical communication system ILD could be increased by a factor of 10-100, permitting either an increase in bit rate for the same span length and receiver, or an increase in span length for the same bit rate and receiver (or both).

Early diodes operated in a multi-longitudinal mode regime, producing light that, while far narrower spectrally than that produced by LEDs, and usable in many applications, was still far from the ideal. Further, the number and amplitude of the longitudinal modes supported changed with modulating current and with its rate of change.

Interest grew in single-longitudinal mode ILDs, and several techniques have proven successful in producing such devices, e.g., the Cleaved Coupled-Cavity (C^3) device, the external, coupled-cavity, and the distributed feedback (DFB) designs. Even when only a single longitudinal mode is supported, however, the spectral line is not arbitrarily pure. Due to thermally induced fluctuations in the cavity dimensions, coupling of spontaneous emission into the laser modes, and fluctuations in the electronic state occupancy and carrier density, the spectral line has finite width.

For very long span lengths (long-haul applications), dispersion limitations can be telling. Further, for coherent detection techniques to be applied effectively, very stable, very narrow spectral width sources are necessary.

With operation at 77 K, the maximum frequency of operation (governed by the resonant frequency), the linewidth, the gain (higher efficiency), noise figure and the life of the ILD all improve [12, 13]. More specifically, the resonance frequency functions as an indicator of speed of operation (an upper bound for small-signal operation); it increases as the square root of the differential gain, which, in turn, increases approximately as $1/T$, so that the speed of operation should roughly double when temperature drops from 300 K to 77 K.

The linewidth is a function of two components, one inversely proportional to power (from the modified Schawlow-Townes formula [14]), and the other independent of power (e.g., due to fluctuations in the electronic state occupancy associated with fast intraband thermalization [15]). The first component drops by a factor of about 8 when the ILD is cooled to 77 K; the other rises by a factor of about 4. At low power levels (a few mW), the first component dominates; at higher levels, the second becomes important.

Relative to power dissipation, several factors contribute favorably at 77 K: the higher efficiency of the device (requiring less dissipation per unit of emitted light power), the higher thermal conductivity of the semiconductor, and the heat removal capability of a liquid nitrogen bath. The external package for a laser typically would have a surface area of several square centimeters; at 10 watts per cm^2, the package could probably dissipate more than 50 watts, especially if heat-sink style packaging were used to increase the surface area. An output of 100 mW (20 dBm) would be very useful in comparison to the more typical 1-5 mW launched in many present systems; taking into account the quantum efficiency and resistive losses, the attendant dissipation should be easily handleable. The upper limit would likely be set more by inducement of fiber nonlinearity than by device dissipation.

The LED: LEDs are known for their relatively low launchable output power and broad spectral output (due to operation in the spontaneous rather than stimulated emission regime). They present problems in efficiently launching power into fiber in surface-emitting configurations (though lenses are helpful), and they are generally very poor for launching into single-mode fiber (with the exception of edge-emitting configurations, which can launch more effectively but are generally of relatively low power).

At present, surface emitting LEDs are used primarily for illuminating large-core fiber (either step-index or graded-index). Modal dispersion effects are so severe in step-index fibers that applications are limited to very short spans. Modal dispersion effects in graded-index fiber are much reduced, so that material or chromatic dispersion effects become significant. The broad spectrum of an LED at room temperature imposes significant limitations upon span lengths, especially when operated at short wavelengths where the fiber is functioning far from its material dispersion zero.

The spectral width of an LED narrows significantly as a function of temperature due to a reduced energy spread in carrier distributions [16]. The ratio $\Delta\lambda/\lambda$ is proportional to T [17], so that the spectral width should drop by a factor of about 4 at 77 K. (Note also that the spectral width is proportional to λ^2, so that the width increases rapidly with longer wavelength.)

For the same reasons as for the ILD, the power output of an LED could be increased by a factor of 10-100 at such temperatures.

2. Light Detecting Devices

Light detectors perform the function of transduction of most of the incident photons into hole-electron pairs which are collected as leakage or photo-current that is sensed (in the case of the p-i-n), or sensed after multiplication via collision ionization (in the case of the APD). Both device types suffer from thermally-induced leakage current (dark current) even when not illuminated. When combined with the photo-current, the dark current constitutes noise which impairs the signal-to-noise ratio of the detecting-receiving function.

Silicon devices are employed for short-wavelength detection, and germanium or III-V compounds are used at long-wavelengths.

Though a function of the device structure and material, a typical diode will have a dark current proportional to $e^{-\Delta/kT}$, with Δ of the order of unity. This current consequently drops by orders of magnitude at cryogenic temperatures. Such reductions in the dark-current noise component can contribute toward improving the span length or bit-rate of a system.

In general, the photocurrent gain of an APD increases as temperature decreases [18], but the effect is dependent upon device structure and is difficult to predict.

3. The Receiver Preamplifier

The photo-current presented by the light detector to the preamplifier is quite feeble (e.g., of the order of a few microamperes). It must be amplified in the presence of noise to levels adequate for treatment as a digital signal (in a digital system). The most widely used preamplifiers are the transimpedance type and the high-impedance or integrating amplifier. The transimpedance amplifier has the advantage of a wide dynamic range and wide bandwidth, requiring equalization only at very high frequencies. The integrating amplifier has low-pass characteristics requiring differentiating equalization. It can, however, display a signal-to-noise ratio superior to that of the transimpedance amplifier.

Both amplifier types suffer from noise associated with the active device used, and with thermal or Johnson noise, flicker noise, etc. The noise associated with silicon FETs decreases by about a factor of 10 at 77 K [19]. The rms Johnson noise is proportional to temperature, so that a reduction from 300 K (room temperature) to 77 K will reduce this component by a factor of about 4. It is not unusual for thermal noise to predominate, so that the signal to noise ratio may improve by about the same factor [20].

Overall Optical Communication System Improvement

The reduction in spectral width of an LED would have the effect of reducing the dispersion and increasing the allowable span length or the allowable bit rate (or both).

The bandwidth quoted by fiber manufacturers is the "laser" bandwidth, measured with a narrow spectrum source. The actual bandwidth that a fiber will provide in a given application depends upon the source used. At short wavelengths, graded-index fiber illuminated by an LED is limited primarily by material dispersion (as opposed to modal dispersion).

It can be shown [21] that the bandwidth of a fiber free of modal dispersion can be approximately expressed as

$$B = \frac{1}{\Delta\lambda\sqrt{A + B\,(\Delta\lambda)^2}}$$

where $\Delta\lambda$ is the rms width of the spectral power density of the source, and A and B are functions of the dispersion at λ and the dispersion slope, respectively. This means that the bandwidth improvement is (approximately) bounded by a $1/\Delta\lambda$ and a $1/(\Delta\lambda)^2$ dependency. For a reduction by a factor of 3 in spectral width, this would translate to an improvement by from a factor of 3 to a factor of 9 in bandwidth.

The narrowed spectrum would also lend itself to increasing the number of channels that can be effectively wavelength-division multiplexed onto a single fiber.

In the case of ILDs, consider the effects of reducing the linewidth of a single-mode laser by a factor of 8. An index-guided laser with a linewidth of 10^{-4} nm (30 MHz) would enjoy a linewidth of only 3.75 MHz. Because coherent detection techniques favor linewidths as narrow as possible [22], such a result would be highly advantageous. For noncoherent detection, the Relative Intensity Noise, which influences the system signal-to-noise ratio, is of interest. Depending upon the power level, cooling with liquid nitrogen can improve the RIN by a factor of up to 15 dB/Hz.

APPLICABILITY TO THE SSC

The SSC will require an extraordinary number of high-data-rate communications links which may well utilize fiber optics. The primary advantage (in this application) of cooling lasers involved would be a significant increase in their reliability; if receiver detectors and preamplifiers were also cooled, the design parameters could be relaxed, perhaps permitting less expensive devices.

The computational needs of the SSC are so great as to suggest that all means for performance improvement should be considered. Cooled CMOS would allow increased speed and density, and cooled GaAs might provide extraordinary results.

The memory needs of the SSC are immense by any measure, and cooled CMOS DRAM-like SRAMs might aid in that segment of the memory needs requiring such capability.

In view of the availability of the cryogen, and the expected small incremental cost of extending its availability, cooling electronics and photonics associated with the SSC data acquisition and reduction activities may be well worth considering.

REFERENCES

[1] R. C. Jaeger and F. H. Gaensslen, "MOS Devices and Switching Behavior," *Low Temperature Electronics*, IEEE Press, 1986, p 92.

[2] S. K. Tewksbury, "N-Channel Enhancement-Mode MOSFET Characteristics from 10 to 300 K," *IEEE Transactions on Electron Devices*, vol. ED-28, Dec. 1981, pp. 1519-29.

[3] S. Hanamura, et al., "Operation of Bulk CMOS Devices at Very Low Temperatures," *1983 IEEE Symposium on VLSI Technology*, Maui, Hawaii, September, 1983, p. 46.

[4] R. W. Keyes et al., "The Role of Low Temperatures in the Operation of Logic Circuitry," *Proceedings of the IEEE*, vol. 58, no. 4, 1970, p. 1914.

[5] B. Lengeler, "Semiconductor Devices Suitable for Use in Cryogenic Environments," *Cryogenics*, vol. 14, no. 8, August, 1974, pp. 439-447.

[6] M. Shoji, *CMOS Digital Circuit Technology*, Prentice-Hall, 1988, p. 427.

[7] P. M. Solomon, "Materials, Devices, and Systems," *Low Temperature Electronics*, IEEE Press, 1986, p. 17.

[8] J. G. Hust, "Thermal Conductivity and Thermal Diffusivity," *Materials at Low Temperatures*, ASM, 1983.

[9] M. Nisenoff, *Superconductor Week*, April 25, 1988, p. 5.

[10] R. K. Kirschman, "Cold Electronics: An Overview," *Cryogenics*, vol. 25, no. 3, March, 1985, p. 118.

[11] R. C. Longsworth and W. A. Steyert, "Technology for Liquid-Nitrogen-Cooled Computers," *IEEE Transactions on Electron Devices*, vol ED-34, no. 1, January 1987, p. 6.

[12] J. Katz, "Low-Temperature Characteristics of Semiconductor Injection Lasers," in R. K. Kirschman, *Low Temperature Electronics*, IEEE Press, 1986, pp. 465-470.

[13] A. K. Jonscher, "Semiconductors at Cryogenic Temperatures," *Proceedings of the IEEE*, vol. 52, October, 1964, pp. 59-60.

[14] K. Vahala and A. Yariv, "Semiclassical Theory of Noise in Semiconductors Lasers - Part 1," *IEEE Journal of Quantum Electronics*, vol. QE-19, 1983, pp. 1096-1101.

[15] K. Vahala and A. Yariv, "Occupation Fluctuation Noise: A Fundamental Source of Linewidth Broadening in Semiconductor Lasers," *Applied Physics Letters*, vol. 43, 1983, pp. 140-142.

[16] J. M. Senior, *Optical Fiber Communications Principles and Practices*, Prentice-Hall, 1984, p. 312.

[17] J. Gowar, *Optical Communication Systems*, Prentice-Hall, 1984, pp. 229-236.

[18] H. Kressel, *Topics in Applied Physics: Semiconductor Devices for Optical Communication*, vol. 39, Springer-Verlag, p. 70.

[19] R. W. Keyes, "Semiconductor Devices at Low Temperatures," *Comments on Solid State Physics*, 1977.

[20] C. C. Lo and B. Leskovar, "Cryogenically Cooled Broad-Band GaAs Field-Effect Transistor Preamplifier," *IEEE Transactions on Nuclear Science*, vol. NS-31, February, 1984, p. 474.

[21] B. E. Briley, *An Introduction to Fiber Optics System Design*, North-Holland, 1988, Chapter 6.

[22] Longsworth, Ibid.

9. Poster Sessions

THE SSC SUPERCONDUCTING AIR CORE TOROID DESIGN DEVELOPMENT

J. A. Pusateri[c.]

P. Bonanos[j], A. Carroll[b.], I.H. Chiang[b.], A. Favale[c.], T. Fields[a.], J. S. Frank[b.], J. Friedman[f.], S. Gottesman[c.], E. Hafen[f.], J. Haggerty[b.], P. Haridas[f.], L. W. Jones[h.], H. W. Kendall[f.], K. Lau[d.], L. Littenberg[b.], J. N. Luton[i.], M. Marx[k.], R. McNeil[e.], W. Morse[b.], L. Osborne[f.], I. Pless[f], B. Pope[g.], L. Rosenson[f.], R. C. Strand[b.], E. Schneid[c.], R. Verdier[c.], and R. Weinstein[d.]

a. Argonne National Laboratory
b. Brookhaven National Laboratory
c. Grumman Aerospace Corporation
d. University of Houston
e. Louisiana State University
f. Massachusetts Institute of Technology
g. Michigan State University
h. University of Michigan
i. Oak Ridge Laboratory
j. Princeton Plasma Physics Laboratory
k. State University of New York at Stony Brook

Abstract: Superconducting air core toroids show great promise for use in a muon spectrometer for the SSC [1,2]. Early studies by SUNY at Stony Brook funded by SSC Laboratory, have established the feasibility of building magnets of the required size[3]. The toroid spectrometer consists of a central toroid with two end cap toroids. The configuration under development provides for muon trajectory measurement outside the magnetic volume. System level studies on support structure, assembly, cryogenic material selection, and power are performed. Resulting selected optimal design and assembly is described.

INTRODUCTION

The air core toroid assembly shown in Figure 1 is the key element in this experiment in the detection and characterization of muons produced at interaction points of the SSC Scientific and engineering collaboration has proceeded to the point where the design shown has been selected for continued refinement[4]. This design places detectors external to the magnetic field. Access to muon measurements inside the magnetic volume is not required, this producing a less complex toroid and magnet system and a simplified precision muon measuring system.

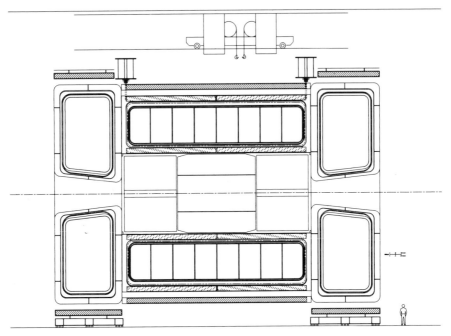

Figure 1.

Designs are based on the following set of initial parameters. All other physical parameters will be chosen to optimize design, i.e. maximize field integral, minimize radiation length, minimize cost and schedule for fabrication and assembly.

SPECIFICATIONS

Central Torus, External Detector:

Basis:

Inside Radius	4 m
Outside Radius	8 m
Half-length	8 m
Bmax	1.5 Tesla

Coil Characteristics:

Material	Alum Stabilized NbTi
Total Ampere Turns (MAT)	32.446
Field Radius Product (T-m)	6.489
Bending Lg at 90o (T-m)	3.561
Conductor Current (kA)	32.189
No of Turns	1008
Cavity Current Density (kA/sq. cm.)	4.000

END TORUS, EXTERNAL DETECTOR

Basis:

Inside Radius, Dewar Corner	1 m 6.3 Degrees
Outside Radius	9 m
Z Start	9 m
Z Length	5 m
Bmax	4.5 Tesla

Coil Characteristics:

Material	Alum Stabilized NbTi
Total Ampere Turns (MAT)	40.803
Field Radius Product (T-m)	8.161
Minimum Detection Angle (Deg)	9.010
Corresponding pseudorapidity	2.541
Bending Lg at Min angle (T-m)	18.583
Conductor Current (kA)	33.666
No of Turns	1212
Cavity Current Density (kA/sq. cm.)	2.200

Important aspects of this design are the consideration of the size, weight, transportability of major assembly elements, and assembly and integration of muon detectors and calorimeter. Site support requirements both during and after final assembly also have a significant effect on subassembly size and support concepts.

AIR CORE TOROID ESTIMATED WEIGHTS

DESCRIPTION	Weight (tonnes)
Central Toroid	
Dewar	82.198
Insulation	117.982
Cond. Structure	156.576
Cond. Spool	45.454
Superconductor	155.058
Web	194.910
TOTAL	752.177
End Toroid	
Dewar	42.706
Insulation	64.146
Cond. Structure	96.418
Cond. Spool	22.949
Superconductor	193.235
Web	112.457
TOTAL	531.910

Weight/Segment	5 deg	60 deg	
Central Toroid	10.447	125	tonnes
End Toroid	7.388	89	tonnes

AIR CORE TOROID ASSEMBLY OPERATIONS

The assembly operations takes place in three phases. The initial phase requires the shop assembly of major elements into safety transportable units. This minimizes the field support services requirements for the superconductor winding, insulation, and testing. The second phase involves the above ground field assembly of these components into larger units which are coordinated with the site requirements and are capable of being lowered into the final assembly area through construction access shafts. The third and final phase takes place in the interaction region 80 meters underground. This portion of the assembly is coordinated with the calorimeter assembly sequence and the detector assembly and alignment requirements.

SHOP MANUFACTURE AND ASSEMBLY

This stage of manufacture is the most complex in that it requires the most specialized support equipment for coil winding, insulating, joining, and testing. Tools and structural details and subassemblies need not be manufactured in the same locality, but careful planning of manufacture of all elements is essential.

Subassembly at this stage, leading to the assembly of five degree elements of the central and end toroids consists of the superconductors, superconductor structure, superinsulation and liquid nitrogen walls, dewar subassemblies, and support structures. Details of a typical five degree assembly are shown in Figures 2 and 3.

ABOVE GROUND FIELD ASSEMBLY

Prior to the start of the above ground assembly operations, an assembly support structure is erected in the detector assembly building. This structure is designed to provide access to the inside and outside of the segments. Welding and various quality control and test facilities are required at this point. Twelve five degree segments are joined together at this stage to produce a sixty degree sector assembly. Weights of these assemblies are approximately 125 metric tons for central toroids and 89 metric tons for the end toroids. This includes dewar walls and insulation. Refer to Figure 4 for central toroid elevation. End toroids assembly is similar in constriction. The eighteen sector assemblies, together with their related support assemblies, are now ready for lowering into the interaction hall area for final assembly.

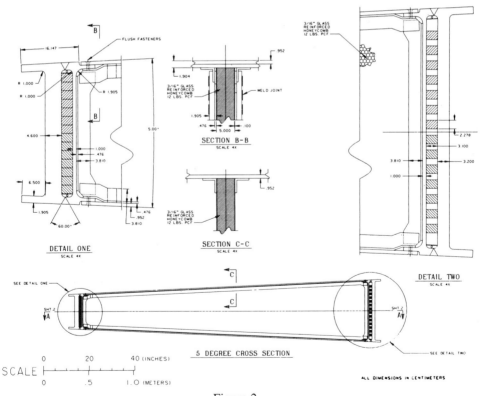

Figure 2.

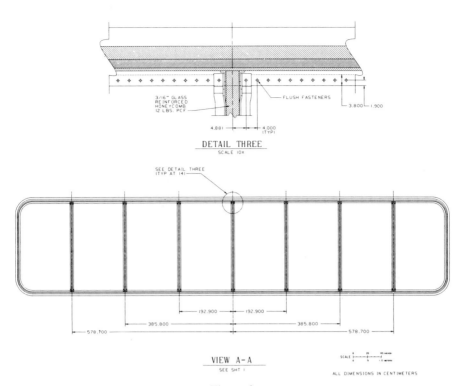

DETAIL THREE
SCALE 10x

3/16" GLASS
REINFORCED
HONEYCOMB
12 LBS. PCF

FLUSH FASTENERS

SEE DETAIL THREE
(TYP AT 14)

VIEW A-A
SEE SHT 1

SCALE

ALL DIMENSIONS IN CENTIMETERS

Figure 3.

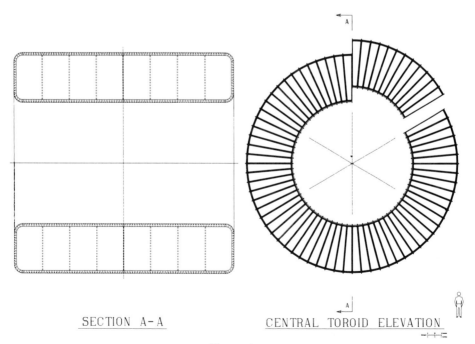

SECTION A-A

CENTRAL TOROID ELEVATION

Figure 4.

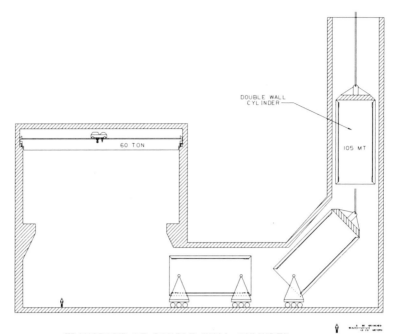

TRANSPORT OF DOUBLE WALL CYLINDER

Figure 5.

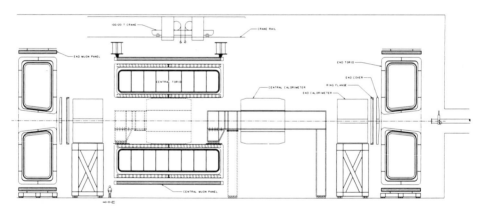

EMPACT ASSEMBLY SEQUENCE

Figure 6.

FINAL ASSEMBLY

Overhead crane and moving supports are required above ground at the construction access shaft to lower the segment assemblies into the interaction hall. A conceptual drawing showing the minimum requirements is seen in Figure 5. In addition to the support structure which remains in place, removable assembly support structure is needed for joining and supporting the 60 degrees sectors. It is anticipated that the central toroid is assembled in its

442

final position and end toroids assembled apart from it so that calorimeter and detectors may be placed into position in the central area. Completely assembled end toroids are capable of being moved into final position when central assembly of all elements are completed. Refer to Figure 6.

REFERENCES

1. M. Marx, SSC/CDG and Physics Department, SUNY, Stony Brook, SSC Report SSC-219 May 1989 "EMPACT - An Alternative Approach to a High Pt SSC Experiment".

2. T. Fields, el al. "Proposal For Development of Superconducting Air Core Toroids as a Precision Muon Spectrometer For SSC Experiments".

3. James N. Luton Jr., and Peter Bonanos, Princeton Plasma Physics Laboratory, Oak Ridge National Laboratory, SSC Report SSCL-N-697, March 1990, "Toroidal Detector Scoping Study".

4. Peter Bonanos, Princeton Plasma Physics Laboratory, February 1990, "External Toroidal Detector Study".

ZEUS MAGNET TESTS

A. Bonito Oliva, O. Dormicchi, G. Gaggero, M. Losasso,
G. Masullo, S. Parodi, R. Penco, P. Valente (1)
Q. Lin, and R. Timellini (2)

(1) Ansaldo ABB Componenti, Genova, Italy
(2) World-Lab. Geneva, Switzerland

ABSTRACT

A thin superconducting solenoid was designed (on the ground of a first conceptual study carried out by INFN [1]) and built [2] at ANSALDO ABB COMPONENTI. This detector was installed in the HERA ring accelerator (DESY, Hamburg) with its relative vacuum plant, 5.5 meters transfer line, current leads cryostat, current leads, power supply, dump resistor, breakers and computer control system. The magnet was cooled down and energized to the nominal field of 1.8 T. In spite of the double layer winding configuration, and the indirect cooling system, no quench occurred. The results of the tests (finished on Febr. '90) and preliminary analysis are reported.

INTRODUCTION

The thin solenoid has been designed to operate in a detector for the ZEUS experiment, in order to study e- p+ collisions.
For this reason a high uniformity magnetic field of 1.8 T is required in the Central Tracking Detector (CTD) volume and also a radiation thickness of 0.9 is necessary. A schematic view of the plant is shown in fig. 1.

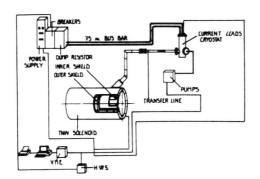

Fig. 1 – Schematic view of Zeus solenoid plant

In Table 1 are reported the main parameters of coil, cryostat and cable. Before starting the construction of this magnet a model was built and tested [3] .

Table 1. Main parameters of ZEUS thin solenoid

Type: double layer solenoid indirectly cooled by two-phase helium

Coil inside diameter:	1849 mm
Coil outside diameter:	1914.22 mm
Coil length:	2487 mm
Number of turns:	907
Number of joints on the internal layer:	2
Number of joints on the external layer:	2
Inductance:	0.844 H
Central field:	1.8 T
Rated current:	4987 A
He pipe length: on coil two parallel paths 40 m each	
Hydraulic diameter of He pipe:	$1.2 * 10^{-2}$ m

Conductor type: Rutherford (10 strands) with 99.996 % Al stabilizer
Al:Cu:NbTi ratio: 14:1.1:1   (4.3 x 15 mm)
                           18:1.1:1   (5.56 x 15 mm)

Short sample critical current (2.3 T, 4.5 K) :	15000 Amps
Stored magnetic energy:	10.5 MJ
Dump resistor resistance:	0.063 Ohm

Dimension of vacuum vessel:

Inner diameter:	1720 mm
Outer diameter:	2220 mm
Length:	2850 mm

Radiation thickness:	0.9

COOL DOWN TESTS

    ZEUS solenoid was assembled in his cryostat and cooled down to liquid
nitrogen temperature in ANSALDO ABB Laboratories in June 1989. In order   to
limit temperature differences between cryogen and coil at less than 40 K   a
mixture of gas and liquid nitrogen was used. Due to the little mass   flow
allowable this cooling had a term of about 110 hours: then the magnet   was
warmed up and shipped to DESY, Hamburg. The installation of the coil inside
the iron yoke, and electrical and hydraulic connections were over on
August, 1989. In fig. 2 is shown a scheme of the connections.

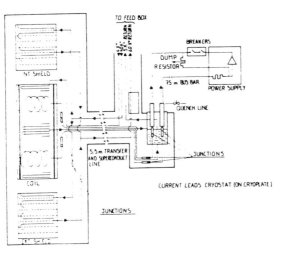

Fig. 2 - Electrical
and hydraulical
connections

On September 1989 the thin solenoid was cooled down to liquid helium temperature, with an average cooling rate of 3 K/h. Then the iron yoke was closed and the coil was gradually excited to the nominal field of 1.8 T. No spontaneous quenches occurred during the charging. This test could not be kept on for a long time due to a programmed shut down of the supply in the DESY line. A second cool-down started on January 1990, the curve with shields and coil temperatures is shown in fig. 3.

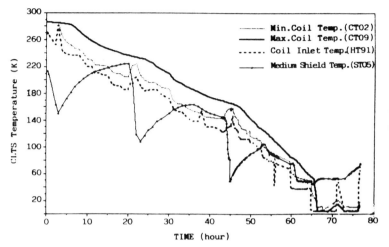

Fig. 3 - ZEUS cool-down curve

The residual pressure of insulation vacuum was less than $1*10^{-5}$ mbar at liquid helium temperatures, starting from $3*10^{-4}$ mbar at the beginning of cool-down. To avoid dangerous stresses on the coil the thermal gradients were kept low ($<$ 40 K) using the maximum mass flow obtainable by mixing 40 K and 300 K helium gas. From 70 K on the coil was cooled with LHe at 4.45 K. The shields were cooled down to 40 K with GHe at inlet pressure of about 15 bar. The measured pressure drop of the shields was about 160 mbar with a nominal flow of 1.4 gr/sec (Pinlet = 14 bar, Taverage = 45 K). The pressure drop of the coil was less than the sensitivity of the gauges ($\leqslant$ 20 mbar). In the flow range 2-10 g/sec no two-phase flow instabilities were detected. The coil thermal losses were about 6 W, not far from the calculated value of 5 W. In Table 2 are reported same measured cryogenic parameters. The stabilizer Al RRR was measured and the value was 1400. Temperatures on the coil were quite uniform, being the maximum difference about 0.01 K. The maximum temperature on the winding was 4.46 K with helium inlet temperature 4.45 K. A maximum temperature of 5.2 K was measured on the top flange of the restraining cylinder.

Table 2. Main cryogenic parameters

Coil pressure drop (Tinlet 4.45 K):	$\leqslant$20 mbar
Shield " " (Tinlet 40 K, Pinlet 14 bar):	$\leqslant$160 mbar
Steady state condition (0 Amps)	
. helium mass flow shields:	1.4 gr/sec
. coil:	$>$ 4 gr/sec
Total current leads mass flow (5000 Amps):	0.7 gr/sec
Minimum flow in coil (0 Amps) without temperature increasing:	2 gr/sec
Minimum flow in shields for average temperature 60 K:	1.3 gr/sec
Thermal loads on coil (4.45 K):	6 W
Thermal loads on shields (45 K):	60 W

ENERGIZATION TESTS

The electric circuit used for the energization includes, for safety reason, two quench detection systems, two breakers and several hard-wired securities. The latter, in case of emergency, gives rise to a fast or slow discharge of the coil, depending on the type of the alarm.
Owing to a room lack in the experimental hall, the power supply, dump resistor and breakers were installed 75 m far from the magnet and control system. This one was connected, through a bus extender, to a IEEE488 interface on the power supply, allowing remote control. ZEUS solenoid was ramped up to nominal current, using 3 Amps/sec up to 3000 Amps and 2 Amps/sec up to 5000 Amps (the maximum nominal one was 1.75 Amps/sec) without quenches. The axial magnetic force acting on the coil due to the detector iron yoke was 2.8 T. In fig. 4 are reported the strains measured on the coil axial tie rods versus current. The average magnetic pressure on Al cylinder was 3.5 Kg/mm2. The yoke magnetization coils had no significant influence on the reported strain values.

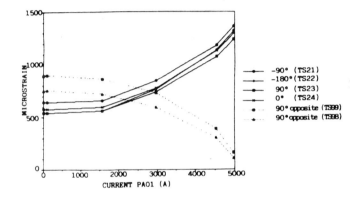

- -90° (TS21)
- -180° (TS22)
- 90° (TS23)
- 0° (TS24)
- 90° opposite (TS99)
- 90° opposite (TS98)

Fig. 4 - Axial tie-rods strains during energization

The resistance of junctions were measured.
The measured temporal current stability was less than $1.4 * 10^{-5} h^{-1}$.
In order to test the safety margin, ZEUS magnet was kept on operation for two hours at the nominal field with an inlet temperature of 4.65 K instead of the nominal value of 4.45 K.

FAST DISCHARGE TESTS

To check the electrical insulation, the maximum temperatures rise of the coil and cylinder, the correct behaviour of safety circuits and the occurring of quench-back effect, about 10 fast discharges were operated at different currents. In Table 3 are shown the main results concerning some of these fast discharges.

Table 3. Some fast discharges results

Fast discharge number	Current (Amps)	Maximum voltage (Volts)	Max. cylinder temperatures (K)	Max. coil temper.(K)
4	3000	168	18	19
7	4000	223	26	32
6	5000	280	32	43

The maximum pressure in CLC during a 5000 Amps fast discharge was about 1.75 bar, and the max temperature of dump resistor was less than 80° C. In

448

the figures No. 5 and 6 are reported the current and voltages decay during a discharge.
No appreciable temperature increasing was observed in correspondence of the junctions, due to Joule heating.

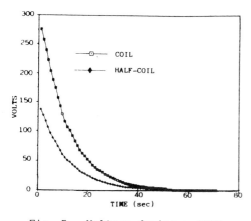

Fig. 5 - Voltage during a 4987
Amps discharge

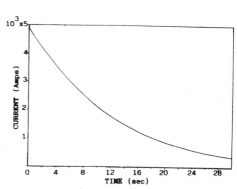

Fig. 6 - Current decay during 4987
Amps discharge

In order to estimate the induced current in the cylinder during the fast discharge, the magnetic field was measured by hall probe. The results for a 4987 Amps fast discharge are shown in fig. No. 7.
The temperature increasing of coil and cylinder is shown in fig. No. 8.

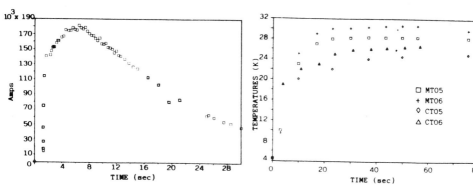

Fig. 7 - Cylinder current (breaker
opening at 4987 Amps)

Fig. 8 - Temperatures increases in
coil and cylinder during a
4000 Amps discharge

CONCLUSIONS

ZEUS thin superconducting solenoid has been installed and tested inside the detector iron yoke. The nominal field of 1.8 T was reached and no quenches occurred. The forces acting on the coil due to the return iron yoke were well below the safety margins.
The temperatures of the coil during induced fast discharges at full current indicate that the maximum value is less than 45 K.
We are actually proceeding to an elaboration of experimental data in order to obtain all possible information on quench-back effect.

Fig. 9 - LN2
cool-down of Zeus
solenoid

Fig. 10- Zeus arrival in DESY
experimental hall

Fig. 11- Zeus installation in the
iron yoke

ACKNOWLEDGEMENTS

We wish to thank for their cooperation DESY and LINDE personnel.
A special thank to Mr. Di Paolo, Mr. G. Bruzzone for their help in the
installation of the solenoid and to Ms. Secci and Mrs. Margiacchi for their
help in preparing this paper.

REFERENCES

1   The ZEUS thin superconducting solenoid
    E. Acerbi, F. Alessandria, C. Baccaglioni, L. Rossi, Paper presented
    at MT-9, 1985 Int. Conf. on Magnet Technology

2   ZEUS magnets construction status report
    A. Bonito Oliva, F. Bordin, O. Dormicchi, G. Gaggero, M. Losasso, R.
    Penco, N. Valle, R. Bruzzese, M. Spadoni, N. Sacchetti, Q. Lin
    Paper presented at MT-11, 1989 Int. Conf. on Magnet Technology

3   Quench behaviour of a thin solenoid model
    A. Bonito Oliva, G. Masullo, O. Dormicchi, G. Gaggero, R. Penco,
    Paper presented at MT-11, 1989 Int. Conf. on Magnet Technology

MANUFACTURE OF 10 T TWIN APERTURE SUPERCONDUCTING DIPOLE MODEL

FOR LHC PROJECT

L. Loche, P. Gagliardi, and A. Martini

ANSALDO ABB COMPONENTI, Genova, Italy

ABSTRACT

Ansaldo ABB Componenti is actually involved in the manufacturing of one model, 1 meter long, of a superconducting twin aperture dipole for LHC Project for CERN Laboratory.
Some constructive solutions are presented together with some mechanical calculations of the dipole ends. The distribution of elasticity modulus of the coil is reported up to 100 MPa of stress.

INTRODUCTION

In the frame of a study on the feasibility of a future hadron collider to be installed in the LEP tunnel, new superconducting magnets must be studied.
The first result of this study was a manufacture of two 8 T prototypes carried out by Ansaldo ABB Componenti in 1988 and 1989.
The first prototype was tested in the laboratories of the CERN. The magnetic field measured have resulted superior at the design maximum value.
A second prototype also was tested in the laboratories of the CERN. The results of the tests showed the good fabrication made by Ansaldo ABB Componenti and confirmed the high values of magnetic field obtained by the first prototype. The main design characteristics of 8 Tesla magnet are represented in Table 1.
In Table 2 are shown the results of the experiences carried out by CERN Laboratories in Geneva. These results are those obtained with the second 8 T model.
After these experiences CERN decided to proceed with conceptual design of a prototype of 10 T twin aperture superconducting dipole model.
The main parameters of the dipole are shown in Table 3.
The CERN charged four Firms in Europe for the construction, design and manufacturing of the model. One of these firms is Ansaldo ABB Componenti. The goal shows the feasibility of the model to obtain the design value of magnetic field and to get some information concerning the future 10 T magnet prototype, 10 m long. Concerning that, Ansaldo ABB Componenti will be charged of the construction of two prototypes of this magnet, 10 m long, commissioned by INFN, the Italian National Institute for Nuclear Physics.

CONSTRUCTION DESIGN

CERN have developed the conceptual design of 10 T superconducting twin aperture model for LHC. Ansaldo ABB Componenti received a technical specification and a set of conceptual design.
Since the design is not complete in all the details, a close collaboration with CERN is necessary to elaborate the best solutions for the fabrication of 10 T twin aperture dipoles.
The model manufactured by Ansaldo will have a stainless steel shrinking cylinder around the yoke, while the other three models have an aluminium alloy shrinking cylinder around the yoke.
Ansaldo ABB Componenti have developed, jointly with CERN, a new solution for the construction of the ends of magnet.
This new solution is supported by a finite elements calculation made with the ANSYS program (see paragraph of mechanical calculations).
Ansaldo have also developed, with CAD program, the design of the vetronite fillers with permits based on the magnetical calculations, to define the geometry of the cross section with the relevant longitudinal spacers.

MECHANICAL CALCULATIONS OF THE ENDS MAGNET

Ansaldo and CERN have decide to redesign the ends of the magnet in order to reduce the problems of assembly and simplify the operations for workmen. The model proposed for the analysis is shown in Fig. 1.

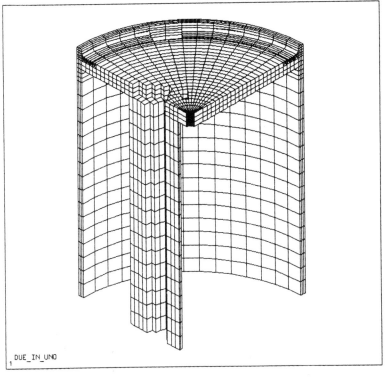

Fig. 1 Element finite model of end magnet

Thanks to geometrical and load symmetry only a eighth of the prototype was studied. The model is a three-dimensional model with a plane of symmetry that cut at half the structure (medium plane).
Two other planes of symmetry are done by two planes perpendicular to the medium plane and between them. The intersection of two planes is the axis of the structure.
In the model was shown the shrinking cylinder around the yoke (thickness 12 mm), the end flange (thickness 40 mm) and a dipole magnet.
The weld between the flange and the cylinder was dimensioned in function of forces act on the magnet.
We suppose that a total vertical force of 100 MPa acts over the ends in direction of the axis of structure.
This force is the result of a prestress, imposed at the dipole to recover the differential contraction between steel and magnet when the magnet is in function at cryogenic temperature, and of the magnetic stress.
In the model this force was distributed on the nodes of the magnet.
The results obtained show that this solution is able to withstand the stress imposed with acceptable deformation.

Table 1. Main characteristics of 8 Tesla model

Nominal field B (2 K) .... 7.5 T
Excitation per dipole (one aperture) .... 988.000 At
Operation current .... 9500 A
Stored energy for both channels .... 5 MJ
Resultant of magnetic forces in the first coil quadrant (see dwg. LHC MTAP 1002.2):
-   Fx' .... 1.8 MN/m
-   Fy' inner layer .... 0.5 MN/m
-   Fy' outer layer .... 0.5 MN/m
Resultant of axial coil forces per magnet end.. 0.55 MN
Coil inner diameter .... 75 mm
Coil outer diameter .... 119 mm
Distance between aperture axes .... 180 mm
Overall length .... 9.100 mm
Overall diameter .... 576 mm
Total weight .... 16 t

ACTUAL DEVELOPMENT

The technicians of Ansaldo ABB Componenti are now carrying out the manufacturing of the model 1 m long. The conductor is substituted by a copper cable with the same geometrical characteristics of superconductor cable. At present, one of the first and one of the second layers of the dipole have been manufactured. More problems arose during this manufacturing regarding the set-up of head fillers before the start-up of winding and the arrangement of these fillers.
After the baking of four first and four second layers, the magnet will be assembled and collared as showed in Fig. 2 .

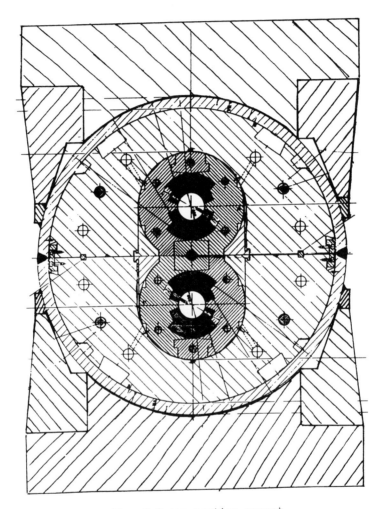

Fig. 2 Cross section magnet

Table 2. Analysis of quenches of 8 Tesla model

Quench No.	Current A	Field T	Pole Trans.	$\int I^2 dt$	Tmax K	Stored Energy	Temperature K
1	9782	8.68	1	–	–	–	1.86
2	10081	8.94	1	9.16 10	80.2	77 %	1.76
3	10039	8.91	2	9.61 10	83.3	80.2 %	1.88
4	10317	9.14	2	9.79 10	84.6	75 %	1.80
5	9689	8.66	2	9.01 10	79.3	79.2 %	2.05
6	9812	8.77	2	8.52 10	76.1	80 %	1.83
7	10119	8.99	2	8.85 10	78.2	78.4 %	1.80
8	10388	9.20	2	9.2 10	81.5	76.1 %	1.76
9	9988	8.88	1	9.03 10	79.4	80.2	1.76
10	10515	9.32	1	9.93 10	85.8	78 %	1.79
11	9733	8.66	1	–	–	–	1.85
12	10660	9.36	1	9.81 10	84.7	75.2 %	1.74
13	10303	9.16	1	9.57 10	83	77 %	1.68
14	10662	9.44	1	–	–	–	1.73
15	9730	8.66	1	–	–	–	1.80
16	10532	9.32	1	–	–	–	1.78
17	10652	9.44	1	–	–	–	1.76

Table 3. Main characteristics of 10 Tesla model

Nominal field Bo (2 K)	10	T
Operation current	15000	A
Turns per beam channel 1st layer	2x13	
2nd layer	2x24	
Peak field in winding 1st layer	10.2	T
2nd layer	8.4	T
Overall current density in compressed and insulated cable 1st layer	357.5	A/mm2
2nd layer	532.9	A/mm2
Operational current density (Jo) in NbTi at 2 K 1st layer	1150	A/mm2
2nd layer	1900	A/mm2
Jo/Jc at 2 K 1st layer	0.77	
2nd layer	0.87	
Coil inner diameter	50	mm
Coil outer diameter	120.2	mm
Length of the straight part	648	mm
Overall length of the coils	1080	mm
Distance between aperture axes	180	mm
Collars outer dimension (horizontal)	380	mm
Iron outer diameter	540	mm
Overall mass	2000	Kg
Stor.energy for both channels combined	684	KJ/m
Self-inductance per single dipole	3.134	mH/m
Mutual inductance between dipoles	7.08	H/m
Longitudinal magnetic forces for both channels combined	700	kN
Resultant of magnetic forces per meter length in the first coil quadrant		
Fx (horizontal)	2276	kN/m
Fy, 1st layer (vertical)	-234	kN/m
Fy, 2nd layer (vertical)	-98	kN/m

ANALYSIS OF COLLAR, YOKE AND SKIN INTERACTION FOR THE

MECHANICAL SUPPORT OF SUPERCONDUCTING COILS

T.K. Heger and J.S. Kerby

Fermilab National Accelerator Laboratory
Box 500
Batavia, IL 60510

ABSTRACT

The transition from warm iron to cold iron superconducting magnets has made the iron and skin assembly an important structural component of the coil support system. The effect of the yoke configuration on the apparent stiffness for the 40mm SSC collider dipole has been reported extensively elsewhere[1]. In this analysis a finite element model has been constructed to investigate the effect of the collar, yoke, and skin interactions for a geometry similar to that proposed for the High Energy Booster dipole. The model allows the evaluation of changes in materials and geometry on the stiffness of the coil support structure. Preliminary results, showing the effects of horizontally and vertically split yokes, skin thickness and prestress, and stainless steel and aluminum skins are reported.

INTRODUCTION

Finite element models have been used extensively in investigations of the cold mass mechanics of the 40mm SSC collider dipole design. Due to differing operating requirements, the High Energy Booster dipole proposed in the Site Specific Conceptual Design Report[2] is a substantially different magnet. The changes center on the larger bore aperture, necessary for injection, while maintaining the same central field as the main collider dipoles. The difference in geometry between models of the two dipoles may also allow for generalization of the effectiveness of coil support systems.

The geometry chosen for initial investigation is based on the HEB dipole report, except that the flattened yoke has been made circular, the collar skin has been removed, and the cooling channels have not been accounted for (Figure 1). The resulting design has a 70mm bore, with an aluminum collar, and 420mm yoke diameter. The analysis consists of two parts: 1) the generation of the coil loads and 2) the support structure model consisting of the collar, yoke, and skin with the coil loads applied. Although the analysis is not expected to give exact predictions of the system, it provides a means to analyze different support schemes quickly for comparison, before proceeding to a complete analysis.

Coil Model

The coil model was used for the calculation of the preload, cooldown, and Lorentz forces applied by the coil to a surrounding rigid cavity. The coil is an offset geometry,

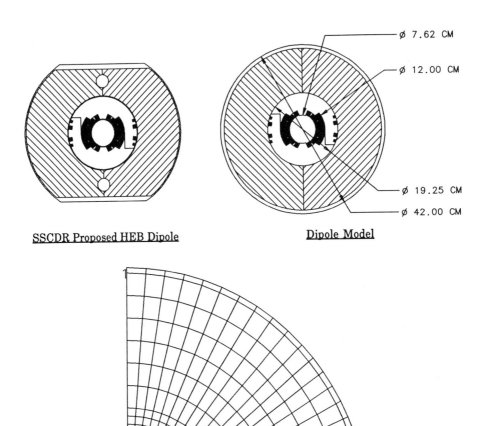

SSCDR Proposed HEB Dipole

Dipole Model

Finite Element 1/4 Dipole Model

Figure 1. HEB Dipole Models

consisting of a 36 turn inner and 30 turn outer winding, with centers offset 4.144 mm, resulting in a 70mm horizontal bore diameter (Fig. 2a). The conductors are assumed to have a linear Young's modulus of 1.25×10^6 psi, and the inner and outer layers are surrounded by a filler material of the same modulus. Frictionless gaps are used between each turn, the turns and the filler material, and to the ground nodes which represent the rigid cavity. The elements surrounding the coil are for the magnetic analysis, and represent the collar, iron and bore vacuum.

The reaction forces at the cavity nodes have been calculated for six cases: preload, cooldown, and central fields of 1.6 T, 3.2 T, 4.8 T, and 6.4 T (Fig. 2b). Preload is applied by the vertical displacement of the coil horizontal plane nodes, until the inner and outer coil are loaded to 9700 psi. Cooldown is simulated by modifying the temperature of the coil, and displacing the cavity nodes radially inward the same amount as an unstressed aluminum collar. Lorentz loads are calculated using the magnetic capabilities of ANSYS, by applying current source terms to the conductor elements for the four central field values above.

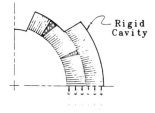

Figure 2a. Preload Displacement

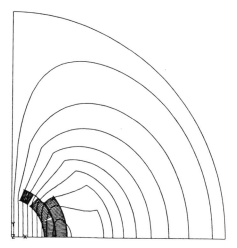

Figure 2b. HEB Magnetic Field Calculation

The resultant forces of the conductors on the cavity nodes are then stored for application to the support system model. Figure 3 shows the inner and outer pole azimuthal loads generated for each of the load steps. The decrease in pole stress due to the Lorentz loads scales to within two percent of the square of the central field.

<u>Support Structure Model</u>

The support system finite element model consists of a complete collar, including front collar, back collar, tab, and key elements, a yoke and a skin. The collar models are similar to that in the 40mm analyses, except that two keys are used per quadrant instead of

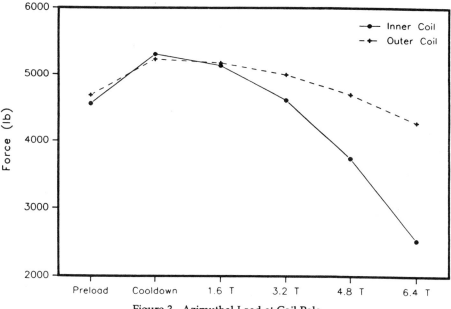

Figure 3. Azimuthal Load at Coil Pole

one. Collar and yoke laminations are assumed to have plane stress. Constraint equations define the azimuthally equal and opposite / radially equal motion of the front and back collar sections at the horizontal and vertical planes. Frictionless gaps are used between the collar and tab, collars and yoke, and collars and keys. A couple defines the spot-weld. The collars are aluminum, while the keys are stainless throughout the analyses.

The yoke can be modeled as either vertically or horizontally split by altering the placement of gap elements and symmetry boundary conditions. Gaps are also used between the yoke and skin. Prestress, when used, is applied to the skin by displacement of skin nodes (nearest yoke split) until the desired stress is achieved.

The coil loads are applied to front and back collars through a set of parallel springs to each pair of coincident nodes on the inner edge of the collars (Fig. 4). The springs are connected to a common 'load node', where the rigid cavity coil load is applied. The spring constant used is calculated from the modulus, azimuthal area and length of coil section adjacent to the force location. Since there is not a one-to-one correspondence between the coil cavity nodes and the collar nodes, area weighted average loads were calculated for application in the support system model. As an example, preload forces are shown in Figure 5.

It is important that the spring rate calculated here is correct since this will determine the proportions of load distributed to each collar. As a check, the spring rate was doubled and halved. The horizontal plane collar displacement results for both cases were within 1.8 percent of the original (spring rate) results.

RESULTS

The cases which were run with this model are given in Table 1. Material properties are listed in Table 2. The results are given in Figs. 6 through 11. The criteria used for the assessment of the effectiveness of the support system is the radial displacement of the front collar at the horizontal and vertical planes. There is some relative motion between front and back collars, however, the front collar is as close or closer to the coil than the back coil. The horizontal collar displacement results are grouped in four graphs: vertically split yoke, horizontally split yoke, stainless steel skin, and yokes with gaps when warm, while the vertical collar displacements are grouped by either vertical or horizontal yoke split.

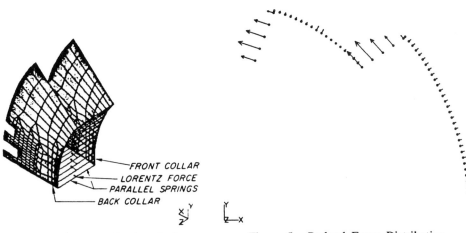

Figure 4. Load Application Springs   Figure 5.   Preload Force Distribution

Table 1. Load Cases

<table>
<tr><td colspan="5">I.   <u>STAINLESS STEEL SKIN</u></td></tr>
<tr><td>Skin Thickness / Prestress</td><td>.1875" / 0 ksi</td><td>.1875" / 30 ksi</td><td>.375" / 0 ksi</td><td>.375" / 30 ksi</td></tr>
<tr><td>Horizontally Split Yoke</td><td>HSH</td><td>HSHP</td><td>HSHT</td><td>-</td></tr>
<tr><td>Vertically Split Yoke</td><td>HSV</td><td>HSVP</td><td>HSVT</td><td>-</td></tr>
</table>

<table>
<tr><td colspan="5">II.   <u>ALUMINUM SKIN</u></td></tr>
<tr><td>Skin Thickness / Prestress</td><td>.1875" / 0 ksi</td><td>.1875" / 30 ksi</td><td>.375" / 0 ksi</td><td>.375" / 30 ksi</td></tr>
<tr><td>Horizontally Split Yoke
Gap = .005"
Gap = .007"</td><td>
HAHG5
HAHG7</td><td>
-
-</td><td>
-
HAHG7T</td><td>
HAHG5TP
HAHG7TP</td></tr>
<tr><td>Vertically Split Yoke
Gap = .005"
Gap = .007"</td><td>
HAVG5
HAVG7</td><td>
-
-</td><td>
-
HAVG7T</td><td>
HAVG5TP
HAVG7TP</td></tr>
</table>

Table 2. Material Properties

	E (psi)	υ	α (for 300 →4°K)
Aluminum	10×10^6	.33	.0040
Iron	30×10^6	.3	.0023
Stainless Steel	28×10^6	.3	.0030

461

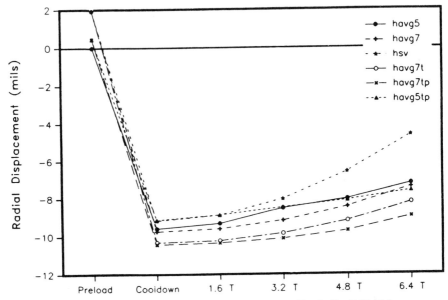

Figure 6. Horizontal Plane Collar Displacement - Vertically Split Yokes

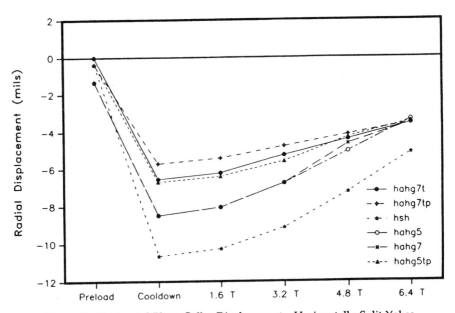

Figure 7. Horizontal Plane Collar Displacement - Horizontally Split Yokes

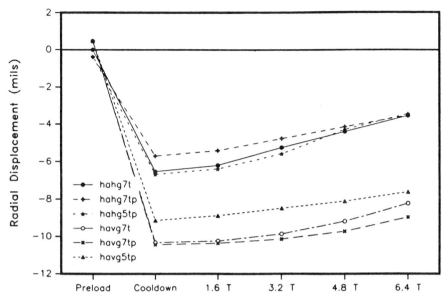

Figure 8. Horizontal Plane Collar Displacement - Horizontally and Vertically Split Yokes with Gaps

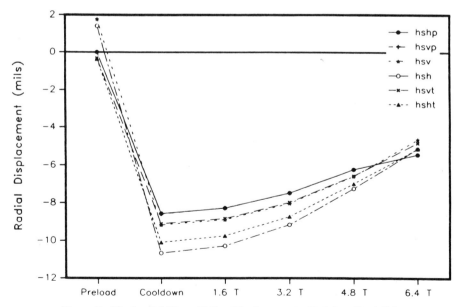

Figure 9. Horizontal Plane Collar Displacement - Stainless Steel Skins

Three mechanisms affect the stiffness of the support structure. The first is contact between the collars and yoke, i.e. are the coil loads restrained by the collars alone or is a portion of the load transferred to the yoke as well. The second is contact between the yoke halves, i.e. is the yoke functioning as two split rings or can loads be transmitted from one half to the other. The third is the direction of the load in comparison to the plane in which the yokes are split , i.e. what portion of the load is resisted by the bending moment of the yoke versus the portion being transferred to the skin as the yokes translate.

Preload

Under preload the collars deflect forming an ellipse with a vertical major axis. The now oval shaped collars are restrained well by the bending moment of the vertically split yoke. The horizontally split yoke, however, has no stiffness in the direction of preload forces; a midplane gap opens up and a portion of the preload is transferred to the skin. In general, for a skin with no prestress, the vertical plane collar displacement for a horizontally split yoke is between .0046" and .0072" radially outward (Figure 10). The results for a vertically split yoke range between .0024" and .0029" radially outward (Figure 11).

Prestressing the skin increases the stiffness of the support structure so that, compared to the unprestressed cases, the vertical plane collar displacement decreases on the order of .001" for a vertically split yoke and .003" for a horizontally split yoke. Increasing the skin thickness decreases vertical plane collar deflection .001" for a horizontally split yoke, but has almost no effect for the vertically split case. During preload the importance of skin variations are most pronounced for the horizontally split yoke because a greater portion of the load is transferred to the skin. In contrast, for the vertically split case most of the load is resisted by the bending moment of the yoke so skin variations have less effect.

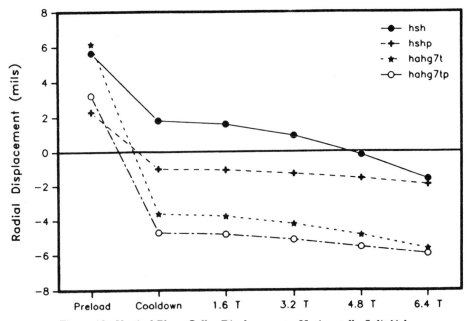

Figure 10. Vertical Plane Collar Displacement - Horizontally Split Yoke

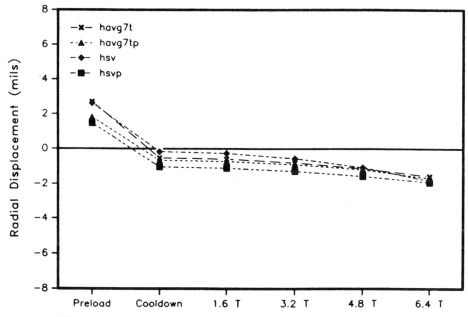

Figure 11. Vertical Plane Collar Displacement - Vertically Split Yoke

Cooldown

During cooldown the difference in thermal contraction of the materials causes the aluminum collars to shrink away from the iron yokes. Therefore contact at the yoke/collar interface decreases and the resulting contribution of the yoke and skin to the structure stiffness is reduced.

For the cases studied here, the smallest horizontal plane gap at cooldown was that of vertically split yokes with gaps between yoke halves when warm. At cooldown the skin forces the yoke halves to follow the contracting collar inward. The initial gap between yokes can be sized to control the amount of yoke/collar and yoke/yoke contact after cooldown. In contrast, a horizontally split yoke maintains contact with the collars at the vertical plane. Because of the location of collar/yoke contact, the vertically split yoke assembly results in collars which are more oval shaped than the horizontally split case.

Lorentz Loads

A preliminary case was conducted with only collars supporting the coil during preload, cooldown, and energization. The collar alone has a linear relationship between Lorentz forces and horizontal plane collar displacement throughout energization. This implies that the horizontal plane collar displacement is proportional to the magnetic field squared. The displacement from its original position at 0T is -.003" and at 6.4T is +.0045", for a total travel of .0075".

Horizontal plane collar displacement in a full model is also proportional to the square of the field until the collar contacts the yoke. The rate of this linear spring can be increased by striving for, optimally, horizontal plane yoke/collar contact after cooldown. In this study, the least total travel of .0015" occurred for vertically split yokes with a .007" gap warm and 0.375" prestressed aluminum skin (HAVG7TP). Horizontal plane yoke/collar contact occurred at magnetic fields between 0T and 1.6T. The largest total travel (.0055") occurred for a horizontally split yoke with no gap when warm and an unstressed 0.1875" steel skin (HSH). Nearly as much travel (.0045") is predicted for a similar case, where the yoke split is at the vertical plane.

The role of the yoke is either to transmit the Lorentz forces to the skin or to resist a portion of the load in bending. The first role is executed by a vertically split yoke; the second role is executed by a horizontally split yoke. Interference between yoke halves increases the effective stiffness of the assembly. For example, in the case of a horizontally split yoke with gaps and with prestressed skin (HAHG5TP), we see that after preload the yokes do not contact (Figure 12a). When the central field has reached the range of 4.8T to 6.4T, the yoke rotates so that there is contact along the entire surface of the split (Figure12b). (Arrows graphically symbolize reaction forces between the yokes at the midplane.) This is evidenced by a decrease in the slope of the horizontal plane collar displacement curve (Figure 7).

CONCLUSIONS

For the cases of this study, the most effective coil support is provided by a gapped vertically split yoke surrounded by an aluminum skin (case HAVG7TP). Comparison of the various cases has quantitively shown importance of the collar/yoke, interference, yoke/yoke interference, yoke orientation, and skin properties on the stiffness of the coil support structure. Specific conclusions from these analyses include:

1. The bending moment of a vertically split yoke results in a stiffer structure for preload.

2. During preload and cooldown the structure's stiffness is also varied by the properties of the skin. Prestressing has more effect than increasing skin thickness.

3. With a line to line fit between the yokes when warm, gaps appear at the horizontal plane between the aluminum collar and iron yoke after cooldown. The collar alone resists Lorentz loads for central fields less than 4.8T.

4. A vertically split yoke with gaps when warm reduces the horizontal plane gap between collar and yoke after cooldown by allowing the yoke to follow the collar contraction. This forms a collar which is more oval.

5. A horizontally split yoke with gaps when warm reduces the horizontal plane gap between collar and yoke after cooldown by compressing the collars at the vertical plane, thereby pushing the collars outward at the horizontal plane. This method does not oppose Lorentz forces directly (as the vertically split case does) and may induce high stress gradients at the coil poles[3].

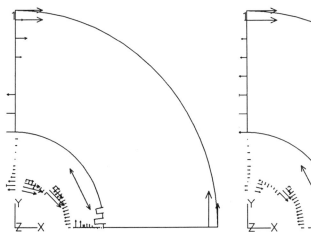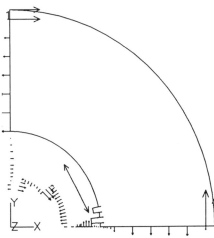

Figure 12a. Yoke/Yoke Contact - Preload   Figure 12b. Yoke/Yoke Contact - 6.4 T

466

6. A vertically split yoke with gaps has greater overall horizontal plane collar motion from preload to 6.4T than a horizontally split yoke. However, under Lorentz loads the collars deflect less with a vertically split yoke than a horizontally split yoke.

RECOMMENDATIONS

In this study, the role of the yoke/collar interference in the support structure was emphasized. A variation which may improve this contact is the use of a different (not line to line) initial fit between the collar and yoke. Further cases could include differing collar materials and differing component dimensions to investigate their effect on the individual and overall structure stiffness.

REFERENCES

1. R. Wands and M. Chapman, "Finite Element Analysis of NC-9 Dipole - Notes #1-#6," SSC-N-530, Fermilab, May 13, 1988.

2. SSC Laboratory, "Site-Specific Conceptual Design Report," November 22, 1989.

3. M.S. Chapman, J.M. Cortella, and R.I. Schermer, "Mechanical Analysis of Different Yoke Configurations for the SSC Dipole," SSC Central Design Group, Lawrence Berkeley Laboratory.

SSC DIPOLE MAGNET MEASUREMENT AND ALIGNMENT USING LASER TECHNOLOGY

A. Lipski, J.A. Carson and W.F. Robotham

Fermilab National Accelerator Laboratory
Box 500
Batavia, IL 60510

ABSTRACT

Advancing into the prototype production stage of the SSC dipole magnets has introduced the need for a reliable, readily available, accurate alignment measuring system which gives results in real time. Components and subassemblies such as the cold mass and vacuum vessel are being measured for various geometric conditions such as straightness and twist. Variations from nominal dimensions are also being recorded so they can be compensated for during the final assembly process. Precision laser alignment takes specific advantages of the properties of helium-neon laser. The laser beam forms a straight line of the greatest accuracy. When combined with an optically produced perpendicular plane, this results in a system of geometric references of unparalleled accuracy. This paper describes the geometric requirements for SSC dipole magnet components, sub and final assemblies as well as the use of laser technology for surveying as part of the assembly process.

INTRODUCTION

The laser alignment system will replace the present optical alignment process. The laser system offers the following advantages:

- Results are in real time

- Eliminates the human factor

- Reduces survey time

- Eliminates the need for a special survey crew

- Repeated measurements for time dependent variations can be obtained with minimal preparations

- Hard copy of results (numerically and graphically) can be obtained at any time

The laser system itself is commercially available from Hamar Laser which will be applied to the alignment procedure. The term "alignment procedure" is defined as the method to obtain a finished magnet which has the magnetic centerline and vertical plane aligned, within specified tolerances to some datum located on the outside of the vacuum vessel. (Most likely - bottom of vessel foot and a fiducial located on foot.) The alignment procedure excludes the alignment of the interconnect region at the magnet ends. In addition, this procedure does not attempt to compensate for inherent errors built into the cold mass assembly, namely the cold mass straightness between its supports is unreconcilable as are variations of the magnetic vertical plane. Only average values (x; y) of the beam tube centerline taken in specified locations along the length of the cold mass and those of the magnetic vertical plane are as reference used in the alignment procedure. Note that due to the crude tolerances of the drawn beam tube its reference center line will be a result of best fitted curve.

For budgetary considerations the adopted construction philosophy of the cryostat has been to minimize the need for tight tolerances in piece parts and assemblies. It is also realized that design of stacked up parts results in tolerance build up which can affect magnet alignment. Part of the alignment process is taking into account the dimensional variance of parts from their nominal dimensions and compensates for it during the assembly process (Figure 1). This part, however, is not going to be covered under the scope of this paper. This paper will only deal with the portion where laser technology is replacing commonly used optical survey instrumentation. The various steps which include laser alignment of cold mass vacuum vessel and the final assembly will be briefly discussed and illustrated (see Appendices).

WHAT IS LASER ALIGNMENT

Precision laser alignment takes specific advantages of the properties of a helium-neon laser. This type of laser produces an intense beam of red light which is a straight line of the greatest accuracy. When bent through a precise 90° angle by a special rotating penta prism a very flat plane is produced. The combination of the straight line and the flat plane results in a laser equivalent of a surface plate, straight edge, cylindrical square, etc. It is a system of geometric references of great accuracy. The laser can be precisely leveled to produce a unique and versatile alignment system.

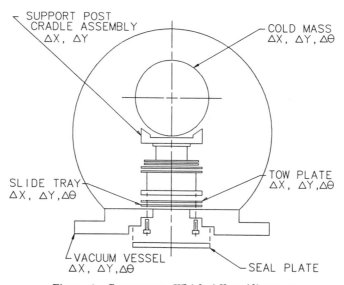

Figure 1. Components Which Affect Alignment

COLD MASS

Tooling Alignment

Prior to the placement of the cold mass on its support tooling it is essential to align all 5 centers of the support tooling with the survey line of sight. The concept used here is to establish the laser beam as the centerline between support tooling #1 and #5 and aligning the rest of the supports centers with the previously established centerline. The 4-axis bore target measuring both the two axis centering and squareness (pitch and yaw) is being held by an adjustable target holder which is supported by two micrometer stick legs and one spring loaded plunger leg (Appendix 1).

Straightness - Horizontal and Vertical

The incorporation of precision skin allows for straightness measurement at any increment along the length of the cold mass. As a first step a survey of the horizontal straightness is taken. The equipment used here will consist of the 4-axis target, the single axis sweep target and the remote scan optical square. The laser unit is mounted in a vertical position on a fixture. The beam which is emitted from the center top laser head is "bucked-in" with the theoretical beam tube centers in both ends of the cold mass using the 4-axis bore target. By turning the thumb lever on the laser head the beam is now emitted from the side of the laser head and parallel to the top of the alignment bar to the center of the remote scan optical square. The vertical sweep plane which is produced will be parallel to the nominal cold mass centerline and perpendicular to the horizontal beam aimed at the remote scan optical square. Moving a single axis sweep target along the cold mass length would indicate the variation of the skin from straightness with respect to the referenced centerline.

Similarly for vertical measurement of straightness the laser is left in its place while the remote scan optical square is moved to a horizontal position right above the laser unit. This will cause the remote scan optical square to sweep a plane which is parallel to the nominal centerline and perpendicular to the beam aimed toward the remote scan unit. Rotation of 180° of the cold mass and repetition of the survey sequence would indicate the effect of gravity on the results (Appendices 2 and 3).

Twist Measurement

As part of the cold mass survey it is important to study the level of twist that has been introduced to the cold mass during its assembly process. The amount of twist will effect the establishment of the vertical plane later on in the process (not within the scope of this paper). Cold mass twist together with deviations of vertical plane measurements coupled with angular deviations of vacuum vessel (Figure 2) and slide tray are being compensated for by rotating out the cold mass prior to being anchored to the center support post (not covered in this paper).

The twist measurement utilizes the laser's scanning capability, thus, forming a horizontal plane. Two single axis sweep targets are placed on both ends of the horizontal surface of a twist fixture. This fixture is locked into the press fit "V" keys which are accurately positioned on either side of the cold mass. Any twist in the cold mass will be translated into an angular displacement which will be read by the targets placed on the fixture. Twist measurements are taken in the five post center line locations (Appendix 4).

Define Geometric Center

A target is installed in the beam tube and moved longitudinally. The target center position is measured with respect to the established sight line and Δx, Δy values are recorded. This procedure is repeated at 24 longitudinal locations as follow:

a. At each support
b. Magnet ends
c. 3 equally spaced positions between supports or supports and ends .

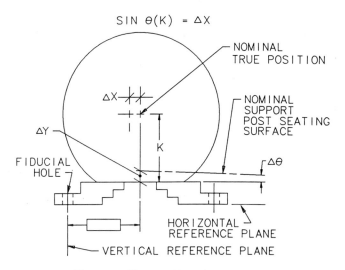

Figure 2. Vacuum Vessel Deviations

This data will be used to define the center of the magnet in the finished assembly as well as measure the straightness and concentricity of the beam tube with respect to the cold mass. The laser target is mounted on the centerline of an air bearing/target transport assembly which is connected to the end of a self propelled in-tube shuttle (inchworm) via double spherical air bearing. The air-bearing/target transport assembly will center itself in the beam tube and will maintain its angular positioning due to the counterweight located on its lower level while the double spherical air bearing will prevent any angular input from the inchworm to the air-bearing/target transport assembly (Appendix 5).

VACUUM VESSEL

Tooling Supports Alignment

Prior to the alignment of the vacuum vessel it is essential to align its tooling supports. In order to compensate for the possible effects of the combined "dead weight" of vacuum vessel and posts on the tooling supports, those are being simulated via an empty vacuum vessel and a suspended "dummy" cold mass which are loading the tooling supports during the alignment process.

The initial alignment will take place before the vacuum vessel is moved into place, yet the final alignment will be done with the vacuum vessel on top of the tooling supports. The laser unit will be in its scanning mode, thus, forming a horizontal plane. The two top support plates are being leveled and elevated to form a horizontal plane using three single axis sweep targets at which time they are locked in place (Appendix 6).

Level Survey of Support Surfaces

The vacuum vessel subassembly, although made to exacting tolerances derived from assembly fixturing, is expected to have geometric variations induced primarily by welding stresses. These variations from the nominal dimensions must be measured so that they could be compensated for during the final assembly of the magnet.

The critical geometric surfaces of the vacuum vessel are the surfaces to which the five support posts will be attached. These surfaces must be measured and deviations from the nominal condition determined. The bottom of the two vacuum vessel support feet serves as the horizontal reference plane while the center of the fiducial holes within the support feet forms the vertical reference plane. The vacuum vessel feet are bolted to a plane tooling surface and "dead weight" load applied to simulate cold mass and supports the weight. Deviation from the ideal geometry are measured at the five support surfaces, thus, establishing a Δy and $\Delta \theta$ for each support location (Figure 2).

The laser is at a scanning mode, thus, forming a leveled horizontal plane. The two single axis sweep targets placed on both ends of the support pad centerline, which is perpendicular to the vacuum vessel axis, will indicate the Δy deviation from nominal (Appendix 7).

FINAL ASSEMBLY

Define Nominal Location of Cold Mass in Vacuum Vessel

At this time it is necessary to establish the survey sight line which coincides with the nominal magnet center with respect to the magnet's horizontal and vertical reference planes (Figures 3). Since it will be very impractical to form a target which will be referenced from the external reference planes (namely the bottom of the vacuum vessel support feet and the fiducial hole within the support feet) it will be assumed that the hole in the support pad within the vacuum vessel was machined per specifications and thus it will serve as datum. A 4-axis target mounted in a fixture and located in the support pad hole over the support tooling will be used to "buck in" the laser beam to where it will form the survey sight line, at which time the laser location will be registered. This location will be used in a later stage to verify the magnet's final position and determine the assembly errors (Appendix 8).

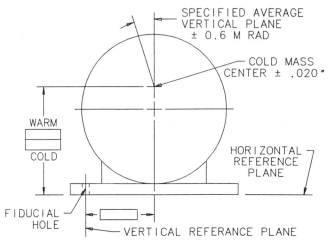

SPECIFIED AVERAGE
VERTICAL PLANE
± 0.6 M RAD

COLD MASS
CENTER ± .020"

WARM
COLD

HORIZONTAL
REFERENCE
PLANE

FIDUCIAL
HOLE

VERTICAL REFERANCE PLANE

Figure 3. SSC Dipole Magnet Geometric Requirements

Define Reference Plane for Support Posts Alignment

After the cold mass/cryostat assembly has been placed in the vacuum vessel, plugs are inserted through access holes in the vacuum vessel and into a hole located in the bottom of each support post. Using a fixture, the plugs are then centered in the support pad hole. The lower portion of each plug has a flat surface machined in it to accommodate the base of a single axis sweep target. The laser beam, via a remote scan optical square, forms a vertically sweeping plane. Most of the alignment will consist of merely rotating the plug and target to its extreme position and then aligning the target in place with the rest of the plugs, thus, forming a reference plane from which the adjustment of each plug in the "x" direction will be initiated.

The variations of all the components which effect the "x" axis position as well as the sagitta values added to or subtracted from the nominal position with respect to the magnet's vertical datum plane will determine the final location of the plugs in the "x" direction (Figures 4 and 5). The posts are then secured to the vacuum vessel (Appendix 9).

Confirmation

To confirm the geometric center a target is reinserted into the beam tube and readings taken at the five support locations. In order to do this, the laser unit is brought back to its registered location previously established, using the same floating 4-axis target (Appendix 8). Deviations from nominal are expected to be within the given tolerance range. Deviations which exceed the range are recognized as assembly errors. Some adjustment can be done to correct these errors. Corrections along the "x" axis can be easily accomplished by loosening the appropriate support post from the vacuum vessel and repositioning it. Errors in the "y" axis are more difficult to correct since these require jacking the post up and adding or eliminating shims through the access hole in the bottom of the vacuum vessel in addition to the loosening of the support post.   Vertical plane errors requires disassembly to correct (Appendix 10).

ALIGNMENT ERRORS

Possible alignment errors can result from two sources:

a. Measurement errors of component parts and assembly process

b. Plastic deformation of components occurring after the final assembly was
    completed .

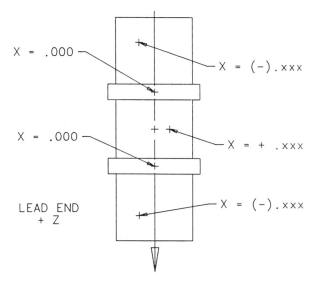

Figure 4. Plan View of

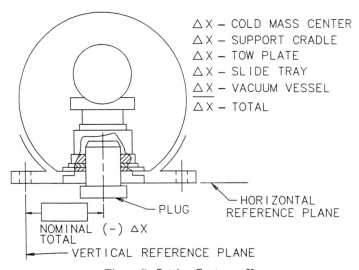

Figure 5. Setting Center on X

Measurement Errors

Applying laser technology to the alignment process has the potential to substantially reduce measurement errors. The red beam from the laser is always visible, consequently, alignment is simplified and the possibility of error minimized. The two-axis readout or the computer screen display geometric errors directly thus, eliminating "eyeball" mistakes. In a short time operators can be trained to use the system and perform tasks which would normally be considered for experts only. In addition the "dynamic testing" allows the operator to have the readout continually displaying the horizontal and vertical variations from the nominal line or plane and make corrections instantaneously. Table 1 denotes the contribution of the various elements to the final assembly total tolerances.

Table 1. Alignment Error Resulting from Components and Assemblies

Obtainable Measurement Accuracy

Component	X(in)	Y(in)	0(mrad)	Laser Alignment
Average cold mass	±.002	±.002	±0.1	Yes
Cradle/support assy	±.001	±.001	-	No
Tow Plate	±.001	±.001	±0.1	No
Slide tray	±.001	±.001	±0.1	No
Vacuum vessel	±.005	±.002	±0.1	Yes
Assembly	±.002	±.002	±0.05	Yes
Total (RMS)*	±.0024	±.0016	±092	

* Root mean square

Plastic Deformation Errors

Alignment errors resulting from plastic deformation involve creep of the cradle/support assembly and stress relaxation of the welded vacuum vessel assembly. These values need to be determined. In addition, handling, storage, and shipping of a magnet can possibly cause the alignment to be disturbed. The physical limits that can be imposed on a magnet (i.e. handling loads, storage temperatures, etc.) will have to be established.

SUMMARY

A considerable effort will be required to develop and demonstrate the alignment accuracy put forth in this paper. Formalization of procedures, gaining experience using the new laser technology, measurement fixtures and a determination of component stability must all be addressed. The goal of this paper was to offer a different approach to alignment via laser technology which promises to be faster, more accurate, reliable and simplified yet practical in a production environment.

ACKNOWLEDGEMENTS

The authors gratefully acknowledge the sincere contribution and professional performance of R. Dixon and his team as well as K. Swanson.

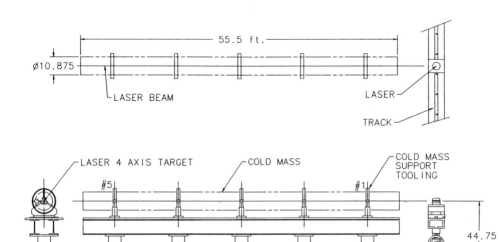

APPENDIX 1. ALIGN ALL 5 CENTERS OF SUPPORT TOOLING WITH SURVEY LINE OF SIGHT.

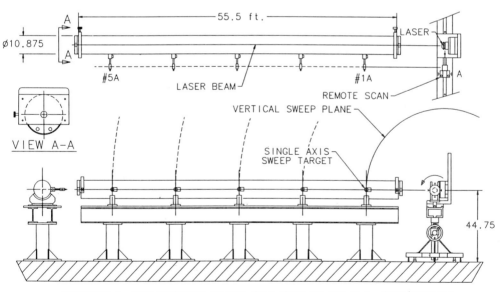

APPENDIX 2. STRAIGHTNESS OF THE COLD MASS ON THE O.D. (HORIZONTAL)

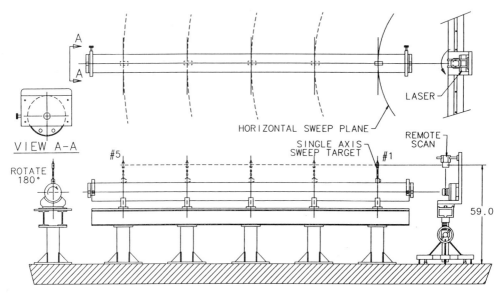

APPENDIX 3. STRAIGHTNESS OF THE COLD MASS ON THE O.D. (VERTICAL)

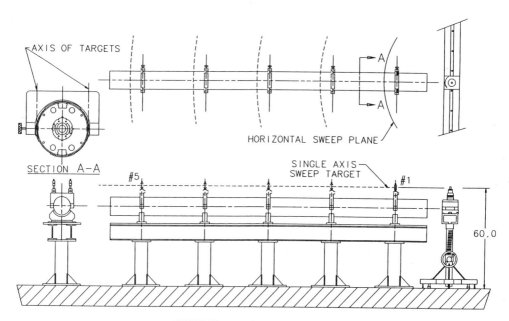

APPENDIX 4. TWIST OF COLD MASS.

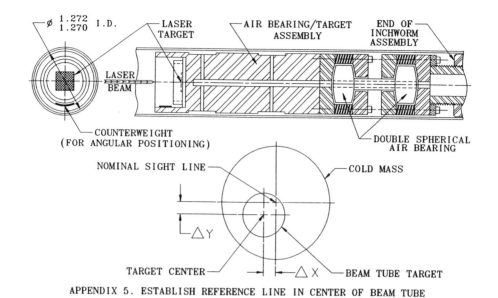

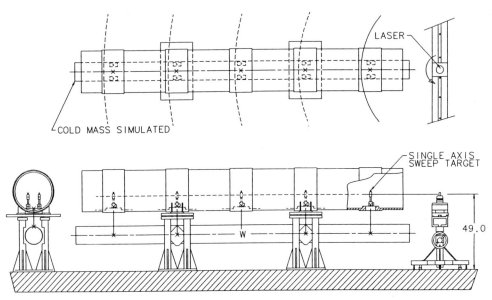

APPENDIX 7. LEVEL SURVEY OF SUPPORT SURFACES.

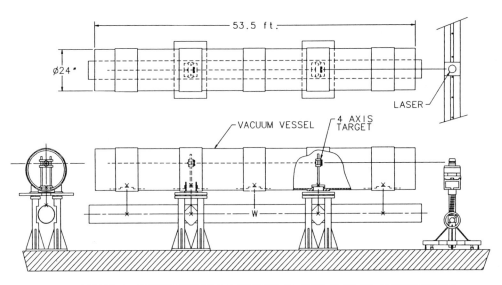

APPENDIX 8. DEFINE NOMINAL LOCATION OF COLD MASS IN VACUUM VESSEL.

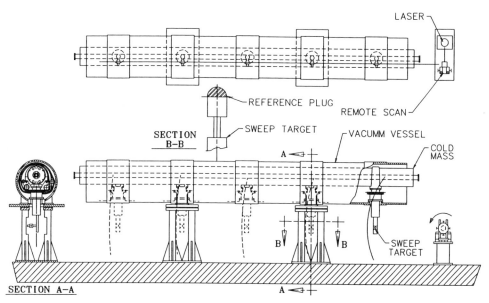

SECTION
B–B

SECTION A–A

APPENDIX 9. DEFINE REFERENCE PLANE FOR SUPPORT POSTS ALIGNMENT.

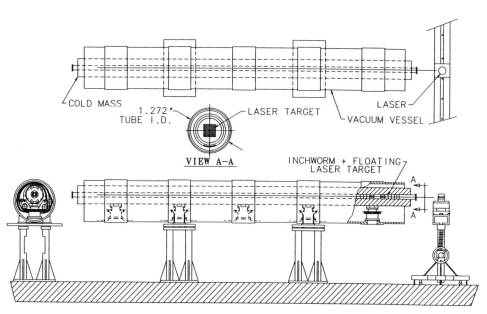

VIEW A–A

APPENDIX 10. ESTABLISH SAGITTA AND CONFIRM.

481

SSC 40mm SHORT MODEL CONSTRUCTION EXPERIENCE

R. C. Bossert, J. S. Brandt, J. A. Carson, C. E. Dickey, I. Gonczy, W. A. Koska, and J. B. Strait

Fermi National Accelerator Laboratory
MS 315, P. O. Box 500
Batavia, IL 60510

ABSTRACT

Several short model SSC magnets have been built and tested at Fermilab. They establish a preliminary step toward the construction of SSC long models. Many aspects of magnet design and construction are involved. Experience includes coil winding, curing and measuring, coil end part design and fabrication, ground insulation, instrumentation, collaring and yoke assembly. Fabrication techniques are explained. Design of tooling and magnet components not previously incorporated into SSC magnets are described.

INTRODUCTION

A series of short model dipoles are being built for the purpose of analyzing the Fermilab SSC magnet design. Changes in design as well as methods of construction can be implemented rapidly by testing them in short models before incorporating them into the more costly and time consuming long magnets. The Fermilab short magnet program attempts a number of design alternatives to the already established "baseline" 40 mm SSC dipole. Manufacturing methods are also incorporated which should facilitate the eventual mass production of SSC magnets. Table 1. lists the design features unique to the Fermilab short models. None of these are currently used in the baseline SSC dipoles. All are being analyzed during the construction of short models for their effect on manufacturability, magnet to magnet consistency and ultimate magnet performance.

EARLY CONSTRUCTION EXPERIENCE

Several SSC short models at Fermilab were built according to a different cross section, the NC9[11], than the baseline. This cross section had the same 40 mm bore diameter and cable sizes, but had slightly different conductor placement and wedge sizes. NC9 magnets had many other design differences, including aluminum collars. As a result many of the design variations cannot be meaningfully compared to the baseline. Nevertheless there are some areas in which the NC9 experience is relevant to the present magnets. Analytically designed coil ends were used in NC9 magnets. Winding and curing techniques are identical. Conductor placement has been observed and analyzed. Various ground wrap systems were attempted. NC9 magnets were built both with and without the addition of teflon as a slip plane for coils in the cross section. They were also built both with and without collaring shims and shoes.

COIL WINDING

All coils are wound on laminated mandrels (see Fig. 1). Steel pole keys made by an EDM process are bolted to the mandrel. Turns are wound around the pole keys and around steel winding keys. The steel winding keys are replaced by G-10 end keys after the coil is cured.

Winding tension is a major concern. Tension must be kept within certain limits. These limits vary depending on the type of coil being wound. Upper and lower winding tension boundaries are dependent on three physical restraints. First, the tension must be high enough that no more than a reasonable amount of effort is required to lay the turns into position as they are wound around the end parts. Second, the cable must be wound tightly enough that the uncured coil can be inserted into the curing mold without an undue amount of pressure. If winding tension is too low the coil will sag by an unacceptable amount making it more difficult to insert into the curing mold. On the 4 cm bore SSC short models it is this cable sag which defines the lower boundary for winding tension. The upper boundary is controlled by the mechanical stability of the cable. If tension is too high, the individual strands will try to take a less stressful shape, causing them to come out of lay. The exact values of these boundaries are usually found by trial and error during the initial coil winding. The winding tension for the SSC 4 cm short models is kept between 70 and 75 lbs. on the inner and 80 and 85 lbs. on the outer coils.

Table 1. Fermilab SSC Short Model Design Features

1. Coils cured using closed cavity mold with hydraulic pressure on ends.

2. Coil ends consisting of:

   a. All parts manufactured from solid G-10.
   b. G-10 keys, saddles and spacers made from an analytically derived geometry.[1] Surface shapes are produced by NC machining using output from computer programs developed at Fermilab.[2,3]
   c. Turns are separated only between current blocks.
   d. Wedges are terminated by solid G-10 spacers.
   e. Spacers incorporate shelves for internal support.
   f. Both "grouped" and "individually determined" options are being tried.[1]
   g. Several manufacturing options are being persued, including machining and molding of different materials.

3. External inner-to-outer coil splice with collet style end clamp system.

4. Kapton only coil insulation and ground wrap system.

5. Elimination of all collaring shims and shoes.

6. Pro-ovalized collar.

7. Square key insertion method.

8. Use of yoke and skinning press.

9. Full length fiducial on skin.

COIL END CONSTRUCTION

Fermilab coil ends are made from an analytically derived geometry.[1,5] The goal in creating these geometries is to cause the cable as it is wound around the end to be subject to as little stress as possible. The need for new end geometries developed because the past SSC ends have been difficult to wind and impossible to maintain conductor placement consistency from magnet to magnet. Some magnets have failed due to end problems.

Two different methods of creating cable paths have been developed at Fermilab. These are called the "ellipse on cylinder"[2] and the "developable surface"[3,4] methods. In either case the surface is created by a computer program which automatically generates a surface from the input parameters given by a designer.

Independent of the method of creating the cable paths is the method of stacking the cables within a current block. These are the "individually determined" and the "grouped" methods (see Fig. 2.).

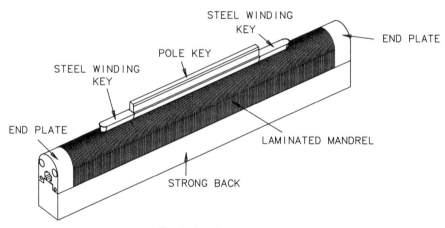

Fig. 1. Laminated Mandrel

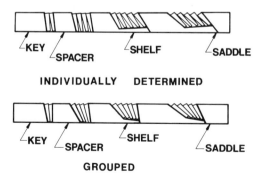

Fig. 2. Cable Stacking Methods

<u>Individually Determined</u>. The minimum stress surface for each turn is calculated individually. This results in the turns not laying directly upon each other. As turns get farther from the pole, the geometry requires that they be stacked at an increasingly larger angle (see Fig. 2.) Small spaces develop between each turn. These spaces typically become filled with epoxy.

<u>Grouped</u>. In the grouped method only the center of each current block is a calculated surface. The turns within each block are layed directly upon each other with no spaces between them. As the turns get farther away from the center or "guiding strip", their internal stresses become progressively higher.

The advantage to the individually determined method is that the stress in each turn is minimized. The advantage of the grouped method is that the spaces between turns are eliminated, further restricting cable movement. There are therefore four different types of ends which can be used on a magnet:

1. Ellipse on Cylinder/Individually Determined

2. Ellipse on Cylinder/Grouped

3. Developable Surface/Individually Determined

4. Developable Surface/Grouped

One objective of the Fermilab short model program is to determine what type of end design is most desirable for the SSC magnet.

Three NC9 magnets were built with Ellipse on Cylinder/Grouped ends. One was cold tested. Two were potted and sectioned. Fig. 3. shows an image of the sectioned end from one of these magnets (F5). These ends were very easy to wind. Conductor placement is consistent between magnets. Some problems still remain. The first turn in each current block has often pulled slightly away from the key around which it is wound, causing gaps as shown. This could be due to an inherent weakness in the grouped method in that it attempts to stack the first cable at a more extreme angle with respect to vertical than a minimum stress condition would dictate. It could also be due to the incompleteness of the model. The model assumes the cable is a homogeneous, infinitely thin strip and not the composite of shapes and materials that really make up a cable. It could also be due to the weakness of the ellipse on cylinder format.[1,5] The developable surface format attempts to correct these weaknesses.

Four magnets have been wound with Ellipse on Cylinder/Individually Determined ends. All four are C358 cross section. A total of five short magnets will be made with these ends of which four will be tested cold. Fig. 4. shows an image of the sectioned end of one of these magnets (DS0307). The end windings do not appear to conform to the individually determined format as readily as they do to the grouped. Conductor placement was poorer than in the grouped case. Winding was also more difficult.

The grouped method of stacking cables appears superior. More work still needs to be done to determine the viability of the grouped format. Developable Surface/grouped ends have been designed which will be incorporated into at least three 40 mm short magnets.

All ends incorporate shelves beneath keys and spacers to radially support the end windings. The shelves function very well as shown on Figs. 3 and 4. End turns no longer protrude into the bore as has been common in past SSC magnets.

When implementing any new end design it is necessary to demonstrate that the coil ends will not have turn-to-turn shorts in operation. Turn to turn hipots were performed on the ends from two magnets (one grouped and one individually determined). Each turn was hipotted successively

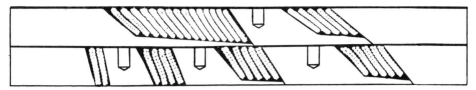

Fig. 3. Ellipse on Cylinder/Grouped End

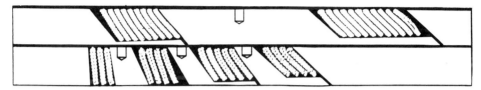

Fig. 4. Ellipse on Cylinder/Individually Determined End

to 100V, 300V, 500V, and then to breakdown. Approximately 100V is required in operation. All turns in F5 passed the 500V hipot and the lowest breakdown voltage was 1700V. On DS0307 all turns passed the 100V hipot. One turn failed at 300V but the next lowest breakdown voltage was over 1200V. The low breakdown voltage was on the first inner coil wound (#101). This coil had been used extensively over a period of several weeks to calibrate and cross check long coil size measurements. It was subject to many compressions in the size measurement fixture and was probably subjected to some mishandling by inexperienced technicians. Just before assembly of the magnet end a turn-to-turn short was found that had not been present immediately after curing. The short was not present, however, after the coil end was clamped and at this point passed the 100V hipot. Because of the unusual handling of this coil we do not believe that its performance is representative of coils manufactured under normal circumstances.[10]

Consequently it does not appear that the Fermilab ends are likely to have a turn-to-turn failure problem. This will be verified by cold testing many magnets.

CURING COILS

Fermilab short coils are cured in a closed cavity mold. Two separate sets of cylinders apply load to the coil (see Fig. 5). The mandrel cylinders apply a radial load to the coil through the mandrel. The platen cylinders apply the azimuthal load to the coil through the sizing bars.[6] The coil is pressed in a two stage process. First the load is applied to the mandrel and then to the sizing bars. The position of the sizing bars is set by stops on the curing mold.

The closed cavity mold is expected to create a uniform coil in both the radial and azimuthal directions. Many short SSC coils have been wound and cured with this process. Six NC9 inner coils and six NC9 outer coils have been produced. Eleven C358 inner coils and eight C358 outer coils have been produced to date.

A problem with the laminated mandrel became evident upon curing. B-stage epoxy from the glass tape extruded into the gaps between the mandrel laminations. This gave the inside surface of both the inner and outer coils a serrated effect. This is unacceptable on the outer coil since it is this surface which contacts the inter-coil ground insulation. A "sheath", or a .015 thick layer of steel was permanently attached to the surface of the outer coil mandrel, giving the cured outer coils a smooth inside surface. This could be done for the inner coil if necessary.

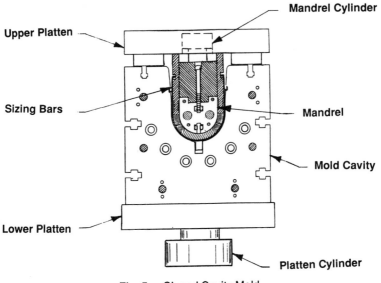

Fig. 5. Closed Cavity Mold

CURING ENDS

When a coil is cured, the curing tooling establishes the inside and outside coil diameters and applies azimuthal pressure to the straight section of the coil. The ends of the coil are also preloaded at this time. End preloading is intended to move the G-10 end parts to their proper positions, thereby compressing the end conductor groups and closing the gaps between the G-10 parts and wedges.

End preloading has been accomplished in two different ways. The long model tooling uses hydraulic force. Fermilab short model tooling will ultimately use hydraulic force as well. Currently three small screws are used to apply end pressure on short coils. Short model experience has shown it to be difficult to achieve proper G-10 end part positioning. The force generated by the small screws has not been sufficient to close the gaps between the G-10 end parts and wedges. Any remaining gap leaves an area in the coil that is not directly compressed by the curing tooling. This also creates end conductor groups that are not completely compressed. This can be seen in coil end sections and is shown in Figs. 3 and 4 where end parts are sometimes not completely pushed back to meet the shelves behind them. Fermilab long model tooling has produced more satisfactory small wedge gaps, although another problem has become evident. The end force is not equally distributed throughout the end parts and does more to compress the last wound end group than the first.

A new approach to these problems is now in the design stage. Each end part will be pushed to its proper position and pinned in place during the winding operation. This will close the wedge gaps and allow the next wound conductor group to be more correctly positioned.

The Fermilab magnet testing program has not shown the lack of end compression or presence of wedge gaps to be a problem. Nevertheless the potential for problems have caused this to be an area for design consideration.

COIL SIZE

All coils are measured before collaring. Coil measurements are used to determine what the expected preload will be. The measurements compare the azimuthal or "arc length" of a quadrant to a steel master of the correct size. Measurements of Fermilab short coils are made on a small portable fixture. The cross section and basic design features of this fixture are shown in Fig. 6. Both coil and master are individually placed in a steel cavity and a load is applied with a hollow bore hydraulic cylinder. The coil size is measured with an LVDT which contacts the back of the cylinder plunger. The master measurement is subtracted from the coil.

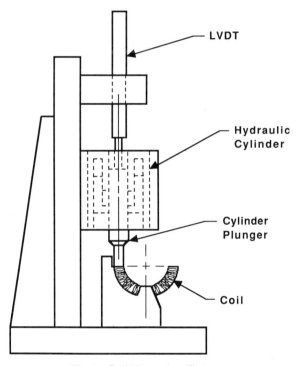

Fig. 6. Coil Measuring Fixture

The coil is divided for measuring purposes into 24 positions (see Fig. 7). Each of these positions is measured at 12000 coil psi. This measurement determines the average coil size as well as the consistency in size within the coil. Eight of these positions are measured at several different pressures (6000, 8000, 10000 and 12000 psi.) This measurement determines the modulus of elasticity of the coil.

Eleven inner and eight outer C358 coils have been cured and measured. Complete measurements of the most recently cured coil are shown in Figs. 8 and 9. Table 2. lists the relevant data for all C358 short coils currently produced.

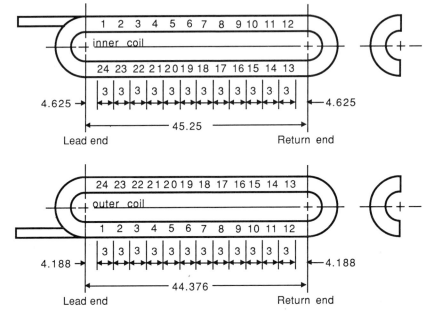

Fig. 7. Coil Measuring Positions (C358 coils)

Table 2. Short Coil Size Measurements

Measured at 12000 psi		Side A			Side B		
Coil No.	Type	Mean Coil size with respect to Master	Range	Std. Dev.	Mean Coil size with respect to Master	Range	Std. Dev.
101	Inner	.0018	+/-.0027	.0019	.0011	+/-.0016	.0013
102	Inner	.0027	+/-.0012	.0008	.0022	+/-.0016	.0001
103	Inner	.0023	+/-.0013	.0008	.0014	+/-.0027	.0016
104	Inner	.0044	+/-.0013	.0008	.0034	+/-.0028	.0015
105	Inner	.0044	+/-.0021	.0013	.0067	+/-.0027	.0019
106	Inner	.0018	+/-.0018	.0010	.0015	+/-.0019	.0012
107	Inner	.0004	+/-.0012	.0008	.0026	+/-.0023	.0012
108	Inner	.0038	+/-.0013	.0008	.0020	+/-.0012	.0007
109	Inner	.0031	+/-.0023	.0011	.0031	+/-.0020	.0013
110	Inner	.0021	+/-.0011	.0006	.0042	+/-.0018	.0010
111	Inner	.0043	+/-.0011	.0007	.0034	+/-.0008	.0005
301	Outer	-.0028	+/-.0013	.0009	-.0010	+/-.0020	.0010
302	Outer	-.0012	+/-.0039	.0020	-.0020	+/-.0021	.0012
303	Outer	-.0011	+/-.0020	.0011	.0012	+/-.0013	.0008
304	Outer	-.0022	+/-.0018	.0011	-.0008	+/-.0016	.0011
305	Outer	-.0016	+/-.0010	.0007	-.0030	+/-.0011	.0007
306	Outer	.0030	+/-.0016	.0008	.0021	+/-.0012	.0007
307	Outer	.0021	+/-.0010	.0006	.0030	+/-.0016	.0010

Short Inner Coil 111

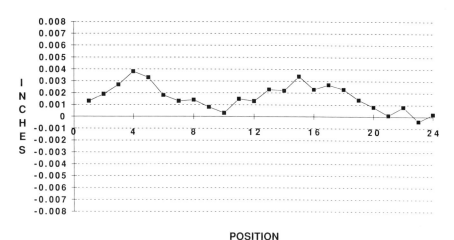

POSITION

Fig. 8. Inner coil Size Measurements (at 12000 psi.)

Short Inner Coil 111 Modulus

PSI x1000

Size with respect to Master, inches

Fig. 9. Inner coil Modulus Measurements

Variations in coil size do not appear to be due to any systematic effects. This could represent a limit on how consistent coils can be made given the component parts (cable, glass tape, wedges and kapton). If so, cable variations of this magnitude will have to be considered in the design of SSC dipoles. Modulus of elasticity for both inner and outer coils is approximately 2×10^6 psi.

It is necessary to know the modulus of elasticity as well as the size of coils when making production magnets. Cured coils in production need to be made in large numbers and matched int pairs according to their size and modulus. Coils with similar sizes are paired together to avoid having errors in the position of the parting plane. This is in contrast to prototype programs in which coils are wound and then immediately put into magnets in the order in which they are wound. For this reason it is necessary to have a production method of measuring coils. The fixture currently being used for short models operates well, but is very labor intensive and could not be used as a production device. Fermilab is therefore developing a measuring system in which load cells and LVDT's are mounted onto the curing press and are used to measure each coil immediately after curing.[13] The short SSC curing press has been instrumented with this equipment. Measurements have been taken on two outer coils and one inner coil.

SPLICES

Inner to outer coil splices in Fermilab short models are located exterior to the coil. The configuration is shown in Fig. 10. This has been done previously in SSC short models built at LBL and Low Beta Quadrupoles made at Fermilab. Locating the splice in this position eliminates the need to break the pole turn away from the rest of the coil when making the splice. Breaking the pole turn away after curing is undesirable because the pole turn is never recured to the rest of the coil. The possibility of damaging the rest of the coil is also decreased since the splice is made farther from the coil body. Quenches in the splice area are also made less likely because the splice is in a lower field area than it would be in the coil body.

The coil end and splice are enclosed in a collet style clamp assembly (see Fig. 11). The lead end configuration is shown. The return end clamp is identical except that there are no splices. The coil is surrounded by a four piece G-10 collet. The G-10 collet is closed by driving on a stainless steel tapered sleeve, thereby compressing the end sections of the coil.

One objective of the Fermilab short model program is to analyze the end clamp system to see if it provides enough preload to the coil. Measurements are taken of the outside diameter of the stainless steel ring before and after the collet is closed. The measured deflections are compared with a finite element analysis of the return end to determine whether the preload in the end is adequate. Fig. 12. shows these measurements for the return end clamp of magnet DS0308. The calculations indicate that these deflections are comparable to a preload of 4000 psi. It appears that the stainless steel shell needs to be made more rigid. Measurements of the deflection of the clamp are also taken during magnet cooldown and powering.

The outer shell, being made of stainless steel, has a relatively low coefficient of thermal expansion. It seems likely that the outer shell will therefore be changed to aluminum and made thicker to maintain higher preloads. A combination of finite element calculations and measurements of end clamps in assembled magnets will be used to determine the thickness used in the final design.

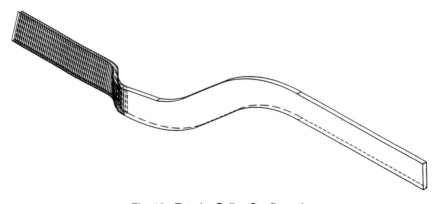

Fig. 10. Exterior Splice Configuration

S.S. TAPERED SLEEVE   G-10 COLLET

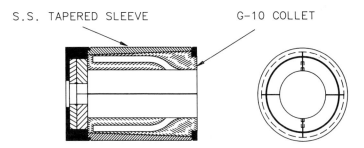

Fig. 11. Collet Style End Clamp

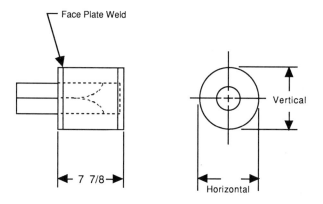

Face Plate Weld

7 7/8

Vertical

Horizontal

<u>DS0308 Return End Clamp Deflections with Magnet in Warm Free State</u>

Distance from face plate weld	Horizontal	Vertical
.5 inches	-.011	+.010
1.5 inches	-.010	+.009
2.5 inches	-.008	+.007
3.5 inches	-.006	+.006
4.5 inches	-.005	+.005
5.5 inches	-.005	+.006
6.5 inches	-.001	+.002

Nominal Diameter = 6.250 inches

Fig. 12. Return End Clamp Deflections

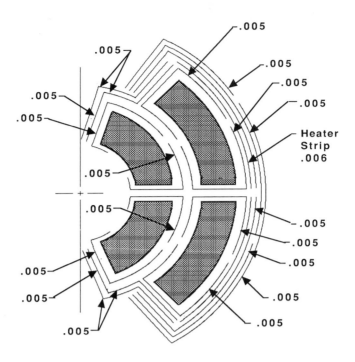

.005
.005
.005
.005
.005
.005
.005
.005
.005
.005
.005
.005
.005
.005
.005
.005
.005
Heater
Strip
.006

Fig. 13. Fermilab Coil Insulation System

Table 3. Coil Insulation Features

1. All insulating material is .005 kapton. This is the most simple configuration. All previous Fermilab magnets have been built this way.

2. No collaring shims are used. Collaring shims have created many problems in the construction of SSC magnets. They frequently fall out of place. Their primary function is to protect the kapton ground wrap at the poles from deterioration due to contact with the serrated edges of the collar. If this contact does not prove to be a problem, then they are an unnecessary extra part and can be eliminated.

3. No collaring shoes are used. The function of the collaring shoes is also to protect the kapton ground wrap from the serrated edges of the collar. If it is unnecessary, then it should also be eliminated.

4. There is no teflon or any other material except kapton used to provide a slip plane between the radial surfaces of the coils. Fermilab has built many magnets with kapton only insulation. No problems have been encountered in these magnets. It is desirable to eliminate as many low modulus materials inside the coil area as possible. There is also some uncertainty about the stability of teflon in a radiation environment.

COIL GROUND INSULATION

The Fermilab short model coil and ground insulation system is shown in Fig. 13. It has several features which are unique to SSC magnets. They are listed in Table 3.

The Fermilab coil insulation has been designed to simplify the configuration as much as possible. Tevatron has been used as an example, with consideration given to special SSC problems when necessary. Three C358 magnets (DS0307, DS0308 and DS0309) and one NC9 magnet

494

(F5) have been built with this insulation system. The NC9 magnet has been cold tested. Questions to be asked concerning the system include:

1. Will the absence of collaring shims allow excessive stress on the kapton insulation at the poles, causing ground shorts during either collaring or powering?

2. Will the lack of collaring shoes allow excessive stress on the kapton insulation at the radial surface of the outer coil, causing ground shorts during either collaring or powering?

3. Will the absence of a teflon slip plane cause any problems related to coil preload or training behavior?

Tests of insulation integrity at the poles and outer surface of the coils without collaring shims and shoes have been completed on one short model (DS0307). The magnet was placed in the press. Full press pressure was applied to the collared coil. At full press pressure the coil was .010 inches beyond closed and the azimuthal coil preload (calculated from pump psi) averaged in excess of 15000 psi in both inner and outer coils. Each coil was hipotted to ground at 5000V. Upper to lower coils were hipotted to each other at 3000V. Hipots were done at the maximum pressure and at several intermediate pressures while the load was being applied. No shorts were detected. This insulation test will be repeated on future magnets.

During cold testing the magnets will be cycled many times to determine if any insulation deterioration will occur during the life cycle of the magnet. One NC9 magnet (F5) has been current cycled 4500 times. No failures developed in the ground insulation.

Experiments are being done to determine whether the kapton only system can provide an adequate slip plane for the layers of ground insulation. One magnet (DS0307) is being assembled and collared three different ways: with kapton insulation only, with teflon added between kapton layers in the same manner as the baseline dipole and with teflon applied directly to the coils. Measurements are taken of two different values: coil psi at the poles and press load. Coil psi at the poles can be measured directly by the strain gages. The press load can be used to determine the approximate parting plane coil psi. Knowledge of these values then allows one to understand the relationship between preload at the parting plane and preload at the poles. This relationship will determine whether the coefficient of friction between the insulation layers is small enough to allow sufficient preload at the poles.

Two magnets (DS0308 and DS0309) have been assembled and will be cold tested with the kapton only insulation system. Another magnet (DS0310) will be tested with teflon slip planes added. Training behavior of the magnets will be compared. If they are significantly different, any of the magnets can be reassembled and retested with an alternate coil insulation system. Results should indicate whether teflon is a necessary part of the ground insulation system.

COLLARING

Fermilab short models are collared using laminated collaring tooling. The tooling consists of several components (see Fig. 14). They are: a laminated structure into which the collared coil is placed, a hydraulic system to drive in the tapered collaring keys, "key supporting bars" to support the tapered keys as they are being inserted into the collars and a transport mechanism to aid in rolling the tooling in and out of the press.

COLLARING METHODS

The tooling provides for two different methods of collaring a coil. Both have been used at various times in SSC magnet fabrication. They are called the "tapered key method" and the "square key method". In the tapered key method a vertical load is applied with the press until the collars are closed just enough to allow the tapered keys to engage. The final portion of closing is

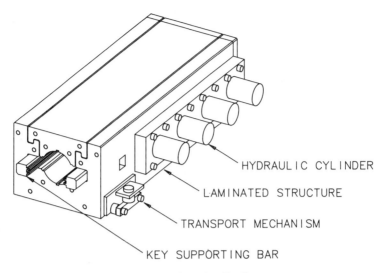

Fig. 14. Collaring Tooling

done by driving in the tapered keys. In the square key method the vertical load is applied with the press until the collars are completely closed. The keys are then pushed in from the side. Very little side force is needed to push in the keys. Either square or tapered keys can be used with this method. The advantage of the tapered key method is that a minimum amount of overcompression of the coils is necessary to collar the magnet. The square key method requires temporary higher preloads during collaring (see Fig. 15). The advantage of the square key method is that the keys are not damaged or "grooved" during collaring. A more consistent relationship between coil size and preload can therefore be achieved.[9] The Fermilab collaring tooling can be used to collar a magnet by either method.

Fermilab short models are collared using the square key insertion method. The collaring procedure is described in Table 4.

Three short models have currently been collared with the square key method. All were overcompressed by .005 before inserting the keys. No turn-to-turn shorts or ground insulation failures have resulted from stresses due to overcompression.

COLLAR/YOKE INTERFACE

All SSC magnets currently use the iron yoke as a mechanical support for the collared coil. Collars alone have proven to be mechanically insufficient to keep the coils adequately contained. Contact between the collar and yoke is necessary to maintain preload, decrease collar deflection under excitation and to transfer the axial component of the Lorentz force to the skin via coil-collar-yoke friction.[8] This can be difficult because the collar material has a higher coefficient of thermal expansion than the iron yoke.

Fermilab short models have a horizontally split yoke with a line-to-line fit between the collar and yoke when cold. This is accomplished by designing the collar configuration such that,

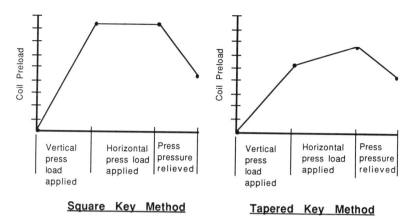

Square Key Method   Tapered Key Method

Fig. 15. Preloads from Square and Tapered Key Methods

Table 4. Collaring Procedure (square key method)

1. Place tapered keys on key support bar. They are held to the bar by a spring clip assembly as shown in Fig. 16.

2. Place the preassembled collared coil in the bottom half of the collaring tooling.

3. Roll the assembled tooling (with the collared coil inside it and the cylinders attached to it) into the press. It is supported vertically by belleville washers and guided horizontally by camrollers.

4. Activate the vertical cylinders. The press closes until it bottoms out against the lower laminations. The collars are now completely closed.

5. Activate the horizontal cylinders. The tapered keys are then pushed in until the horizontal cylinders reach their stops. Since the collars are already completely closed, it takes minimal force to place them in their slots. The collared coil is then complete.

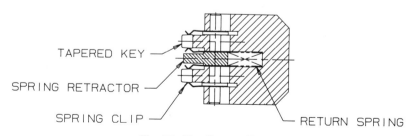

Fig. 16. Key Support Bar

when the collared coil is in the undeflected state it is vertically oval. When the collars are in an undeflected state the centerlines of the two collar halves are separated by .004 (see Fig. 17). This allows the collared coil, after deflections caused by coil preload and changes in size caused by thermal contraction of components, to still contact the yoke when cold.[9] Two magnets (DS0308 and DS0309) have been completed with this design. Several more will be built.

A vertically split yoke has also been designed. It will be used in some short models. The vertically split yoke has certain advantages relative to the horizontal.[7] Since the iron laminations are drawn onto the collars from the horizontal direction, the collared coil can be inserted easily into the yoke while still maintaining the appropriate horizontal interference necessary at room temperature. The vertically split yoke requires a different collar design to achieve the same line-to-line fit when cold.

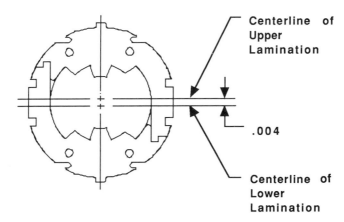

Fig. 17. Pro-ovalized Collar Configuration

YOKE AND SKIN DESIGN

The baseline SSC magnets are contained in a 3/16 inch thick 304 stainless steel skin. Alignment is achieved by use of intermittent fiducial balls.

The Fermilab yoke and skin system is designed to improve the straightness and angular alignment of the magnet as well as completely close the parting plane gap in the iron. To properly close the parting plane gap requires that the skin must also be longitudinally straight. If not, the skin will merely contact the laminations at the high points leaving the intermediate laminations unsupported.

Neither the collared coil nor yoke assembly offer either beam or torsional stiffness. It is therefore necessary to index each and every lamination during the assembly process. This is accomplished by using a full length alignment key. The key indexes to each yoke lamination. It also indexes to the assembly tooling to properly align the angularity of all laminations. The tooling plays an important roll in accomplishing the design goals. It must precisely define the geometry of the assembly prior to welding. Once welded, the shape is fixed.

One SSC short model (DS0308) has been completed with this yoke design. Welding of alignment keys was done by hand. Fermilab's internally developed requirement for twist over the iron length is a maximum of 1 milliradian. Measurements of the fiducial bar indicate that twist in DS0308 is approximately 5 milliradians. This was caused by an out-of tolerance condition in the alignment key. This condition will be corrected by the next short model. Similar magnets produced at Fermilab with alignment keys that are within the specified tolerance have no more twist than the design specifications.

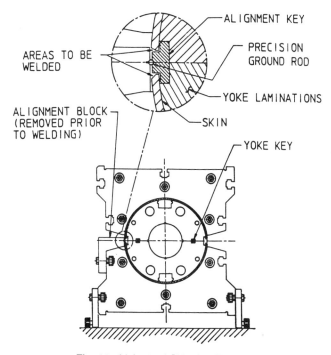

Fig. 18. Yoke and Skinning Tooling

INSTRUMENTATION

Fermilab short models are instrumented with resistive strain gage load cells to measure the coil preload at the poles both during collaring and cold testing. The strain gage system is identical to that used in the baseline design at BNL.[12]

Capacitance load cells are also being mounted in the collars.[14] The capacitance of these cells changes by a measurable amount when they are subjected to an external load, hence a correspondence between pressure and capacitance can be established and used to measure the load applied to these devices. These gages are still in the developmental stage and their inclusion in the short magnets has been primarily to learn what difficulties their use presents. If a reliable capacitance load cell can be produced, they will have the advantage of being very thin (less than 0.030 inches with insulation), and therefore can be used to measure coil stress in places which are currently unreachable, such as inside wedges or at the parting plane.

Strain gages are applied to the skin on magnets to be tested cold. These gages will measure stresses in the skin during many phases of construction and testing. Knowledge of skin stresses during cooldown and coil excitation is needed to make decisions involving the yoke/collar interface design.

CONCLUSION

The Fermilab short model program has been implemented to achieve two objectives: to prove that the Fermilab tooling can produce a working magnet and to test design alternatives which could improve the performance of the established baseline SSC dipole. Several magnets have been built and tested. Continuing development is necessary concerning both design of magnet components and production manufacturing methods. The short magnet program is equipped to continue this process and proceed from it into the next SSC dipole design.

REFERENCES

1.   R. C. Bossert, J. S. Brandt, J. A. Carson, H. J. Fulton, G. C. Lee, and J. M. Cook, Analytical Solutions to SSC Coil End Design, in: " IISSC, Supercollider 1", p. 387, M. McAshan, ed., Plenum Press, New York, NY (1989).
2.   G. C. Lee, 1989, Autoend Program, (a program creating surfaces by wrapping ellipses on cylinders), Fermi National Accelerator Laboratory.
3.   J. M. Cook, 1989, Program Bend, (a program creating surfaces using elasticas and developable surfaces), Argonne National Laboratory.
4.   J. M. Cook, An Application of Differential Geometry to SSC Magnet End Winding, Argonne National Laboratory (Preprint).
5.   R. C. Bossert, J. S. Brandt, and J. M. Cook, End Designs for Superconducting Magnets, presented at Breckenridge Workshop, Colorado, August 14-24, 1989.
6.   J. A. Carson, E. J. Barczak, R. C. Bossert, J. S. Brandt, G. A. Smith, SSC Dipole Coil Production Tooling, in: "IISSC, Supercollider 1", p. 51, M. McAshan, ed., Plenum Press, New York, NY (1989).
7.   J. B. Strait, K. J. Coulter, T. S. Jaffery, J. Kerby, W. A. Koska, and M. J. Lamm, Experimental Evaluation of Vertically versus Horizontally Split Yokes for SSC Dipole Magnets, presented at the IISSC, Miami Beach, Florida, March 14-16, 1990.
8.   J. B. Strait, Prestress, Ovality and Multipoles in DSS Magnets, presented at the July 1989 MSIM, Fermi National Accelerator Laboratory.
9.   J. B. Strait, R. C. Bossert, J. A. Carson, P. M. Mantsch, SSC Cold mass Design Review, presented to SSCL at Fermi National Accelerator Laboratory, November 8-9, 1989.
10.   J. B. Strait, R. C. Bossert, J. A. Carson, P. M. Mantsch, Response to Action Items from Fermilab SSC Cold Mass Design Review, submitted to SSCL, December 1989.
11.   M. S. Chapman, and R. H. Wands, A Finite Element Analysis of the SSC Dipoles, in: "IISSC, Supercollider 1", p. 659, M. McAshan, ed., Plenum Press, New York, NY (1989).
12.   C. L. Goodzeit, M. D. Anarella, and G. L. Ganetis, Measurement of Internal Forces in Superconducting Accelerator Magnets with Strain Gauge Transducers, SSC Note SSC-MAG-R-7312.
13.   C. E. Dickey, Coil Measurement Data Aquisition and Curing Press Control System for SSC Dipole Magnet Coils, in: "IISSC, Supercollider 1", p. 415, M. McAshan, ed., Plenum Press, New York, NY (1989).
14.   C. E. Dickey, and D. L. Kubik, Bulk Modulus Capacitor Load Cell for SSC Model Dipole Measurements, presented at the IISSC, Miami Beach, Florida, March 14-16, 1990.

PROCESS CONTROL ORIENTED

QUENCH ANALYSIS OF A SSC MAGNET

T.J. Fagan and P.W. Eckels

Westinghouse Electric Corporation
Science and Technology Center
Pittsburgh, PA

ABSTRACT

Control of the refrigeration system for the SSC magnets will
require a very advanced process control system. To be effective, the
control system must be able to recognize magnet quench behavior and
respond to mitigate the event and recover stable operating conditions.
The purpose of this quench analysis is to provide quench characteristics
in the format of manifold pressure and temperature rise with time as
well as to evaluate the outflow that must be accommodated by the
refrigeration system. The method of analysis and results of the
analysis are described.

INTRODUCTION

The most technically advanced facet of the SSC project is the
helium management system process control program and apparatus. In
order to achieve a lead in the SSC technology of discrete event process
control, Air Products, Inc. formed a PA initiative Benjamin Franklin
group in process control. It consists of teams from Westinghouse,
Carnegie-Mellon University, Lehigh University, and Air Products. The
Westinghouse contribution to that group, the subject of this work, is to
provide an analytical model of the worst case discrete event that the
process control system must manage, a magnet quench. Our objective,
then, was to provide the helium temperature, flow rate and pressure rise
imposed on the helium cooling system by the magnet as a function of time
after the initiation of a quench, i.e., a massive loss of
superconductivity in the magnet winding. We are concerned with the
worst case event for the cooling system to mitigate, the quench at peak
operating current. It has been shown[1] that current levels in the range
of 4000 A produce the highest winding temperature because of their
reduced quench propagation velocities.

Figure 1 is a picture of a polished section through the 17 m long
dipole winding and stainless steel collar lamination. Figure 2 shows
the insulation and slip plane (anti-friction) system placed around the
winding. The arrangement is more or less two-dimensional except that
every six inches a lamination is missing to ventilate and mitigate the
winding pressure rise. The ventilation feature is a relatively recent
design change.

A quench occurs in the SSC magnets when the superconductor temperature exceeds the critical temperature, that temperature above which the superconducting state cannot exist. Because of the very low heat capacity of most materials at LHe temperature and the relatively high thermal resistance of the conduction cooling system used in the SSC magnet windings, a quench can be caused by heat generation of tens of mJ/cc of conductor. This means that heating resulting from friction due to winding slippage, mild ac losses, acoustic pressure waves, and Ohmic heating can produce a quench.

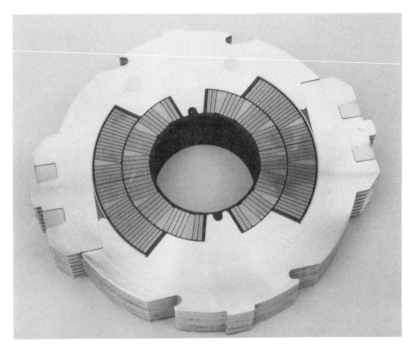

Figure 1 - Transverse polished section through the collar and winding.

When a winding quenches, the energy stored in the magnetic field of a magnet is dissipated as Ohmic heating. That heat may be localized as in a "locked quench zone" or distributed throughout the winding by a rapidly propagating quench zone. Rapid propagation of the quench zone distributes the heat over a large fraction of the winding reducing the local temperature rise of the winding; however, rapid propagation increases the pressure rise within the winding. Between these two limiting conditions, it is expected that there is an optimal propagation rate which does not produce a damaging pressure rise, yet is sufficiently rapid to distribute the heating over the whole winding. To achieve this optimum, a quench protection system is designed into the SSC magnet which consists of a spatially distributed heater energized by capacitor discharge when a resistive voltage is detected. The energy

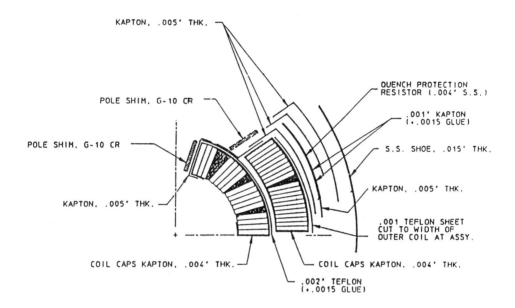

KAPTON, .005' THK.

POLE SHIM, G-10 CR

POLE SHIM, G-10 CR

KAPTON, .005' THK.

COIL CAPS KAPTON, .004' THK.

QUENCH PROTECTION RESISTOR (.004' S.S.)

.001' KAPTON (+.0015 GLUE)

S.S. SHOE, .015' THK.

KAPTON, .005' THK.

.001 TEFLON SHEET CUT TO WIDTH OF OUTER COIL AT ASSY.

COIL CAPS KAPTON, .004' THK.

.002' TEFLON (+.0015 GLUE)

EXPLODED VIEW
COIL QUADRANT-NO SCALE
(COLLAR REMOVED FOR CLARITY)

Figure 2 - Lay of the insulation, quench heater, and slip planes.

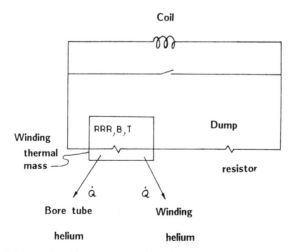

Coil

RRR,B,T

Dump

Winding
thermal
mass

resistor

$\dot{q}$ $\dot{q}$

Bore tube   Winding

helium   helium

Figure 3 - Schematic diagram of the electromagnetic quench circuit.

delivered by the quench protection system is designed to cause the
entire winding to become resistive within 79 ms.

The present analysis began by assuming a non-propagating, or
uniform quench of the whole winding. Although conductors do not
naturally quench uniformly, the protection system imposes a nearly
uniform quench. The validity of the uniform quench assumption is then
examined within the context of the results of the prediction. It is
found that the pressure rises are sufficient to produce a quench by
adiabatic compression temperature rise of the helium and winding via an
acoustically propagating pressure wave. It is also evident from the
results that the rate of field change produced by the acoustically
propagating quench can produce an ac loss temperature rise of the helium
and winding of sufficient magnitude to cause an electromagnetic quench.
The time constants of these events are so short that high pressures are
anticipated to exist within the insulation layers around the
superconducting winding unless protective measures are taken. These
pressures are confined to the outer winding layer which is enclosed in
Kapton and Teflon. The cooling system and piping do not experience
these large pressure rises because they are large enough to offer little
resistance to the quench flow surge.

Although the high pressure within the outer winding layer might be
expected to offer some structural difficulties such as reducing the
winding preload, the peak in hydraulic pressure occurs while the
magnetic pressure from the inner layer is still higher than the
hydraulic pressure and that magnetic pressure prevents any reversal of
loading within the winding.

In the following text, we shall describe the finite difference
model of the magnet hydraulic system, the input data and its source, the
results of the analysis, and some recommendations for design
improvements to eliminate the quench pressure rise problem.

MAGNET QUENCH MODEL

To model the quench process, a finite difference technique was
utilized that marched forward in time from initiation of the event. The
continuity, energy, and constitutive equations were discretized using a
central difference technique. The problem of pressure interaction in
the momentum and energy difference equations was handled by using an
intermediate mesh overlay. The base mesh chosen was based on major
geometric characteristics of the magnet and the overlay mesh had the
same geometric characteristics, but was displaced one-half time step
from the base mesh as is the general practice. Central difference must
be used for the intermediate mesh also.

The magnet quench model is subdivided into electromagnetic and
thermohydraulic models. The electromagnetic model includes a time,
temperature, and field dependent computation of winding resistance, heat
capacity, heating, and current. It provides the winding temperature at
each time step to the thermohydraulic model. The thermohydraulic model
computes the fluid and thermal potentials and fluxes using temperature
dependent properties for solid materials and temperature - pressure
dependent properties of helium.[2] The heat fluxes from the winding are
provided to the electromagnetic computation in an iterative, converging
computation process that achieves a balance according to the
conservation equations.

504

Electromagnetic Model

The schematic diagram of Figure 3 shows the elementary circuit from which the electromagnetic model is derived. Opening the switch corresponds to the quench event and sends current through the normal metal in the winding and through the energy dump resistor, if there is one. Magnetic field is presumed linear with current and the normal metal resistance is computed using the appropriate field and temperature values in the Kohler equation for copper resistivity.[3] The winding is subdivided into "n" series elements. The current decay increment is computed by Equation 1.

$$\delta(I) = 2(\Sigma_n R_n/L) \; \delta(\theta) \tag{1}$$

The Ohmic heat generated within the winding is $I^2 R \; \delta(\theta)$, or including it in the energy balance of a winding element, we have Equation 2.

$$\delta(T_n) = (I^2 R_n \; \delta(\theta) - Q_n)/(C_n) \tag{2}$$

Here Q_n is provided by the thermohydraulic computation and is the heat removed from the winding by conduction and convection. Both the temperature rise and current decrement are summed to maintain a current value at each time step.

The electromagnetic program can be checked independently of the thermohydraulic program by setting the heat transferred from the winding to zero, an adiabatic quench. This was done and the results are presented below. The computed temperature rise of the winding agreed with hand computations very well.

Thermohydraulic Model

At low temperatures, solid materials have very low thermal capacity and during the initial phase of the transition to the normally conducting mode, the bulk of the heat generated in the windings will be absorbed by the helium filling the winding voids, the spaces between the laminations in the yoke and collar, and the helium in the bore tube flow passages and the coolant channels in the yoke. The addition of heat to the helium will result in a significant increase in temperature and pressure. The coolant supply system and the winding structure must be designed to accommodate this pressure rise without damage and to permit a rapid return of the system to the superconducting operating mode.

Therefore, the transient pressure and temperature distribution internal to the magnet structure during the "quench" process must be realistically modeled.

The following briefly describes the model developed to predict the thermo-hydraulic characteristics of the supercollider dipole magnets during a transition to the normally conducting mode, typically referred to as a "quench."

Technical Approach

The magnet has been divided into a series of inter-connected fixed volumes. Figure 4 shows the basic thermo-hydraulic element. The heat transfer and fluid flow processes occurring in this basic element can be analyzed by applying the principles of conservation of mass, conservation of energy and conservation of momentum as follows.

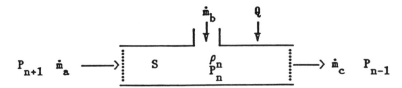

Figure 4 - Thermo-hydraulic element.

Conservation of Mass

Element n with volume S is assumed to be initially filled with a fluid at temperature T_n and pressure P_n. From the equation of state for the fluid, the density of the fluid ρ_n can be determined and the initial mass of fluid in the element $\rho_n S$ can be calculated. During time step Δt fluid flows into element n from element n+1 at a flow rate $\dot{m}_a$. A second stream of fluid flows into the element through openings in the wall of the element in a direction normal to the main stream velocity at a flow rate $\dot{m}_b$. Fluid flows from element n to element n-1 at flow rate $\dot{m}_c$. At the end of time step Δt the density of the fluid remaining in element n is ρ_n'. Application of the principle of conservation of mass gives Equation 3:

$$\dot{m}_a \Delta t + \dot{m}_b \Delta t + \rho_n S = \dot{m}_c \Delta t + \rho_n' S \qquad (3)$$

Conservation of Energy

The internal energy of the fluid initially stored in volume n is $\rho_n S u_n$. Streams $\dot{m}_a$ and $\dot{m}_b$ add energy to volume n in the form of internal energy, pressure-volume work and kinetic energy. Energy Q is added to volume n in the form of heat transferred from the walls of the element. Energy is removed from volume n by stream $\dot{m}_c$ in the form of internal energy, the pressure-volume product and kinetic energy. The internal energy of the fluid remaining in volume n at the end of time step Δt is $\rho_n' S u_n'$. The sum of internal energy and the pressure-volume product is commonly referred to as enthalpy. For stream $\dot{m}_a$:

$$\bar{h}_a = \bar{u}_a + \bar{P}_a / \bar{\rho}_a \qquad (4)$$

Similar relationships are used for streams $\dot{m}_b$ and $\dot{m}_c$. Application of the principle of conservation of energy to volume n gives Equation 5:

$$\dot{m}_a \Delta t \left(\bar{h}_a + \frac{\bar{V}_a^2}{2} \right) + \dot{m}_b \Delta t \left(\bar{h}_b + \frac{\bar{V}_b^2}{2} \right) + \rho_n S u_n + Q =$$

$$\dot{m}_c \Delta t \left(\bar{h}_c + \frac{\bar{V}_c^2}{2} \right) + \rho_n' S u_n' + \frac{S}{2} \left(\rho_n' V_n' V_n' - \rho_n V_n^2 \right) \qquad (5)$$

Conservation of Momentum + Friction

The flow rate between element n and element n-1 is a function of the average difference in static pressure between the elements. This pressure difference must balance the pressure drop due to friction between the fluid and the walls of the passage and the change in momentum of the fluid. The friction pressure drop has been defined using the Moody friction factor approach:

$$\Delta P_f = f \frac{L_x}{D_h} \bar{\rho}_n \frac{\bar{V}_c^2}{2} \qquad (6)$$

Where L_x is the length of the flow path, D_h is the hydraulic diameter of the flow passage and f is the Moody friction factor. The magnitude of f is a function of the Reynolds number of the flow through the passage and the ratio of the rms surface roughness of the passage walls to the passage hydraulic diameter.

The pressure difference due to the change in momentum of the main stream is given by:

$$\Delta P_m = \frac{\dot{m}_c}{A_f} \bar{V}_c - \frac{\dot{m}_a}{A_f} \bar{V}_a + \frac{S}{A_f \Delta t} \left(\rho_n' V_n' - \rho_n V_n \right) \qquad (7)$$

Where A_f is the flow area of the passage. Note that since stream b flows at right angles to the main stream, it has no initial momentum in the main flow direction. However, the kinetic energy of stream b has been included in the conservation of energy equation.

The definition of mass flow rate gives:

$$\dot{m}_a = \bar{\rho}_a \bar{V}_a A_f \qquad (8)$$

$$\dot{m}_c = \bar{\rho}_c \bar{V}_c A_f \qquad (9)$$

The average pressure difference between elements n and n-1 must equal the sum of ΔP_f and ΔP_m. Substituting Equations 6 and 7 into the combination of Equations 4 and 5 gives:

$$\frac{P_n + P_n'}{2} - \frac{P_{n-1} + P_{n-1}'}{2} = \left(2 + f\frac{L}{D}\right)\left(\frac{\rho_n + \rho_n'}{2}\right)\frac{\bar{V}_c^2}{2} - \bar{\rho}_a \bar{V}_a^2$$
$$+ \left(\rho_n' V_n' - \rho_n V_n\right)S/A_f \qquad (10)$$

The simultaneous solution of Equations 1, 3, and 8 determines the quasi-steady state coolant properties in element n at the end of time step Δt. Due to the complex nature of the equation of state for helium, a closed form solution does not appear to be possible and an iterative method has been used.

Iteration Method - Single Element

Several iteration parameters were considered for a single element. The parameter selected, defined as X_f, is the ratio of the flow out of the element $\dot{m}_c \Delta t$ to the sum of the flows into the element, $\dot{m}_a \Delta t$ and $\dot{m}_b \Delta t$ and the mass stored initially in the element $\rho_n S$.

$$X_f = \frac{\dot{m}_c \Delta t}{\dot{m}_a \Delta t + \dot{m}_b \Delta t + \rho_n S} \qquad (11)$$

A description of the iteration method used for a single element is shown in Table 1.

Table 1 — Iteration Method for Single Element.

---

1. Assume value of X_f.

2. Calculate ρ_n' from conservation of mass.

3. Calculate P_n' from conservation of momentum plus friction.

4. Find T_n' from P_n' and ρ_n'.

5. Find u_n' from P_n' and T_n'.

6. Find u_n'' from conservation of energy.

7. Compare u_n' and u_n''.

8. If within desired error limit stop.

9. Alter estimate of X_f.

   If u_n'' is greater than u_n' increase X_f.

   If u_n'' is smaller than u_n' decrease X_f.

   If sign of error has changed decrease step size.

Iteration Method - Multiple Elements

In order to model the entire magnet, a number of single elements must be combined. Since the flow into and out of element is a function of pressures and temperatures in the elements upstream and downstream from it, and the temperatures and pressures in the elements connected to it through openings in the passage walls, the combination of elements is a highly iterative procedure. The methods used to perform this iteration are outlined in Table 2.

Table 2 — Iteration Method for Multiple Elements.

---

1. To start, assume no flow between elements.

2. Calculate pressure distribution at end of time step using estimated values for downstream pressures and flows between elements.

3. Re-calculate pressure distribution at end of time step using estimated values for downstream pressures and flows between elements.

4. Compare old and new estimate of pressure at end of time step for all elements.

5. If within desired error limit move to new time step.

6. Revise estimates of downstream pressures using:

$$P_{n-1}'' = P_{n-1}' + X_c(P_{n-1}' - P_{n-1})$$

Computer Code

The single element model described above has been incorporated into a FORTRAN 77 subroutine known as PFLOW. PFLOW is called iteratively, using the method outlined in Table 2, by a main program known as DPFLO. Currently the thermodynamic and transport properties of the helium coolant are evaluated by a series of subroutines and functions extracted from the NBS HELIUM package. The code was developed and tested on a DEC PDP-11/73 minicomputer using the RSX-M11+ operating system.

The PDP-11 is clearly too slow for production runs and the code was transferred to one of the VAX computers located at the R&D Center and an IBM PS/2 model 70-121 computer. Care was taken to write the source code in standard FORTRAN 77 so that it will be fully portable. The NBS HELIUM subroutines and functions may be replaced by simplified models to reduce computing time. A table look-up approach is much faster.

RESULTS

The quench prediction program has predicted the pressure, temperature, and mass flow for the cases of adiabatic and non-adiabatic with 1989 modifications. While the majority of the parameters are directly set by the dimensions, those parameters associated with the sizing of flow channels through the Kapton enclosing the winding layer can only be estimated. We have assumed that 0.025 mm (1 mil) spacings between the Kapton layers will exist with a length given by the minimum Kapton overlap. This leads to a very large L/D ratio for these passages. It is safe to say that the better the magnet construction (insulation fit and lay), the smaller the flow passages will be. These are presented below.

Adiabatic Quench

This case describes the condition of no escape of heat or fluid from the confines of the winding. The helium volume fraction within the winding is 3% and it is initially at 4.3°K and 4.0 atm. Adiabatic conditions may describe the very early (< 0.1 ms), high current (100% operating) quench process reasonably well. Figures 5, 6, and 7 show the current decay with time, the magnet terminal voltage with time for the current previously mentioned, and the conductor temperature as a function of time. These are not necessarily the highest values of temperature, voltage, nor current decay that can be encountered. A propagating quench will very likely produce the highest values. Still, it is of interest to examine the implications of these results relevant to our predictions of the worst discrete event that the cooling system control program must handle.

Adiabatic compression of the helium to 8 atm will produce the 0.6°K temperature rise necessary to quench the winding. Adiabatic processes are constant entropy processes and entropy is a property of the helium fluid. Thus, the process end points are independent of the means used to get between the points. Physically, this all means that if a process is adiabatic (or quickly accomplished in this case), a 0.6°K temperature rise will produce 6 atm of pressure or, vice versa, pressurization to 6 atm will produce a 0.6°K temperature rise.

In a quench situation, the initial winding rate of temperature rise is about 10°K/ms (Figure 7) due to the low specific heat of the winding pack below 10 K. A pressure relief wave traveling at 120 m/s from the edge of a disturbance can move only 0.6 cm before the pressure rise is sufficient to produce a propagating quench wave.

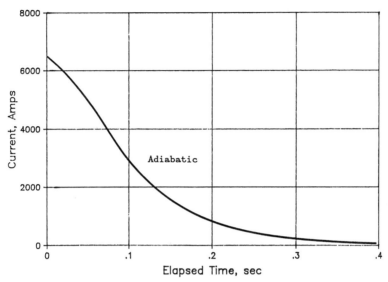

Figure 5 - Decay of the magnet current with time after an adiabatic, uniform quench.

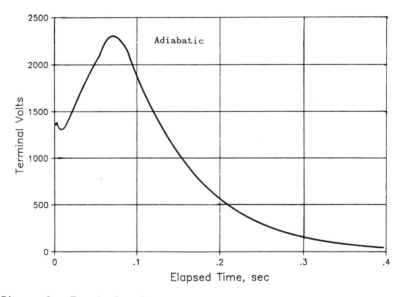

Figure 6 - Terminal voltage as a function of time after a quench.

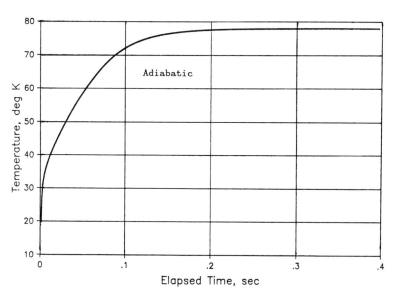

Figure 7 - Temporal variation of conductor temperature during an adiabatic, uniform quench.

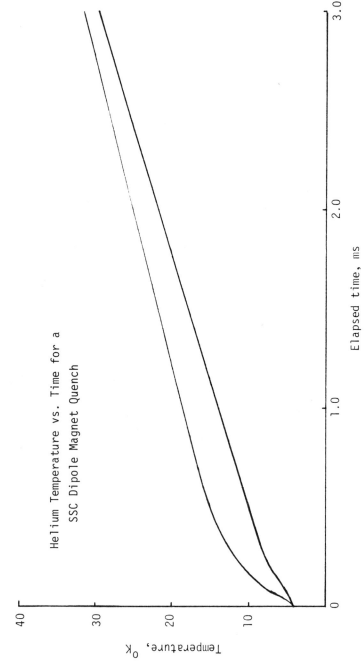

Helium Temperature vs. Time for a
SSC Dipole Magnet Quench

Figure 8 - Peak helium temperature in the winding and bore tube efflux temperature
following a quench.

512

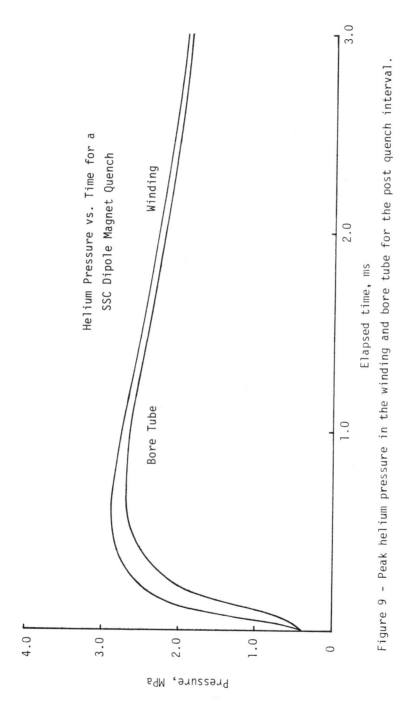

Figure 9 - Peak helium pressure in the winding and bore tube for the post quench interval.

513

It is also interesting to note that the current decay is in the range of 35 kA/s which is a field change of approximately 35 T/s. In the uninsulated strands of the cable, such high values of field change are likely to produce a quench from ac losses. Thus, our uniform quench model for predicting the maximum disturbance which the helium cooling system will encounter is entirely realistic.

Non-adiabatic, Ventilated Collar Quench

This case describes the outrush of helium due to a quench with a transverse or radial vent (missing lamination) through the collar every six inches. Figures 8, 9, 10, and 11 show the helium temperature, helium pressure, helium outflow rate and helium outrush velocity as a function of time. Note that the transient results with flow are only significant for short intervals because of the small helium fraction in the windings and so the computation is terminated after 3.0 ms. After that time, the adiabatic results indicate winding temperature well enough to provide a basis for cooldown computations which are beyond the scope of this work.

Figure 9, helium temperature in the winding and bore tube, again shows the rise to be roughly 10 K/ms. The winding temperature is several degrees warmer than the helium leaving the bore tube as would be expected.

The pressure rise is very significant early in the process but decays to acceptable levels very quickly. The pressure peaks in the winding at 29 atm in 0.6 ms and in the bore tube at 27 atm in the same time scale. In 3.0 ms the pressure has decreased to about 20 atm. Any magnet component of high density is unlikely to respond to such a pressure application.

Helium flow from the winding and bore tube is shown in Figure 10. The peak flow is 400 kg/s and again decreases quickly to 100 kg/s. Integration of the area under the curve yields 0.77 kg expelled within 3 ms, which is 85% of the helium initially in the winding and bore tube.

Helium velocity, Figure 11, shows potentially damaging velocity levels within the magnet. When temperature is considered, the Mach number is typically 0.5 or less. This is important from the standpoint that the equations used in the development are compatible with these Mach number levels.

The results of this work are clearly within our expectations for a magnet with a very low helium mass contained within the winding. We find the computer computations to be in agreement with overall mass and energy conservation considerations. Convenience of use is excellent, the program runs on an IBM PS 2 model 70 in about 1.5 hrs, compilation having been done by the Ryan-MacFarland Fortran Compiler. Run time can be greatly reduced by evaluating properties via table interpolation routines rather than the direct NBS property routines.

Due to the small helium fraction within the winding, the quench event helium surge is not a major problem for the refrigeration system and there is no danger from excessive temperature rise in these magnets.

CONCLUSIONS

The quench analysis computer program is a very useful and convenient tool.

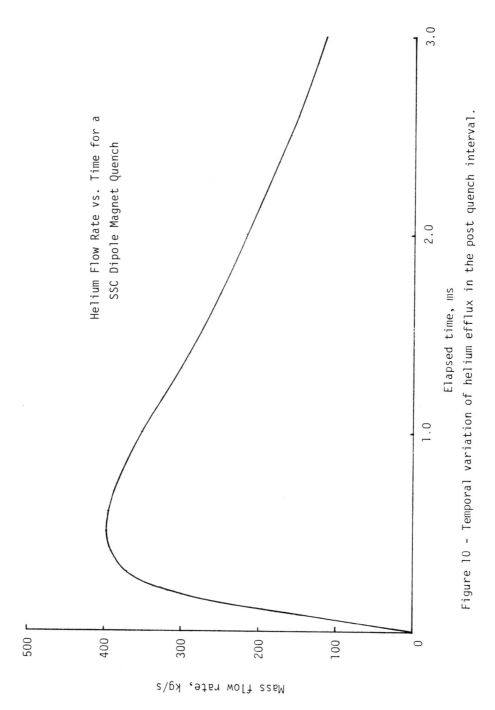

Helium Flow Rate vs. Time for a
SSC Dipole Magnet Quench

Figure 10 - Temporal variation of helium efflux in the post quench interval.

515

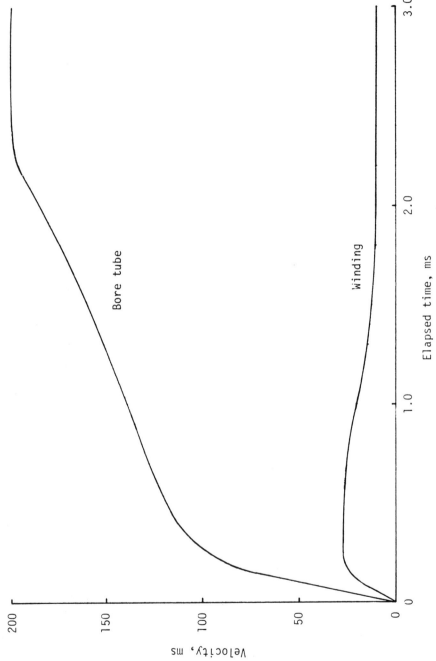

Figure 11 - Helium velocity in the winding and bore tube of an SSC dipole magnet after a quench.

The quench event does not produce damaging pressure, temperature, or flow levels for the structure or refrigeration system.

The results show that quenches will propagate both thermo-acoustically and electromagnetically.

REFERENCES

1. A. Devred, et al., "Quench Characteristics of Full-Length SSC R&D Dipole Magnets," Presented at the 1989 CEC at UCLA, July 27, 1989.
2. Vincent D. Arp, "Helium Properties for the PC, HEPROP.EXE," Personal Communication at the 1989 CEC at UCLA.
3. Handbook on Materials for Superconducting Machinery, Revised, Jan. 1977, Battelle Labs., MCIC-HB-04.
4. E. A. Ibrahim, M. A. Hilal, and S. D. Peck, "Quench Pressure Analysis of Adiabatically Stable Magnets", IEEE Transactions on Magnetics, Vol. MAG-23, No. 2, March 1987.

BULK MODULUS CAPACITOR LOAD CELLS

C. E. Dickey

Superconducting Super Collider Laboratory*
2250 Beckleymeade Avenue
Dallas, TX 75237

Abstract: Measurement of forces present at various locations within the SSC Model Dipole collared coil assembly is of great practical interest to development engineers. Of particular interest are the forces between coils at the parting plane and forces that exist between coils and pole pieces. It is also desired to observe these forces under the various conditions that a magnet will experience such as: during the collaring process, post-collaring, under the influence of cryogens, and during field excitation. A twenty eight thousandths of an inch thick capacitor load cell which utilizes the hydrostatic condition of a stressed plastic dielectric has been designed. These cells are currently being installed on SSC Model Dipoles. The theory, development, and application of these cells will be discussed.

INTRODUCTION

The development of the SSC Collider Dipole and virtually all superconducting magnet projects worldwide, would be greatly aided by the realization of load measuring instrumentation which could be installed at various locations within superconducting magnet structures. To be of true value, such instrumentation must not perturb the superconducting magnet structure itself. To minimize perturbations, it would be necessary to develop an extremely thin load cell. If a thin, accurate load measuring device could be developed, such a device might also have significant commercial applications.

THEORY OF MODULUS OF ELASTICITY CAPACITOR LOAD CELLS

As you know, the capacitance of an idealized parallel plane capacitor is given by:

---

*Operated by the Universities Research Association, Inc., for the U.S. Department of Energy under Contract No. DE-AC02-89ER40486.

$$C = \frac{\epsilon A}{d}$$

Where C is the capacitance in Farads, epsilon is the electric permitivity of the dielectric material, A is the plate area and d is the plate separation.

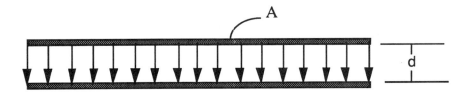

A load cell, a device which measures mechanical load, may be designed based on the parallel plate capacitor.

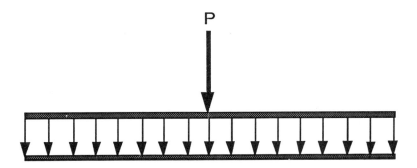

In a capacitor load cell, the capacitance will change as the load cell is loaded. Primarily the change in capacitance corresponds to the change in plate separation and plate area. The change in plate separation is related to the modulus of elasticity of the dielectric material;

$$\delta = \frac{\sigma}{E}$$

where delta represents the resultant strain, sigma represents the applied stress, and E represents the modulus of elasticity.

Taking the strain of the dielectric material into account, we may rewrite the capacitance equation:

$$C = \frac{\epsilon A}{d(1 - \frac{\sigma}{E})}$$

Taking into account the Poisson ratio effect on the plate area;

$$C = \frac{\in A \left[1 + 2v_p \frac{\sigma}{E_d}\right]}{d \left[1 - \frac{\sigma}{E_d}\right]}$$

where nu represents Poisson's ratio of the plate material. It should be noted that the load axis strain and the Poisson ratio both tend to increase the capacitance as the gage is loaded. In addition, the above equation does not consider fringe field effects or the possible relationship between applied stress and the electric permitivity of the dielectric material.

THEORY OF BULK MODULUS CAPACITOR LOAD CELLS

It should be possible to construct a capacitor gage based on the bulk modulus effect. This could be achieved by causing a condition of spherical or hydrostatic stress to exist within the dielectric material. This condition could be established by using a constrained fluid or a plastic dielectric.

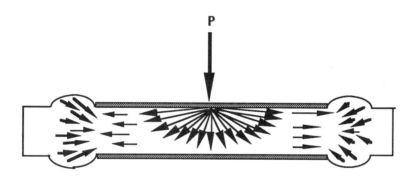

The resultant strains in a spherically stressed system are equal in all directions and are given by;

$$\delta = \frac{\sigma}{E} (1 - 2v) \; .$$

Substituting these into the capacitor equation;

$$C = \frac{\in A \left[1 + 2v_p \frac{\sigma}{E_p}\right]}{d \left[1 - \left[\frac{\sigma}{E_d} [1 - 2v_d]\right]\right]} \; .$$

This is the equation for the Bulk Modulus Capacitor Load Cell.

THEORETICAL DISCUSSION

The equation for the modulus of elasticity capacitance gage is linear up to the elastic limit of the dielectric material where the equation displays a singularity. In fact, experimentation with various dielectric materials has shown that the capacitance continues to increase beyond the elastic limit until the dielectric material fails. However, operation in the dielectric material's plastic region usually results in significant permanent capacitance offset which is no doubt due to plastic flow of the dielectric material.

$$C = \frac{\in A \left[1 + 2v_p \frac{\sigma}{E_p} \right]}{d \left[1 - \frac{\sigma}{E_d} \right]}$$

The case of the bulk modulus capacitance gage is somewhat different:

$$C = \frac{\in A \left[1 + 2v_p \frac{\sigma}{E_p} \right]}{d \left[1 - \left[\frac{\sigma}{E_d} \left[1 - 2v_d \right] \right] \right]}$$

Here we see that the singularity condition is;

$$\left[\frac{\sigma}{E_d} \left[1 - 2v_d \right] \right] = 1$$

this suggests that the bulk modulus capacitance gage may be able to operate beyond the elastic limit of the dielectric material. Experimentation has shown that bulk modulus gages do indeed operate linearly well beyond the dielectric materials elastic limit.

DEVELOPMENT OF BULK MODULUS CAPACITOR GAGES

Fermilab has been developing the Bulk Modulus Gage over the past year. During this time, we have achieved a more thorough understanding of the gage's properties. In addition, we have gone far toward developing the readback electronics and making the gage's more convenient and reliable to use.

Our initial work was conducted with a classic parallel plate configuration. Since our gage's displayed fairly low capacitance, on the order of a few hundred picofarads, we noticed a significant signal contribution due to the existence of an external shunt capacitance circuit.

To eliminate this

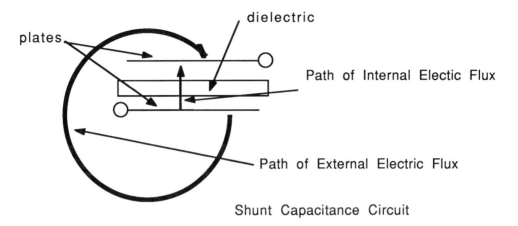

dielectric

plates

Path of Internal Electic Flux

Path of External Electric Flux

Shunt Capacitance Circuit

shunt capacitance we developed the Y-Gage. the Y-Gage is effectively a shielded capacitor and all significant flux paths are contained within the gage.   In addition,   the Y-gate displays

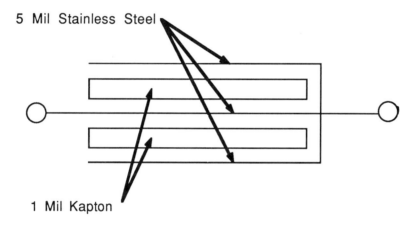

5 Mil Stainless Steel

1 Mil Kapton

The Y Gage

double the capacitance of a classic parallel plate design. This increase in capacitance results effectively in an increase in the gages sensitivity and it reduces the demands on metering electronics.

Gage physical dimensions have varied somewhat. But typical dimensions are approximately 1.5 –2" × 0.375" × 0.028". One of the significant advantages of this type of load cell is that its physical dimensions can be easily varied to match the specific case geometry. Curved surfaces or even thinner load cells are probably possible with careful development.

The electronic design of the metering circuit has undergone development and we now feel that we have a reliable and linear circuit.

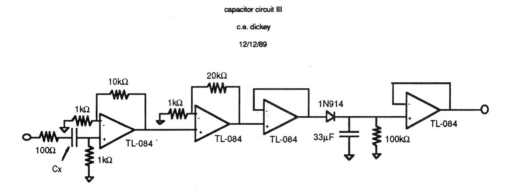

Linear performance of the metering circuit is quite important. With a linear circuit, we have the opportunity to merely adjust gain to calibrate the circuit to the individual gage. This should eliminate the necessity of using the same individual electronics channel for reading a specific gage throughout the various phases of the gage's life. However, it may be necessary to measure and record circuit gain after calibration. Another advantage of using a linear circuit is the elimination of uncertainties caused by the combination of the various nonlinear subcircuit transfer characteristics. Researchers will be presented with the opportunity of studying the load cell directly. Looking at the circuit transfer plot, it should be noted that the point where the plot goes horizontal corresponds to the onset of clipping in the final gain stage of the circuit.

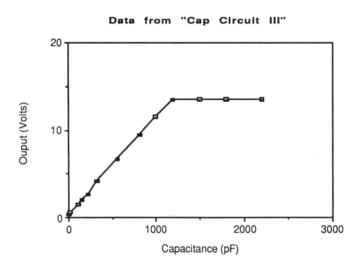

Figure 1 is a chart recorder output of a typical Bulk Modulus Capacitor Load Cell which has been loaded and unloaded ten times. Its repeatability appears acceptable for the purpose of making quantitative measurements within a collared coil assembly. However, these measurements were made in a fixture. The load cell was not removed from the fixture between cycles. Changes in output which result from repositioning the load cell are currently under study.

Clearly, two areas should be carefully considered. First, thought should be given to the possibility of alternative methods of construction. Precision machining, vapor deposition of the dielectric material onto the plates, and packaging should all be carefully considered. Second, thought should be given to the interface between the magnet and the load cell. These interfaces should be precisely formed and uniform from interface to interface. This does not seem impossible since the coil surfaces are filled with an epoxy layer.

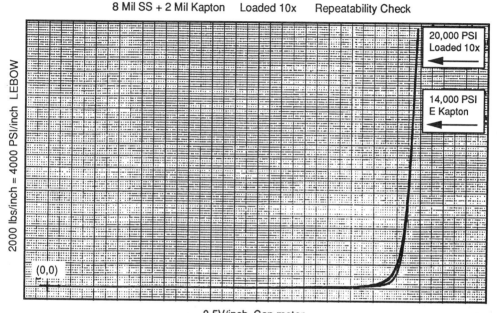

Figure 1.

CONCLUSION

Bulk Modulus Capacitor Load Cells are apparently a new type of transducer technology. They offer the possibility of providing instrumentation to study superconducting magnet structures. In addition, important commercial applications are possible. However, more work needs to be done.

RADIATION EFFECTS ON INTEGRATED CIRCUITS USED IN HIGH ENERGY

RESEARCH

Alvin S. Kanofsky

Department of Physics, Lehigh University
Bethlehem, Penn.18015

Bert Yost and Werner Farr

LeCroy Corporation
Chestnut Ridge, N.Y. 10977

We report here on radiation effects on two amplifiers used in high energy experiments. These are standard devices that are produced by LeCroy . We describe each of the devices and the experimental techniques. Finally, we present and discuss the results of our measurements.

MONOLITHIC MODEL TRA402 - 4 CHANNEL TRANSRESISTANCE PREAMPLIFIER

The TRA402 is a monolithic four channel, fast, low noise preamplifier. Its principal application is amplification of wire chamber signals for time-resolved measurements. Its high density makes it a practical solution, especially if space at the detector is at a premium, yet chamber mounting is a necessity.

The amplifier provides a gain of 25 mV/ microamp with a risetime of 3 nsec. The low input impedance of <100 ohms and low chamber noise of < 100 nA RMS is ideally suited to chamber applications, providing little integration of the current pulses. The special input geometry of the TRA402 makes it useful even with high capacitance detectors such as liquid argon calorimeters and wire chamber strips and pads. Because the risetime of the wire input is maintained by the TRA402, the user has greater freedom in selecting the RC coupling to the subsequent circuitry, allowing the risetime/noise trade-off to be optimized.

The TRA402 provides both inverting and non-inverting inputs for all four channels. This allows use of the amplifier for both positive and negative inputs provided by the cathodes and anodes of a wire chamber respectively.

MODEL TRA1000 - CHARGE/CURRENT PULSE PREAMPLIFIER

The LeCroy Model TRA1000 monolithic preamplifier is a versatile, economical, low-noise device which can be used either as a current-to-voltage preamplifier or as a charge-to-voltage preamplifier. The device has been designed for use with negative input signals;

however, it may also be configured for operation with positive inputs. The various options are selected through the use of external components.

The TRA1000 has been designed for direct connection to a variety of detectors. Its low noise and low input impedance make it ideal for use with proportional wire chambers even when resistive wire is used for position measurements. The device also finds application with photomultipliers when the economy of low-gain tubes is factor or dynamic range considerations are important. When used in conjunction with both the last dynode and the anode signals, exceptional dynamic range can be achieved by selecting different gains for the two devices. In this way, high and low sensitivity channels are configured.

The linearity of the preamplifier is excellent for a wide range of input risetimes and selected gains. The linear range of the output is 0.0 to 1 V, even for 50 ohm loads. Linearity has been measured in conjunction with a LeCroy 2280-Series 12 bit ADC. The result is typically <0.5% of reading with proper compensation.

The equivalent input noise of the preamplifier is as low as 2 pA in some configurations. Noise measurements are quoted in rms pC, referred to the preamplifier input. The LeCroy 2280-Series current-in tegrating ADC's employing a 500 nsec wide gate, were used to characterize the current-to-voltage configurations. For the charge-to-voltage configurations a peak sensing 2280 - Series ADC was used.

The low price, compact packaging, and low power dissipation of the Model TRA1000 make it an ideal choice for use in large-scale systems applications. It is particularly suited for use as a current-sensitive preamplifier for MWPC analog position measurements. The Model TRA1000 is a low-cost, high performance answer to a wide range of charge-sensitive and current sensitive preamplifier needs.

IRRADIATION TECHNIQUES

The 3 MeV High Voltage KN3000 Van De Graaff at Lehigh University was used for the irradiations. This machine provided beams of 2.5 MeV electrons which were directly used. We looked for slowly changing time effects using a continuous beam as well as transient effects with a pulsed beam arrangement. Figure 1 shows the experimental arrangement.

The electron beam dosage was measured by determining the beam profile using the luminescence from a glass plate at the exit window, placing the collimator hole at the beam location with the sample over the collimator hole, and measuring the total charge collected from the beam on a Faraday cup behind the sample.

TRA1 000 RESULTS

We bench tested the gain for the TRA1000 amplifier to check the linearity and waveforms under normal conditions. Figure 2 shows the test set-up. Figure 3 shows typical waveforms, as obtained on a LeCroy 9450/20 digital oscilloscope.

We then irradiated the circuit directly in the beam, with the circuit in its package and mounted on the LeCroy test board. In the first three second separate irradiations, we observed a widening of the leading and trailing edge overshoots while irradiating, but no change in the plateau voltage. Each second of irradiation corresponded to 1/3 Mrad of radiation. There was about half a second between exposures. With the fourth exposure, which lasted for 5 seconds, the gain went to zero after three seconds. This corresponded to 1 Mrad additional exposure.

With the fifth exposure, the output returned to the original form after about a second of irradiation, and then the overshoots and the output waveform decreased with further exposure for the next sixty seconds of exposure - an additional 20 Mrads of exposure. The chip then returned to a normal signal with the next irradiation, and finally the overshoots decreased to zero and the output stabilized at 12 mV. flat output with no overshoots for the next 120 seconds. During the next two minute exposure, the output returned to its original waveform for the next 6 minutes of exposure, although at the end of the exposure, a small high frequency component with a DC baseline shift was observed.

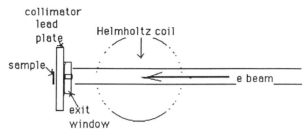

Figure 1. Electron Beam Irradiation.

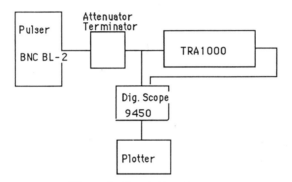

Figure 2. Circuit Test Set-up.

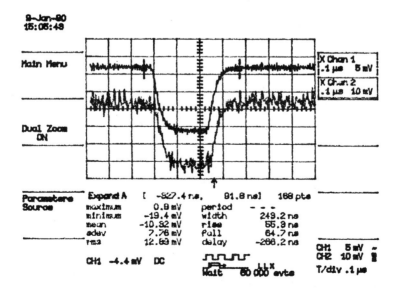

Figure 3. Typical Waveforms.

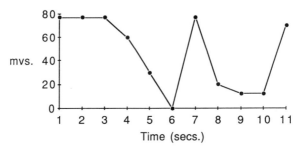

Figure 4a. TRA1000 output vs. seconds (0.3Mrad/sec).

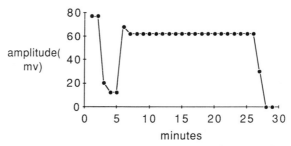

Figure 4b. TRA1000 output vs. minutes (20 Mrad/min).

The chip was then further irradiated at a higher beam current to determine if it could then be rendered completely inoperative. The beam current was 10 microamps (3.33 Mrads/sec dose) instead of 1 microamp. The circuit operated for the next 120 seconds without an appreciable change in output, and then failed within 5 seconds, at which point it no longer gave any output. This indicates that a radiation hardening process has occurred. The output for the short time and long time irradiations are shown in Figures 4a and 4b.

TRA402 RESULTS

The TRA402 amplifier was set up identical to the TRA1000 amplifier, using the LeCroy test board. The gain was checked to be nominally the same as the specifications, using the same method as that of the TRA1000. In order to obtain a more uniform sampling, the output waveforms were stored on a LeCroy 9450/20 Digital oscilloscope every 1 second. The beam current was set at 4 microamps, which corresponds to a dose rate of 1.3 Mrads/sec.

The TRA402 amplifier showed no decrease in the output amplitude with no waveform distortion until 80 seconds into the irradiation with a dose rate of 1.3 Mrads/sec. when the signal amplitude rapidly decreased without distortion over four a seconds to provide no output. The circuit was continuously irradiated for another 10 seconds. with no signal output. The beam was then off for a couple of minutes. It was then resumed and there was a partial restoration of the output waveform for several seconds, but with severe rounding of the leading and trailing edges. The output signal then totally disappeared and no longer appeared with further application of the beam and with removal of the beam. The performance of the TRA402 with time is given in Figure 5.

Thus, there was no indication of a radiation hardening process occurring, as with the TRA1000 amplifier, but rather a radiation tolerance up to about 100 Mrads, and then a uniform deterioration for the next 5 Mrads.

530

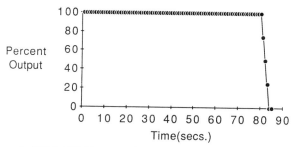

Figure 5. TRA402 Percent Output vs. time(1.3 Mrad/sec).

CONCLUSIONS

The two amplifier chips showed markedly different behavior. Up to the first Mrad of radiation, we notice no effect on the circuits. Beyond this, the TRA1000 showed erratic behavior initially with radiation up to the first 100 Mrads of irradiation and then stabilized for a period of an addition 200 Mrads, before ceasing to function. This indicates the possibility of a radiation hardening process occurring.

The TRA402 maintained its performance for about 100 Mrads and then deteriorated uniformly for the next 5 Mrads, indicating no radiation hardening, but rather a tolerance to radiation. Thus it seems the performance under radiation was different for the two circuits, and that initial radiation hardening in an electron beam may be a way to obtain radiation tolerances in circuits that initially are susceptible to radiation.

We are investigating these phenomena further both at the microscopic and gross circuit level, as well as obtaining further measurements on these and additional circuits.

ACKNOWLEDGEMENTS

This work was partially supported by the D.O.E., Bell Laboratories, and the Navy. We also thank Mr.Walter LeCroy for his support and encouragement.

SENSOR ARRAY FOR CHARACTERIZATION OF MAGNETIC FIELDS

IN COLD BORE DIPOLE SSC MAGNETS

Piotr M. Starewicz

Resonance Research, Inc.
46 Manning Road
Billerica, MA 01821

INTRODUCTION

A new method for characterization and qualification of the Superconducting Super Collider (SSC) dipole and multipole magnets prior to installation, including testing of the magnitude, stability, and homogeneity of the magnetic field is proposed. This paper describes a mechanically simple, stationary magnetic field sensor capable of measurements within the extended length and the narrow bore of the SSC design. The main feature of the proposed method is the ability to perform fast, precise measurements of magnetic field at cryogenic temperatures using a stationary NMR antenna array capable of operation over a wide range of excitation currents (magnetic field range from 0.2 to 7 tesla). The array is constructed of elongated antenna elements, distributed evenly around a hollow cylinder, centered and fixed within the beam tube. Each element filled with helium-3 acts as a self-contained NMR sensor resonating at frequency corresponding to its external magnetic field. Each sensing element has active cross-section small with respect to the highest spatial frequency magnetic inhomogeneity of interest. Based on the sensor array design, the measurement system will permit virtually instantaneous magnetic field characterization over the integrated magnet length with high transverse resolution and without need for mechanical motion of the sensor array.

Target Specifications

- Ability to test magnets from 0 to 0.01 T at high temperatures and 0.2 to 7 T at low temperatures.
- Measurement of the integrated field along the beam axis.
- Cold bore operation (liquid helium temperature).
- Field mapping up to 18th order in Fourier coefficients.
- Resolution better than 1 ppm.
- Simultaneous acquisition of data from the entire volume.
- Real time field drift measurement and compensation.
- Instantaneous data availability and display.

Technical Approach

- Operation over a wide range of magnetic fields.
- Use of nuclear magnetic resonance (NMR) for field measurements method between 0.2 and 7 tesla at cryogenic conditions.
- Use of electron spin resonance (ESR) for measuring low fields (below 0.01 T) at ambient conditions.
- Use of a stationary antenna array.
- Minimal mechanical interface requirements.
- Single array capable of uniform simultaneous measurements over the entire length of the magnet.
- Information of the axial inhomogeneities can be obtained from the resonance line-shape.

Significant Features

- Use of stationary sensor array.
- Frequency insensitive design using impedance matched but untuned antenna elements.
- Number of independent elements within the array determines the highest spatial frequency component of the magnetic field which can be measured.
- Electrically independent elements allow for simultaneous or sequential scanning.
- Each element filled with helium-3 acts as a self-contained NMR sensor.
- Each element filled with a stable free radical acts as a self-contained ESR sensor.

Sample Selection

Sample chosen for low temperature NMR mapping system must fulfill three conditions:

1. Must permit NMR observation at liquid helium-4 temperature.
2. Must have relatively short T (1) relaxation time in order to enable fast measurement cycle without saturation.
3. Must present a single NMR line.

The only substance which fulfils all three conditions is helium-3 (^{3}He). 3He gyromagnetic ration of (γ_{3He}) 32.433 MHz/T results in resonance frequency of 227 MHz at 7T. In comparison to hydrogen-1, a common NMR standard, it offers a relative sensitivity of 44% at constant field (γ_{1H} = 42.576 MHz/T).

Sample chosen for ESR mapping system must fulfill less stringent conditions:

1. Compound must contain a stable-free radical.
2. Must present an uncomplicated ESR spectrum.

The gyromagnetic ratio of an electron is approximately 657 times that of proton. As a direct consequence of this substantial useful magnetization can be observed at low fields at similar frequency ranges to the NMR (280 MHz at 100 gauss). A number of compounds containing stable-free radicals are commercially available.

Non-Resonant NMR Sensing Probe

Sensor element requirements
- Uniform RF performance
- Minimum crosstalk
- Sample confinement
- B_1 orientation orthogonal to B_0
- Mechanical integrity
- Ease of interfacing
- Ease of manufacture

Factors influencing the absolute signal amplitude
- Sensitivity. A constant for each nucleus (or electron).
- Absolute temperature. The available signal is proportional to the difference in absolute populations of the excited and ground states of spins under investigation. This distribution, as described by the Boltzman equipment, $N = \exp(-hv/kT)$, is strongly favored at low absolute temperatures.

  There is approximately a two order of magnitude signal gain at 4.2oK over ambient conditions.
- Receiving antenna Q-factor. Signal available for detection is proportional to Q. Conventional lumped circuit element resonant detection coils used in routing NMR and ESR applications have receiving antenna Q-factors on the order of 100.

  The proposed design has Q of approximately unity.

Signal strength for the non-resonant design will be equivalent to that available for a conventional tuned system at ambient temperature.

Transmission Line as NMR Sensing Probe

Several types of transmission lines may be used. Shielded balanced pair or single coaxial lines appear most applicable. In the dipole field a substantial part of the volume contains the B_0 and B_1 flux lines substantially orthogonal to each other. This solution presents a satisfactory geometry to induce and receive NMR or ESR signals.

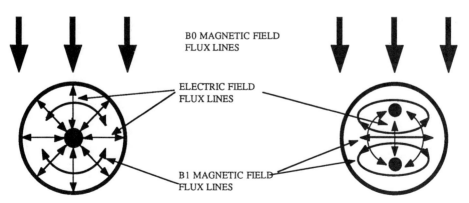

Figure 1. Transmission Lines

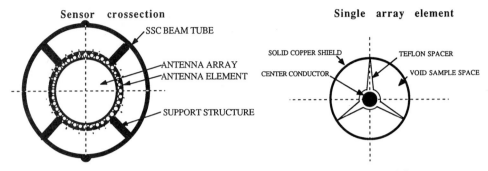

Figure 2.   NMR Sensor Probe Array (36 element assembly)

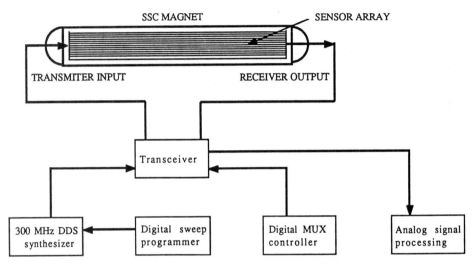

Figure 3.   Block Diagram of SSC NMR Field Mapping System

Feasibility Questions

-   Practicality of non-resonant NMR sensor under cryogenic conditions.
-   Practicality of NMR sensor of linear dimension significantly exceeding wavelength of the RF required for resonance.
-   Practicality of use of coaxial transmission line as NMR probe.
-   Selection of NMR and ESR sensitive samples for ambient and cryogenic conditions.
-   Method of supplying the excitation and receiving RF signals with low thermal losses.
-   Method of receiving signal from the linear element without significant local cancellations.
-   Electronic design of simultaneous vs. sequential scan RF transceiver for multi-element array.
-   Selection of pulsed RF vs. continuous wave excitation methods.

Reference

1.   W.G. Clark, J. Appl. Phys., 63(8), 4185 (1988).

THE LSU ELECTRON STORAGE RING, THE FIRST COMMERCIALLY-BUILT

STORAGE RING

Richard Sah

Brobeck Division
Maxwell Laboratories, Inc.
Richmond, California

ABSTRACT

The Brobeck Division of Maxwell Laboratories, Inc., is building the first industrially-produced storage ring. It will be located at Louisiana State University (LSU) at the Center for Advanced Microstructures and Devices (CAMD) in Baton Rouge. The purpose of this electron storage ring is to provide intense beams of x-rays to advance the state-of-the-art in lithography and to permit research in a broad area. This facility consists of a 1.2 GeV, 400 mA electron storage ring with a 200 MeV linac injector. The magnet lattice is a Chasman-Green design (double-bend achromat), and the ring circumference is 55.2 meters. There are four 3.0 meter, dispersion-free straight sections, one for injection, one for the 500 MHz RF cavity, and two for possible future insertion devices. The storage ring construction project is in the detailed-design stage, and many systems are in the initial stages of fabrication.

INTRODUCTION

In 1988, the Brobeck Division of Maxwell Laboratories, Inc., (MLI) entered into an agreement to design, build, and commission an electron storage ring for Louisiana State University (LSU). This 1.2 GeV ring represents the first instance in which an American industrial firm has assumed sole responsibility for a project of this kind. MLI will supply an integrated storage ring system whose design stresses performance, cost effectiveness, and flexibility. The purchase agreement is a fixed-price contract.

Since the technology of storage rings has been developed primarily in national laboratories in the United States and in other countries, the LSU Electron Storage Ring Project has presented significant challenges in technology transfer from national laboratories to industry. MLI is meeting these challenges by adopting the following three strategies, listed in descending order of importance: (1) building an experienced technical staff with key personnel who have had substantial work experience in national laboratories, (2) relying on consultants from several national laboratories to provide expert advice and review, and (3) using (at full cost recovery) actual technology developed at national laboratories. We believe that the most effective form of technology transfer is the transfer of trained personnel from one institution to another. To the greatest extent possible, the storage ring design utilizes proven technology which has been developed in the national laboratories.

PURPOSE OF THE RING

The Center for Advanced Microstructures and Devices (CAMD) at LSU is being established as a leading center for research in the VLSI (very large scale integration) electronic devices of the future. The production of such devices requires writing line features in semiconductors with a resolution of approximately 0.25 micrometers. The research which is to be carried out requires intense short-wavelength photon beams with excellent spatial stability. At present, the most promising approach for carrying out this work is with the soft x-rays emitted by the bending magnets of an electron storage ring. Therefore, MLI is designing, manufacturing, and commissioning the first industrially-produced electron storage ring in the United States, specifically for the CAMD application at LSU. The inclusion of 3 meter straight sections in the ring lattice greatly increases the flexibility of this facility by permitting the future addition of insertion devices (wigglers and undulators).

PERFORMANCE REQUIREMENTS

The most important performance requirements for the LSU Electron Storage Ring, as spelled out originally in the Invitation to Bid, are that the critical wavelength of the synchrotron light be in the range 6-12 angstroms and that the maximum beam current be 400 mA. In the MLI Proposal for the storage ring, the selection of 1.2 GeV beam energy and 1.4 Tesla magnetic field provides a critical wavelength of 9.5 angstroms. Furthermore, the storage ring subsystems have been designed for operation at 1.4 GeV and 200 mA, in order to provide a reduction of the critical wavelength to 6 angstroms (i.e., higher photon energy).

STORAGE RING DESCRIPTION

Specific performance aspects stressed in the design of the storage ring include (1) warm magnet technology, (2) 9.5 angstrom critical wavelength, (3) 400 mA capability, (4) large access to synchrotron radiation, (5) low beam emittance, and (6) utility straight sections. A Chasman-Green lattice was selected, and a quadrant of the storage ring is shown in Figure 1. Table 1 gives the major parameters of the ring.

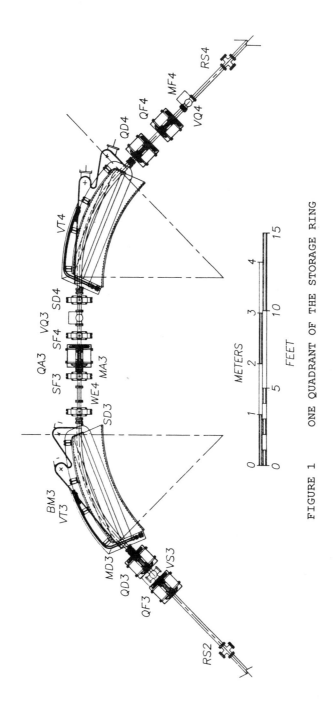

FIGURE 1   ONE QUADRANT OF THE STORAGE RING

539

TABLE 1
STORAGE RING PARAMETERS

Injection Energy	200 MeV
Nominal Operating Energy	1.2 GeV
Peak Operating Energy	1.4 GeV
Beam Current at 1.2 GeV	400 mA
Beam Current at 1.4 GeV	200 mA
Number of Superperiods	4
Circumference	55.2 m
Radiofrequency	499.65 MHz
Horizontal Betatron Tune	3.26
Vertical Betatron Tune	1.168
Natural Emittance	2×10^{-7} pi-m-rad

The cost-optimized injection system consists of a 200 MeV, S-band travelling wave linac. It will provide short filling times and multiple filling modes.

DESIGN DETAILS

This section contains a brief description of the design details for the technical systems of the storage ring.

Figure 2 shows a cross-sectional view of a dipole magnet including the vacuum chamber and an ion pump. The magnet is constructed of steel laminations with welded steel straps, and the end plates will be constructed of glued laminations. The gap of the dipole magnet is 62 mm. A substantial fraction of the dipole magnet laminations have been manufactured and delivered, and the assembly of a prototype will begin soon. The quadrupole magnets are also laminated, and they are assembled from four quadrant subassemblies. The quadrupole aperture is 78 mm.

The vacuum system is designed to provide an operating pressure of 1×10^{-9} Torr. Ion clearing electrodes will be incorporated within the vacuum system. Figure 3 shows plan and elevation views of the dipole vacuum chamber. The dipole chambers are brazed stainless-steel assemblies. Water-cooled copper absorbers are incorporated, and dipole chambers include two output ports for photon beams and two ion pumps. The other vacuum chambers consist of 3 inch stainless steel tubing and are pumped by a system of distributed ion pumps.

The purpose of the radiofrequency system is to add energy to the injected beam during ramping and to replace beam energy lost to synchrotron radiation. About 20 kW of power is required, plus an additional 20 kW to make up for resistive losses in the cavity walls. The amplifier is rated for 60 kW.

The architecture of the control system, shown in Figure 4, is based on an online database which is distributed over Ethernet. The computers are microVax's operating under VMS with DecWindows. The system will support CAMAC, VME, Allen-Bradley, and GPIP interfaces to the accelerator hardware. Almost all of the computer and

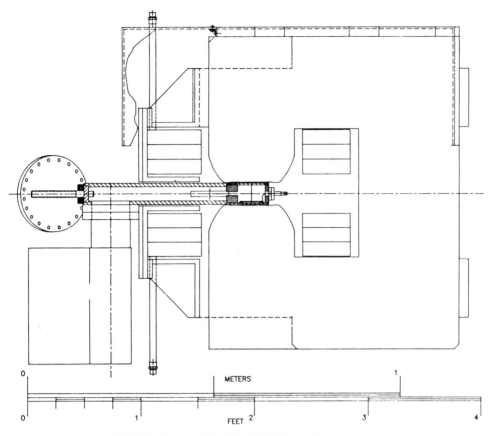

FIGURE 2   DIPOLE MAGNET CROSS SECTION

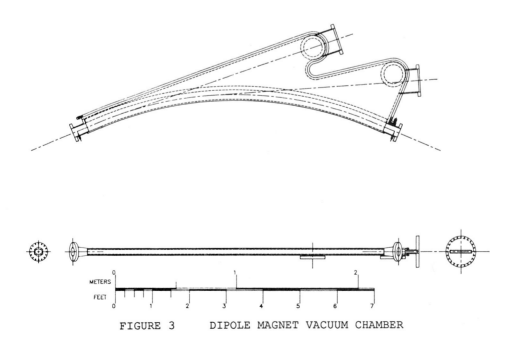

FIGURE 3   DIPOLE MAGNET VACUUM CHAMBER

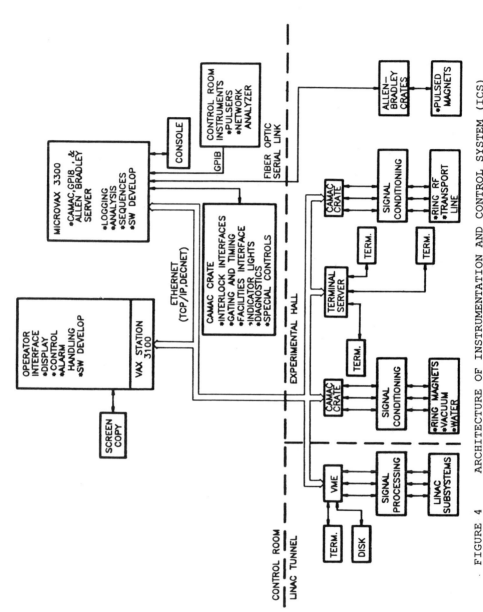

FIGURE 4   ARCHITECTURE OF INSTRUMENTATION AND CONTROL SYSTEM (ICS)

542

interface hardware and a majority of the software are commercially supplied. This will minimize development cost and maximize serviceability.

CONCLUSION

The LSU Electron Storage Ring is the first industrially-produced storage ring in the United States. The various technical systems of the storage ring are in detailed design and fabrication at this time. When completed, this highly flexible research facility will be the centerpiece of CAMD in Baton Rouge, and it will contribute to the development of advanced electronic devices for many years to come. The design, manufacture, and commissioning of this ring depend critically on an effective technology transfer from national laboratories to industry, so the successful completion of this project will represent a significant advance in the ability of industry to support major accelerator construction projects in the United States.

10. Program Schedules and Challenges

PROJECT ORGANIZATIONS AND SCHEDULES*

Richard J. Briggs

Superconducting Super Collider Laboratory[†]
2550 Beckleymeade Avenue, Dallas, Texas 75237

Abstract

The Superconducting Super Collider Laboratory (SSCL) faces the challenge of simultaneously carrying out a large-scale construction project with demanding cost, schedule, and performance goals; and creating a scientific laboratory capable of exploiting this unique scientific instrument. This paper describes the status of the laboratory organization developed to achieve these goals, and the major near-term schedule objectives of the project.

Organization

The SSCL has grown rapidly during the first year of its existence. The laboratory staff started with a core group of less than 100 employees located in the Central Design Group in Berkeley, California, and grew to over 600 employees located in the temporary quarters in Dallas, Texas. The challenge faced by the laboratory is how to combine the execution of a large scale construction project—with demanding cost, schedule and performance goals—with the creation of a laboratory infrastructure that will support and promote creative science activities and its associated educational benefits. Models of other scientifically oriented megaprojects, such as those in NASA, combined with the successful approaches used in the past in high energy physics projects provide guidance, but it is clear that new ground is being broken by the SSC.

The most recent organization chart is shown in Figure 1. The roles and responsibilities of the key leadership positions are as follows. The Director establishes the goals and policies for the laboratory, and he develops the strategic plan for the project and the laboratory in the context of the world high energy physics program. The Project Manager reports to the Director, and he is responsible for the execution of the construction project, consisting of the accelerator technical systems and the conventional facilities. The development and acquisition of the superconducting magnets, because

---

*Presented at the International Industrial Symposium on the Super Collider, Miami Beach, Florida, March 14–16, 1990.

†Operated by the Universities Research Association, Inc., for the U.S. Department of Energy under Contract No. DE-AC02-89ER40486.

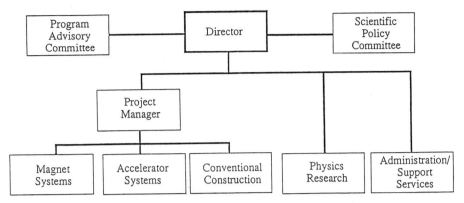

Figure 1. Organizational Chart

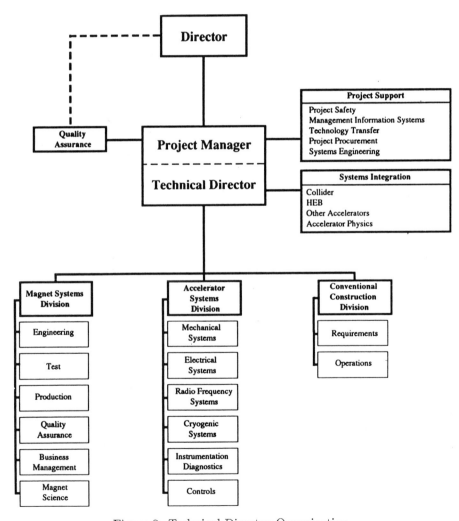

Figure 2. Technical Director Organization

of the magnitude of this accelerator component task, is the responsibility of a separate division (Magnet Systems). All other technical systems in the accelerator complex are the responsibility of the Accelerator Systems Division. The Conventional Construction Division is responsible for the conventional facilities, including the underground tunnels and the surface facilities.

A new feature of the present organization is the addition of a Technical Director supporting the Project Manager, as shown in more detail in Figure 2. The Technical Director, supported by system integration group leaders for the major portions of the accelerator, leads the physics design of the accelerator and has responsibility for all the technical requirements and interface definitions.

The Technical Services and Administrative Services Divisions provide support to the construction project, and they are also part of the laboratory infrastructure supporting scientific research. These scientific activities will, of course, become the mainstream of the laboratory when the construction is completed and the operating phase begins.

The Physics Research Division reports to the Director, separate from the construction project organization. The Physics Research Division is responsible for the detectors, and these components are designed and constructed by scientific collaborations involving many universities and other research organizations. The management and coordination of these collaborations is of a different character from the rest of the construction project, which is a more traditional project activity under the direct line authority of the SSCL project manager.

Schedules

The near-term priorities of the SSC project are:

1. Baseline cost/schedule/performance validation.

2. Development of industrial capability to produce large numbers of superconducting magnets.

3. System tests of a prototype collider segment by September 1992.

The development of the near-term schedule is driven by these priority items, which are critical to establish the readiness of the SSC project to move forward. The current definition of the baseline design of the SSC complex has been documented in a preliminary draft of the Site-Specific Conceptual Design, and a final version reflecting major design decisions is being prepared. Other papers at this symposium provide details of this design. The cost and schedule is also being finalized, and it will be subjected to a major review in early Summer 1990.

One major change from the earlier designs is the choice of aperture in the collider dipole magnets (CDM)–namely, 50 mm rather than 40 mm. This change led to restructuring of the superconducting magnet development. Strong support from Fermilab, Brookhaven, and Lawrence Berkeley Laboratories will continue (Figure 3), with the industrial contractor working in partnership with the laboratories after the selection has been made (Figure 4). Assembly of about a dozen collider dipole magnets by the industrial contractor at Fermilab is part of the technology transfer plan. These magnets, assembled by the industrial team, will be used to form a segment of the collider system for system testing by September 1992 (item 3 on the above list).

The first part of this "string test" will take place at a surface facility where five superconducting dipole magnets will be connected together. The major goals of the string test are:

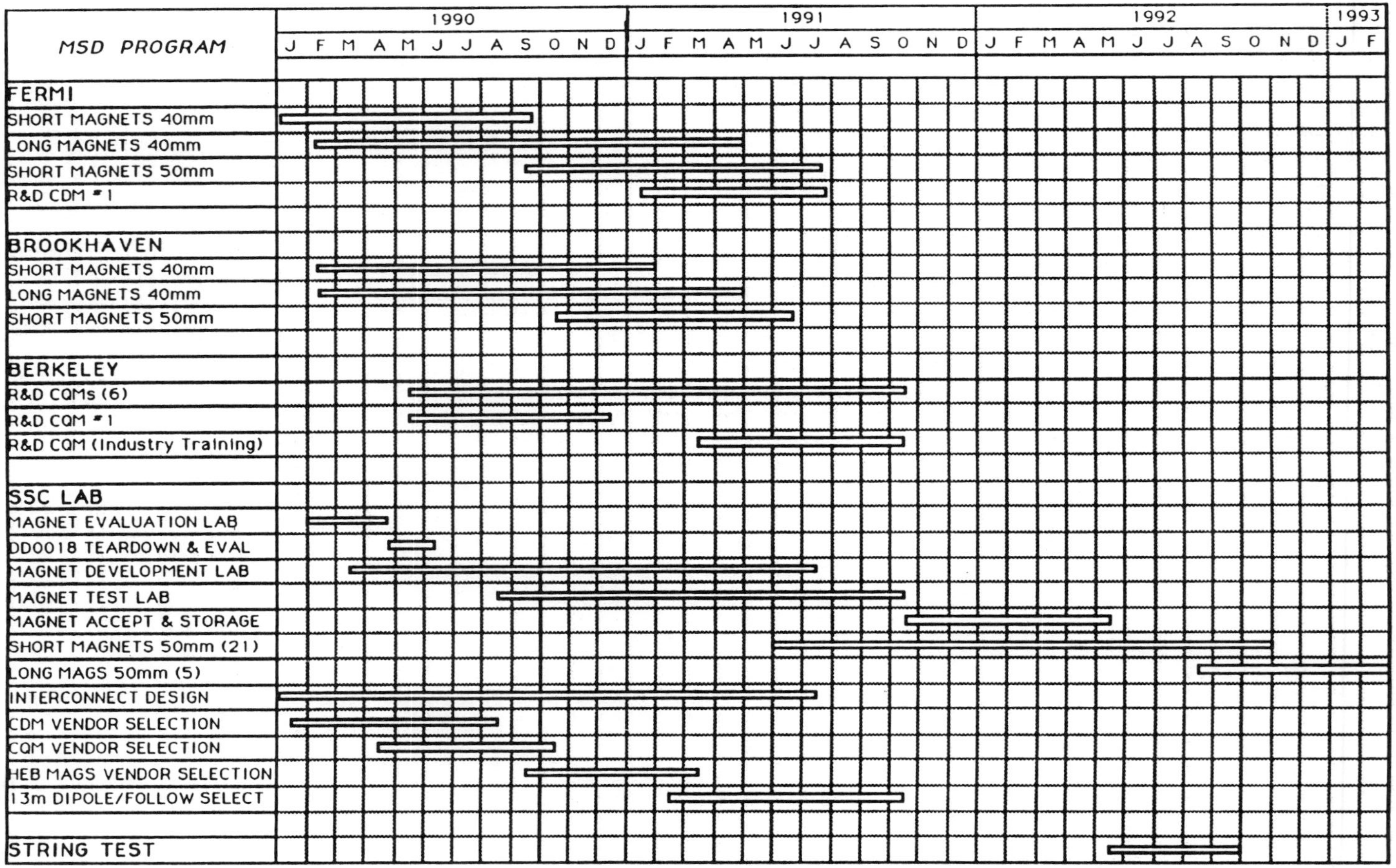
MSD PROGRAM
1990
1991
1992
1993
J F M A M J J A S O N D J F M A M J J A S O N D J F M A M J J A S O N D J F
FERMI
SHORT MAGNETS 40mm
LONG MAGNETS 40mm
SHORT MAGNETS 50mm
R&D CDM #1
BROOKHAVEN
SHORT MAGNETS 40mm
LONG MAGNETS 40mm
SHORT MAGNETS 50mm
BERKELEY
R&D CQMs (6)
R&D CQM #1
R&D CQM (Industry Training)
SSC LAB
MAGNET EVALUATION LAB
DD0018 TEARDOWN & EVAL
MAGNET DEVELOPMENT LAB
MAGNET TEST LAB
MAGNET ACCEPT & STORAGE
SHORT MAGNETS 50mm (21)
LONG MAGS 50mm (5)
INTERCONNECT DESIGN
CDM VENDOR SELECTION
CQM VENDOR SELECTION
HEB MAGS VENDOR SELECTION
13m DIPOLE/FOLLOW SELECT
STRING TEST

Figure 3. Lab Support

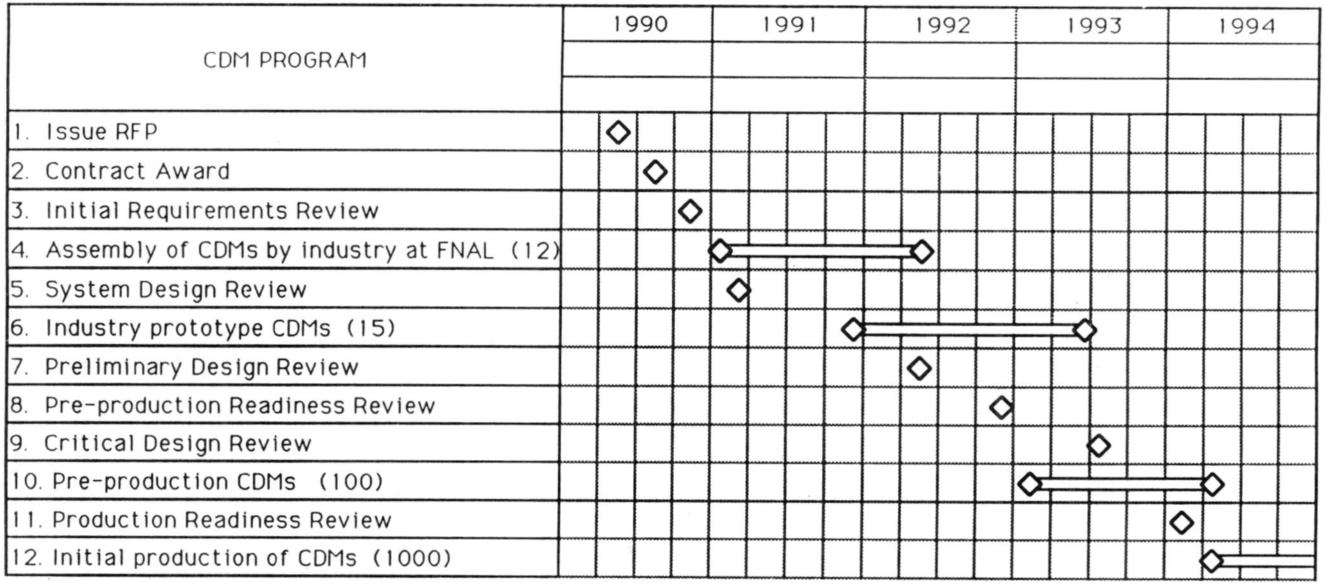

CDM PROGRAM	1990	1991	1992	1993	1994
1. Issue RFP	◇				
2. Contract Award	◇				
3. Initial Requirements Review	◇				
4. Assembly of CDMs by industry at FNAL (12)		◇———◇			
5. System Design Review		◇			
6. Industry prototype CDMs (15)			◇———◇		
7. Preliminary Design Review			◇		
8. Pre-production Readiness Review				◇	
9. Critical Design Review				◇	
10. Pre-production CDMs (100)				◇———◇	
11. Production Readiness Review					◇
12. Initial production of CDMs (1000)					◇

Figure 4. Industrial Contractor Selection Criteria

551

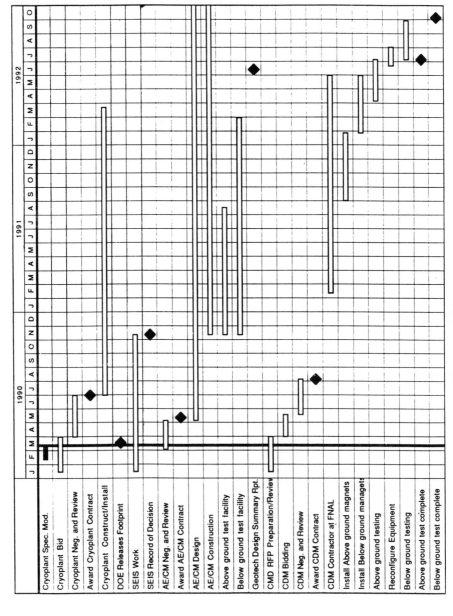

Figure 5. Schedule of Tasks

MAJOR MILESTONES FOR SSC	PRELIMINARY DATE
AE/CM Award	May 90
Baseline Validation Complete	July 90
Collider Dipole Magnet (CDM) Contract Award	Aug. 90
SEIS Record of Decision	Nov. 90
CDM Preliminary Design Review	July 92
Collider String Test	Sept. 92
First Collider Sector–Start Installation of Major Components	Jan. 94
Start CDM Production	April 94
Linac–Start Commissioning	Sept. 94
MEB–Start Commissioning	Oct. 95
HEB–Start Installation	Jan. 96
MEB–Test Beams Available	April 96
HEB–Start Commissioning	Oct. 97
SSC–Start Commissioning	April 98
SSC–Beam to Experiments	Oct. 98

Figure 6: Top-Level Milestone Chart

- The installation of the magnets to test the design of the magnet interconnects: electrical, cryogenic, and vacuum connections.

- The operation and integrity of the vacuum system.

- The ability to cool the string to liquid helium temperatures, verifying operation and control of the refrigerator system.

- The demonstration of the power supply, quench protection and controls systems, as well as the superconducting magnets, by ramping the string under conditions similar to those expected in operation.

These tests involve the first implementation of the major systems necessary to operate the collider and are the next important step in the SSCL program following successful tests of individual magnets. Following the surface facility tests, a similar test will be made below ground in a tunnel segment to test the operation of the support systems (cryogenics, electrical, and vacuum) in a realistic environment. The schedule of tasks involved in these system tests is shown in Figure 5. Results from these system tests are essential before the preproduction collider dipole manufacturing begins, and the production design of interconnections and spool pieces is finalized.

The top-level milestones over the full duration of the SSC construction project are shown in Figure 6. These milestones are preliminary versions; finalization of the baseline schedule will occur in late-Summer 1990 according to our present plan.

11. Materials and Conductors II

SUPERCONDUCTING CABLE FOR HERA

Stefan L. Wipf

Deutsches Elektronen-Synchrotron, DESY
Notkestrasse 85
D-2000 Hamburg 52

INTRODUCTION

HERA (Hadron Electron Ring Accelerator) consists of an accelerator
storage ring for electrons at 30 GeV and one for protons at 820 GeV, both
located in a tunnel of 6.3 km circumference, the proton ring being above the
electron ring[1]. The ring has four quadrants each with a 90° bend and a 360 m
straight section. The straight section can accomodate colliding regions for
the electrons and protons circulating in opposite directions. So far two
such regions are being equipped with experiments ZEUS and H1.

Along the curved sections are dipole magnets for bending and quadrupole
magnets for focussing[2]. For the proton ring these are superconducting, and
the required dipole field strength is 4.68 T. The total number of magnets
and the required superconducting cable quantities are listed in Table I.

The cable was procured from industry under contracts which were issued
at the end of 1984. Preparatory to the request for proposals was a period of
development and testing in collaboration with industry in order to
establish the state-of-the-art. The experience from Fermilab (Tevatron),
Brookhaven Natl. Lab and Saclay was also made available for drawing up the
specifications. The quadrupole cable was produced by VAC (Vacuumschmelze,

Table I. HERA superconducting cable requirements

	Necessary		Spare	Total
DIPOLES	416 (8 x 52)		29	453
	6 (vertical bending 3 collision zones)		2	
cable	874 km (1930 m/dipole)		24 km	898 km
QUADRUPOLES	224		22	246
cable	109 km		6 km	115 km

Germany), the dipole cables half each by ABB[3] (Asea Brown Bovery, Switzerland) and EM-LMI (Europa Metalli - La Metalli Industriali, Italy). ABB subcontracted the manufacturing to the Swiss superconductor consortium (Metalworks Dornach, Huber & Suhner, ISOLA Breitenbach, Swiss Metalworks SELVE). The three firms finished delivery of the cables by 1987, 1988, 1989 respectively.

I have been associated with this task at DESY since the end of 1987. Consequently this report covers work by many other people, notably S. Wolff who was in charge of the magnet development at DESY, A.F. Greene, BNL, who helped in setting up the specifications and introduced me to working with the manufacturers, M. Garber, BNL, who was responsible for the critical current measurements on all cable probes, H. Bathow, DESY (now retired), who set up the geometrical quality control devices and to Hongda Ma, visitor to DESY (1986-1988) from Academia Sinica, Beijing, who handled the quality control[4].

As to the contents of this paper:

After a brief discussion of basic considerations governing the specifications of the cable geometry and of the critical current at high fields come a few remarks on how cables are made and on the working principles of cabling machines.

The main topic will be quality control. First, the critical current of the dipole cables, followed by an example of a serious problem of cabling degradation, that is, a critical cable current smaller than the sum of the individual critical currents of the constituent wires. It was found that too much shear stress in cabling has to be avoided. Second, the control of the cable thickness and keystone angle. A surprising observation was an increase in critical current after a second pass through the roller dies to reduce a slight oversize.

The most important consideration is the successful performance of the magnets containing the cables. A secondary consideration of interest concerns the cost of the cables which at approx. 20 M$ represents 3-4 % of the total cost of the HERA project. A controlled improvement of only 1 % in performance of the cable is already eminently worthwhile.

BASIC CONSIDERATIONS

Specification: Geometry

The cross section of the dipole cable is specified by width and thickness of thick and thin edge, see Fig. 1. As these measurements are difficult to establish because of the rounding of the edges, an average thickness of 1.475 mm and a keystone angle of 2.234° were used as specifictions; the allowable error in the measured quantities has to be calculated from the specified errors (see below, Fig. 22).

The cable is made of 24 strands (23 strands for the quadrupole cable) with a copper superconductor ratio of 1.8 +0.2/-0.1 and filament diameter to be chosen between 12 and 16 μm. The chosen diameter to be kept within ± 1 μm for the whole production. The choice was 14 μm, 1230 filaments and 16 μm, 912 filaments for ABB and LMI, respectively. The strands are twisted to a filament twist pitch of 25 ± 3 mm.

A second twist pitch is for the strands in the cable. This is to be chosen between 75 and 95 mm and then to be kept within ± 2 mm. The cable is

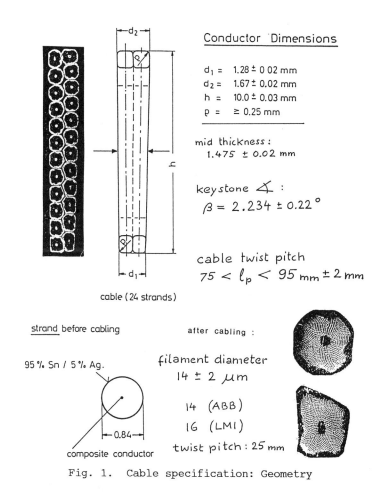

Conductor Dimensions

$d_1 = $ 1.28 ± 0 02 mm
$d_2 = $ 1.67 ± 0.02 mm
$h = $ 10.0 ± 0.03 mm
$\rho = $ ≥ 0.25 mm

mid thickness:
1.475 ± 0.02 mm

keystone ∢ :
$\beta = 2.234 \pm 0.22°$

cable twist pitch
$75 < \ell_p < 95 \text{ mm} \pm 2 \text{ mm}$

cable (24 strands)

strand before cabling

after cabling :

95 % Sn / 5 % Ag.

filament diameter
14 ± 2 μm

14 (ABB)

16 (LMI)

0.84

composite conductor

twist pitch : 25 mm

Fig. 1.   Cable specification: Geometry

formed as a lefthanded screw, the filaments in the strands have a right-
handed pitch. Thus the finished cable lies almost flat, with little inherent
twist, because the two opposite twist pitches balance each other.

The insulation of the cable consists of 0.025 mm thick Kapton tape,
12 mm wide, wrapped with 58 % overlap (right-hand spiral) and over this a
0.12 mm thick, 9 mm wide glass fibre tape impregnated with B-stage epoxy (to
be cured at 160 °C), wrapped as a lefthanded helix with 3 mm gap between
turns.

The dipole coil consists of inner and outer half coils, held together
by aluminum collars under a pre-stress of approx. 50 MPa (see Fig. 2).
During assembly of the coils, the pre-stress has been measured to be up to
110 MPa. The cable cross sectional tolerances allow a smallest keystone
angle which is correct for the inner coil, whereas the largest keystone
angle would be correct for a coil with an even smaller aperture (smaller by
8 mm in radius). Thus the danger is alleviated of crushing the insulation
between the narrow edges of the cable. Instead, the wide edges receive more
pre-stress and since they are less compacted they are capable of deforming
before the insulation is damaged. For these reasons the same cable is also
used for the outer coil where the keystone angle mismatch is even higher.

559

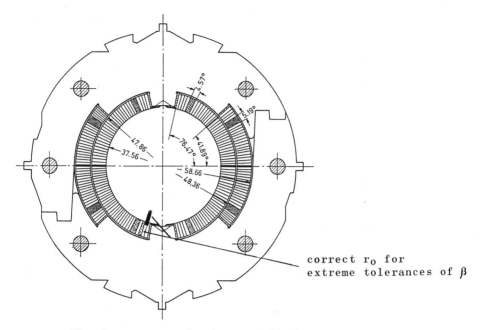

correct r_0 for
extreme tolerances of β

Fig. 2.   Superconducting HERA dipol coil with collar

The pre-stress is needed to fix the cable against any motions which would lead to a quench. Motion occurs more easily in the less compacted parts of the cable, another reason for preferring higher pre-stress near the wide edges.

It is easier to make cable with a less compacted narrow edge. Cabling degradation is usually caused by the severity of the narrow edge deformation. Limits for the widest tolerable narrow edge want careful reasoning, therefore. Pre-stressing of the winding in the finished coil is one of the important features which contribute to reaching a high operating current with little or no training, with a given cable.

Specifiction: Critical Current

At a temperature of 4.6 K and an applied field of 5.5 T the dipole cable must be able to carry at least 8000 A (Quadrupole cable 7000 A). It was agreed that for each cable length (usually 1910 m comprising 2 x 585 m and 2 x 372 m for the two inner and outer coils of a complete magnet) one piece should be tested. Because of the large currents and Lorentz forces such tests are not trivial[5] and require a massive rig as shown schematically in Fig. 3. The Lorentz force is large enough that a motion of 2 μm releases energy of > 1mJ per cm length of cable, sufficient for a temperature rise of several K. Therefore, a strong restraining force is necessary similar in size to the pre-stress in the coil. The bifilar arrangement leads to self fields which modify the applied field by approx. ± 6 %. The chosen arrangement has the field maximum at the narrow edge of the cable (as in Fig. 3). The specified current above refers to the critical current under these conditions, the test procedure being part of the specification. This means that the "true" critical current of the cable in a uniform field of 5.5 T may be higher by approx. 11 %.

There is no easily convincing way to correct for the non-uniform field[6,7]. The difficulty is in the transition between superconducting and

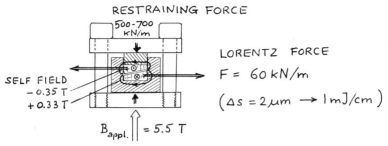

Fig. 3.   Method of measuring cable critical current

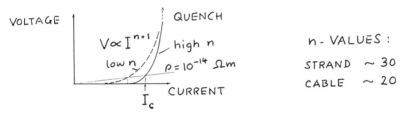

Fig. 4.   Definition of critical current and quality factor n

resistive state. The critical current is defined (somewhat arbitrarily) as the current where the resistivity $\rho = 10^{-14}$ Ωm. Resistivity and voltage are approximated by an exponential increase with current: $\rho \propto I^n$ and $V \propto I^{n+1}$ (see Fig. 4).

The exponent n is also known as the "quality factor" because among other things it depends on the quality of the filaments, such as sausaging, breaks and metallurgical defects. Cabling degradation is caused by damage to filaments, mostly at the narrow edge, and would reduce n. But n is also affected by flux creep and especially by non-uniformity of the field. (Typical n-values are: for strands ~30, for cable ~20).

The most important aim in making a good cable is to achieve the highest possible current density in the finished coil. The critical current density in filaments is determined by the state-of-the-art for flux pinning and has to be considered as given. Only two parameters affecting the critical current are still open for design decisions. These are the copper to super-conductor ratio (CSR) and the compaction of the cable. Both enter the over-all packing factor $\lambda = Asc/Awdg$ where Asc is the cross-section of all the superconducting filaments in the conductor and Awdg the cross-section available for the conductor, including superconductor, copper, insulation, voids and other materials, if any. Fig. 5 illustrates the relative impor-tance of filament critical current density jsc and λ. The usual accelerator dipole is a so-called cosϕ-magnet. The volume of superconductor needed, Vsc, is expressed in terms of the required field B_o, the aperture radius r_1 and λ and jsc. Because superconductor is the most expensive ingredient, Vsc is a good measure for the total cost of the magnet[8]. (Vsc = 4 NAsc, N: number of turns each half coil). Using the data for HERA, we learn that a 1 % increase in jsc leads to a 1.2 % decrease in Vsc, whereas a 1 % increase in λ gives only 0.2 % decrease in Vsc. That is, an increase in λ is 6 times less efficient than an increase in jsc. Increasing λ either by decreasing CSR or

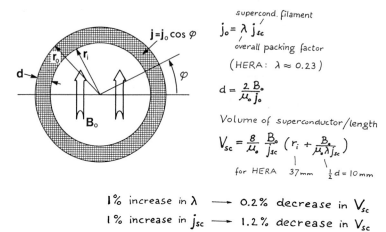

$$j = j_0 \cos \varphi$$

supercond. filament

$$j_0 = \lambda \, j_{sc}$$

overall packing factor

$$(HERA : \lambda \approx 0.23)$$

$$d = \frac{2}{\mu_0} \frac{B_0}{j_0}$$

Volume of superconductor/length

$$V_{sc} = \frac{8}{\mu_0} \frac{B_0}{j_{sc}} \left(r_i + \frac{B_0}{\mu_0 \lambda j_{sc}} \right)$$

for HERA   37mm $\frac{1}{2} d = 10$ mm

1% increase in λ ⟶ 0.2% decrease in V_{sc}

1% increase in j_{sc} ⟶ 1.2% decrease in V_{sc}

Fig. 5.   Importance of critical current density

by increasing cable compaction is dangerous: on the one hand lack of copper leads to instabilities, degraded current and training, while on the other hand cabling degradation is easily caused by compaction. Both instabilities and cabling degradation reduce jsc and hence directly negate any advantage of increased λ.

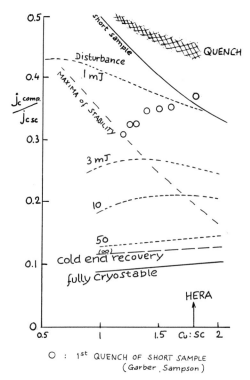

O : 1st QUENCH OF SHORT SAMPLE
(Garber, Sampson)

Fig. 6.   Copper: Superconductor ratio and stability

Copper to Superconductor Ratio (CSR)

The question is: How is stability and consequent reduction in operating current dependent on CSR? Theoretical and experimental results in Fig. 6 illustrate the trend. The ratio of critical current of the composite to that of the filament is plotted against CSR. Changing CSR from 2 to 1 increases the short sample current from 0.33 to 0.5 in the ratio 2 to 3 but the fully cryostable current drops from 30 % to 18 % of short sample current. The dotted lines give calculated limits for creating minimum propagating zones by localized deposition of the indicated "disturbance" energy in a conductor representative of a HERA cable in a dipole winding. These stability limits have maxima and it would be sensible in designing for a given disturbance to choose a CSR near the maximum (actually, to the right of the maximum, because copper is cheaper than superconductor). Between CSR 2 and 1 the disturbance energy of the maximum changes by a factor > 20. The design disturbance not being well known, a good safety factor is recommended, but this safety factor rapidly diminishes towards smaller CSR values.

The experimental points[10] are first quenches of cable samples measured according to the procedure in Fig. 3. The HERA sample, with a CSR of 1.8, reaches short sample. The others reach short sample performance after a number of training steps. This number increases with decreasing CSR to about 10 at 1.25; at CSR = 1.1 short sample performance can no longer be reached even after many training steps. (The difference between theory and experiment, not to be discussed here, may have several reasons: different disturbances, lack of computational sophistication, etc.).

This discussion demonstrates that, even with more accurate theoretical and experimental data on disturbances in dipole coils, the advantages of a move towards lower CSR will be marginal.

Cabling: Geometrical Considerations

Two things can be shown by elementary geometrical arguments: the limits to compaction and the unavoidable cabling degradation due to the cabling twist.

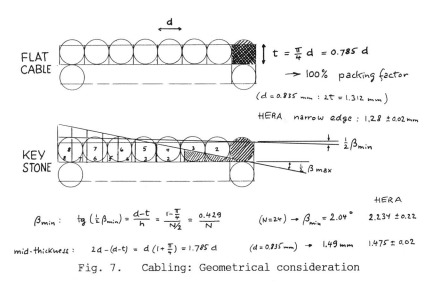

Fig. 7.   Cabling: Geometrical consideration

The assembled wires, before cabling, have a packing factor of 78.5 %. As a flat cable they can easily be compacted to 100 %, as seen in Fig. 7. For a keystoned cable one can define the minimum keystone angle ßmin as the one where the narrow edge is compacted to 100 % while the wide edge is still at 78.5 %.

The maximum angle is given by 100 % compaction throughout and narrow edge compressed to zero width. To achieve anything approaching ßmax also requires a compaction of the cable towards the wide edge; in the compaction process all the strands have to be moved towards the wide edge (see Fig. 7). In principle this is possible but would require a sophisticated effort not to damage the superconducting filaments in the process. In my opinion the effort would be wildly out of proportion to the possible gain. It is seen that the HERA specification is very close to ßmin.

It should be mentioned that there is always a small amount of lateral compaction at the wide edge because the uncompacted assembled strands require a width h + 2Δh*. This means that even at ßmin the strands are slightly deformed also at the wide edge and are, therefore, held by the precompression in the coil.

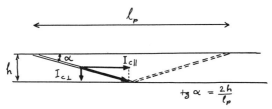

$$tg \, \alpha = \frac{2h}{l_p}$$

$$I_{c\,\|} = I_{c\,STRAND} \cdot \cos \alpha \quad \rightarrow \text{useful for dipol}$$

$$I_{c\,\perp} = I_{c\,STRAND} \cdot \sin \alpha$$

$$I_{c\,CABLE} = \sum_{}^{N} I_{c\,\|} = \cos \alpha \sum_{}^{N} I_{c\,STR.}$$

SUPERCONDUCTOR NEED

without cabling: $(l_p \rightarrow \infty)$ $\quad L_{CABLE} \cdot I_{c\,CABLE} = L_{CABLE} \cdot N \, I_{c\,STR.}$

with cabling: length of strands: $L_{CABLE}/\cos \alpha$
current $I_{c\,STR} = I_{c\,CABLE}/N\cos \alpha$ $\left. \right\}$ $\quad \dfrac{L_{CABLE} \cdot N \, I_{c\,STR.}}{\cos^2 \alpha}$

Spec. $\quad 75\,mm < l_p < 97\,mm$

$\quad 14.93° > \alpha > 11.65°$

$\quad 0.966 < \cos \alpha < 0.979 \rightarrow 2\%$ reduction of I_c

$\quad 1.071 > 1/\cos^2 \alpha > 1.0425 \rightarrow 4\%$ more superconductor

Fig. 8.   Cable twist pitch

---

* for N even: h + 2Δh = (N/2-1+√3)d; for HERA cable 2Δh = 0.69 mm, see Fig. 17 (top), and Fig. 18, cable cross section between 2.58 and 1.68 mm curves

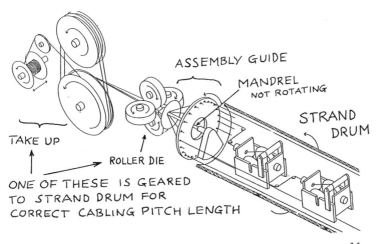

ASSEMBLY GUIDE

MANDREL
NOT ROTATING

STRAND
DRUM

TAKE UP

ROLLER DIE

ONE OF THESE IS GEARED
TO STRAND DRUM FOR
CORRECT CABLING PITCH LENGTH

Fig. 9.   Example of cabling machine (from R. Coombs[11])

The cable is twisted for two reasons: electrically the strands must be transposed against external fields; mechanically the cable must be able to bend (in both directions) without loosing its integrity.

The twisting (see Fig. 8) introduces a current component $I_{c\perp}$, not useful for producing the dipole field, that is, the useful component $I_{c\parallel}$ is reduced from I_c strand by the factor $\cos\alpha$. This is equivalent to a cabling degradation because the critical current of the strands has to be larger by $1/\cos\alpha$. There is a further penalty: the strands have to be longer than the cable by $1/\cos\alpha$. Naturally, the manufacturers chose the largest pitch length which the specification allowed.

Cabling Machines

There are many versions of cabling machines. For a specific cabling task specific solutions have to be developed, details of which are usually considered proprietary. However, all have the same basic functional components (see Fig. 9). A rotating strand drum accommodates N sufficiently large strand reels* and supplies the wire strands under equal controlled tension to an assembly guide which has to lead the N strands without cross-overs into the roller die. The strands must arrive at the rollers spread into a flat cross- section. The strand tension tends to collapse the assembled strands into a circular cross-section and means are needed to keep the correct shape. Usually, an internal mandrel ("core pin", "lamella") tapering to a shape like the head of a screw driver fulfills this function; it must extend as close as possible to the rollers and it must not rotate. Sometimes an external shroud serves a similar purpose. Naturally, assembly guide and roller die are most important and sensitive components influencing the quality of the cable.

It is also important that the forces are controlled. The pitch is given

---

* As a rule the strand reel holders are geared so as to keep the reel axis parallel to itself while the drum rotates, in order not to impart any extra twist[12] to the strands

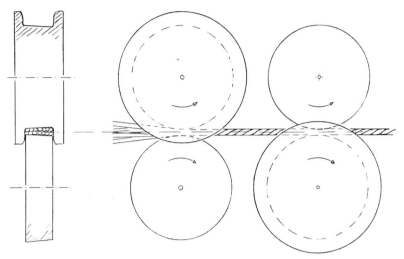

Fig. 10.   Two-roller set (UNK)

by either roller die or <u>take-up</u> being geared to strand drum; the advance per
rotation gives the pitch length. If both are geared the forces on the cable
are ill determined.

The roller die usually has the shape of a turk's head. But there are
other solutions, e.g. the two-roller set (see Fig. 10) is a two-stage die.
One roller pair shapes the upper face, the second one the lower, giving the
cable the final shape. I was told by a visitor from UNK that this arrangement produces excellent results.

QUALITY CONTROL

The accelerator operates most of the time as storage ring at high
field. It is, therefore, important that all the magnets perform equally well
at high fields - the critical current of the cable is very important. Also
important is the dimensional precision, to avoid disturbances by lack of
pre-stress (cable too thin) or damaged insulation by too much pre-stress
(too thick).

Quality Control: Critical Current

In order to obtain a narrow distribution of the cable critical current
- the goal was ± 2 % - the manufacturer was requested to select the strands
composing cables whose sum of the measured strand critical currents is as
nearly equal as possible. The effect is illustrated in Fig. 11.

In Fig. 12 is a plot of critical cable currents versus the cable identifying number (cables were sequentially numbered as they were produced *).

Two things are seen immediately:
1. There is a learning curve showing asymptotically increasing I_c
2. There is a case of increasing cabling degradation, indicated by the
   arrowed curve.

---

* Not each number represents the same cable length; some numbers have no
  critical current samples.

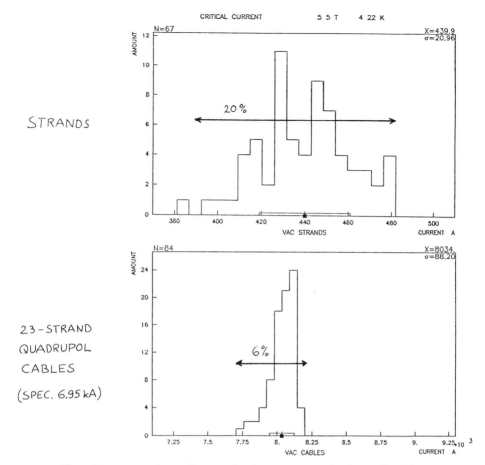

STRANDS

23-STRAND
QUADRUPOL
CABLES

(SPEC. 6.95 kA)

Fig. 11.   Sorting of strands improves variation of cables

The learning curve indicates progress in improving strand critical currents; indeed, the highest results are among the world leaders at the time, suggested in the insert by the bold open circle corresponding to a cable I_c = 9.2 kA.

The five points of progressively decreasing I_c were alarming. Microscopic analysis showed broken filaments at the narrow cable edge, a typical telltale of severe cabling degradation. It was found that the screws holding the mandrel in place had become loose, allowing the mandrel, which normally extends very close to the turk's head rollers, to jam between the roller and thus damaging the strands by excessive pull[3].

Usually the time lag between cable production and I_c measurement (at least 10 days) was too large to control the cabling machine by feed-back from the I_c results. In this case cabling was done in batches on a non-dedicated machine which was periodically re-arranged for other jobs. As luck would have it, the cable with the lowest I_c was the last in the production batch and the loose mandrel was noticed during re-equipment of the machine. The incident taught how better to control against cabling degradation, also contributing to the learning curve.

Is the opposite of cabling degradation possible, i.e. a critical current enhancement caused by the cabling process? The critical current of

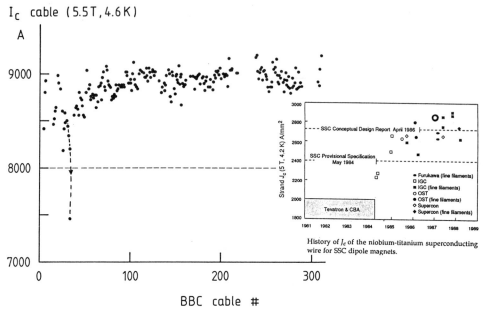

Fig. 12. Critical current of dipol cable (ABB)
Insert: o = j_{sc} (5 T, 4.2 K) of filaments in best cable

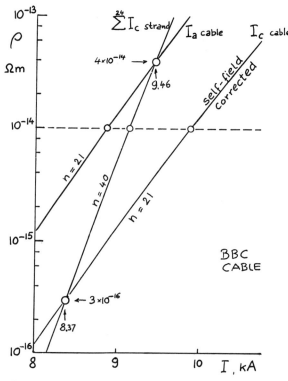

Fig. 13. Cabling degradation or enhancement?

a strand is increased by cold work up to a certain value; further cold work then reduces I_c again. By using strands with cold work just short of the optimum, cabling enhancement should be possible[13]. We think that examples with the highest cable I_c show such an effect, but it is difficult to prove. This has to do with the definition of the critical current.

An example[3] is illustrated in Fig. 13 where ρ vs I_{14} is given for the cable labelled as "I_a cable" (8.9 kA at specified 10^{-14} Ωm) as measured according to the procedure described above (Fig. 3) in the non-homogeneous field, composed of external field (5.5 T) and self field of measuring current; the curve fits $\rho \propto I^n$ with n = 21. The sum of the individually measured strand critical currents (9.1 kA at 10^{-14} Ωm) is also presented. We can shift the critical cable curve (I_c cable, self-field corrected, 9.9 kA at 10^{-14} Ωm) to indicate the situation where the narrow edge is at B = 5.5 T, the rest of the cable at lower field but otherwise equally non-homogeneous as for the I_a curve. Because the strands have n = 40, the strand critical current is equal to I_a at 4×10^{-14} Ωm and equal to I_c at 3×10^{-16} Ωm. Unequivocally, the critical current of the cable, defined as $\rho = 10^{-14}$ Ωm, is superior to the sum of the strand critical currents. But this does not mean that the strands have improved their flux pinning strength through the cabling process. The difficulty is that we compare two measurements, the one on the strands and the other on the cable, both with different field inhomogeneities for neither of which we can make a "correction".

It has been suggested that, rather than making the difficult cable sample measurement, the cable is taken apart and the individual strands be measured for comparison to the strand measurements before cabling. This may indeed be better. The change in mechanical stresses, especially if the deformed strand is bent into a straight shape, would introduce a non-negligible but easier to argue correction. So far we cannot give accurate qualitative values for cabling degradation or enhancement.

Other cable test arrangements have been suggested with seemingly better

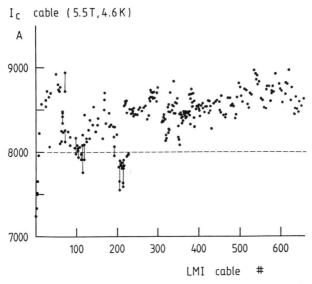

Fig. 14.   Critical current of dipol cable (LMI)

field distributions[14,15]. But none of these methods has been developed to the level of reliability and precision comparable to the BNL setup used here.

The learning curve for the other producer, (see Fig. 14) looks less tidy. At the beginning there were real problems. After a good start the critical current performance became erratic. The first question was: are the cable samples representative of the whole length? Points connected by vertical lines are measurements on different samples from the same cable. The scattering of these points is much too big considering the measurement error being rather smaller than ± 100 A. Obviously, something was out of control by comparison with the initial period where I_c values close to 9 kA were reached. After almost 1/3 of the production run this was a very serious problem.

Because of the time lag between cable production and I_c results, one needed criteria that could be checked, preferably on the shop floor.

Cabling Degradation and Pitch Length

Under the microscope the cabling degradation revealed itself as broken filaments near the narrow edge. For ordinary inspection, however, a degraded

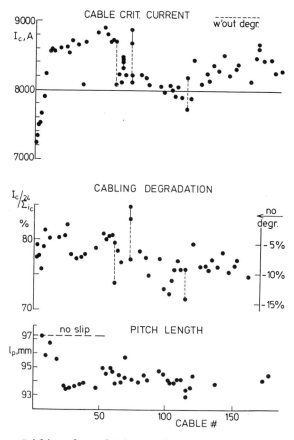

Fig. 15.   Cabling degradation and pitch length of LMI cable

and a healthy cable look more similar than two violins. Yet the violins may have a price difference a factor 10 or more and an expert can tell even without playing them; we want to develop similar expertise for cables.

Measurement of the pitch length gives a hint of a possible connection between pitch length and cabling degradation (see Fig. 15). The design pitch length is 97 mm, that is, the surface of the turk's head rollers moves forward by this amount per rotation of the strand drum. If the cable pitch length is shorter than 97 mm, the cable in effect "slips" back against the turk's head rollers. A relative measure for cabling degradation is the ratio between the cable critical current and the sum of the strand critical currents ($\sum^{24} i_c$, measured at 4.2 K, 5.5 T). Because of the difficulties discussed above, we calibrate the degradation by taking the highest I_c value as undegraded and obtain a degradation scale (on the right of Fig. 15) sufficient for the purpose of this discussion.

As a further peculiarity a "pitch anisotropy" was observed for the samples with a deficient pitch length. Normally, the pitch angle α should be the same on both sides of the cable. Here, α is larger on the upper surface where in rolling the as yet undeformed strands lead towards the narrow edge. On the other surface α is smaller. This is illustrated in Fig. 16 with the help of cable foot prints (made with carbon paper). The prints of upper and lower side are laid parallel to each other with the narrow edge in exact register. Equal sections are cut and exchanged. If α on the two surfaces is the same, the strands of the exchanged sections match accurately; by a mismatch, progressively worsening with distance from the narrow edge, even very small differences in α are accurately revealed.

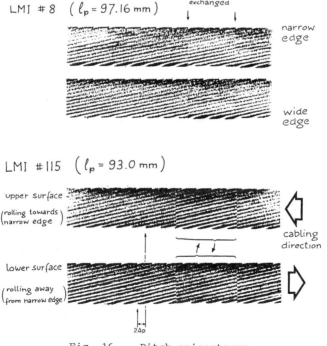

Fig. 16.   Pitch anisostropy

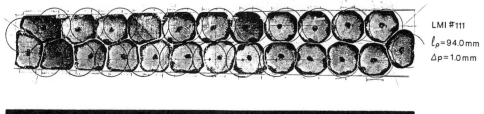

LMI #111

$l_p = 94.0\,mm$

$\Delta p = 1.0\,mm$

Fig. 17.   Cable cross-section with (above) and
without (below) pitch anisotrophy

Naturally, on the side with the smaller α the strands are squeezed together more than on the other side. This is seen clearly in a cross sectional micrograph in Fig. 17. For comparison, the cross section in Fig. 1 shows a symmetric pitch; the difference is most obvious by looking at parallel strands pairs at narrow and wide edge.

To understand what happens during deformation we need to cast an analytic look at the space between the turk's head rollers. In Fig. 18 the view is from inside the turk's head roller (as if it were transparent) onto the deformation zone. Lines of equal distance between the two conical roller surfaces are dashed parabolas (cone sections). The one marked 2.58 mm indicates the line beyond which the tip of the mandrel cannot be moved without starting to grab. The one marked 1.68 mm indicates the limit of the

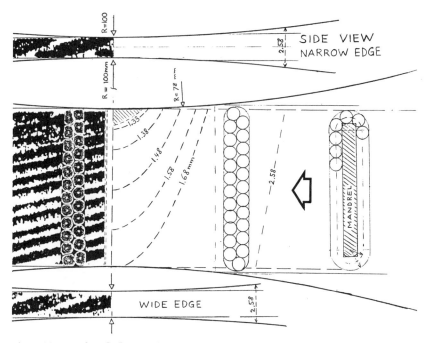

Fig. 18.   The deformation zone between the turk's head rollers

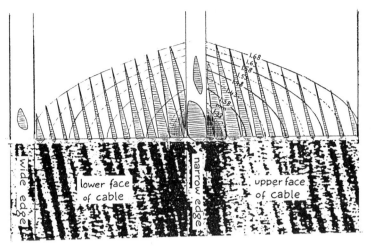

Fig. 19.   How the cable touches the rollers

deformation zone; between this line and the line of closest approach of the rollers (which is the actual die-cross section) all the deformation of the strands takes place. There follows a small zone; approx. 0.5 mm, of elastic relaxation, and the finished cable emerges, no longer contacting the rollers. The indicated foot print of the cable shows the side marked as "upper" in Fig. 17. The side rollers start to touch and thereafter squeeze the cable at a marked distance between the 2.58 and the 1.68 mm line.

Slip between cable and roller surface could only occur against frictional forces in the areas where the deforming cable touches the rollers. In these areas the pressure must be just above the yield pressure of copper ($\geq$ 200 MPa). The touching areas on all four sides of the cable are indicated in Fig. 19. The estimated total area is 30-40 mm^2. With a friction coefficient of 0.3, the driven turk's head rollers can, therefore, transmit a friction force in excess of 2 kN to the cable. Suppose this friction force is used to pull the cable against the strand tension from the strand drum spool brakes. As soon as the individual strand force exceeds the rather small friction force possible for strands near the wide edge of the cable, such strands will be pulled by forces transmitted from other strands downstream of the deformation zone. The resulting asymmetry of forces will create the described pitch asymmetry: the cable will be distorted in a way that the wide edge lags behind the narrow edge by a small amount Δp, or, the pitch angle on the upper face will be larger than on the lower face as can be seen from inspection of Fig. 19.*

Transient forces during stopping and starting of the machine deviated sufficiently from running forces to give unrepresentative test samples. These were often cut from the beginning or end of cables where the cabling machine was in transition between running and standing still. This most likely explains the large variation observed between samples from the same cable.

The largest force against this friction force is estimated at < 1800 N

---

* Pitch anisotropy can also have other reasons, such as alignment of the incoming cable relative to the turk's head roller (up-down, sideways, parallel or angular misalignment). Nevertheless, lack of pitch anisotropy or at least lack of variation along the cable of pitch anistropy are signs of good control of the many cabling variables.

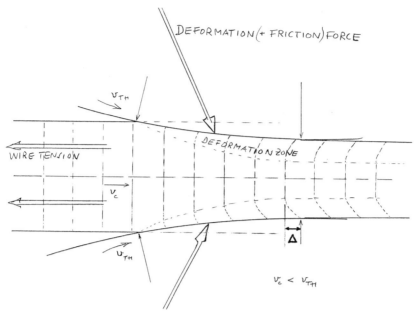

Fig. 20.   Deformation with longitudinal shear

consisting of the strand tension of up to 60 N per strand from the strand
drum brakes, the friction on the mandrel, approx. 100 N, and the horizontal
component of the deformation force, < 200 N. Normally this largest force is
reduced by the take-up force. It is unlikely, that the cable can slide
against the friction force on the turk's head roller surface.

The observed slip is not due to sliding, it is due to shear deformation. Longitudinal shear deformation* can explain both pitch length deficit

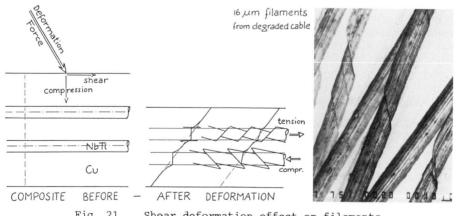

Fig. 21.   Shear deformation effect on filaments
           (SEM micrograph, M. Garber et al.[16])

---

* There is also the almost entirely transverse shear required for the
necessary plastic deformation by compression. Here we talk about an
additional, unneccessary shear deformation.

and cabling degradation. The deformation process under the turk's head roller is illustrated in Fig. 20. Adjacent to the surface in touch with the roller extends a zone in a state of plastic deformation. This deformation zone increases in depth up to the point of closest approach between the rollers where the finished cable leaves the die. If a friction force is to be transmitted, it creates a shear stress in the cable. In the deformation zone this shear stress leads to a shear deformation. The incoming wires in Fig. 20 are marked with equidistant dashed lines. These markers move with the cabling speed V_c through the rollers. It is seen that the roller surface moves faster to make up for the total shear Δ.

The design pitch length is given by the speed of the rollers; the cable pitch length is shorter by Δ/length of deformation zone (here about 5 mm) which is about 4 %, the pitch length being 93 instead of 97 mm, from this we conclude that $\Delta \approx 0.2$ mm.

Shear deformation can lead to reduced critical currents as explained with the help of Fig. 21. The dashed marker line would in the deformed state still be straight if the filaments were equally as soft as the Cu matrix. With the filaments being much harder and not capable of plastic deformation, the marker line within the filament remains as before and continues into the Cu matrix where a region of highly enhanced shear compensates for the missing filament deformation. The enhanced shear stresses at the filament surface eventually cause shear fractures with an appearance depending on tension or compression as indicated. Indeed in severely degraded samples the filaments near the narrow edge of the cable have a similar appearance prior to being completely broken.

Once the situation as discussed here was recognized, incoming strand tension and take-up force were more nearly balanced and thereafter the cabling degradation remained in the range of 6 ± 5 % leading to the results shown in Fig. 14 for numbers > 220. Towards the end (# > 500) cabling degradation remained better than 6 %.

Quality Control: Cable Geometry

The cable behaves sufficiently like a spring, so that in its relaxed state neither width nor thickness nor keystone angle correspond to the values valid for the cable under the stresses encountered in the winding. The cable width is most efficiently measured while it is under tension still in the cabling machine.

The thickness is measured under a pressure of 30 MPa*. Ten samples about 15 cm long cut adjacently from the cable samples are stacked, alternating thick and thin edge, between parallel steel brackets a distance h apart. The thickness is then measured to a precision of ± 2.2 μm or better.

The keystone angle is measured on a single 150 mm long sample under a pressure of 10.5 MPa. Again, the specimen fits into a precisely machined steel mold[4]. The measurement precision is better than ± 0.02°.

The thickness tolerances, given in Fig. 1, must be translated into

---

* Below approx. 10 MPa the thickness is pressure dependent because of springiness. During the first pressure cycle the stack of samples settles by 10-20 μm, after about 4 pressure cycles, the settling has become negligible and between 10 and 30 MPa we find an elastic deformation with an effective modulus of approx 30 GPa. This value compares with a tensile modulus of the cable of effectively ~ 80 GPa, and the modulus of the finished coil winding (compressive) of approx. 20 GPa.

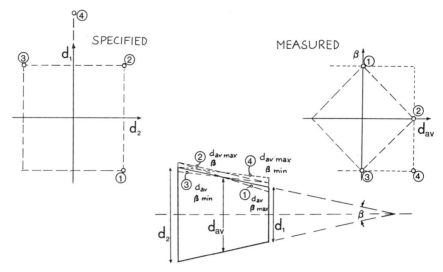

Fig. 22.   Geometrical tolerances

tolerances in the measured quantities average thickness, d_{av} and keystone angle, ß. This is demonstrated in Fig. 22. Tolerances of constant ± Δß and ± Δd$_{av}$ would allow cases such as Example 4 in Fig. 22, where the tolerance on the narrow edge is doubled. Instead the correct tolerances obey the

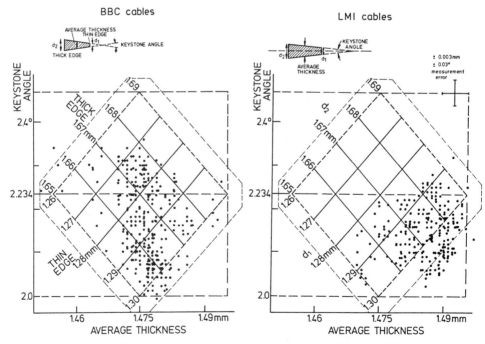

Fig. 23.   Measurements of all dipole cable samples

Table II. I_c enhancement after reducing thickness of narrow edge

Cable #	I_c before	I_c after	ΔI_c	n before	n after
399					
400	8520	9030 A	+ 510 A	19	24
419		8920			
420	8540	8990	+ 450	19	23
435	8270	8900	+ 630	18	22
466	8390	8950	+ 560	18	22
470	8440	8970	+ 530	19	23
476	8640	8930	+ 290	16	20
502	8760	8870	+ 110	21	23
527	8690	8910	+ 220	18	24
531	8680	8800	+ 120	19	22
I_{cav}	8548	8927			
σ	151	61			

equation $|\Delta\beta/0.2355°|+|\Delta d_{av}/0.02mm| \leq 1$. In practice the measurement error is added to the tolerance. Acceptable specimens have measurements within the border indicated by dashed lines.

The allowed error is larger than the actual measurement error. A re-measurement of the same sample even-re-stacked is routinely done with errors $< \pm 0.5$ μm in thickness and $< \pm 0.02$ in angle. Different samples cut from a 2 m piece of cable show variations in thickness and angle of $\lesssim 7$ μm and $\lesssim 0.1°*$. The thickness difference between the beginning and end of a cable rarely exceeds 10 μm. Such a difference, incidentally, corresponds to the effect of a temperature change of the turk's head rollers of 5 °C.

Displayed in Fig. 23 are all the thickness and angle measurements of the cables from the two manufacturer. Towards the end of the production both producers could cluster their values to better than ± 10 μm in both d_1 and d_2, but they both gradually shifted towards the maximum of d_1 and towards small values of β.

In one production many cables exceeded the tolerance for d_1 and had to be re-rolled before being acceptable. Such cables were treated like new cables, and new samples were subjected to the normal measurements. In many such cases, and quite against expectations, noticeably higher critical currents were measured, as listed in Table 2.

A detailed analysis of this effect has not been done yet. In the re-rolling process the cable came straight off the spool and therefore ran through the rollers in the opposite direction to the original cabling direction. The plausible assumption is, that some of the cabling degradation is reversible**.

---

* Difference between maximum and minimum of 10 measured samples; thickness samples measured in pairs of 2.
** The influence of stress on superconductors has been studied by Ekin[17], who finds a (reversible) shift in critical current of more than 30 % up to below the breaking stress. In the degradation process discussed above, local stresses close to breaking must be reached.

SUMMARY, CONCLUSIONS, RECOMMENDATIONS

After ending on an optimistic - if tentative - note, we summarize

- Industry produces superconducting cables of high quality. The high field critical current displayed in Fig. 12 over the whole production (not counting rejected cable) has an average value I_{cav} = 8874 A, standard deviation σ = 175 A; towards the end of the learning curve (cable # > 170) I_{cav} = 8970 A, σ = 95 A, a truly remarkable result. Similarly in Fig. 14 we find I_{av} = 8490 A , σ = 219 A and (cable # > 500) I_{cav} = 8730 A, σ = 118 A.

- Cable cross section profile, specified to ± 20 μm in thickness of narrow and wide edge, can be produced to even better tolerance. Once the cabling process was routinely established, tolerances close to ± 10 μm were reached.

- Neither critical current nor thickness and keystone angle measurements with the required precision are trivial. Well developed methods are needed. Many disputes can be avoided if specifications are related to measuring methods in agreement between producer and customer. This is particularly true for the critical current with its dependence on magnetic field variation and the "quality" exponent n, or keystone angle, which depends on pressure.

- Cabling degradation can be a serious problem. It is not sufficiently understood. In the case discussed here it seemed to be caused by excessive shear deformation.

- Shear deformation also caused a shortening of the cabling pitch length and a pitch angle difference between the two cable faces (pitch aniso-tropy), pitch length and pitch anisotropy can be checked along the cable and can be used to control the cabling machine and thus the cable quality.

- Critical cable current is important. The temptation to increase it by larger cabling compaction or by smaller copper/superconductor ratio is balanced by the danger of increased cabling degradation.

- A dedicated cabling machine is desirable. All the forces need to be controlled very well. The assembly guide should allow fine adjustment of alignment. Temperature of roller die should be controlled. The aim: to reduce variations along the cable sufficiently to make the test sample representative of the cable, and to reduce cabling degradation altogether.

Acknowledgement

It is a pleasure to thank my colleagues mentioned in the introduction for much help and useful critical comments. The thickness and angle measure-ments were made by D. Habercorn and F. Esume. I benefitted from many critical discussions with S. Rossi which aided in clarifying and focussing the technical arguments. My thanks go to S. Ceresara, G. Donati, R. K. Maix and D. Salathé for fruitful collaboration with industry.

References

1. B.H. Wiik, Progress with HERA, IEEE Trans.Nucl.Sci. NS-32: 1587-91(1985)
2. S. Wolff, Superconducting HERA Magnets, IEEE Trans.Magn.24: 719-22(1988)
   S. Wolff, The Superconducting Magnet System for HERA, Proc. 9th Int.

Conf. Magnet Technology, MT-9,1985, pp. 62-67, C. Marinucci, P. Weymuth ed., SIN, Zürich

3. R.K. Maix, D. Salathé, S.L. Wipf, M. Garber, Manufacture and Testing of 465 km Superconducting Cable for the HERA Dipole Magnets, IEEE Trans.Magn.25: 1656-1659 (1989)

4. Hongda Ma, The Quality Control of the Superconducting Cable in the HERA Project, Proc.Int.Low.Temp.Mat.Conf., June 1988, Shen-yang, China. also: Report DESY HERA 88-01, Jan. 1988

5. M. Garber, W.B. Sampson, M.J. Tannenbaum, Critical Current Measurements on Superconducting Cables, IEEE Trans. Magn., MAG-19: 720-723 (1983) updated and more detailed: M. Garber, W.B. Sampson, Test Methods for Cable Critical Currents and Normal State Resistance, Report AD/SSC/Techn. No. 70, March 3, 1988 (15 p.)

6. M. Garber, A.K. Ghosh, W.B. Sampson, The Effect of Self Field on the Critical Current Determination of Multifilamentary Superconductors, IEEE Trans.Magn.25: 1940-44 (1989)

7. L.F. Goodrich, S.L. Bray, Current Capacity Degradation in Superconducting Cable Strands, IEEE Trans.Magn. 25: 1949-52 (1989)

8. S.L. Wipf, Dipole Aperture and Superconductor Requirements, pp. 167-170, in "Accelerator Physics Issues for a Superconducting Super Collider", ed. M. Tigner, Univ. Michigan, Ann Arbor (1984), UMHE 84-1 S.L. Wipf, High Field Dipoles for Future Accelerators, LA-10219-MS Sept. 1984 (45 p.)

9. S.L. Wipf, in Proc. of Informal Workshop on Superconductor Stability, ed. A. Greene, P. Thompson, BNL Magnet Div. Notes, 268-19 (RHIC-MD-72) Apr. 1988 (unpublished). See also LA-7275, (1978)

10. W.B. Sampson, private comm. (1988); see also A.K. Ghosh, M. Garber, K.E. Robins, W.B. Sampson, Training in Test Samples of Superconducting Cables for Accelerator Magnets, IEEE Trans.Magn.25: 1831-34 (1989)

11. R.C. Coombs, Cabling Machine Modification to Manufacture 23 Strand "Keystoned" Superconducting Cable, RAL Report SMR/50

12. J. Grisel, J.M. Royet, R.M. Scanlan, R. Armer, A Unique Cabling Machine Designed to Produce Rutherford-Type Superconducting Cable for the SSC Project, IEEE Trans.Magn.25: 1608-10 (1989)

13. R.K. Maix, D. Salathé, Practical Scaling Formulae for the Determination of Critical Currents of NbTi Superconductors, Proc MT-9 (loc. cit. ref. 2) pp. 535-538

14. P. Fabbricatore, A. Matrone, A Parodi, R. Parodi, C. Salvo and R. Vaccarone, A Multiple Sample Holder for J_c Measurements on HERA Cables, Proc. ICEC 12, R.G. Scurlock, C.A. Bailey, ed. Butterworths, Guildford (1988) pp. 903-907

15. H.H.J. ten Kate, H. Boschmann, B. ten Haken, L.J.M. van de Klundert, A Test Facility for High-current Superconducting Cables up to 25 kA at 7 Tesla, in Proceedings of ICEC-11, 1985, pp. 500-504

16. M. Garber, M. Suenaga, W.B. Sampson, R.L. Sabatini, Critical Current Studies on Fine Filamentary NbTi Accelerator Wires, Adv. Cryog. Eng. Mat. 32: 707-714 (1986)

17. J.W. Ekin, Mechanical Properties and Strain Effects in Superconductors, in "Superconductor Materials Science. Metallurgy, Fabrication and Applications". S. Foner, B.B. Schwartz ed. Plenum, N. York (1981) pp. 455-510

DESIGN OF MULTIFILAMENTARY STRAND FOR SUPERCONDUCTING

SUPERCOLLIDER (SSC) APPLICATIONS -- REDUCTION OF MAGNETIZATIONS

DUE TO PROXIMITY EFFECT AND PERSISTENT CURRENT

E. W. Collings*, K.R. Marken Jr., and M. D. Sumption*

Battelle Memorial Institute, Columbus, Ohio, U.S.A.
* Also affiliated with Ohio University, Athens, Ohio, U.S.A.

ABSTRACT

Nonsuperconducting saddle magnets can in principle be designed to produce an undistorted dipolar magnetic field. But if the coils are wound from superconducting strands, residual magnetization, M_R (i.e. "persistent current") resident in the filaments is responsible for multipolar distortions of the desired field. Recognizing that the height (or thickness) of the M(H) hysteresis loop -- $\Delta M(H) \equiv (M_{R+} - M_{R^-})$ -- is proportional to the product of critical current density, $J_c(H)$, and the filament diameter, d, there is a strong interest in producing, on a commercial scale, multifilamentary strands with smaller and smaller filaments. In order to preserve filament quality in small filaments, some authors have deemed it necessary to confine the ratio of filament spacing (s) to filament diameter (d) to s/d $\leq$ 0.15±0.02. The combination of small d with low s/d results in interfilamentary spacings sufficiently close to proximity-effect couple the filaments. But if the interfilamentary matrix is alloyed with ~0.5 wt.% Mn, coupling is barely perceptible even with 1 μm diameter filaments. But having disposed of excess, interfilamentary, magnetization one is still faced with the intrafilamentary magnetization of the NbTi filaments themselves. Since during the operating cycle (field-increasing) of the SSC magnet this magnetization is diamagnetic (throughout most of the winding) it can be neutralized by including in the superconducting strand a material with a large positive magnetization, such as Ni. Possible methods of deploying and administering the Ni are discussed.

INTRODUCTION

Nonsuperconducting saddle magnets can in principle be designed to produce an undistorted dipolar magnetic field. But if the coils are wound from superconducting strands, residual magnetization, M_R (i.e. "persistent current") resident in the filaments contributes a multipolar distortion to the desired field. Recognizing that the height (or thickness) of the M(H) hysteresis loop -- $\Delta M(H) \equiv (M_{R+} - M_{R^-})$ -- is proportional to the product of critical current density, $J_c(H)$, and the filament diameter, d, there is a strong interest in producing, on a commercial scale, multifilamentary strands with smaller and smaller filaments. In order to preserve filament quality in small filaments, it has been suggested necessary to confine the ratio of filament spacing (s) to filament diameter (d) to s/d $\leq$ 0.15±0.02. The combination of small d with low s/d results in interfilamentary spacings sufficiently close to proximity-effect couple the filaments. The proximity effect contributes an unwanted excess magnetization to at least a portion of the M(H) hysteresis loop thereby counteracting to some extent the advantage that would otherwise accrue from taking the strand through its final stages of reduction[1,2,3].

It has been shown that interfilamentary coupling can be suppressed by alloying the matrix with a low concentration of Mn[4]. In the strands prepared for this study, an interfilamentary alloy of Cu-0.5wt.% Mn was used. As a result, filaments as small as d = 2.5 μm (hence s = 0.5 μm, at the design s/d of 0.19 in this case[3]) were successfully decoupled. But even in the absence of proximity-effect coupling one is still faced with the inherent magnetization of the NbTi filaments themselves. During the operating cycle (field-increasing) of the SSC magnet, this magnetization is mostly diamagnetic; accordingly it can be neutralized by associating the superconducting strand with a material of large positive magnetization, such as Ni. To be sure, Ni barriers have been incorporated into multifilamentary strands to eliminate proximity-effect interfilamentary coupling[5], and bulk Ni inserts have been recommended for magnetization compensation in SSC dipoles[6], but the present idea of associating Ni directly with the strand for local magnetization compensation is new.

EXPERIMENTAL

Magnetization Measurements

Magnetization was measured as function of temperature up to the T_c of NbTi with field sweep amplitudes of from a few tens of gauss up to 15 kG. A computerized PAR-EG&G vibrating-sample magnetometer (VSM) was used, in association with a 17 kG iron-core electromagnet powered by a $\pm$ 65 A field-controlled bipolar power supply. In completing a full hysteresis loop, including the initial branch from the origin, the instrument records 1,023 data pairs. Thus the field resolution in any experiment is about 1/200th of the field-sweep amplitude, which enables all fine structure associated with coupling magnetization to be fully recorded.

Materials and Magnetization Samples

Multifilamentary Cu-base/NbTi composites were prepared by commercial vendors from Nb-Ti rods clad with a thin barrier-layer of Nb (whose presence was ignored in the data analyses). Three groups of strands were studied -- Series-I: A Cu-matrix, 6,000-filament series, with s/d $\simeq$ 0.15, designated RHIC (indicating a class of strand intended for *Relativistic Heavy Ion Collider* application). Series-II: An 23,000-filament series, designated CMN, with an *interfilamentary* Cu-0.5wt.%Mn matrix and an s/d of 0.19 [3]. Strand-III: An 11,000-filament strand which was selected for the Ni-compensation study. In each case the strand design called for an annular filamentary bundle, encased in Cu and surrounding a Cu core; see, for example, Figure 1. Further details concerning the sample materials are listed in Table 1.

Magnetometer-Sample Preparation

The samples for magnetization measurement consisted of cylindrical bundles, about 3 mm in diameter and 6 mm in length, of parallel multifilamentary strands imbedded in epoxy. Depending on the strand diameter, the number of strands in the bundle varied from 1 (CMN-115) to 200 (CMN-5) so as to keep the volume of superconductor roughly constant at about 0.01 cm[3]. In the present study, the sample orientation was axis-normal to the applied field. In general it was customary to prepare two sets of samples: one consisting of the as-received composite strands, and the other of bare NbTi filaments -- all the Cu having been removed by etching. By comparing the magnetizations of the clad and bare materials, the influence of proximity effect could be clearly observed.

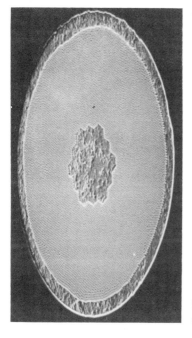
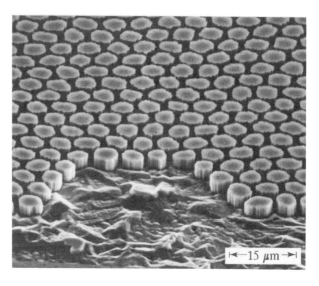

Figure 1.
Scanning electron micrograph of SSCNI. The source
of the strand was IGC Advanced Superconductors, Inc.

PROXIMITY EFFECT

If a superconductor, s, is in intimate contact with a normal conductor, n, superelectron pairs will leak through the s/n interface. The probability of finding a superelectron pair in n at a distance x from the interface is

$$P \; \alpha \; \exp(-k_n x) \tag{1}$$

where, for high-conductivity (so-called "clean") normal conductors, k_n^{-1}, the characteristic decay length, is given by

$$k_n^{-1} = \bar{h} \, v_f/2\pi k_B T \tag{2}$$

in which $\bar{h} \equiv h/2\pi$ and v_f is the Fermi velocity in n. The latter is judged to be clean if its electronic mean-free-path, mfp, ℓ, is very much greater than a "coherence length", ξ_n, given by

$$\xi_n = (\bar{h} \, v_f \ell/6\pi k_B T)^{1/2} \tag{3}$$

Consider "pure" unalloyed Cu. At 4.2 K the coherence lengths of coppers with residual resistance ratios (RRR) of from 75 to 200 lie between 0.69 and 1.13 μm; the corresponding mfps are 3.15 to 8.45 μm, respectively. By the above criterion these coppers would be regarded as clean with decay lengths as prescribed by Eqn. (2), which for T = 4.2 K, yields k_n^{-1} = 0.45 μm. Thus, for example, in Cu/NbTi composites stabilized by Cu of normal purity, the filaments will be coupled by proximity effect as soon as the interfilamentary spacing drops below about 0.9 μm; i.e. when the filament diameter (at s/d = 0.15) decreases below about 6 μm.

Four factors will reduce, and eventually destroy, proximity effect coupling: (i) an increase in the temperature; (ii) an increase of the applied magnetic field strength; (iii) a decrease of the matrix mfp, ℓ; (iv) a decoupling of the superelectron pairs in the matrix through local-moment scattering.

Table 1 Specifications of the Three Classes of Sample

Sample Code	Strand Diam.,D, 10^{-2} cm	Fil. Diam.,d*, μm	Magnetization Test Sample**		
			No. of Strands	Length, mm	NbTi Filament Volume*, 10^{-3}cm^3
Series-I Cu-Matrix Strands: 6,108 filaments (heat treated)					
RHIC-009	2.54	2.106	80	6.31	10.74
RHIC-013	3.47	2.890	48	6.31	12.13
RHIC-026	6.57	5.490	14	6.44	13.03
Series-II Cu-0.5wt.%Mn-Matrix Strands: 22,902 filaments (not heat treated)					
CMN-5	1.27	0.500	200	5.88	5.28
CMN-10	2.75	1.068	58	5.74	6.83
CMN-15	3.85	1.495	33	5.60	7.43
CMN-21	5.29	2.051	21	6.15	9.77
CMN-25	6.35	2.459	15	5.64	9.20
CMN-115	29.99	11.556	1	5.892	14.15
Strand-III Cu-matrix strand: 11,000 filaments†					
SSCNI††	8.23	5†	8	4.98	8.35

*   Obtained by etching-and-weighing using separately measured density of bulk Nb-46.5Ti (= 6.097).

** Referring to clad samples only.

†   Nominal.

†† Data supplied by H. Kanithi of IGC Advanced Superconductors, Inc, the source of the strand.

Nonmagnetic Scattering

Decrease of the matrix mfp can be achieved by solid-solution alloying; the Cu is then characterized as "dirty", in which case k_n^{-1} is identical to ξ_n. Thus in response to alloying with nonmagnetic solutes

$$k_n^{-1} = 9.96 \times 10^{-15}/\sqrt{\rho_{\Omega cm}} \tag{4}$$

which in the case of Ni (at concentrations below the ferromagnetic limit) can be written

$$k_n^{-1} = 9.50 \times 10^{-2}/\sqrt{c} \qquad (\mu m) \tag{5}$$

where c is the Ni concentration in at.% (approximately equal to wt.% when the Ni is dissolved in Cu). By way of illustration, let us assume that s = 0.1 μm and it has been decided to reduce k_n^{-1} to one-third of s/2, hence to 0.017 μm. Eqn. (5) indicates that the amount of Ni needed would be 31 at.%.

Magnetic Scattering

It is well known by now[7] that localized magnetic moments are potent decouplers of superelectron pairs. Thus Mn dissolved in Zn reduced its T_c at the rate of about 300 K/at.%. Let us assume that the same rate also holds for Mn in Cu, and that $T_{c,Cu}$ itself is 0.07 K[8]. It then follows according to arguments advanced in Ref. 4 that the above-mentioned decay length of 0.017 μm can be achieved in Cu by alloying it with 0.9 at.% of Mn. The coupling-reduction effectiveness of Mn will be demonstrated below.

Relative Coupling-Reduction Efficiencies of Mn and Ni

In order to establish a decay distance of $k_n^{-1} = 0.017$ μm in Cu, 31 at.% of Ni was required as compared to 0.9 at.% for Mn. Thus, on an at.% basis, Mn is 30 times more potent than Ni. Furthermore, taking into account the specific scattering strengths, in Cu, of Ni ions (viz. 1.1×10^{-6} Ωcm/at.%) and Mn ions (viz. 2.9×10^{-6} Ωcm/at.%) it follows that for a given reduction in coupling the penalty, in terms of matrix resistivity, is a factor of 13 less if Mn is the solute rather than Ni.

VISUALIZATION OF PROXIMITY-EFFECT COUPLING

It is well known that the total height of the magnetization loop, $\Delta M_v(H)$ (emu/cm^3 of the NbTi component of the composite) is related to the filament's critical current density, J_c (A/cm^2), and diameter, d (cm), by[9]

$$\Delta M_v(H) = (0.4/3\pi) J_c d \qquad (6)$$

It follows that any magnetization enhancement due to proximity effect coupling can be expressed in terms of an effective filament diameter, say d_{eff}[10], which is larger than the actual diameter, d, as indicated in Figure 2. The onset of the $d_{eff} > d$ regime can be visualized through application of either of the following two methods:

The $\Delta M_v/J_c$ Method

From Eqn. (6) a plot of $\Delta M_v/J_c$ against d, in the absence of coupling, should be linear through the origin. Otherwise at small values of d, positive departures from the line should be observed, heralding the onset of coupling[1]. If it can be assumed that J_c is independent of d over the range concerned, the same should be true for a plot of ΔM_v *versus* d, as indicated in Figure 2.

The $\Delta M_{v,clad}/\Delta M_{v,bare}$ Method

As insurance against the possible variation of J_c with d, and then to avoid the difficulty of having to measure a possibly large low-field J_c, the $\Delta M_{v,clad}/\Delta M_{v,bare}$ method was devised. Referring again to Eqn. (6), and after associating $\Delta M_{v,clad}$ with a coupling-enhanced d_{eff}, we obtain simply that $\Delta M_{v,clad}/\Delta M_{v,bare} = d_{eff}/d$, the latter being constant and close to 1 when coupling is absent. With decreasing d, departure from the constant value indicates the onset of coupling. Figure 3 shows that, for the series of Cu-matrix RHIC-type strands, coupling sets in below d = 5.5 μm (i.e. s $\simeq$ 0.8 μm).

Figure 3 also demonstrates that, for the CMN series of strands with the Cu-0.5 wt.%Mn interfilamentary matrix, d needs to be reduced below 1.1 μm (s $\leq$ 0.2 μm, in this case) before we begin to see the effect of coupling.

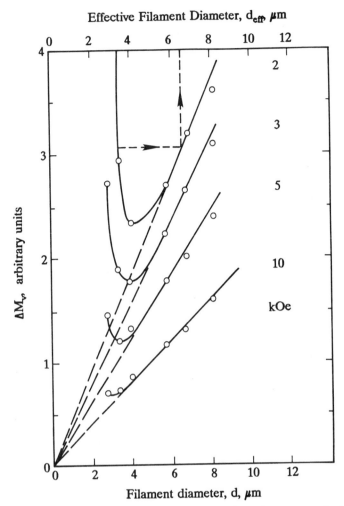

Effective Filament Diameter, d_{eff}, μm

ΔM_v, arbitrary units

Filament diameter, d, μm

Figure 2.   Height of the M(H) hysteresis loop plotted *versus* nominal filament diameter, d -- derived from Reference 1 (Ghosh *et al.*).

MAGNETIZATION COMPENSATION

Introduction

It has been shown that in the filament range of relevance to the SSC, proximity effect coupling can be eliminated through the incorporation of Cu-Mn as interfilamentary matrix. But as indicated elsewhere[11,12], practically all of the coupling magnetization shows up along the field-decreasing or *trapping* branches of M(H), whereas the SSC magnet is nominally operated along the field-increasing or *shielding* branch. Nevertheless in the SSC dipole, even when the central field has been increased up to its injection value, a decreasing field is still being experienced by some parts of the winding. The magnetization of these components thereby remains susceptible to proximity-effect enhancement unless the interfilamentary spacing is sufficiently large or Mn is alloyed into the interfilamentary Cu. In either case, having disposed of the coupling problem one is still faced with the inherent magnetization of the filaments themselves.

586

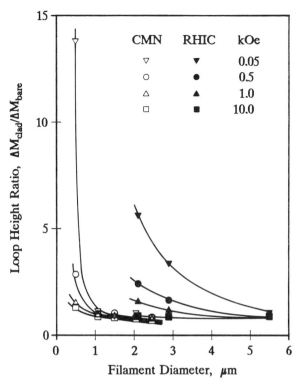

Figure 3.   Relative heights of the M(H) loops at 4.2 K and the fields specified for clad and unclad (bare) samples of RHIC and CMN, respectively, strands.

Preliminary Compensation Study

    Since during the operation of the SSC magnet this magnetization is diamagnetic over most of the winding, it can be compensated locally by associating the strand with a small volume-fraction of ferromagnetic material such as Ni. In our preliminary study the magnetization of a typical sample was compensated at a field of about 4-5 kG by a short length of Ni wire.

CONCEPTUAL DESIGN OF MAGNETIZATION-COMPENSATED RHIC AND CMN STRANDS

    Data needed to magnetically compensate RHIC and CMN strands in fields of 3 to 4 kG are provided in Table 2(a).   The strand magnetizations referred to have been obtained from the field-increasing branches of their M(H)s; the data for Ni were acquired during a magnetization measurement at 10 K [11,12]. Listed in Table 2(b) are: (i) the volume percentages of Ni that need to be associated with the strand for compensation within 3-4 kG; (ii) the thickness of a ferromagnetic Ni layer to be plated onto the surface of the strand for compensation at 3.3 kG; (iii) the number of Ni filaments that would need to be included in the strand, again for compensation at 3.3 kG, should it be decided to replace some of the NbTi filaments with Ni.

    Since Table 2 was first published[11], Hitachi Cable Ltd has demonstrated the technical feasibility of the Replacement-Filament Method [Item (iii) above] both from a metalworking standpoint and as a method of magnetization compensation. A paper in this proceedings, authored by G. Iwaki et al., describes the configuration of a Ni-filament-compensated strand and the magnetic results obtained from it.

Table 2. Conceptual Design of Magnetization-Compensated Strands
Based on Series-I and Series-II Starting Materials

(a) Specific Magnetization of Filamentary NbTi (at 4.2 K) and Ni (at 10 K)

Sample Material	Strength of the *Increasing* Applied Field, kG		
	3.0	3.3	4.0

Magnetization of NbTi, M_{SC}, emu/cm^3

RHIC-009	-10.758	-10.176	- 9.124
RHIC-013	-13.073	-12.391	-11.185
RHIC-026	-20.881	-19.917	-18.164
CMN-5	- 6.366	- 6.036	- 5.178
CMN-10	- 6.223	- 5.796	- 5.048
CMN-15	- 6.865	- 6.472	- 5.685

Magnetization of Ni, M_{add}, emu/cm^3 *

Ni	+ 467.4	+ 478.1	+ 496.8

(b) Volume Percentage and Actual Volume of Ni Needed for Compensation
at Various Fields

Sample Code	Vol. Pct. Ni, $100R_C \equiv 100A_{add}/A_{SC}$**			No. of Ni Filaments†	Thickness of plating†† t, μm
	3.0 kG	3.3 kG	4.0 kG		
RHIC-009	2.302	2.128	1.837	127	0.6
RHIC-013	2.797	2.592	2.251	154	1.0
RHIC-026	4.468	4.166	3.656	244	3.0
CMN-5	1.362	1.263	1.042	286	0.1
CMN-10	1.331	1.212	1.016	274	0.3
CMN-15	1.469	1.354	1.144	306	0.5

*   In normalization to unit volume, a density of 9.04 was taken.

**   If M represents a material's specific magnetization and A its cross-sectional area, while subscripts "SC" and "add" denote NbTi and the compensating addendum (i.e. Ni), then the fractional amounts of addendum material required for compensation are simply $R_C = A_{add}/A_{SC} = -M_{SC}/M_{add}$.

†   Number of Ni-replaced NbTi filaments, for a total of 6,108 in the case of RHIC and 22,902 in the case of CMN, needed for compensation at 3.3 kG.

††   Plated layer applied to the outside of the strand (appropriate to 0.33 T operation) computed from the relationship $t = (A_{SC}/\pi D)R_C$.

DESIGN AND PERFORMANCE OF Ni-PLATED STRANDS

Design of Ni-Plated Strands

A length of arbitrarily chosen multifilamentary strand -- designated SSCNI in Table 1 -- was taken for the electroplating study. Its diameter, D, and moment per unit length (M_ϱ, emu/cm) at "design fields" of 3 kG and 6 kG were measured. Next, the desired plating thicknesses, t, were calculated according to

$$t = |M_\varrho/\pi D\sigma| \tag{7}$$

where σ is the moment per unit volume of Ni at the design field. In practice, σ is close to the saturation moment of Ni at liquid-He temperatures. In the present study, D = 8.226×10^{-2} cm, M_ϱ = -3.9309×10^{-2} emu/cm (3 kG) or -2.7603×10^{-2} emu/cm (6 kG), and $\sigma(4.2K)$ = $+494.64$ emu/cm^3 (3 kG) or $+515.92$ emu/cm^3 (6 kG). Eqn. (5) then called for plating thicknesses of 3.08 μm and 2.07 μm, respectively.

Chemically (Electrolessly) Plated Strands

Electroless plating was first recommended to us as a method of administering a uniform Ni layer onto a thin Cu wire. Accordingly, Ni layers intended to range in thickness from 4.1 μm to 1.5 μm were deposited from a calibrated chemical bath. Some magnetization results for the strand most thickly coated with Ni are presented in Figure 4. There it can be seen that the as-deposited electroless layer was nonferromagnetic, which is characteristic of the expected amorphous-Ni film. The figure also indicates that a mild anneal of 3 hours at 300°C restores some of the ferromagnetism as partial recrystallization takes place.

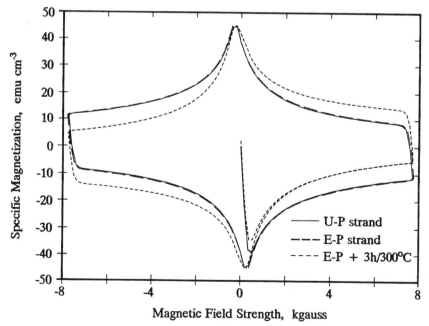

Figure 4.   Magnetization per unit volume of NbTi *versus* field at 4.2 K for the SSCNI strand in the various conditions: (i) as-received (U-P); (ii) chemically (electrolessly) Ni-plated (E-P); chemically plated and annealed (partially recrystallized) for 3 hours at 300°C.

Electroplated Strand

Electroplated Ni layers are known to be crystalline, and hence ferromagnetic, in the as-deposited state. Using a calibrated electroplating bath an attempt was made to deposit a ferromagnetic-Ni layer, 2.1 μm thick, for magnetic compensation at 6 kG. Estimated by weight change, the actual thickness of the layer deposited was 2.2 μm. The degree to which compensation was achieved is indicated in Figure 5 which shows the shielding magnetization crossing the H-axis at 7.35 kG.

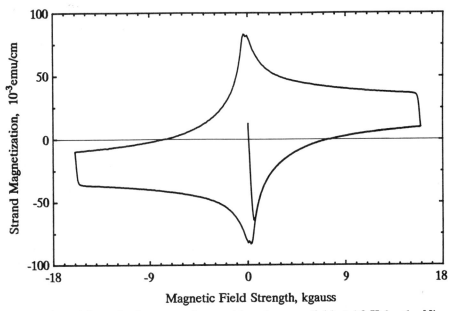

Figure 5.   Magnetization per unit strand length *versus* field at 4.2 K for the Ni-electroplated SSCNI strand. Note that moment compensation (+H increasing) takes place at 7.52 kgauss.

Recommendations for a Continuous Strand-Plating Process

Using a simple laboratory Ni-plating bath, it was determined that a successful 2-μm-thick Ni layer could be deposited on a typical SSC strand -- sample SSCNI in this case -- in 10 minutes at a "current density" of about 90 mA for a 33-cm length of strand. In round numbers this is equivalent to treating 6,000 feet of material with 100 A for 60 minutes, which leads to the process scheme depicted in Figure 6.

As indicated in Figure 6, SSC-type strand passed continuously along the axes of a series of cylindrical anodes each 100 feet in length, could be electroplated with Ni at a rate of about 6,000 ft/h -- a rate that is compatible with those of other strand-processing steps, such as the twisting operation.

Conclusion

A few percent of Ni added to a superconducting strand can offset most of its shielding magnetization over a limited magnetic field range. Pure Ni adds little to the existing magnetic hysteresis, and once it reaches saturation its effect is to bodily shift the wings of the M(H) loop in the +M direction when H is positive and in the -M direction when H is negative.

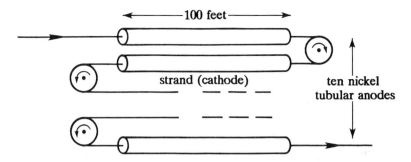

Figure 6.   Recommendation for continuous strand electroplating. Using such a scheme
it should be possible to coat a typical strand with about 2 μm of Ni at a rate
of about 6,000 feet per hour.

The addition of Ni to the strand may relieve the SSC magnets's need for fine filaments, one of the purposes of which was also to minimize conductor magnetization.

Early studies have suggested that magnetization compensation could be achieved through the insertion of bulk Ni into the dipole "wedge"[6]. But in this work we are proposing that the Ni could be associated with the composite strand itself -- either in the form of replacement filaments or as a coating on the outside surface. The feasibility of the replacement-filament idea is demonstrated elsewhere in these proceedings. In this paper the effectiveness of the electroplating approach has been demonstrated, and a method for continuously administering the ferromagnetic layer, at a rate commensurate with other strand-processing operations, has been recommended.

ACKNOWLEDGEMENTS

The clad and unclad epoxy-potted magnetization samples were prepared by R. D. Smith, Battelle. Dr G. Iwaki, Hitachi Cable Ltd, provided us with a preprint of his "Ni-filament" paper in time for it to be mentioned in association with this work. The research has benefitted from discussions with R. L. Scanlan and M. A. Green of the Lawrence Berkeley Laboratory, and R. F. Steining of the SSC Laboratory. It was sponsored by the U.S. Department of Energy, Division of High-Energy Physics.

REFERENCES

1.   A. K. Ghosh, W. B. Sampson, E. Gregory, and T. S. Kreilick, "Anomalous low field magnetization in fine filament NbTi conductors", IEEE Trans. Magn. MAG-23, 1724 (1987).

2.   A. K. Ghosh, W. B. Sampson, E. Gregory, T. S. Kreilick, and J. Wong, "The effect of magnetic impurities and barriers on the magnetization and critical current of fine filament NbTi composites", Tenth Int. Conf. Magnet Tech., Boston, MA, Sept. 21-25, 1987.

3.   E. Gregory, T. S. Kreilick, J. Wong, E. W. Collings, K. R. Marken Jr., R. M. Scanlan, and C. E. Taylor, "A conductor with uncoupled 2.5 μm diameter filaments designed for the outer cable of SSC dipole magnets", IEEE Trans. Magn. MAG-25, 1926 (1989).

4. E. W. Collings, "Stabilizer design considerations in ultrafine filamentary Cu/NbTi composites", Sixth NbTi Workshop, Madison, WI, Nov. 12-13, 1986; see also Adv. Cryo. Eng. Materials <u>34</u>, 867 (1988).

5. T. S. Kreilick, E. Gregory, and J. Wong, "Geometric considerations in the design and fabrication of multifilamentary superconducting composites", IEEE Trans. Magn. <u>MAG-23</u>, 1344 (1987); see also Ref. 2.

6. M. A. Green, "Control of higher multipoles in SSC dipole magnets due to superconductor magnetization using ferro-magnet material in the dipole wedge", Lawrence Berkeley Laboratory Report LBID-1533, SSC-MAG-661, September 1989.

7. C. Rizzuto, "Formation of localized moments in metals: experimental bulk properties", Rep. Progr. Phys. <u>37</u>, 147-229 (1974).

8. E. W. Collings, *Applied Superconductivity, Metallurgy and Physics of Titanium Alloys, Volume 1*, Plenum Press, New York, 1986, p. 435.

9. W. J. Carr Jr., and G. R. Wagner, "Hysteresis in a fine filament NbTi composite", Adv. Cryo. Eng. Materials <u>30</u>, 923 (1984).

10. S. S. Shen, "Magnetic properties of multifilamentary Nb_3Sn composites", in <u>Filamentary A15 Superconductors</u>, ed. by M. Suenaga and A. F. Clark, Plenum Press, New York, 1980, pp. 309-320.

11. E. W. Collings, K. R. Marken Jr., and M. D. Sumption, "Design of coupled or uncoupled multifilamentary SSC-type strands with almost zero retained magnetization at fields near 0.3 T", Proc. CEC/ICMC Conf. Los Angeles, CA, July 1989, Adv. Cryo. Eng. Materials <u>36</u>, to be published.

12. E. W. Collings, K. R. Marken Jr., and M. D. Sumption, "Magnetization studies of multifilamentary strands for superconducting supercollider (SSC) applications -- methods of controlling proximity-effect coupling and residual magnetization", Proc. Int. Atomic Energy Agency Symposium, Tokyo, Japan, September 1989, to be published.

DEVELOPMENT OF NbTi SUPERCONDUCTING CABLE FOR THE SSC

BY SUMITOMO ELECTRIC INDUSTRIES, Ltd.

S.Saito, T.Sashida, G.Oku, K.Ohmatsu, and M.Yokota

Superconductor Project Group
Sumitomo Electric Industries, Ltd.
1-1-3, Shimaya, Konohana-ku, Osaka, 554 Japan

ABSTRACT

NbTi superconducting cable to be used for the SSC project has been developed by Sumitomo Electric Industries, Ltd. We have developed and manufactured various types of superconducting wires which are fabricated by using the single stacking technique in a 260 mm diameter full size billet. 30,000 NbTi filaments can be embedded in this billet. The critical current density (Jc) is as high as 3,300 A/mm^2 at 5 Tesla in the laboratory scale samples. We have achieved 2,800-3,100 A/mm^2 at 5 Tesla in the industrial scale wires with a unit length of 12 km from a 250 kg full size billet.

We also have developed and manufactured many kinds of Rutherford-type cables. Fabrication techniques for a cable with precise dimensions, less critical current degradation and a high residual resistance ratio (RRR) value were studied.

INTRODUCTION

The superconducting wire for the SSC is required to have very fine multi-filaments 6μm in diameter and high critical current density of over 2,750 A/mm^2 at 5 Tesla. In order to fulfill such requirements, it is necessary to assemble regularly a very large number of filaments and maintain a clean surface and uniformly-sized filaments during the manufacturing process.

Sumitomo Electric has been developing superconducting wires of very fine multi-filaments with high critical current density for use in devices such as the SSC and superconducting generators. We usually use a large diameter Cu/Nb-Ti composite billet 260 mm in diameter, which makes it easy to assemble a large number of Nb-Ti filaments. Now we have succeeded in manufacturing wires which have a very large number of filaments for the SSC cable.

The superconducting cable for the SSC is required to have very strict tolerances of ±6 μm. Sumitomo Electric, which is the largest company for electric conductor and cable in Japan, has developed many kinds of superconducting cable for use in fusion, accelerators, superconducting generators, and the electro-magnetic thrusters (EMT) for ships. In this paper, we report on the fabrication and properties of the wire and cable for the SSC.

FABRICATION OF SUPERCONDUCTING WIRE AND CABLE FOR THE SSC

Fig. 1 shows the process used at Sumitomo Electric to manufacture NbTi superconducting wire and cable[1]. A typical feature of this process is the application of a large diameter Cu/NbTi composite billet by using a 5,000 ton hot extrusion press, which is the biggest press in Japan. The billet size is 260 mm in outer diameter and 900 mm in length, with a total weight of over 400 kg. The large diameter makes it easy to assemble a very large number of Cu/Nb-Ti single-core hexagonal segments. A wire with the largest number of NbTi filaments of all our products was fabricated from a billet which was constructed of 13,000 Cu segments and 17,000 Cu/NbTi segments with a size of 1.2 mm. We made these segments straight and stacked in a large Cu can by our original method. The packing factor of this billet was over 98%. We made billets for the SSC by the same billet making method.

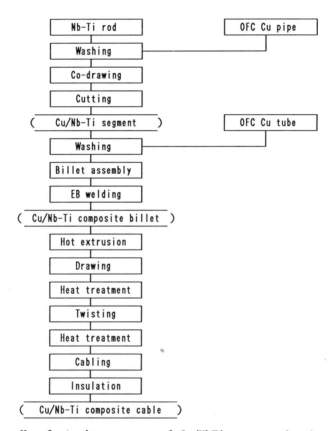

Fig. 1. Manufacturing process of Cu/NbTi superconducting cable

Fig. 2 shows the cross section and filament surface of superconducting wires for the inner cable of the SSC. The arrangement of NbTi filaments is very uniform and the filament surface is smooth.

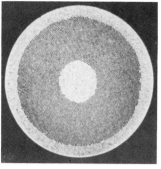

wire for the inner cable   NbTi filament surface

Fig. 2. Cross section of wire and NbTi filament surface

The multiple heat treatment was adopted to increase the critical current density[2]. Fig. 3 shows the dependence of the critical current density of the wires, whose filament diameter is 6μm, on the external magnetic field. The criterion on the critical current density was at a sensitivity of $1\times10^{-14}\Omega$-m for wire area. The quantity of the laboratory
scale is about 2~3 kg, and the quantity of the industrial scale is about
50 kg. The critical current density of the industrial scale wire is 2,800
~3,100 A/mm^2 at 5 Tesla, and the value for the laboratory scale wire is
as high as 3,300 A/mm^2 at 5 Tesla.

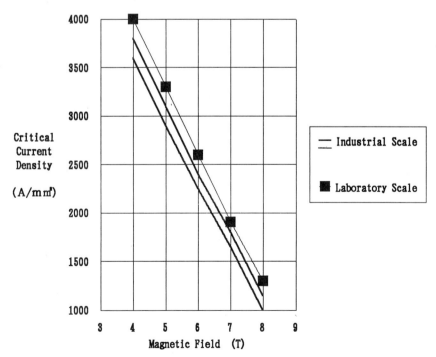

Fig. 3. Dependence of the critical current density
on the external magnetic field

In our experience, the drawability of wire with very fine multi-filaments (～ 10,000) is inferior to that of wire with fewer filaments (～ 1,000). We assumed that this depends on the quantity of surface area between the construction materials (Cu, NbTi). It is necessary to carefully clean the segment surfaces and take great care to maintain the cleanness during the billet making process. In order to improve the metallurgical bonding between the construction materials, the extrusion condition must be strictly controlled. Furthermore, the mechanical strength of wire with high critical current density (～3,000 A/mm^2 at 5 Tesla) is very high and the drawability is inferior to wire with lower critical current (～2,000 A/mm^2 at 5 Tesla). Suitable die designs and die reduction schedules were also investigated. As a result of these developments, a unit length of 12 km (about 50 kg) at 0.8 mm diameter could be obtained without wire breakage from a 250 kg full-size billet in industrial scale.

The Rutherford-type cables for the inner and outer windings of the SSC were fabricated using the planetary-type cabling machine. Fig.4 shows the cross-section of the inner and outer cables. We adopted a dimension measurement tool as shown in Fig.5 in order to satisfy the precise dimensional requirements.

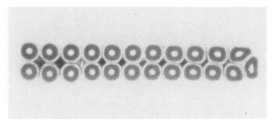

inner cable

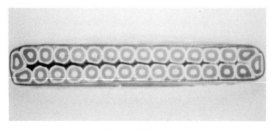

outer cable

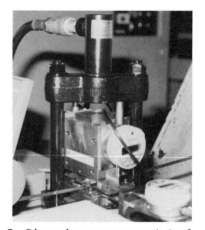

Fig. 4. Cross section of the inner   Fig. 5. Dimension measurement tool
and outer cable

PROPERTIES OF SUPERCONDUCTING WIRE AND CABLE FOR THE SSC

AC Loss

If the residual magnetic field of the cable is large, it causes the deformation of magnetic field distribution which affects the orbit of particles. The residual magnetic field mainly depends on the proximity effect of NbTi filaments in the strands. It is necessary to investigate the relationship between the proximity effect and the filament size and spacing in order to design the conductor. In the measurements, the samples are the wires for the outer cable and these sizes are 0.64, 0.47, and 0.20 mm in diameter. The change of magnetic field ($\triangle$B) is -0.5 Tesla→0→+0.5 Tesla and the changing ratio ($\dot{B}$) is 0.2 T/s.

Fig. 6 shows the magnetization loops of these wires. The peak corresponding to the proximity effect begins to be generated in the 0.47 mm diameter wire whose filament diameter is 4.4 μm and filament spacing is 0.92 μm. In the 0.2 mm diameter wire, there is a clear peak of the proximity effect, and the magnetization loop is the largest.

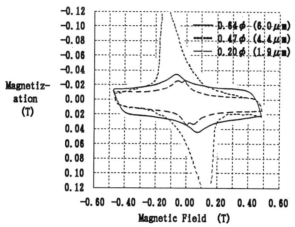

Fig. 6. Magnetization loops of wires for the outer cable

Fig. 7 shows the relationship between the AC losses and the designed filament size and filament spacing. The AC losses increase in the 0.2 mm diameter wire. There is no proximity effect in the 0.6mm diameter wire, which is the specified size for the outer cable. This demonstration certifies that our fabrication methods are suitable to produce wire for the SSC.

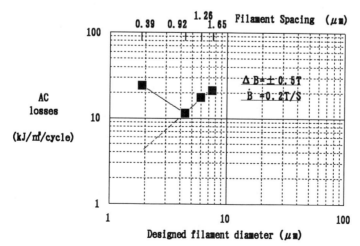

Fig. 7. Relationship between the AC losses and the designed filament size

Annealing Properties

Fig. 8 shows the dependence of the critical current and the residual resistance ratio of the wires, which were sampled from the inner cable, on the annealing temperature. The critical current begins to degrade at the higher temperature above 250 ℃ for two hours. The residual resistance ratio value is 73 after cabling and begins to rapidly increase above 200℃ and reaches more than 200 at the higher temperature above 240 ℃. These values are controllable by annealing after cabling.

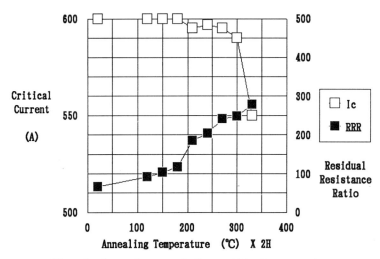

Fig. 8. Dependence of the critical current
and the residual resistance ratio
on the annealing temperature

Degradation After Cabling

It is important to develop a highly compacted cable which makes it possible to wind the coil tightly. However, degradation is generated if the compaction is too high. We investigated the degradation of the critical current when the cable is compacted. The samples were from the inner cable but these shapes were not keystone. The cables were compacted by two ways. One was to press on the wide side, and the other was to press on the narrow side by the Turk's head roll. The critical current values were measured for the strands picked out of the cables. The packing factor was calculated by the following formula (1):

$$P = \{(\pi \cdot d^2 \cdot n) / (4 \cdot w \cdot t \cdot \cos\theta)\} \times 100 \quad \cdots \cdot (1)$$

P:packing factor	n:number of strands
d:strand diameter	θ:cabling angle
t:thickness	w:width

Fig. 9 shows the dependence of critical current on the cable packing factor. The degradation becomes large when the packing factor is over 95%. Strand breakage occurs at the corner of the cable when the packing factor is 99% in pressing on the wide side and 97% in pressing on the narrow side. It is more degraded by pressing on the narrow side than by pressing on the wide side. The maximum degradation ratio is 8.0% which occurs by pressing on the narrow side when the packing factor is 95.4%.

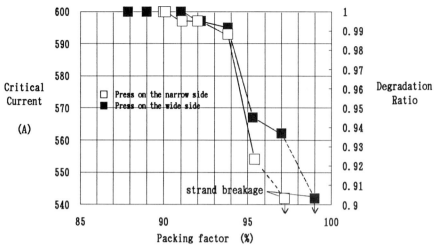

Fig. 9. Dependence of the critical current on the packing factor

CONCLUSIONS

1. Wires for the SSC with very fine multi-filaments of 6 μm diameter were fabricated by using the single stacking technique. The critical current densities of these wire was 2,800-3,100 A/mm^2 at 5 Tesla in the industrial scale and 3,300 A/mm^2 in the laboratory scale.

2. Wire and cable for the SSC were developed and fabricated in the industrial scale. The test results of these products met the specifications of the SSC.

3. The proximity effect of NbTi filaments was not observed in the 0.6 mm diameter wire, which was for the outer cable.

4. The residual resistance ratio of the cable could be increased by annealing, without degradation of the critical current.

5. Degradation of the critical current by cabling was less than 2% in the dimensions which met the specifications.

6. In the same compaction cable, there was more degradation by pressing on the narrow side than by pressing on the wide side.

ACKNOWLEDGMENTS

The authors would like to express their thanks for the invaluable assistance and support of their many colleagues.

REFERENCES

[1] S.Saito, et al.,   "Recent Activities for NbTi Superconducting wire at SEI", Sumitomo Electric Technical Review, No.29, January 1990, pp 51-59

[2] S.Saito, et al.,   "Superconducting Properties of Fine Filamentary Superconducting Wires", Proc. of ICFA, New York U.S.A., May 1986

A STATUS REPORT ON THE DEVELOPMENT OF INNER AND OUTER CONDUCTORS

FOR THE SSC DIPOLE AND QUADRUPOLE MAGNETS

H. Kanithi, M. Erdmann, P. Valaris, F. Krahula, R. Lusk, E. Gregory
and B. Zeitlin

IGC Advanced Superconductors Inc.
1875 Thomaston Avenue
Waterbury, CT 06704

ABSTRACT

Results obtained on SSC strand made from both 309 mm and 355 mm diameter billets at IGC are described. In the interest of meeting very stringent quality assurance criteria, and to ensure high current densities and long piece lengths, many elaborate fabrication steps have been introduced in recent years. As full scale production approaches, the economic impact of these many steps becomes of increasing importance and they are now being reexamined. Scale-up is one of the more obvious ways in which cost reduction can be accomplished and the plans which have been made to affect such scale-up are briefly indicated.

INTRODUCTION

The present strand and cable specifications[1] for the Superconducting Super Collider (SSC) dipole and quadrupole magnets have evolved over the past years. The evolution can be traced to the increased understanding for improved magnet performance. Two aspects of the SSC conductor have remained as the focal points during all these years. Namely, high critical current density and low magnetization at low fields. As the J_c started to increase in prototype manufacturing trials so did the specification target. It is now set at 1600 A/mm^2 at 7T for the SSC inner strand and at 2422 A/mm^2 at 5.6T for the outer strand prior to cabling. These two values roughly correspond to the frequently quoted 2750 A/mm^2 at 5T. High J_c not only reduces the final cost of conductor (due to a smaller volume fraction of NbTi alloy required) but more importantly, it increases the safety margin for the magnets. At beam injection field, on the other hand, the magnetization has to be as low as possible to minimize field distortions. Fine and ultrafine filaments result in low magnetization and thus became a requirement. During the early stages of prototype strand development, filament sizes in the range of 9 to 2.5 μm were being considered[2]. For practical limitations in the manufacturing technology that has evolved, the filament size has now been set at 6 microns for the dipole and quadrupole magnet conductors. It should be noted, however, that there is continued interest in 2.5 μm filaments which are expected to eliminate the need for active correction coils. This smaller filament size has brought about an added problem of proximity effect coupling which results in increased magnetization. The use of dilute alloys of Cu with Mn instead of pure copper as the matrix has been found to effectively decouple the filaments[3]. Work in this area is ongoing at various U.S. manufacturers including IGC/ASI. In the last 1-2 years, many U.S. and foreign manufacturers have been able to produce quality strand and cable meeting essentially all the SSC specification requirements. This was possible due to: a) An extensive microstructural development effort by Larbalestier and coworkers[4] and b) The realization of certain key design and process parameters by the conductor manufacturers. These have been reported in the literature.

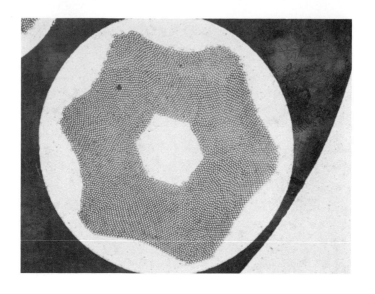

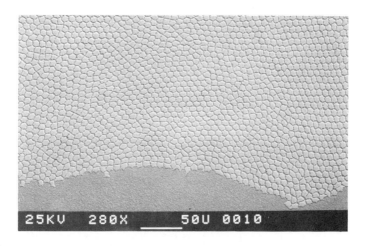

25KV   280X   50U 0010

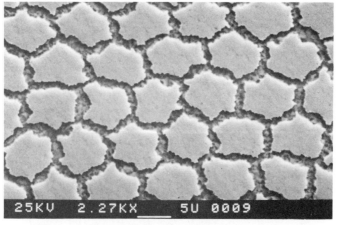

25KV   2.27KX   5U 0009

Figure 1. Photomicrographs of 0.808 mm dia. inner grade strand from billet 5264R-B0309.

The combined effort led to the identification and control of extrinsic factors which were limiting the achievement of the high J_c which is truly intrinsic to NbTi. Besides high homogeneity of NbTi alloy and heat treatment details, there are many factors influencing the final conductor characteristics. These include local Cu/NbTi area ratio in the filament region, the adequacy of the diffusion barrier between NbTi and copper, the bond integrity at various interfaces and billet geometry.

During the period 1986 to 1988, we have produced both inner and outer grades of wire with NbTi filament diameters of 2.5 to 9 μm[5-7]. These materials have been successfully used in the early prototype magnet development. Economic considerations for large scale wire production played an important role during that phase of development. We have now developed a manufacturing technology that results in very long lengths of a more reliable and consistent product. One of the requirements that is recognized as important but is not yet incorporated into the present SSC specifications is the consistency in critical current performance of the cables over the production period. The variation in conductor Ic from beginning to the end of production should be minimal in order to maintain the magnetization relatively constant. To that end, preselection of strands based on individual I_c is an option to exercise during cable strand mapping. However, this is an expensive option as the number of short sample tests increases. We have recently produced two billets each of inner and outer grade strand from nominal 309 mm diameter billets. The details of processing are different in these two types of billets. The data presented in the following sections address the statistical nature of the characteristics of the two types of conductor produced. Our main goal here is to begin generating a sufficient database which will guide us in the right choice of process parameters for both strand and cable.

INNER GRADE CONDUCTORS

The design approach for the two billets of inner grade strand was based on the so-called single round restack rod. The identities given to these billets are 5264R-B0309 and 5264R-B0310. Monofilament rod was produced by a hot extrusion process using high homogeneity grade Nb-47 wt % Ti alloy ingot made by Teledyne Wah Chang. Having realized that the Nb barrier thickness in earlier pilot billets was inadequate, we have doubled it in this run. Although attempts were made to reduce the stacking faults present in earlier billet assemblies, they were only partially successful. The processing of these two billets followed conventional manufacturing operations including hot isostatic pressing, extrusion and cold-drawing with intermediate heat treatments. Figure 1 shows the strand cross section and the shape of filaments in 5264R-B0309 strand. The filament geometry is typical of this stacking method. The lack of Cu-Ti intermetallic formation on the filament surface indicates the adequacy of the Nb diffusion barrier.

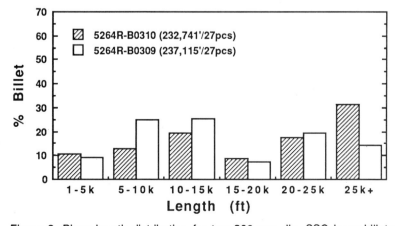

Figure 2. Piece length distribution for two 309 mm dia. SSC inner billets.

Table 1. Performance statistics for inner strand

	5264R-B0309			5264R-B0310		
	Cu/SC	I_c(7T)A	Jc,A/mm^2	Cu/SC	I_c(7T)A	Jc,A/mm^2
No. of Samples	26	26	26	27	27	27
Mean	1.498	365	1777	1.513	351	1722
Standard Deviation	.058	9.9	20.7	.046	10.2	22.7
Data Minimum	1.400	349	1719	1.403	330	1670
Data Maximum	1.591	381	1815	1.619	373	1764

Table 2. Inner Cable (5264R) Performance

I_c (7T, 4.22 K)	7791 A
I_c Margin	7.7%
J_c (7T, 4.22 K)	1670 A/mm^2
Cu/SC Ratio	1.53
RRR	33
I_c Degradation	2.7%

The final strand diameter was continuously gauged with a state-of-the art laser micrometer. Samples were cut from each continuous piece and tested for critical current (I_c) at 7 tesla. The billet yield, in terms of piece lengths, is analyzed and presented in the customary fashion in Figure 2. The fraction of the total yield versus the piece length range is plotted here for the two billets involved. The yield of each billet in terms of total strand length and number of pieces is also noted. While the present data show an improvement in the longer piece lengths, compared to the past[8], the overall results are less than desirable for a large production run. In Table 1 we list the statistics for each billet. The data include the mean, the minimum value, the maximum value and the standard deviation for I_c, Cu/SC and J_c. For the two billets under study, the standard deviation in I_c is 2.7 to 2.9%. The standard deviation in J_c, however, is only 1.2 to 1.3% indicating that the I_c variability is coming from factors such as the strand size and/or Cu/SC ratio. The standard deviations in I_c and J_c could serve as a measure of the variation in strand magnetization and, therefore, the magnetization of the cable.

Approximately 6700 m of cable consisting of 23 strands was made using the Dour Metal cabler[9] at New England Electric Wire Co., in Lisbon, N.H. Representative cable samples were tested at the Brookhaven National Laboratory (BNL) and the typical results are summarized in Table 2. The critical current at 7T and 4.2K for this inner grade cable is approximately 7790 A which gives us a margin of 7.7% over spec value of 7231 A. The I_c degradation due to cabling alone is estimated at 2.7% based on independent stand testing at BNL. This is believed to be typical for such cables.

OUTER GRADE CONDUCTOR

Two 309 mm outer grade billets were designed to be assembled with hexagonal clad monofilament rod. The monofilament rod was produced using techniques slightly different from the previously described inner grade billets with a primary goal of achieving longer piece lengths. Figure 3 shows the strand cross section and shape of filaments in the two outer grade strands. It is evident from these pictures that the hexagonal restack assembly leads to a more regular arrangement of filaments than in the round restack method used in the 5264R set of billets. The 6 micron diameter filaments also look uniform and are free of any intermetallics. Two billets designated as 5345-B0359 and 5345-B0397 were designed to have a final Cu/SC ratio of 1.8:1. Pertinent conductor data was gathered and analyzed. Both the outer billets produced extremely long piece lengths as illustrated in Figure 4. Several of the longest pieces had to be cut to fit on to the shipping spools. This type of piece length performance represents a process that is well under control from start to finish.

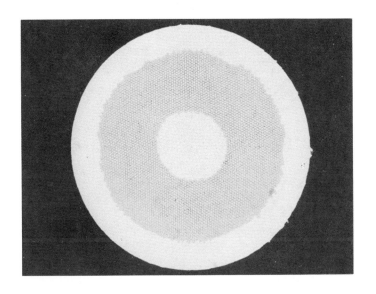

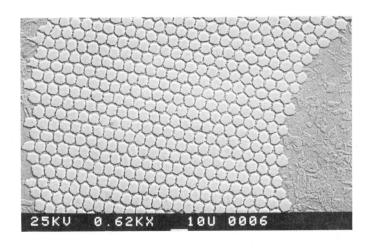

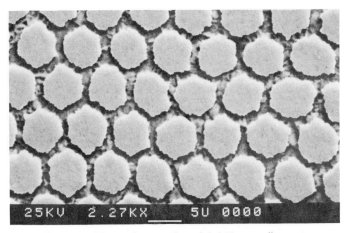

Figure 3. Photomicrographs of 0.648 mm dia. outer
grade strand from billet 5345-B0359.

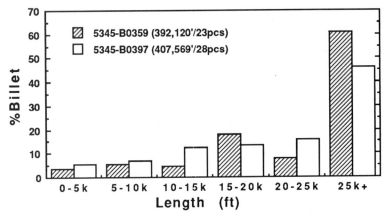

Figure 4. Piece length distribution for two 309 mm dia. SSC outer billets.

Table 3 is a listing of statistics on the I_c(5.6T), Cu/SC and J_c of the outer strand. An average J_c(5.6T) of 2584 and 2574 A/mm^2 is achieved in the two billets B0359 and B0397, respectively. These values correspond to a J_c(5T) of ~2940 A/mm^2. The distribution of I_c appears narrow as indicated by the standard deviation of only 1.1 to 2.4%. In Figure 5, we have compared the variation of I_c and Cu/SC ratio along the billet length as approximated by the piece number. It is interesting to note that the I_c variation is roughly a mirror image of the Cu/SC ratio. This clearly demonstrates that the variation in strand Ic is more due to variation in Cu/SC ratio than due to changes in the intrinsic J_c. Statistical values such as those in Tables 1 and 3 and Figure 5 provide a quantitative method of tracking the production billets at least as far as the strand is concerned.

The strand from the two billets was cabled to produce a 30 strand outer grade cable. The electrical properties of the cable are summarized in Table 4. The cable I_c as measured by BNL is 8975 A at 5.6T which provides a margin of 14.2% over the specification minimum. The high J_c of the magnitude obtained here is more than necessary to offset variations in raw materials and in conductor process that are anticipated for a large scale production.

Table 3.   Performance statistics for outer strand

	5345-B0359			5345-B0397		
	Cu/SC	I_c(5.6T)A	Jc,A/mm^2	Cu/SC	I_c(5.6T)A	Jc,A/mm^2
No. of Samples	23	23	23	27	27	27
Mean	1.769	309	2584	1.704	315	2564
Standard Deviation	.020	7.4	52.2	.031	3.6	29.1
Data Minimum	1.721	298	2460	1.653	308	2504
Data Maximum	1.801	324	2673	1.768	323	2624

Table 4.   Outer Cable (5345) Performance

I_c (5.6T, 4.22 K)	8975 A
I_c Margin	14.2%
J_c (5.6T, 4.22 K)	2540 A/mm^2
Cu/SC Ratio	1.84
RRR	50
I_c Degradation	1.6%

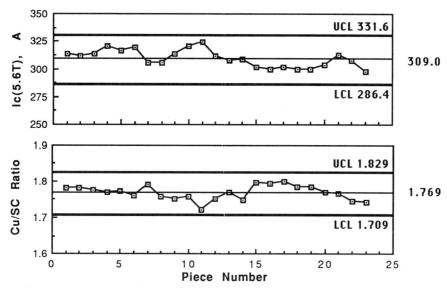

Figure 5. Variation of Ic and Cu/SC ratio along billet length for 5345-B0359.

Although raw material specifications and the general process prescription are expected to be fixed at the beginning of the SSC production campaign, there are enough variables with their respective tolerances which will lead to a certain variation in the final conductor characteristics. Our challenge will be to minimize the variation by tight process control.

BILLET SIZE SCALE UP

As a part of a development exercise, directed by LBL and SSCL, we have participated in the evaluation of one 355 mm dia. outer grade billet. LBL had supplied copper clad monofilament composite rod at 25 mm dia. and unfinished 355 mm dia. billet can materials[10]. The manufacturing approach was similar to that of 309 mm dia. outer grade billets reported earlier. Proportionally larger sized hexagonal restack rods were assembled. After welding and consolidation the billet was extruded. The details of wire drawing and heat treat schedules were also kept essentially identical to the 309 mm dia. outer type billets. However, the results in terms of piece lengths (Figure 6), I_c and J_c (Tables 5 and 6) were disappointing when compared to the 309 mm billets.

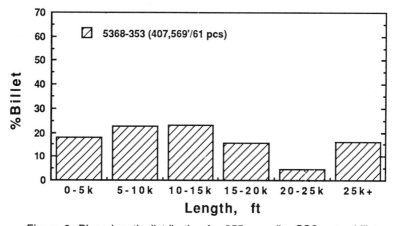

Figure 6. Piece length distribution for 355 mm dia. SSC outer billet.

Table 5. Performance statistics for outer strand from 355 mm dia. billet

	Cu/SC	$I_c(5.6T)A$	Jc,A/mm^2
No. of Samples	36	36	36
Mean	1.588	303	2378
Standard Deviation	.027	9.3	65.0
Data Minimum	1.512	282	2191
Data Maximum	1.641	318	2473

Table 6. Outer Cable (5368) Performance

	Run No. 2501	Run No. 2500
I_c (5.6T, 4.22 K)	8414 A	8537 A
I_c Margin	7%	8.6%
J_c (5.6T, 4.22 K)	2211 A/mm^2	2243 A/mm^2
Cu/SC Ratio	1.60	1.6
RRR	52	52
I_c Degradation	4.5%	3.1%

After an extensive investigation, the primary cause was identified to be the inadequate thickness of Nb barrier and its irregular/nonuniform deformation. There could also be other reasons such as the quality of copper used in the monofilament rod, and the integrity of the bond at the Cu/Nb and Nb/NbTi interfaces. The combined effect of these factors was the formation of intermetallics on the filament surface, some filament sausaging, low J_c and low normal transition index (n) values. The strand also exhibited low ductility in sharp bend tests. An average strand J_c (5.6T) of 2378 A/mm^2 with a standard deviation of 2.7% was achieved for this billet. When cabled the J_c(5.6T) dropped by approximately 4%. As a result of a low Cu/SC ratio (1.6:1 instead of 1.8:1) and low I_c degradation due to cabling, the cable still exhibited an acceptable critical current in excess of 8400 A at 5.6T. Because of the inferior performance of this 355 mm dia. billet material as compared to the 309 mm dia. billets described earlier, this cable should be more thoroughly characterized prior to its use in magnets. The 355 mm dia. billet exercise is a partial success. Further prototype 355 mm dia. billet fabrication is expected in the near future, at which time more attention will be given to the problems identified in this first attempt.

SUMMARY

We have successfully produced 309 mm diameter billets to make both SSC inner and outer grade strand and cables meeting all performance requirements of SSC specifications. The I_c margins in the final cable were approximately 8% for the inner grade and 14% for the outer grade. The first 355 mm diameter developmental billet produced an unacceptable number of piece lengths and a slightly lower J_c as a result of inadequate Nb diffusion barrier thickness. Extensive strand data including diameter, I_c, Cu/SC ratio and J_c have been gathered and analyzed. Further collection of such data for a large number of billets that are uniformly produced should provide useful statistics for the materials and manufacturing processes involved, as large scale production begins.

REFERENCES

1. Material Specification No. SSC-MAG-M-401 for NbTi Superconductor Wire and No. SSC-MAG-M-402 for Cable for SSC Dipole Magnets, issued June 9, 1987.

2. R. Scanlan, J. Royet and C. E. Taylor, "Superconducting materials for the SSC", in Adv. in Cryo. Eng. Mat'ls., Vol. 32, Plenum Press, New York (1986) pp. 697-706.

3. T. S. Kreilick, E. Gregory, J. Wong, R. M. Scanlan, A. K. Ghosh, W. B. Sampson, and E. W. Collings, "Reduction of coupling in fine filament Cu NbTi composites by addition of manganese to the matrix", in Adv. in Cryo. Eng., A.F, Clark and R. P. Reed, eds. Plenum Press, New York 1988, vol. 34, pp. 895-900.

4. P. J. Lee and D. C. Larbalestier, "Development of nanometer scale structures in composites of Nb-Ti and their effect on the superconducting critical current density", Acta Metall, 35 (1987) p.2523.

5. K. Hemachalam, C. G. King, B. A. Zeitlin and R. M. Scanlan, "Fabrication and characterization of fine filaments of NbTi in a copper matrix", in Adv. in Cryo. Eng. Mat'ls., Vol. 32, Plenum Press, New York (1986) pp.731-738.

6. C. King, K. Hemachalam and B. Zeitlin, "Prototype fabrication of ultrafine filament NbTi conductor for the SSC", IEEE Trans., MAG-23, 2, pp. 1351-1354, 1987.

7. H. C. Kanithi, C. G. King, B. A. Zeitlin, "Fine filament NbTi Conductors for the SSC", IEEE Trans., Vol 25, 2, pp. 1922-1925, 1989.

8. H. Kanithi, F. Krahula, M. Erdmann, R. Schaedler and B. Zeitlin, "Superconducting wire and cable for the SSC-progress at Intermagnetics General Corporation towards production", Proc. of the International Industrial Symposium on the Super Collider, Feb. 8-10, New Orleans, M. Mc Ashan, Ed., Plenum Press, New York(1989) pp. 223-229.

9. J. M. Royet, R. Armer, R. Hannaford and R. M. Scanlan, "An Industrial Cabling Machine for the SSC", ibid., pp.273-276.

10. Under separate contracts from LBL, Supercon Inc., Shrewsbury, M.A. had produced the monofilament rod to 25 mm dia. and Oxford Superconducting Technology, Carteret, N.J. had made the copper can parts.

EVALUATION OF SSC CABLE PRODUCED FOR THE MODEL DIPOLE PROGRAM

DURING 1989 AND THROUGH FEBRUARY, 1990*

D. Christopherson, R. Hannaford, and R. Remsbottom, SSC Laboratory
M. Garber, Brookhaven National Laboratory
R. M. Scanlan*
Lawrence Berkeley Laboratory
1 Cyclotron Road
Berkeley, CA 94720

ABSTRACT

During 1989 and the beginning of 1990, approximately 150,000 feet of cable was manufactured for use in the SSC Model Dipole Magnet Program. The wire for the cable was made to SSC specifications by three different manufacturers. The cable was made at New England Electric Wire on the SSC Production Cabling Machine, under supervision of either SSC Laboratory personnel or the wire manufacturer's representative. All the cable produced for SSC model dipoles was subjected to rigorous inspection in order to insure that the magnet construction and performance would be predictable. The cable dimensions were measured at intervals of 10 feet or less with a cable measuring machine. Electrical properties were measured on samples from one end of each cable length. Critical current degradation due to cabling was checked by measuring the critical currents of the wires used to make the cable and comparing these with the cable critical current. The results of the dimensional and electrical measurements will be discussed and compared with the SSC specification requirements.

INTRODUCTION

During 1989 and through February 1990, a considerable amount of experience and knowledge has been gained in the fabrication of SSC type cable. The use of the newly developed cabling machine and the in-line measuring system has allowed the production of SSC cable to be more accurately and continually monitored. This development, together with the existing test procedures, has resulted in an expanded database[1] and better understanding of the cabling process and parameters.

CABLE TESTING

Electrical testing of SSC cable is done at Brookhaven National Laboratory in a specialized testing facility. A detailed description of the electrical testing of SSC cable is beyond the scope of this paper; only a skeleton account of the procedures are included here. We refer the reader to Reference 2 for a more complete discussion of the methods used in testing the electrical properties of SSC cable.

*This work is supported by the Office of Energy Research, Office of High Energy and Nuclear Physics, High Energy Physics Division, Dept. of Energy under Contract No. DE-AC03-76SF00098.

Samples of the cable and strands incorporated into the cable are sent to BNL. Representative samples of uncabled strands are tested for I_c at various fields. The results of these strand tests are used in the determination of cabling degradation and can also exhibit a reasonable estimation of the performance of the cable[2]. Calculation of the cable degradation was done somewhat differently in this report due to the retrospective availability of wire data. Cable degradation is discussed at length in a later section of this paper. Because production scale testing allows only for a few measurements of the cable specimen, the full I_c vs. B curve of the cable is fitted using the slope of the wire curve (averaged from all wire samples tested). Previous experiments have showed that the two slopes agree well and the resulting accuracy is about $\pm$ 150A.[2] All critical current measurements are done at a temperature of approximately 4.35K because of the increased pressure due to the forced flow cooling throughout the test magnet. All reported values of the I_c are calculated for the reference temperature of 4.22K.

Resistance measurements are made on the cable at room temperature (R295) and at 10K (R10), and the Residual Resistance Ratio calculated (RRR = R295/R10). The copper to superconductor ratio is determined from these resistance values, and in turn the Cu/sc is used in calculating the critical current density (J_c) of the cable.

The mechanical testing of cable is just as important as the electrical measurements to ensure a reliable magnet. The dimensional control of the cable is monitored by the Cable Measuring Machine. The CMM is an in-line device that periodically (currently 10 ft. intervals) measures the cable width, mid-thickness, and keystone angle as the cable is being produced. A series of tests and spot-checks are performed before each cable run. These tests examine the residual twist of the cable, as well as the surface quality, lay pitch, lay direction, and bare filament integrity.[3,4]

SSC WIRE

Superconducting wire is the single most important component of the SSC cable fabrication process, therefore a short outline of the wire properties and parameters are included.

(Table 1a) SSC Outer Wire

Billet #	Total Length	Length > 10K	%L > 10K	Norm I_c (5.6T)
2071-1	264,889	227,419	85.90	292.85
2071-2	231,260	150,273	65.00	294.65
2128-1	260,767	230,094	88.20	296.15
2128-2	183,216	183,216	100.00	298.81
2128-3	340,232	331,417	97.40	302.68
B0359	392,120	332,120	84.70	310.00
B0397	388,581	337,882	87.00	313.99
2301-1	392,151	360,285	91.90	281.07

(Table 1b) SSC Inner Wire

Billet #	Total Length	Length > 10K	%L > 10K	Norm I_c (7T)	Type
2127-1	216,231	168,601	78.00	334.37	1.5
B0309	237,115	155,934	65.70	365.90	1.5
B0310	232,753	165,464	71.00	353.37	1.5
2300-1	183,652	138,414	75.40	355.51	1.5
2346-1	261,218	240,046	91.90	364.28	1.3
2346-2	264.232	224,173	84.80	366.07	1.3
2346-3	160,694	35,568	22.10	361.41	1.3

Tables 1a and 1b list the data used in constructing Figures 1a and 1b and Figures 2a and 2b. Total length is the length of wire shipped for cabling after meeting SSC specifications. The I_c reported here for these billets is the sum of the critical currents of each piece normalized with its length. This means that a long piece will have a larger contribution to the total normalized billet I_c than a short piece. Mathematically, the normalized I_c for a billet is:

$$I_{ctot} = \Sigma \, I_{cx} l_x / L$$ where L = total length
I_{cx} = piece critical current
l_x = piece length
I_{ctot} = normalized billet critical current

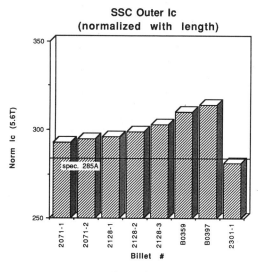

Figure 1a

Figure 1a shows the illustration of how well each SSC outer billet has performed against the SSC critical current specification. In all cases except one this requirement was comfortably met. Although billet 2301-1 fell slightly below the I_c specification for SSC wire, a low degradation during cabling allowed this material to produce acceptable cable.

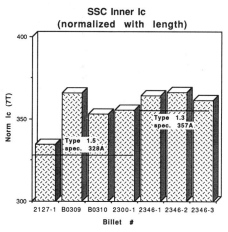

Figure 1b

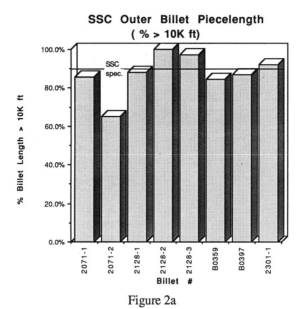

Figure 2a

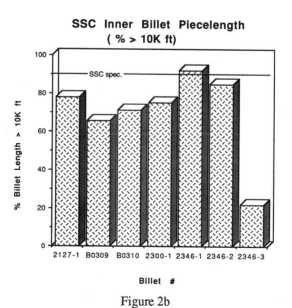

Figure 2b

Figure 1b is the equivalent of 1a for SSC inner billet data. All seven billets represented here meet the SSC specification. However, the four with a Cu/SC ratio of 1.5 exceed the minimum specified I_c by a larger margin.

Figures 2a and 2b show the piece length performance of acceptable strand from each billet. It is obvious that very few billets actually meet the proposed SSC requirement of 90% > 10,000 feet. As more experience is gained in making SSC conductor, it is expected that this piece length problem will be rectified. Piece length is only one facet of the picture however. To get a better idea of billet performance, the overall yield (total length) must also be considered. For example, billet 2128-2 has 100% of its acceptable lengths greater than 10,000 feet. Yet the sum of all acceptable lengths only reaches 183,216 feet. Depending upon the initial size of the assembled multifilament billet, this yield could be as low as 50%. Further information must be acquired from the wire manufacturer in order to clarify this point.

SSC PRE-PRODUCTION CABLE

Pre-Production cable is considered to be cable made with the "main line" SSC conductor. This is the wire fabricated to meet the existing SSC wire specification document.[3] Wire designed to address a specific issue for research purposes is considered R&D and is briefly discussed in a later section.

(Table 2a) SSC Outer Pre-Production Cable

Cable #	Length	I_c (5T)	I_c (5.6T)	Deg.	Material Inc.
SSC22-00004	13,075	9516	8338	7.10%	2128-1,2,3
SSC22-00005	13,623	9800	8608	3.40%	2071, 2128
SSC22-00006	11,000	9661	8488	4.70%	2071, 2128
SSC22-00007	4,400	9247	8138	7.60%	2071
SSC22-00008	11,091	9216	8135	3.50%	2301-1
SSC23-00001	11,800	NA	NA	NA	B0359
SSC-O-I-00002	11,267	NA	NA	NA	B0397
Total Length =	76,256				

(Table 2b) SSC Inner Pre-Production Cable

Cable #	Length	I_c (7T)	Deg.	Inc. Material	Type
SSC12-00001	5,336	7465	2.90%	2127-1	1.5
SSC12-00005	3,613	7407	3.70%	2127-1	1.5
SSC12-00007	7,100	7822	4.30%	2300-1	1.5
SSC-I-S-00008	7,894	8094	3.30%	2346-1,2,3	1.3
SSC-I-S-00009	8,935	NA	NA	2346-1,2,3	1.3
SSC-I-S-00010	10,800	8368	0.00%	2346-1,2,3	1.3
SSC-13-00003	8,572	7840	3.50%	B0310	1.5
SSC-I-I-00003	NA	NA	NA	B0309	1.5
Total Length =	52,250				

Table 2a shows the SSC outer type pre-production cable fabricated in 1989 and 1990 to date. The minimum I_c at 5.6T is specified to be 7860A. The column labeled "Material Inc" shows the material incorporated into each cabling run, thus the cable performance can be referenced back to the appropriate billet and comparisons can be made. Similarly, Table 2b lists the SSC inner type pre-production cable for the same period. The minimum I_c at 7T is 7231A for type 1.5 cable, and 7860A for type 1.3 cable[3] (Type 1.5 designates the use of wire with a 1.5:1 copper to superconductor ratio, and likewise type 1.3 uses 1.3:1 Cu/sc strand). Also included in Tables 2a and 2b is the total footage of production outer (76,256) and inner (52,250) cable made during this period.

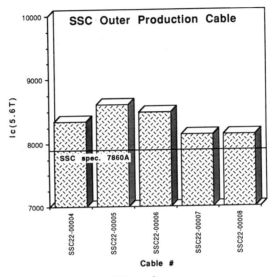

Figure 3a

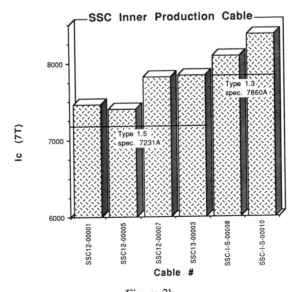

Figure 3b

616

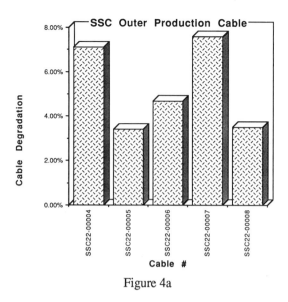

Figure 4a

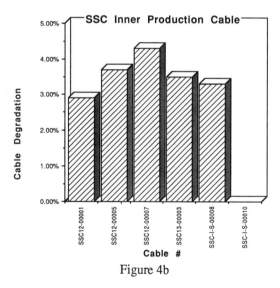

Figure 4b

Figures 3a and 3b show a bar chart plot of the I_c for the outer and inner cable that have been measured to date.

The degradation values listed in Tables 2a and 2b, and illustrated in Figures 4a and 4b, show the approximate decrease in current carrying properties of the cabled strands after experiencing the deformation introduced in the cabling operation. These values are computed by using the <u>normalized I_c</u> numbers of the material involved in the corresponding cabling run. This is multiplied by the number of strands in the cable (23 for inner, 30 for outer) and divided into the reported I_c of the cable. The result of this quotient is subtracted from 1.0 giving the percentage of degradation induced by the cabling process. Depending upon the method used in calculating degradation, values can vary between sources. Because of the limited availability of piece length and I_c data for billets, historically the 'before cabling' I_c used in figuring degradation has been an approximation based on a few strand tests. Using the normalized I_c as is done here takes into consideration each strand used in the cable.

Cable degradation can be influenced by numerous factors, some of which can be controlled in the cabling operation. Certainly a properly aligned and properly set-up machine will help to lower degradation. This seems to be the case for cable SSC-I-S-00010 where the degradation is small enough to be lost in the error factors of the measurements. Compaction is the largest parameter affecting cabling degradation. The higher the compaction, the higher the degradation. Therefore, it is desirable to make cable at the higher end of the midthickness specification limit as long as this limit is not exceeded. This is demonstrated by a simple comparison between the degradation in cables SSC22-00005 and SSC22-00007. SSC22-00007 experienced a consistently higher compaction throughout the run which explains the higher degradation measured.

Sources of cabling degradation can also be a result of traits inherent in the strand itself. The shape and proximity of the filament array in relation to the outside diameter of the wire has been known to contribute to the amount of cable degradation. Higher degradation has been observed in strand with an insufficient amount of copper surrounding the matrix. This deficiency of copper cladding fails to protect the filaments and absorb the deformation introduced during cabling. In Figure 5 we compare the degradation in cables 12-00007, 13-00003, and 12-00001, and we can see a relative fit to the factors described

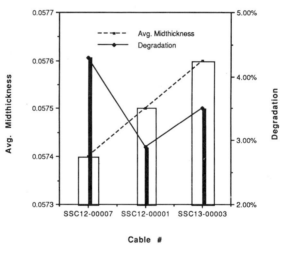

Figure 5

above. The high degradation in 12-00007 can be attributed to the higher compaction (lower average midthickness). However, if we compare 13-00003 and 12-00001, we see a reverse result. Although 13-00003 has a lower compaction, the degradation is higher. This may be attributed to the fact that strand used in 13-00003 has a cross section that is undesirable in terms of cabling degradation. The overall filamentary array has a hexagonal pattern in which the points of the hex come relatively close to the outside diameter of the strand. The problems associated with such a cross section are mentioned above.

Dimensional measurements and statistics of the cable are monitored by the Cable Measuring Machine (CMM) as the cable is being produced. Using data from the CMM, we can produce charts such as Figure 6 which shows the thickness variation throughout the entire run for cable SSC12-00005. Similar charts can be produced for keystone angle and cable width. Table 3 lists the statistics gathered by the CMM during the run.

(Table 3) SSC # 12-00005 In-line Cable Data

	PSI	Angle	Width	Thickness
Average	2547.04132	1.60452	0.36651	0.05741
Minimum	2442.00000	0.00200	0.36596	0.05727
Maximum	2467.00000	1.62700	0.36707	0.05763
Standard/Dev	2.91589	0.08448	0.00009	0.00006

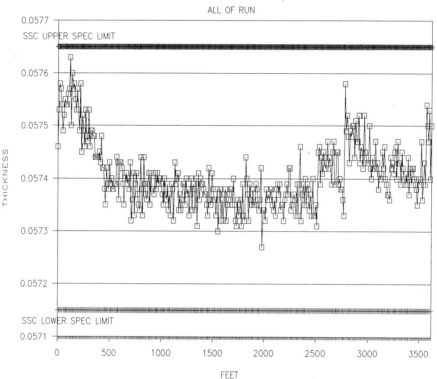

Figure 6

(Table 4) SSC R&D Cable

Cable #	Type	Length	I_c	Fil. Dia.	Notes
SSCIGC14RD	1.8 Outer	8495	NA	6um	14" dia. billet
SSC12-00003	1.5 Inner	1562	7464 (7T)	9um	Unannealed NbTi
SSC12-00004	1.5 Inner	2054	6950 (7T)	6um	Unannealed NbTi
SSC13-00001	1.5 Inner	2750	7968 (7T)	9um	0.45 LAR
SSC14-00003	1.5 Inner	1821	7464 (7T)	6um	Annealed NbTi

SSC R&D CABLE

Table 4 lists some of the R&D cable manufactured during 1989 and 1990 to date. Of considerable interest is SSC-IGC14RD. This cable is SSC outer type material incorporating strand from a 14" diameter single stack extrusion billet. This is a scale-up from the standard 12" diameter extrusion billet, and in design will improve yields and lessen fabrication costs. Problems involving the diffusion barrier and Cu/sc ratio led to less than optimum associated properties. This first attempt at a 14" billet yielded low J_c's, piece lengths and overall yield, as well as bend test failures. It is generally believed that the reasons for this poor strand performance are known and there is confidence that future 14" billets will behave much better. Somewhat surprising, the cable Ic measurements showed less than 2% degradation with cabling.

The other R&D cables listed in Table 4 are all SSC inner type with various parameters differentiating them from production material. Cables SSC12-00003, 12-00004, and 14-00003 are comprised of strand with different filament diameter and/or different NbTi raw material history. Cable SSC13-00001 was fabricated from wire with closely spaced 9um diameter filaments. Table 4 lists the basic characteristics of these R&D cables and the length of each run.

SUMMARY

Of the total cable made during 1989 and early 1990, 76,256 feet was SSC production outer cable, 52,250 feet was SSC production inner, 8495 feet was R&D outer, and 8187 feet was R&D inner cable. In most cases the electrical values of the wire and cable fell amply above the SSC requirements. The piece lengths and yields of many of the billets, however, do not meet the present goals of the SSC project. These shortcomings are being addressed in the present manufacturing scale-up program. Cable degradation numbers have been relatively low, and increased understanding of the degradation parameters can lead to degradations consistently under 4% or better.

Throughout the past year a considerable amount has been learned about the cabling process and an initial foundation has been set for future programs, specifically the newly proposed 50mm SSC Dipole Program. The experience and knowledge developed in the 40mm magnet program will help promote a successful transition to a new cable design if required.

REFERENCES

1. M.J. Baggett, R. Leedy, C. Saltmarsh, and J.C. Thompkins, "The Magnet Component Database System", paper III-B-3, IISSC, Miami Beach, FL, March 14-16, 1990.

2. M. Garber and W.B. Sampson, "Quality Control Testing of Cables for Accelerator Magnets", IISSC, New Orleans, LA, Feb. 8-10, 1989.

3. SSC Specification SSC-MAG-M-4142, NbTi Superconductor Cable for SSC Dipole Magnets.

4. A.F. Greene and R.M. Scanlan, "Elements of a Specification for Superconducting Cable and Why They Are Important for Magnet Construction", IISSC, New Orleans, LA, Feb. 8-10, 1989.

CURRENT DEVELOPMENTS OF THE Cu/Nb-Ti SUPERCONDUCTING CABLES

FOR SSC IN HITACHI CABLE, LTD.

G. Iwaki*, S. Sakai*, Y. Suzuki**, H. Moriai**, and Y. Ishigami*

* Metal Research Laboratory, Hitachi Cable, Ltd.
** Tsuchiura Works, Hitachi Cable, Ltd.
   Tsuchiura, Ibaraki, 300, Japan

ABSTRACT

Extended research and development programs in national laboratories, universities and companies are now at a point where it can be said that the specifications for Cu/Nb-Ti superconducting cables are almost settled. Nevertheless, the need for superconducting cables with higher performances: higher current carrying capacities, lower magnetization and stabler productivity is more apparent than ever. A lot more studies have to be carried out.

In Hitachi Cable, Ltd., we are obtaining results which we feel could make a useful contribution to the SSC project. They are as follows: (1) 3460 A/mm^2 at 5 T in critical current density J_c obtained on a laboratory scale, (2) Low magnetization properties of wire compounded pure Ni confirmed, (3) 3000 A/mm^2 at 5 T in J_c on wires obtained on an industrial scale, (4) Possibility of stranded cables with large key-stone angle and 2750 A/mm^2 at 5 T in J_c confirmed, (5) Excellent oxygen free copper with the high RRR value of 500 developed.

INTRODUCTION

The specifications for Cu/Nb-Ti superconducting cables for the dipole magnets have been modified several times since the SSC project started. The modifications have mainly concerned the critical current density J_c and the dimensions of the Nb-Ti filaments. Cables with a J_c more than 2750 A/mm^2 at 5 T and 6 μm Nb-Ti filaments in diameter are the requirement at the present time.

Active investigation on critical current density has been carried out since 1984. Recently, a conductor with a J_c more than 3800 A/mm^2 at 5 T was obtained[1]. Small filament diameters creat a spacing problem with increase in conductor magnetization. Some arrangements have been designed to reduce this magnetization. Hitachi Cable, Ltd. has been concentrating on these problems. We have developed a system to produce and supply cables to meet all specifications.

In this paper, studies on increasing the J_c values and the results of a trial manufacture of low magnetization superconducting wires compounded Cu-0.5wt%Mn alloy[2] and pure Ni[3] on a laboratory scale, are described. The process conditions for cables with a J_c of 3000 A/mm^2 and the results of a trial manufacture of cable with a large key-stone angle on an industrial scale, are also described. Finally, an oxygen free copper with a high RRR value developed by Hitachi Cable, Ltd., is described.

CRITICAL CURRENT DENSITY

The factors which influence the critical current density (J_c) are numerous but the influence of intermediate heat treatment received is particularly well known. There are three parameters involved. These are temperature, heat treatment duration and the number of steps. It is very difficult to optimize the conditions for intermediate heat treatment because of the influence of the degree of cold working received between the steps and after the final heat treatment. The dependency of temperature and the number of intermediate steps on the J_c was clarified by serial experiments and the J_c raised. Cu/Nb-Ti wires were manufactured to obtain Nb-Ti filaments of 4-6 μm diameter. Time durations of intermediate heat treatment was constant 50 hours in all the cases. The specific resistivity of wires, a value of $10^{-14}\Omega$m, was used to define the superconducting state. Figure 1 shows the transition of optimum temperature at which the maximum J_c has been obtained at each step of heat treatment, up to four steps. All curves in Fig. 1 have a maximum point at which the maximum J_c shows a remarkable increase according to the increment of treatment steps. In the case of a treatment duration of 50 hours, it turns out that the J_c does not exceeds the value of the SSC specifications until after three steps.

The optimum temperature in each curve is different. Decrease is according to increment of treatment step with temperature intervals of 15 K. If multi step intermediate heat treatment is performed at the temperature corresponding to each of these optimum temperatures, that is, the temperature lower at steady intervals of 15 K, starting at 708 K, with the second step at 693 K, the third at 678 K etc, a higher J_c can be obtained. Fig. 2 shows the change in J_c when multi step intermediate heat treatment up to six steps has been performed under the conditions mentioned above. Based on the result in Fig. 1, The optimum temperatures in the fifth and the sixth are 648 K and 633 K respectively.

Comparing these results with that of Fig. 1, in a case of four steps, the J_c of 3230 A/mm^2 at 5 T obtained becomes slightly higher than the 3180 A/mm^2. A J_c of 3460 A/mm^2 at 5 T obtained after six steps treatments is the highest value obtained so far in Hitachi Cable, Ltd., but the conditions of cold woking have not necessarily been optimized yet and it is expected that an even higher value can be obtained after optimizing.

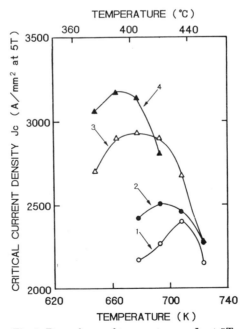

Fig. 1. Dependency of temperature on Jc at 5T.

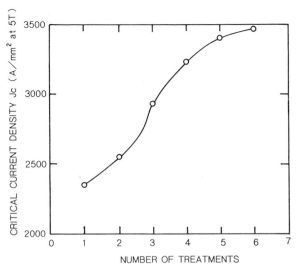

Fig. 2.   Relationship between Jc at 5T and number of steps in intermediate heat treatment. Temperature is 708 K at first, and decreases at constant intervals of 15 K.

LOW MAGNETIZATION WIRES

The smaller the Nb-Ti filaments become, the less the magnetization. However when the copper spacing between filaments is less than 1 μm magnetization increases due to the proximity effect. From this point of view, spacing in the SSC cables with copper matrix are required to be more than 1 μm.

Methods for Low Magnetization

Several methods have been proposed to reduce the magnetization. One is to substitute Cu-Ni or Cu-Mn alloys for the copper in the interfilamentary matrix[2]. Another is to composite with pure Ni[3]. When substitution of Cu-Ni alloys is applied to SSC cables, high Ni content alloys have to be used but these may cause Ni diffusion into the stabilizer copper during the intermediate heat treatment steps and hence a degradation of RRR. Because the SSC cables with Ni alloys have to receive the multi step intermediate heat treatments to obtain a high J_c. But in the case of using Cu-Mn alloys, it is possible to keep the Mn content as low as 0.5 wt%. In our study on SSC type superconducting wires using Cu-0.5wt%Mn, Mn did not diffuse even after 3 steps of heat treatment[4]. It appears therefore that this method is suited for low magnetization cables for the SSC. In the case of pure Ni composite, it is unnecessary to substitute pure Ni for the interfilamentary matrix, but simply to compound a proper quantity[3].

Experimental Details

To evaluate the effectiveness of the methods with respect to magnetization, we made three types of wires: conventional copper matrix wire, copper and Cu-0.5wt%Mn alloy matrix wire and pure Ni composite wire. We carried out measurements of the magnetization at low applied field for all three.

The specifications for the wires are shown in Table 1. CU sample refers to the wire using copper matrix, CML and CMS are the wires with Cu-0.5wt%Mn alloy as the interfilamentary matrix, and NI is the wire compounded with the pure Ni. The diameter of the wires were all 0.385 mm and the ratio of the filament spacing to the filament diameter s/d was 0.2 except for CMS which was 0.05. These wires were made by the double-stacking method with the number of Nb-Ti filaments in CU, CML and CMS being 5610 (=85x66), and in NI, 5478 (=83x66) as two Nb-Ti filaments are replaced by the same size pure Ni filaments in sub-bundles. The ratio of the pure Ni to Nb-Ti in NI was therefore 0.024 (=2/83). This value was determined after reference to Collings's paper[3]. The purity of the Ni used in the experiment was 99.7wt%. The nominal Nb-Ti filament diameter was about $3\,\mu$m for all, i.e. the filament spacings were about $0.6\,\mu$m except for CMS which was $0.16\,\mu$m, calculated from the measured ratio of the matrix to the Nb-Ti filaments.

Table 1. Specifications for wires for magnetization measurement

Sample name	CU	CML	CMS	NI
Wire diameter (mm)	0.385	0.385	0.385	0.385
Filament spacing/filament diameter (s/d)	0.2	0.2	0.05	0.2
Number of Nb-Ti filaments	5610	5610	5610	5478
	(85x66)	(85x66)	(85x66)	(83x66)
Number of Ni filaments	-	-	-	132
				(2x66)
Ratio of matrix to Nb-Ti*	1.72	1.66	1.73	1.76
Ratio of Ni to Nb-Ti	-	-	-	0.024
Nb-Ti filaments diameter$(\mu$m)**	3.1	3.2	3.1	3.1
filament spacing$(\mu$m)**	0.62	0.64	0.16	0.62
Strain after final heat treatment	4.2	4.2	4.2	4.2

*:measured values by etching-and-weighing method
**:nominal values

Fig. 3.   Cross-sectional photos of wires. (a) CU, copper matrix wire, s/d=0.2; (b) CML, Cu-0.5wt%Mn matrix wire, s/d=0.2; (c) CMS, Cu-0.5wt%Mn matrix wire, s/d=0.05; (d) NI, pure Ni compounded, s/d=0.2.

The wires were made under the same process conditions in order to get similar J_c, because the J_c would closely relate to the conductor magnetization. The intermediate heat treatment was done under conditions of temperature of 678 K for 50 hours in three steps. The strain after final heat treatment was 4.2 in all the wires.

Cross-sectional photos of the wires are shown in Fig. 3. It can been seen that the cross-sections are relatively good except for some of the outer filaments in the sub-bundle which are nonuniform due to the influence of double-stacking in the process. In CMS the Nb-Ti filaments are very close, and in NI it can be observed that two pure Ni filaments are sited symmetrically in each sub-bundle. The measurements of conductor magnetization were carried out on a high accuracy device equipped with SQUID. The measurement temperature was 4.5 K, and the external field was applied stepwise perpendicularly to the drawing axis of the samples.

Results

The hysteresis loops of CU, CML and CMS are shown in Fig. 4 to evaluate the effect of the Cu-0.5wt%Mn alloy. The hysteresis loop of CU, indicated by a solid line, shows a remarkable magnetization peak near the applied field of 0 T due to the proximity effect. To the contrary, in that of CML with the same s/d as CU, indicated by a large pitch dashed line, phenomena such as CU do not appear. It seems that the effect of Cu-0.5wt%Mn alloy appears. However magnetization at the beam injection field of 0.3 T is the same. In that of CMS, indicated by a small pitch dashed line, the magnetization peak appears near 0 T because of the very small spacing. However, the peak height is lower than that of CU, so Cu-0.5wt%Mn alloy is effective against the magnetization peak near 0 T. Nevertheless the magnetization of CMS is higher than in CU in all applied field ranges except for near 0 T. When the applied field becomes higher the effect of Cu-0.5wt%Mn disappears, as can been seen from the results on CML.

Figure 5 shows the hysteresis loop of NI in comparison with that of CU. The degree of magnetization of NI is remarkably low even if the difference in the number of Nb-Ti filaments is taken into account. The magnetization at applied field of 0.3 T is positive, and the absolute value is small. This result indicates that it is possible for conductor magnetization at a beam injection field to become extremely small by applying this method. When the width of the hysteresis loop is checked it can be seen that in NI it is fairly narrow. This result is contrary to the primary expectation that the hysteresis loop would have been a simple summation of that of pure Ni and Nb-Ti. The causes for this will be investigated in the future.

Table 2 shows the J_cs of wires used in experiments at several external applied fields. In spite of the intermediate heat treatment conditions mentioned above, J_c at 5 T was only about 2600 A/mm^2 except for CMS which was 2700 A/mm^2. However, the J_c of CU, CML and NI were almost the same. It is therefore not necessary to take the difference in J_c into account in the interpretation of these phenomena, and it can be seen that they are inherent phenomena.

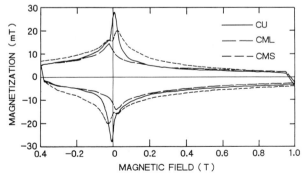

Fig. 4.   Comparison of hysteresis loops of copper matrix wire CU and Cu-0.5wt%Mn matrix wire CML, CMS. The ratios of filament spacing to filament diameter s/d are 0.2 except for CMS of 0.05. Diameter of Nb-Ti filaments is about 3 μm in all wires. Magnetization values are magnetic moments per unit wire volume.

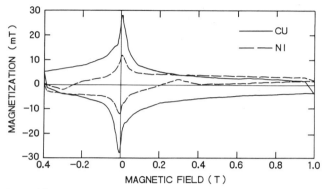

Fig. 5.   Comparison of hysteresis loops of copper matrix wire CU and pure Ni compounded wire NI. The ratios of filament spacing tio filament diameter are 0.2. Diameter of Nb-Ti and pure Ni filaments is about 3 μm. Magnetization values are magnetic moments per unit wire volume.

Table 2. Critical current density of wires for magnetization measurement (A/mm^2)

Sample	Applied field (T)				
	0.3*	4.0	5.0	6.0	7.0
CU	11100	3150	2560	2000	1465
CML	10760	3185	2590	2040	1515
CMS	11870	3335	2700	2090	1530
NI	11190	3150	2570	1990	1465

∗: values without self field correction

DEVELOPMENTS ON INDUSTRIAL SCALE

3000 A/mm^2 at 5 T Class Cable

3000 A/mm^2 at 5 T in J_c on a pre-stranded wire has been obtained on an industrial scale, but this value is not reproducible. The J_c values on wires on an industrial scale which are obtainable have been in the range from 2800 to 2900 A/mm^2 at 5 T. Intermediate heat treatment is done in three steps. Based on the indication that J_c increases according to the increment of the heat treatment steps as mentioned above, four intermediate heat treatment steps were performed. The conditions for heat treatment were at 678 K for 50 hours each step. The results are shown in Fig. 6 alongside the results of three step heat treatment for comparison. The pre-treated wires are the same. After four steps, the J_c at 5 T is 200 A/mm^2 higher than in the three with the result that the J_c is slightly over 3000 A/mm^2. This corresponds to the findings on a laboratory scale.

This result indicates that it is possible to obtain even higher J_cs if the intermediate heat treatment steps are increased. However, four steps is the maximum number of steps acceptable on an industrial scale. In the future, processing conditions such as in the intermediate heat treatment, cold working and others will be optimized to raise the J_c even more.

Cable with Large Key-stone Angle

In the dipole magnets in SSC, there is a problem with the coherency between the wound cables and wedges inserted between them. Recently a dipole magnet with a bore diameter of 50 mm but with no wedges was developed by Sintomi et al[5]. The report so far indicates that the dipole magnet is working well. A wedgeless magnet is desirable not only from the standpoint of magnet stability but also facility of winding.

The cables needed for the 50 mm bore dipole magnet are different from existing cables. The most important is the large key-stone angle of 3 deg. contrary to the 1.6 deg. in existing cables. The degradation of J_c during stranding can be expected to increase even more. To prepare for this, a trial manufacture of a cable with a key-stone angle of 3.07 deg. as shown in Table 3 was carried out. Pre-stranded wires with a J_c of 2800 A/mm^2 at 5 T after three step intermediate heat treatment were used.

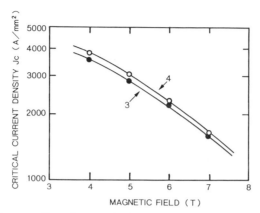

Fig. 6. Relationship between Jc and magnetic field in industrial wire. The numbers indicate the number of steps of heat treatment. Treatment conditions are at 678 K for 50 hours in all the cases.

Table 3. Specifications of cable with a large key-stone angle

Items	Values
Pre-stranded wire diameter (mm)	0.808
Jc at 5T of pre-stranded wire (A/mm^2)	2800
Number of wires in cable	23
Key-stone angle of cable (deg.)	3.07
Lay pitch of cable (mm)	79
Mid-thickness of cable (mm)	1.39
Width of cable (mm)	9.34

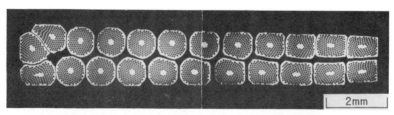

(a) Trial manufactured cable (key-stone angle : 3.07deg.)

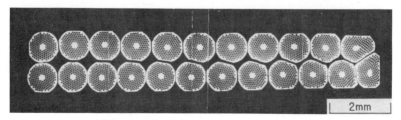

(b) Existing cable (key-stone angle : 1.6deg.)

Fig. 7. Cross-sectional photos of cables for the inner. (a) Trial manufactured cable with key-stone angle of 3.07 deg. ; (b) existing cable with key-stone angle of 1.6 deg.

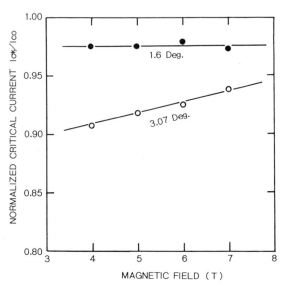

Fig. 8. Relationship between normalized critical current Ic*/Ico. Ic*: critical current after stranding Ico: critical current before stranding

The cross-section of a cable obtained in this trial manufacture is shown in Fig. 7 with an existing cable alongside for comparison. The photo shows that there is no anomalous deformation even at the narrow side, and the cable is good. However, a comparison of the critical current Ic on the wires before and after stranding shows that Ic degradation increased as expected. Figure 8 shows the relationship between the normalized critical current Ic*/Ico (Ic*:after stranding, Ico:before stranding) and the external applied field in a comparison of large and small key-stone angle cables. It can been seen in the figure that the degradation of this trial manufactured cable is 6 % at 5 T and 3% at 7T i.e. greater than in existing cable.

Hitachi Cable, Ltd. are in a position to provide cables with large key-stone angles and J_c of 2750 A/mm^2 at 5 T for 50 mm bore dipole magnets though the Ic degradation may not always be satisfied in a system to produce cables with J_c of 3000 A/mm^2 at 5 T for the above mentioned reasons. Our next step will be to develop even more excellent stranding techniques to achieve an Ic degradation similar to existing cable.

HIGH RRR OXYGEN FREE COPPER

Cables for SSC dipole magnets also require high RRR (Residual Resistivity Ratio) values (more than 66 for the inner and 63 for the outer) to achieve high stability in magnets.

Hitachi Cable, Ltd. is one of the few companies which supply mass produced oxygen free copper (OFC). The high performance of our OFC products are recognized all over the world. In addition, other uses of OFC are being investigated in ongoing projects. An excellent OFC for cryogenic purposes (CG-OFC) with an RRR value of 500 in bulk has been developed[6].

The CG-OFC can be used for stabilizers in superconductors. Figure 9 shows the relationship between the thermal conductivity, a very important property for stabilizers, and temperature alongside Class-1 OFC ASTM grade (OFC-1) with an RRR value of 250. At extremely low temperatures (< 20 K), CG-OFC has roughly twice the thermal conductivity of OFC-1. For experiment, we made two types of superconducting wire, with CG-OFC and OFC-1 as shown in Fig. 10, and measured the RRR values. The measurement results are shown in Table 4 accompanied with the specifications of both wires. It can be seen that CG-OFC matrix wire has a much higher RRR value.

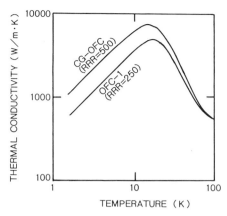

Fig. 9. Relationship between thermal conductivity and temperature.

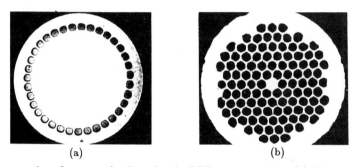

Fig. 10. Cross-section of superconducting wires for RRR measurements. (a) High matrix ratio wire; (b) Low matrix ratio wire.

Table 4. Specifications and RRR values of superconducting wires with CG-OFC and OFC-1

	High matrix ratio wire	Low matrix ratio wire
Wire diameter (mm)	0.47	0.47
Number of Nb-Ti filaments	36	108
Nb-Ti filament diameter (μm)	28	30
Ratio of matrix to Nb-Ti	7.1	1.32
RRR values in CG-OFC	450	272
RRR values in OFC-1	220-250	180-210

Wires with a high matrix ratio have higher RRR values than wires with a low matrix ratio as the filament spacing becomes about 6 μm. The RRR values of the interfilamentary matrix degrades. In the case of SSC cables, the interfilamentary matrix does not contribute to the RRR values because the interfilamentary spacings are nearly 1 μm, and the RRR values are governed by the small area excepting the interfilamentary matrix. It is needless to say that materials with high RRR values are indispensable for stabilizers in SSC cables. CG-OFC seems best suited for the job.

CONCLUSIONS

(1)   The critical current density J_c increases with the number of intermediate heat treatment steps. There are optimum conditions corresponding to each step. The optimum temperatures decrease depending on the number of steps, under conditions of 50 hours. Intervals between the optimum temperatures are 15 K. Based on these results, a J_c of 3460 A/mm^2 at 5 T has been obtained after six step intermediate heat treatment.

629

(2) Wires have been manufactured in trial by compounding with Cu-0.5wt%Mn alloy as the interfilamentary matrix and substituting Nb-Ti filaments with pure Ni filaments. A comparison with conventional wire with an all copper matrix has shown that Cu-0.5wt%Mn alloy is effective against the hysteresis peak which appears near 0 T in an applied field. However, the magnetization of a wire compounded with Cu-0.5wt%Mn alloy at a beam injection field of 0.3 T is similar to that of conventional wire. In a pure Ni composite, the magnetization at 0.3 T is not only small but also the area of the hysteresis loop is remarkably reduced. This confirms that the pure Ni composite is very effective against problems of conductor magnetization.

(3) On a industrial scale, based on experimental results, if the number of intermediate heat treatment steps is increased from three to four production of cables with a stable J_c of 3000 A/mm^2 at 5 T is possible.

(4) Cable with a large key-stone angle for 50 mm bore dipole magnets has been manufactured in trial. The Ic degradation of the cable obtained was larger than expected. However, it is possible to produce cables with a large key-stone angle and J_c of 2750 A/mm^2 at 5 T because of pre-stranded wires with a J_c of 3000 A/mm^2 were obtainable.

(5) An oxygen free copper having an RRR value of 500 has been developed. Using this excellent oxygen free copper as a stabilizer for superconductors the RRR value can be modified further, that is, the stability of superconducting magnets can be increased.

ACKNOWLEDGEMENTS

The authors wish to express thanks to Professors H. Hirabayashi and T. Shintomi of KEK for their useful discussions concerning the cables in the dipole magnets. Mr. R. Saito, Mr. T. Suzuki and Mr. K. Asano of Hitachi Ltd. should also be thanked for their valuable suggestions.

REFERENCES

1. K. Matsumoto and Y. Tanaka, "High Critical Current Density in Multifilamentary NbTi Superconducting Wires",6th US-Japan Workshop on High Field Superconductors, Boulder, Colorado, Feb. 22-24 (1989)

2. E. W. Colings, "Stablizer Design Considerations in Fine-Filament Cu/NbTi Composites", Advances in Cryogenic Engineering Materials, vol.34, 867 (1988)

3. E. W. Collings, K. R. Marken Jr. and M. D. Sumption, "Design of Coupled or Uncoupled Multifilamentary SSC-type Strands with Almost Zero Retained Magnetization at Fields Near 0.3 T", International Cryogenic Materials Conference, Los Angeles, California, Jul. 24-28 (1989)

4. S. Sakai, G. Iwaki, Y. Sawada, H. Moriai and Y. Ishigami, "Recent Developments of the Cu/Nb-Ti Superconducting Cables for SSC in Hitachi Cable, Ltd.", IISSC'89, New Orleans, Louisiana, Feb. 8-10 (1989)

5. T. Shintomi, A. Terashima, H. Hirabayashi, H. Miyazawa, T. Kawaguchi and S. Murai, "Development of Superconducting Dipole with Ideal Arch Structure Using Large Keystone Angle Cable", 11th International Conference on Magnet Technology, Tsukuba, Japan, Aug. 28-Sep. 1 (1989)

6. Y. Nagai, S. Sakai, K. Sugaya and H. Moriai, "Development of High Purity Oxygen Free Copper and It's Application for Superconducting Wires", to be published in Hitachi Cable Review No.9 (1990)

DEVELOPMENT AND LARGE SCALE PRODUCTION OF

NbTi AND Nb$_3$Sn CONDUCTORS FOR BEAM LINE AND DETECTOR MAGNETS

Helmut Krauth

Vacuumschmelze GmbH
Grüner Weg 37
D6450 Hanau, F. R. Germany

ABSTRACT

Fine filament NbTi and Nb$_3$Sn conductors for accelerator magnets were developed and manufactured. With NbTi current densities above 3000 A/mm^2 at 5 T and 4.2 K were achieved in industrial scale production of conductors with 5 μm diameter. The conductors exhibited negligeable filament coupling effects even with 2.5 μm filaments in a pure Cu matrix. Bronze route Nb$_3$Sn conductors with small effective filament diameters (about 7 μm) and non-Cu critical current density of 800 A/mm^2 were developed. Keystoned cables with both, NbTi and Nb$_3$Sn, were manufactured with strand numbers of 23 to 36. The capabilities of large scale production were demonstrated during the fabrication of the cable for the HERA quadrupoles. NbTi and Nb$_3$Sn cables produced for CERN were used in model magnets for LHC verifying the feasibility of dipole magnets with magnetic fields of 9 T to 10 T. Aluminium stabilized NbTi conductors were produced by co-extrusion for detector magnets of several projects in US and Europe. The process of co-extrusion was demonstrated to be applicable also to reacted Nb$_3$Sn conductors.

INTRODUCTION

For the next generation of hadron colliders like SSC in US, UNK in Soviet Union and LHC in Europe cabled conductors are needed for the beam line magnets consisting of strands with high critical current density and fine filaments. Although different in detail, the specifications call for NbTi conductors with a current density of 2750 A/mm^2 or higher at 5 T and 4.2 K at filament diameters of 6 μm or below (down to 2.5 μm). This represents a further enhancement of required conductor performance with respect to the HERA accelerator, where 2400 to 2500 A/mm^2 were achieved in conductors with 12 to 20 μm diameter [1].

In case of Nb_3Sn a non-Cu critical current density of 1300 A/mm^2 at 10 T and 4.2 K is aimed for. This j_c-level is achievable in principle by using specific manufacturing processes, but represents a difficult task in connection with the requirement of having small effective filament diameters. Despite the relatively low non-Cu critical current density the bronze route seems to be suitable to come closest to the specified goal.

Besides the beam line magnets large superconducting solenoids are needed in the detectors installed in the colliding regions. In case that part of the detector is outside the solenoid, the magnets have to be radiation transparent. Aluminium stabilized NbTi conductors were developed for this purpose and were fabricated for example for the detectors ALEPH (CERN), CLEO II (Cornell University), H1 (DESY). The use of Nb_3Sn may be advantageous in the next generation of detectors because of the larger stability margin. The co-extrusion process used for NbTi was therefore adapted for its use also with brittle reacted Nb_3Sn conductors.

NbTi AND Nb_3Sn CONDUCTORS FOR BEAM LINE MAGNETS

The conductors of the dipoles and quadrupoles for high energy hadron accelerators are keystoned flat cables made from 20 to 40 individual strands of NbTi or Nb_3Sn. Of prime importance are high critical current density to achieve high fields and small effective filament diameter to reduce field distortions due to persistent currents in the filaments. The current status of critical current densities in fine filament conductors produced on an industrial scale are summarized in Fig. 1. Effective filament diameters as determined by magnetization measurements are 5 μm (equal to nominal diameter) for NbTi and about 7 μm (at 3 μm nominal diameter) for Nb_3Sn, respectively. For NbTi the j_c-values are also given at 2 K, because the magnets of LHC in the regime of 9 T to 10 T most probably will be built with NbTi and operated at 2 K. These data are extrapolated from 4.2 K by applying a shift of $\Delta B = 3$ T. This rule of thumb has been veryfied to be applicable in general, but has to be confirmed of course before starting large scale production. In case of Nb_3Sn non-Cu critical current density as well as Nb_3Sn layer critical current density is given. The latter values are calculated from the measured matrix to Nb ratio and assuming a volume increase of 38 % during the reaction heat treatment. The data are needed for the evaluation of magnetization measurements and estimation of effective filament diameters.

Transport critical current densities were determined by standard resistive 4 terminal measurements at a sensitivity of 10^{-14} Ωm for NbTi and 0.1 μV/cm for Nb_3Sn, respectively. DC magnetization measurements were performed in a vibrating sample magnetometer with a short piece of wire perpendicular to the magnetic field. Effective filament diameters were estimated from magnetization by the generally accepted method [2] and using the measured transport critical current densities for evaluation.

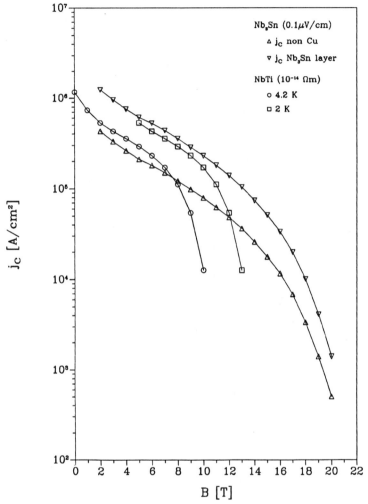

Fig. 1 Critical current density as a function of external
field for fine filament conductors (without self
field correction):
- NbTi: 4.2 K and 2 K (wire diameter 0.65 mm;
nominal and effective filament diameter 5 μm;
see Fig. 2)
- Nb_3Sn bronze route: non-Cu j_c and Nb_3Sn-layer
j_c (wire diameter 0.92 mm; filament diameter:
3 μm nominal, about 7 μm effective; see Fig. 5)

NbTi-Conductors

The scale-up of conductor technology from HERA type conductors to the enhanced performance turned out to be rather straight forward. The main changes were the use Nb barriers around the filaments and the use of more sophisticated stacking techniques in the billet production.

As for the billet design, single stacking techniques of monofilamentary elements is the preferred solution for filament numbers below about 10000, as required for UNK and the main line SSC design with 6 μm filaments. Higher filament numbers as needed for the back-up design of SSC with 2.5 to 3 μm or for LHC conductors are accommodated best by a double stacking technique.

Stacking of hexagonal elements turned out to yield the optimum geometrical uniformity of the filamentary array. Stacking of round elements is also feasible [3] but turned out to lead to increased probability for stacking faults and distortions of the filamentary array. Special toolings were developed to produce hexagonal elements with low tolerances and to facilitate bundling of hexagons with face-to-face dimensions below 2 mm. Full size billets according to SSC specifications were produced by this stacking methods, both, for conventional indirect extrusion (250 kg yield) and hydrostatic extrusion (100 kg yield). Both techniques allow warm extrusion with temperatures well below 600°C and yield very uniform filament arrays. Hydrostatic extrusion even allows cold extrusion offering the opportunity to make available cold work prior to extrusion for j_c optimization. A detailed experimental programme is presently being pursued to explore these opportunity further, in supplement of own previous experience and published results [4].

Single stack conductors with 6000 filaments and a nominal Cu:NbTi ratio of 1.8 were drawn to wires and j_c optimized for different wire and filament diameters. A cross section of a wire according to the SSC specification of the outer cable strand is shown in Fig. 2 together with a SEM micrograph of the 5 μm filaments. Typical critical current densities and n-values at 5 T and 4.2 K are:

$$d = 8\text{--}10 \ \mu m \qquad j_c = (2850\text{--}3100) \ A/mm^2 \qquad n = 40\text{--}50$$

$$d = 5\text{--}6 \ \mu m \qquad j_c = (2750\text{--}3000) \ A/mm^2 \qquad n = 40\text{--}50$$

$$d = 2.5\text{--}3 \ \mu m \qquad j_c = (2450\text{--}2550) \ A/mm^2 \qquad n = 25\text{--}30$$

The conductor described above has a pure Cu-matrix. It was expected therefore, that some proximity coupling would occur in the low field region leading to increased magnetization and to a large effective filament diameter [5]. Therefore magnetization measurements were performed in a vibrating sample magnetometer. The results are shown in Fig. 3. The width of the magnetization curve normalized to the superconductor volume 2 $M_s = 2 \ M(\alpha + 1)$, with α = Cu: NbTi ratio, (e. g. 16 mT at 0.66 T external field) is consistent with a filament diameter of 2.6 mm as expected from the billet design at 0.332 mm wire diameter. Only at very low fields well below 0.1 T coupling effects are seen.

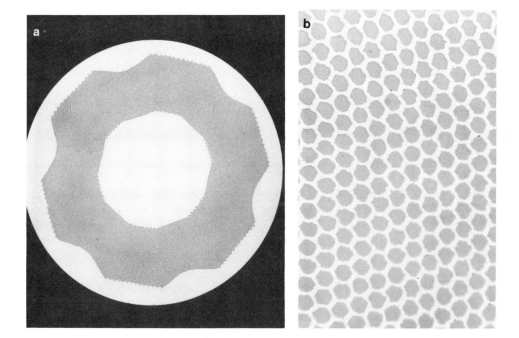

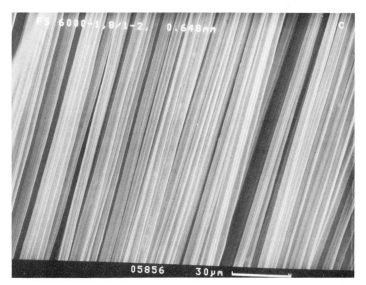

Fig. 2 NbTi conductor corresponding to SSC outer cable strand
with 6000 filaments at final diameter of 0.65 mm with
a critical current density of 3000 A/mm^2 at 5 T and
5 μm diameter filaments (Cu : NbTi = 1.8 : 1)

a) conductor cross section
b) close-up view of the filamentary array
c) SEM micrograph of filaments

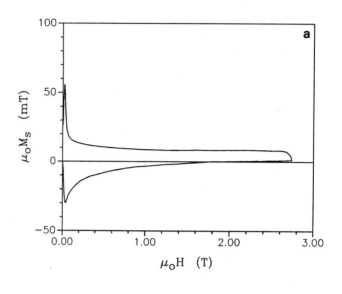

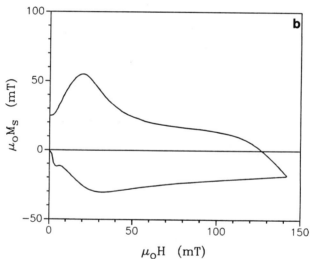

Fig. 3 Magnetization curves of a NbTi conductor with 6000
filaments with a nominal diameter of 2.6 μm in a
pure Cu matrix (M_S normalized with respect to NbTi
volume)

a) high field region (in accordance with $d_{eff} \lessgtr 3$ μm)
b) low field region, showing filament coupling effects
only at fields well below 0.1 T in the virgin run
as well as in the return curve.

The initial slope of the virgin curve is much steeper than expected from the NbTi volume. This is due to proximity coupling and the filament bundle acting as a single filament shielding the total volume of the bundle. Also, in the return path indications of coupling can be seen leading to a typical double hump characteristics in case the full curve would have been drawn [6].

Although filament coupling at injection condition (0.4 to 0.7 T, depending on the project) can be avoided with 2.5 μm filaments in a pure Cu-matrix by a proper conductor design and processing, work is going on to verify the effect of a CuMn-matrix in the filamentary area.

Other important wire parameters besides j_c and magnetization are residual resistance and mechanical properties. For the SSC outer cable strand a RRR value of $\geq$ 200 was measured in annealed condition. The spring back test required by SSC yielded a very good value of 720° in comparison with the specified value of $\leq$ 1090°.

NbTi keystone cables. Different keystone cables were produced from NbTi with strand numbers varying from 23 to 36. This includes cables for the HERA dipoles (24 strands) and quadrupoles (23 strands). Cables for LHC model coils (30 and 36 strands) and test cables for SSC (23 and 30 strands) Examples of cables are shown in Fig. 4. Typical degradation was between 3 and 5 % as measured at wire samples prior to and after cabling. This was veryfied by cable measurements on the HERA cable at BNL [1]. Strand measurements were shown to be in general a simple methods to determine degradation due to cabling [7].

Large scale fabrication capabilities were demonstrated both for strands and cables during production of the HERA quadrupole cable within a relatively short period although, of course the total quantity was much less than required for the present projects. The quality control measures established in our plant like ultrasonics and multifrequency eddy current testing of semifinished products and final conductors lead to a uniform production of conductor. As an example I_c of the strands varied by only $\pm$ 4 %. By sorting of the strands the cable I_c varied by only $\pm$ 1 %. This translated into a variation of magnetization at 0.3 T of also only 1.1 % [1].

Magnet performance The HERA quadrupoles reached design fields with virtually no training and showed excellent field quality [8]. The first two 1 m model dipoles built for LHC showed also little training and reached central fields of 9.37 T and 9.45 T, respectively [9, 15].

Nb₃Sn Conductors

The j_c values of the Nb₃Sn phase is much higher than that of NbTi, as can be seen from Fig. 1, especially in high fields above 8 T. How much of this current density can be made available for magnet producers depends on the wire fabrication route and on the constraints of a specific applications.

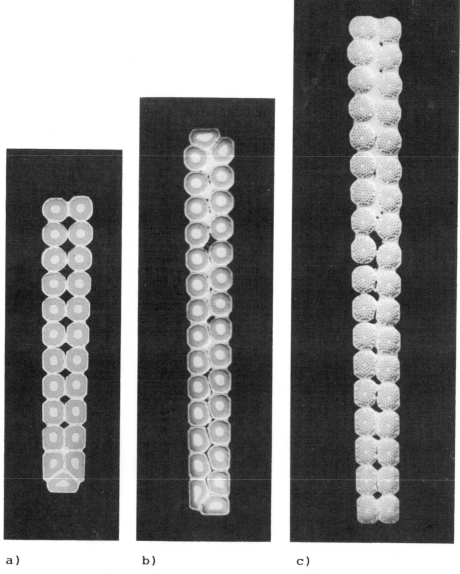

a)   b)   c)

a) 23 strands à 0.84 mm diameter (HERA quadrupoles)
b) 30 strands à 0.84 mm diameter (LHC 9 T model dipole)
c) 36 strand à 0.9 mm diameter (test cable)

Fig. 4 Examples of keystoned NbTi cables

An updated comparison of the bronze route with other Nb_3Sn manufacturing techniques will be presented elsewhere [10]. With respect to accelerator applications it can be stated that conductor versions with high j_c values tend to exhibit very large effective filament diameters due to inherent properties of the approach (e. g. tube techniques, jelly roll techniques) or due to metallurgical coupling by filament bridging after reaction heat treatment. The best compromise seems to be possible with bronze processed Nb_3Sn wires [2,11]. Here, the large fraction of CuSn bronze required to get a sufficiently large Sn reservoir can be redistributed between the filament for decoupling without sacrificing overall critical current density.

An example of an externally stabilized bronze route Nb_3Sn conductor with a local CuSn: Nb ratio of 1.5 is shown in Fig. 5. The measured critical current density (e. g. 800 A/mm^2 at 10 T and 4.2 K) is given in Fig. 1. The nominal filament diameter after reaction is about 3 μm.

Magnetization measurements performed on this wire revealed an effective filament diameter of about 7 μm, which is quite satisfactory and suitable for magnet application. Cross sectional micrograph showed no obvious filament bridging, such that occasional bridges along the conductor length seem to be responsible for the increase in magnetization. Details and ways for further improvements will be discussed elsewhere [12].

Nb₃Sn keystone cables were manufactured for the Nb_3Sn development programme for LHC from externally stabilized Nb_3Sn bronze route conductors. An example is shown in Fig. 6. Measurements of j_c revealed no degradation with respect to strand measurements [13].

Magnet performance. A mirror coil and a 1 m model dipole were built from the described cables. Both magnets went well beyond 9 T reaching a peak field of 10.2 T (mirror) and 10.05 T corresponding to a central field of 9.5 T (model) [14,15], with the limitation in the latter case most probably not being found in the critical current of the cable.

Work is progressing to enhance the performance of bronze route conductors closer towards the envisaged LHC specification.

CONDUCTORS FOR DETECTOR MAGNETS

Large superconducting solenoids are widely used in the particle detectors of colliders. In case the solenoid is incorporated into the detector it has to be radiation transparent. For this purpose aluminium stabilized conductors are used. A continuous co-extrusion process has been developed in which completed NbTi cables are clad with pure aluminium yielding a highly stabilized, radiation transparent conductor. Typical conductors delivered for several projects are shown in Fig. 7 [1].

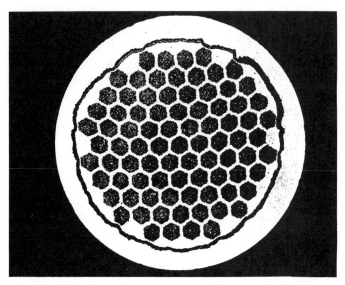

Fig. 5 Nb$_3$Sn bronze route conductor with 20000 filaments and local matrix to Nb ratio of 1.5 (wire diameter 0.92 μm)

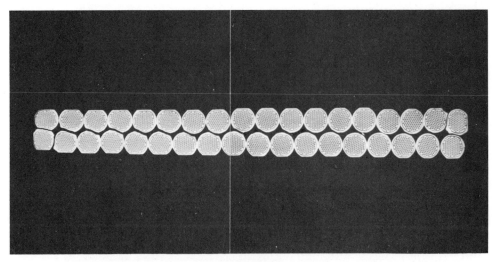

Fig. 6 Keystoned Nb$_3$Sn cable with 36 bronze route strands of 0.92 mm diameter containing 20000 Nb filaments with 2.5 μm diameter and 38 % of Cu-stabilizer

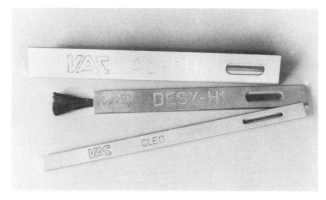

Fig. 7 Aluminium stabilized NbTi conductors used
in detector magnets
- ALEPH at LEP (CERN), 35 x 3.6 mm^2
- CLEO II at CESR (Cornell University), 16 x 5 mm^2
- H1 at HERA (DESY), 26 x 4.5 mm^2

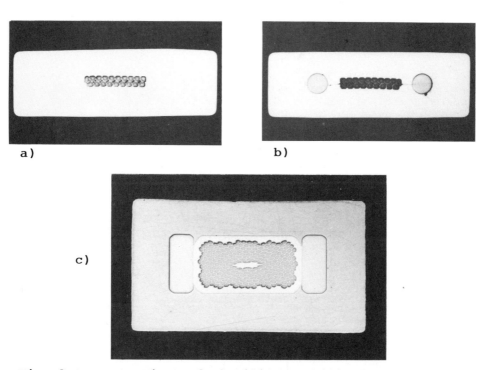

a)

b)

c)

Fig. 8 Cross sections of aluminium stabilized Nb$_3$Sn
conductors with different amounts of Al with and
without mechanical reinforcement
a), b) Nb$_3$Sn cable, dimension of finished conductor
16 x 5.4 mm^2
c) monolithic strand, dimension of finished
conductor 6.6 x 3.6 mm^2

In future large solenoids Nb_3Sn conductors may be of interest because of the larger stability and/or temperature margin. Therefore the co-extrusion process was also qualified for reacted Nb_3Sn conductors [16]. It could be demonstrated that the co-extrusion process does not damage the brittle Nb_3Sn conductor and that the j_c degradation is acceptable. In addition to co-extrusion of aluminium on Nb_3Sn, also reinforcing members were incorporated to strenghten the conductor mechanically. Typical examples of conductor geometries are shown in Fig. 8.

CONCLUSION

Fine filament conductors based on NbTi are now available allowing the construction of dipoles and quadrupoles with fields of 5 to 7 T at 4.2 K (SSC, UNK) and 9 to 10 T at 2 K (LHC). At 5 μm filament diameter current densities of 3000 A/mm^2 can be achieved with standard heat treatment schedules. Work is in progress to demonstrate similar values also for wires with 2.5 μm filament (4.2 K and 5 T).

With bronze route Nb_3Sn conductors a similar regime of 9 T to 10 T could be achieved. Also, the wires used have small effective filaments of about 7 μm. Nevertheless a further increase of critical current density seems to be needed to increase the margin and to allow for more Cu-content for protection reasons.

The conductor technology for large detector solenoids based on NbTi conductors with aluminium is available. Mechanical reinforcement can be incorporated if necessary. Extrapolation to Nb_3Sn conductors is possible.

Although it seems certain that the High-T_c-Superconductors (HTSC) will not be incorporated in accelerator magnets in the near future, they may be used for certain components like current leads and interconnections to reduce heat losses or allow more temperature margin, respectively. It could be demonstrated recently that HTSC wires based on BiSrCaCuO can carry significant current in very high fields at 4.2 K and also at intermediate temperatures (20-40 K) even in the field of several T [17,18].

ACKNOWLEDGEMENT

The author is very much indebted to his collegues at Vacuumschmelze who were responsible for and carried out the work reported here. He also would like to thank Mr. M. Kemper, Institut für Angewandte Physik, Gießen, for performing the magnetization measurements.

REFERENCES

1. H. Krauth, "Recent developments in NbTi Superconductors at Vacuumschmelze", IEEE Trans. MAG 24 (1987) 1023

2. A. K. Gosh, K. E. Robins and W. B. Sampson, "Magnetization measurements on Multifilamentary Nb_3Sn and NbTi Conductors", IEEE Trans. MAG 21 (1985) 328

3. H. Krauth "NbTi Superconductors for Accelerator Magnets: Large Scale Production for HERA and Development of Fine Filament Conductors", Proc. ICFA Workshop on Supercond. Magnets and Cryogenics, BNL, May 12-16, 1986, BNL report 52006

4. E. Gregory, T. S. Kreilick and J. Wong, "Innovations in design of Multifilamentary NbTi Superconducting Composites for the Supercollider and other applications" in "Supercollider 1", M. Mc Ashan, editor, Plenum Press, New York, 1989, p. 277

5. A. K. Gosh, W. B. Sampson, E. Gregory and T. S. Kreilick, "Anomalous Low Field Magnetization in Fine Filament NbTi Conductors" IEEE Trans. MAG 23 (1987) 1724

6. E. Gregory, T. S. Kreilick, J. Wong, E. W. Collings, K. R. Marken, R. M. Scanlan, C. E. Taylor "A Conductor, with Uncoupled 2.5 μm Filaments, Designed for the Outer Cable of SSC Dipole", IEEE Trans. MAG 25 (1989) 1926

7. M. Garber, A. K. Gosh and W. B. Sampson, "The Effect of Self Field on the Critical Current Determination of Multifilamentary Superconductors", IEEE Trans. MAG 25 (1989) 1940

8. H. R. Barton et al. "Performance of the Superconducting Magnets for the HERA Accelerator", loc. cit. 14

9. R. Perin, D. Leroy, G. Spigo "The First, Industry Made Model Magnet for the CERN Large Hadron Collider", IEEE Trans. MAG 25 (1989) 1632

10. H. Krauth, A. Szulczyk, M. Thöner, "Nb_3Sn Multifilamentary Conductors: An Updated Comparison of Different Manufacturing Routes", to be presented at 1990 Applied Supercond. Conf., Snow Mass, CO., Sept. 24-28, 1990

11. A. K. Gosh, M. Suenaga "Magnetization of Internal-Tin Processed Nb_3Sn Wires", Proc. 6th Japan-US Workshop on High-Field Superconducting Materials, Boulder, CO, February 22-24, 1989

12. H. Krauth, A. Szulczyk, M. Thöner, K. Heine, "Properties of NbTi and Nb_3Sn fine filament conductors", to be presented at Europ. Particle Acc. Conf., Nizza, June 12-16, 1990

13. S. Wenger, F. Zerobin, A. Asner, "Towards a 1 m Long High Field Nb_3Sn Magnet of the ELIN-CERN Collaboration for the LHC-Project" IEEE Trans. MAG 25 (1989) 1636

14. A. Asner, R. Perin, S. Wenger, F. Zerobin, "First Nb_3Sn, 1 m Long Superconducting Model Magnets for the LHC Break the 10 T Field Threshold" Proc. 11th Magnet Technology Conf., Tsukuba/Japan, Aug. 28-Sept. 1, 1989

15. R. Perin for the CERN LHC Magnet Team, "First Results of the High-Field Magnet Development for the Large Hadron Collider" loc. cit. 14

16. M. Thöner, H. Krauth, J. Rudolph, A. Szulczyk, "Aluminium Stabilized Nb_3Sn Superconductors" Adv. Cryo. Eng. 34 (1988) 507, Plenum Press, New York

17. K. Heine, J. Tenbrink, M. Thöner, "High-Field Critical Current Densities in $Bi_2Sr_2Ca_1Cu_2O_{8+x}$/Ag wires" Appl. Phys. Lett. 55 (1989) 2441

18. K. Heine, J. Tenbrink, H. Krauth, "Temperature Dependence of Critical Currents in Bi-based High-T_c Superconductor Wires", to be presented at ICMC '90 Topical Conference on High-Temperature Superconductors, Material Aspects, Garmisch-Partenkirchen, May 9-11, 1990

TESTS OF DC CABLE-IN-CONDUIT SUPERCONDUCTORS

FOR LARGE DETECTOR MAGNETS

P.G. Marston, J.R. Hale, J. Ludlam and A. M. Dawson

Plasma Fusion Center, Massachusetts Institute of Technology
Cambridge, MA 02139

Abstract

At the previous IISSC conference, a novel cable-in-conduit conductor (CICC) was proposed that offered not only the many advantages of traditional CICC such as helium containment within the conduit and integral structural support, but also increased protection characteristics. The latter is achieved by separating the protection and stabilization functions in the conductor so that the protection function is incorporated in a conduit with relatively low electrical resistivity, and the stability function is retained by the cable. An embodiment of this approach was the addition of a braided dacron sleeve surrounding the cable, acting as an electrical barrier between cable and conduit. The results of comparison tests performed on two samples, one without the barrier, one with, have been dramatic, indicating clearly the improved protection capabilities of the latter conductor. This new configuration of CICC, with improved protection characteristics should be desirable for all types of detector magnets, but especially for those requiring radiation-thin windings.

Proof-of-Concept Conductor Tests

Two test samples were prepared by winding single-layer noninductive coils on mandrels of G-10, on the OD of which had been machined two V-grooves in a double helix pattern. The total length of subscale CICC on each sample was approximately 9.2 meters; of this length, approximately 1 m was incorporated into the two end-termination assemblies. Every reasonable effort was made to ensure that the sample coils would be physically, electrically, and thermally identical, save for the one difference that was the prime object of the performance comparison, the braided dacron sleeve that insulated the cable from the conduit in one of the samples. Hereinafter, this sample will be referred to as the sleeved conductor, to distinguish it from the conventional CICC sample, in which the cable is not insulated from the ID of its conduit. A cross-sectional photomicrograph of the conductor is shown in figure 1.

A resistive heater measuring 0.001 m in diameter and 0.3 m in length was soldered with a generous fillet along the outer surface of the conduit at one end of each sample coil. The heater is made so that the internal resistance elements are electrically isolated from its stainless steel jacket by compacted ceramic powder, thereby permitting the user to solder the stainless jacket directly to the sample with no need to interpose an insulating barrier.

Two experimental runs were carried out, one for each of the test coils. During each experimental run, a pressure transducer was connected at the sample termination that was nearest the heater. Figure 2 is a schematic of these instrumentation connections. Background magnetic field was generated by a 10 MW, 10" bore Bitter

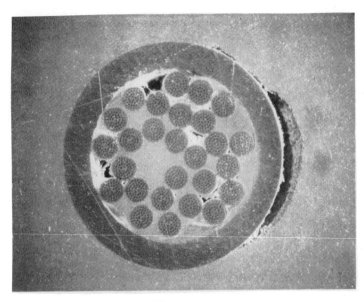

Figure 1
Photomicrograph of CICC

Conduit material	DHP copper
Conduit OD, cm	0.673
Number of strands	27
Cu:SC of strands	1.35
Void fraction	0.38
I_c, kA, @7T	6.5

solenoid at MIT's Francis Bitter National Magnet Laboratory. Measurements were made at three values of field, 5, 6, and 7 T. Transport current was supplied to the sample by a 10 kA constant voltage dc power supply.

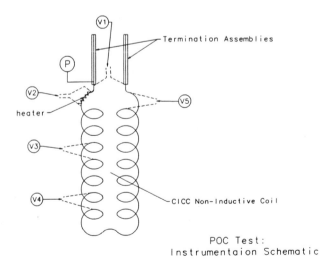

Figure 2. Electrical schematic of test coil, showing the location of heater, the pressure transducer, and connection points for five voltage tap pairs. The coil was wound as a single-layer noninductive double helix on a grooved G-10 mandrel.

Table 1
Data Acquisition Summary

Channel	Name	Source
1	V1	Coil terminal voltage
2	V2	Voltage across heated length of conduit
3	V3	Voltage across turn 4
4	V4	Voltage across turn 7
5	V5	Voltage across last turn
6	P	Pressure transducer
7	I	Transport current
8	Ih	Heater current
9	Vh	Voltage across heater terminals

Data acquisition was organized into nine channels, summarized in Table 1.

The experiment was conducted as follows:

- Current in the background field magnet was set to generate the desired field strength at the conductor;

- the current in the sample was set to a desired value;

- the heater pulse supply (a simple capacitive discharge circuit) was charged to a selected voltage;

- the heater pulse supply was triggered.

- If a propagating normal zone ensued, the transport current supply was interrupted, and the previous steps retraced, but with a smaller heat perturbation.

- If not, the heater was again pulsed with increasingly higher currents until the sample did suffer a quench.

The amplified raw signals were digitized by a 32-channel 12-bit A/D converter, sampling at 2 kHz. The resulting digitized data was read out to a DEC VAXstation II for storage on magnetic disk, and displayed on the workstation monitor.

The next three figures, 4 through 6, are presented here as representative of reduced data, and are useful for making comparisons between the two samples, and between data traces taken under different operating conditions for the same sample. Three of the plots in each figure show the potential drop along the length of conductor spanned by the voltage taps, as a function of time following the onset of the heater pulse (note the expanded time scale in the upper right plot). The fourth shows the trace of the pressure transducer output, relative to the initial pressure. Table 2 is a summary of the three representative data sets.

Table 2
Parameters of Selected Shots

Shot No.	Sleeved	Background Field (T)	Transport Current (kA)
193	no	7	2.96
209	yes	7	3.00
240	yes	6	4.50

These data sets display the behavior of the conductors during the evolution of a nonrecoverable normal zone initiated by the application of a heat pulse along a 0.3 m length of the outside surface of the conduit. The time constant of the capacitive discharge pulse, $RC \approx 0.005$ s. The conductor length for V3, V4, and V5 was one turn, or about 0.56 m. The length for V2, the heated length, was 0.30 m. V1 is the total potential difference from one termination assembly to the other, about 8.2 m.

The most obvious characteristic difference between the behavior of the two different conductors can be seen by comparing the plots for shots 193 and 209, taken with the same values of transport current ($I = 3$ kA, $I_{op}/I_c \approx 0.46$), and background field (7 T):

- In shot 193 (Fig. 3), the progressive onset of current sharing along the length of the sample is clearly evident in this unsleeved CICC, reaching the last turn of the coil more than two seconds following the heater pulse. This corresponds to an average normal zone propagation velocity of about 3.7 m/s. Current sharing grows rapidly and uniformly, as evidenced by the initial slopes, steep and featureless, of the curves V3—V5. The curve V1, the voltage across the entire sample length, is also smooth and featureless, showing a 'knee' at the same value of time that corresponds to the knee in curve V5, the time at which the entire sample is fully normal, i.e., current-sharing has ceased.

- In shot 209 (Fig. 4), by contrast, voltage signals appear throughout this sleeved sample immediately after the heat pulse. As is more clearly evident in the expanded plot in the upper right of the figure, the last turn has begun current-sharing after an elapsed time of less than half a second, as revealed by the sudden increase in the slope of the trace, and has reached T_c at about $t = 0.6$ s.

  These voltage traces are characterized by a more complex shape than those in the previous figure, displaying an early modest slope, followed by a steeper middle section, and finally bending again as the turn becomes fully resistive at the critical temperature. The voltage across the sample terminations, curve V1, also displays this more complex shape; the final slope change in V1 shows that the temperature of the full length of the sample is at least at the critical temperature when $t \approx 0.6$ seconds after the heater pulse. In shot 240 (Fig. 5), taken for different initial conditions, these features occur even more quickly.

The concept of propagation velocity would seem nebulous, at best, in CICCs with sleeved cable, for the transition to the normal state evolves nearly simultaneously (although perhaps not at the same rate) over the entire length of the conductor. During these tests, the current-sharing mode appeared to last longer at distances more remote from the site of the initial disturbance. This may have been because the dacron sleeve was not woven densely enough to preclude occasional 'soft' electrical shorts that reduced the conduit current farther downstream. However, as is the case for any CICC, the most crucial parameters during a quench are the peak temperature and pressure, and for given initial conditions, these depend ultimately on the total time elapsed before the stored energy is dissipated.

This comparison suggests that the insulating dacron sleeve has a dramatic effect on the electrical and thermal behavior of this CICC following a perturbation that exceeds the stability margin. Good electrical contact apparently exists only at the two terminations, as was intended, and as a resistive zone grows in the cable, the resulting potential difference between the two ends of the sample drives current through the full length of the conduit. By this mechanism, the entire length of cable is driven into the current-sharing mode. In an actual magnet, the stored energy would thereby be dissipated over the whole conductor volume, limiting the maximum temperature excursion to a safe value.

Quantitative Analysis of POC Test Data

This section will describe in more detail some of the quantitative aspects of shots 193 and 209.

In the case of shot 193, the cable and the conduit were in intimate electrical and thermal contact along the entire length of the sample. Along any length in which

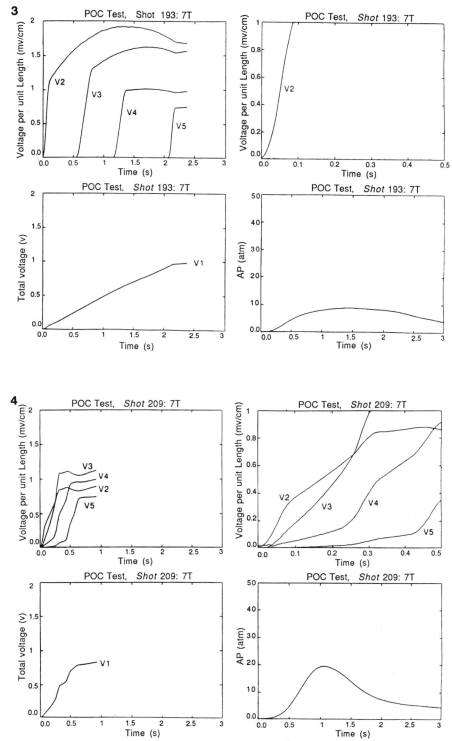

Figures 3 and 4. Voltage traces 2 through 5 are plotted together in these two figures in order to easily discern their relative positions on the Time axis. Pressure data is redrawn relative to the initial pressure; that is, $\Delta P(t) = P(t) - P(0)$. Comparison of these two figures reveals the essential differences in behavior between the two samples.

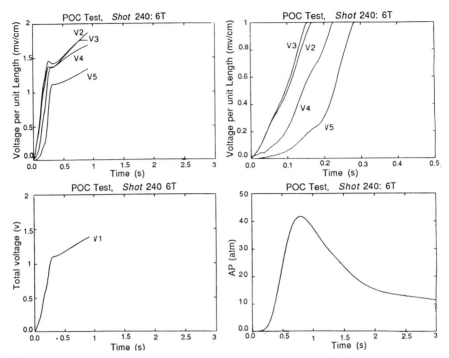

Figure 5. Data from shot 240. The higher slopes in the higher current case are clearly evident, as is the higher peak pressure. As indicated by knee in trace V5, the temperature of the entire length of conductor reached T_c in about 250 ms after the heat pulse.

the temperature of the strands, $T > T_{cs}$, that portion of the current that could not flow within the superconducting filaments transferred to the copper portion of the strands and to the adjacent section of copper conduit, giving rise to the measured potential drop. Note that no voltage is measured unless and until $T > T_{cs}$ along at least some portion of the length of conductor spanned by a given pair of voltage taps. A power vs. time curve was calculated by multiplying the transport current and potential drop point by point. The cumulative energy deposition was calculated by appropriate numerical integration and summing of the power curve data, and plotted in figure 6. All values are presented in terms of a unit length of conductor.

Table 3 summarizes the relevant physical parameters of the unsleeved CICC. (The listed values of enthalpy per unit length are estimates only, inasmuch as the pressure and temperature of the helium during the course of the event are not precisely known).

Table 3
Conductor Data: Shot 193

$A_{strands}$, cm^3/cm	0.123
void fraction	0.38
A_{helium}, cm^3/cm	0.075
m_{helium}, g/cm	9.4×10^{-3}
I_{op}/I_c	≈ 0.4
T_{cs}, K	5.25
T_c, K	6.1
$H_{He}(5.25\text{-}6.1)$, J/g	5
$H_{He}(5.25\text{-}6.1)$, J/cm	0.047

The enthalpy in the range only from T_{cs} to T_c is of interest here, because no voltage is measured along a given length until the temperature of the cable in that section reaches T_{cs}.

The abrupt change in slope of the voltage traces in figure 3 occurs when the temperature of the strands reaches T_c. For trace V4, for example, this happens at $t \approx 1.3$ s. From the plot of cumulative energy deposition in figure 6, one can see that by that time, ≈ 0.13 J/cm has been dissipated at that location in the coil. This amount is a factor of three more that the rough estimate of 0.047 J/cm required to raise the helium temperature from T_{cs}, when the voltage first appears across this section, to T_c. The discrepancy could be the result of some heat conduction through the insulating cocoon-like shell that was intended to isolate the coil from the helium bath.

The values for the comparable shot with the sleeved conductor are listed in Table 4.

Table 4
Conductor Data: Shot 209

$A_{strands}$, cm^3/cm	0.123
void fraction	0.33
A_{helium}, cm^3/cm	0.066
m_{helium}, g/cm	8.2×10^{-3}
I_{op}/I_c	≈ 0.4
T_{cs}, K	5.25
T_c, K	6.1
$H_{He}(4.2\text{-}5.25)$, J/g	28.1
$H_{He}(4.2\text{-}5.25)$, J/cm	0.23

For this shot, the enthalpy of interest is that required to raise the temperature from 4.2 K to 5.25 K, that is, the energy required to drive the strands into a non-recoverable current-sharing mode. For, as explained in an earlier section, once a resistive region develops anywhere in the conductor, the resulting terminal potential difference drives some of the current through the entire length of conduit, and a voltage is detectable at every one of the voltage taps simultaneously, whether or not the underlying cable is resistive at that particular location or not.

In this case, the resistivity of the conduit material must be known in order to calculate the power that it dissipates in any given length. For this purpose, a spare piece of the conduit that had the same history of cold work as the sample conduit was instrumented for resistivity measurements. Data were taken at room temperature and at 77 K. These results, combined with published data on this alloy of copper, yielded a best estimate of 0.5 $\mu\Omega$-cm for the resistivity of the conduit at 4.2 K. The magnetoresistance is believed to be small compared to other uncertainties in this analysis, as is the temperature dependence of the resistance over the range of interest here, 4.2 K — 10 K. The conductor length for V3, V4, and V5 each correspond to one turn, or about 0.56 m. The length for V2 is the heated length, 0.30 m. V1 is measured across the entire sample length from one termination assembly to the other, a length of about 8.2 m.

The resistance for each length spanned by voltage taps, then, was calculated, and made constant for purposes of calculating the power dissipation during the initial quench transient; the power was obtained by taking the quotient V^2/R for each section, data point by data point. It is important to emphasize the difference between this case and the previous one: the only joule heating that occurs at sites remote from the location of the initial perturbation, during the early phases of the quench, is that due to current flowing in the conduit. That current is a comparatively small fraction of the total transport current, a maximum of about 300 A of a total 2.5 kA in shot 209, and flows in the conduit because of the growing potential drop across the entire sample length generated by current sharing that began at the site of the initial perturbation.

Inspection of voltage trace V4 in figure 4 shows two distinct slope changes. The first change occurs at the onset of current-sharing, when $T = T_{cs}$, and the second, when the strands are fully normal, when $T = T_c$. The first slope change occurs at t ≈ 0.22 s. Referring to the cumulative energy plots in figure 7, one finds that it is not possible to read a value that early in the transient, except to estimate that $E < 5 \times 10^{-4}$ J/cm. If the energy generated by joule heating in the conduit is absorbed by the interstitial helium, which in turn would raise the temperature of the strands to T_{cs}, then, as listed in Table 4, 0.23 J/cm would be required. Clearly, current sharing has begun long before enough energy has been generated in the conduit to raise the temperature of the helium to T_{cs}.

However, only about 2.2×10^{-4} J/cm is sufficient to raise the temperature of the conduit and the strands to the current-sharing temperature. This suggests that while the dacron sleeve is effective in electrically isolating the strands from the conduit, the intimate physical contact between the sleeved, compacted cable and the ID of the conduit, an inevitable consequence of the method of manufacture, leads to comparatively effective heating of the strands by conduction from the conduit, despite their being immersed in cooler helium. The cable and subcable twist sequence ensures that all strands will be at the OD of the cable at frequent intervals, and hence they all are able to accept heat conducted from the conduit through the dacron barrier.

Stability

Even though the chief thrust of these tests was not a comparison of energy margins of the two configurations, some stability data were taken. The data are shown in figure 8. Heat pulses that did not produce a nonrecoverable normal zone are indicated by ◇, while those which initiated a quench are drawn as ×.

One of the intended features of this sleeved cable CICC was to enhance the stability of the conductor against small friction-induced heat inputs. By introducing a helium-permeable thermal barrier between the conduit—where the frictional heating arises—and the strands, with their very small heat capacity, this design would enable the interstitial helium to play a major role in absorbing externally-applied heat before the vulnerable strands that were immersed in it were driven into current sharing. Evidence that this feature was not achieved was presented in the previous section; the stability data further bears this out.

Two aspects of these data are noteworthy. First, the stability margin of the unsleeved sample is *larger* than that of the sleeved sample, the reverse of prior expectations. Second, the value of energy margin for the higher field cases of the sleeved sample is apparently a rather weak function of current over the range of I_{op}/I_c covered in the test.

Apparently, the barrier sleeve did not impede the conduction of heat to the strands as much as had been hoped. The lower energy margin of the sleeved conductor can be partly explained by writing the following equations describing the heating rate in the section of conductor that was driven into current sharing. (Cooling mechanisms in the two samples may have been slightly different from one another, but these are not considered in this discussion.) In both cases, the joule power is the sum of two components – heating in the cable, and heating in the sheath. In both cases, this can be expressed as

$$P = I_c^2 R_c + (I_0 - I_c)R_{sh},$$

where I_c is the current flowing in the cable, R_c is the normal-state resistance of the section of cable that is current sharing, I_0 is the total transport current, and R_{sh} is the resistance of the section of conduit that overlies the resistive section of cable.

But, the current that flows in the cable is different in the two cases. In the unsleeved conductor, the copper portion of the cable is electrically in parallel with

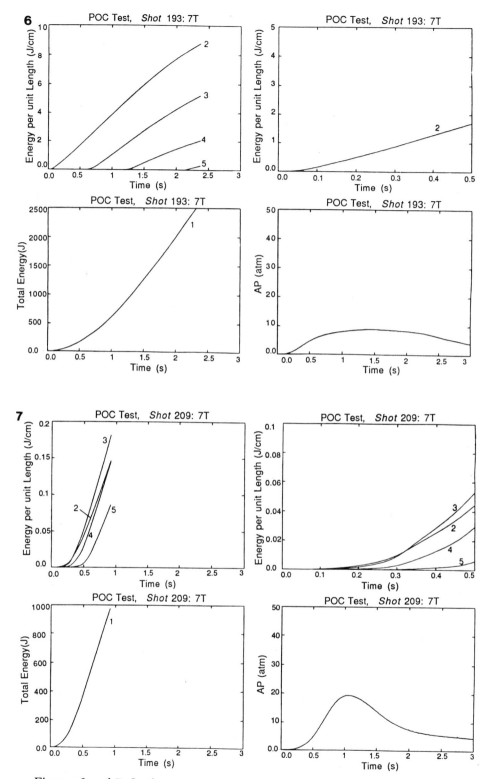

Figures 6 and 7. In these two figures the data for the same three shots has been replotted as cumulative energy deposition per unit length rather than as voltage per unit length, as described in detail in the text.

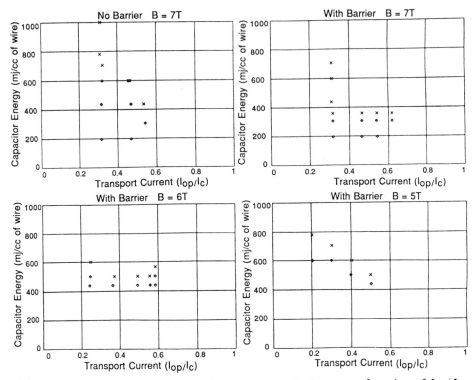

Figure 8. Plots that indicate quench and no-quench shots as a function of I_{op}/I_c.

the length of conduit that overlies it, whereas in the sleeved conductor, it is in parallel with the entire length of conduit. That is, for the unsleeved conductor,

$$I_c = I_0 \left(\frac{R_{sh}}{R_{sh} + R_c} \right),$$

while in the case of the sleeved conductor,

$$I_c = I_0 \left(\frac{\mathcal{R}}{\mathcal{R} + R_c} \right),$$

where $\mathcal{R}$ is the resistance of the entire length of the conduit.

Combining and rearranging, we have

$$P = I_0^2 \left[r^2 (R_c + R_{sh}) + (1 - 2r) R_{sh} \right],$$

where

$$r = \frac{R_{sh}}{R_{sh} + R_c}$$

for the unsleeved case, and

654

$$r = \frac{\mathcal{R}}{\mathcal{R} + R_c}$$

for the sleeved conductor.

Calculations utilizing these equations, with numerical data specific to these conductors, show that other factors being equal, the ratio of joule heating during current sharing in the sleeved conductor to that in the unsleeved sample is about $2.5 - 3$, a result that is consistent with the observed lower energy margin of the sleeved conductor.

However, a further subtlety must be considered in order to account for the apparent reduction of this ratio with increasing current, that is, the comparatively weak dependence of the energy margin on current. The explanation lies, at least in part, in the fact that the resistance ratio, r, in the unsleeved case is essentially constant during early stages of the evolving normal zone, because the resistivity of the heated strands and of the section of conduit that overlies them increase at the same rate as the temperature rises. On the other hand, for the case of the sleeved sample, the ratio r tends to *decrease* as the temperature rises because $\mathcal{R}$ stays relatively constant while R_c, in the denominator, tends to increase. This tendency is enhanced at higher values of transport current because the greater joule heating leads to a faster increase in the normal-state resistivity of the heated cable.

Conclusion

Comparative testing of two samples of Cable in Conduit Conductor have demonstrated a proof of concept for the novel design described herein. The primary characteristics of this design are:

- Strands with a comparatively small Cu:SC ratio;
- Conduit made with material that has a comparatively low electrical resistivity;
- An electrically insulating, helium-permeable sleeve or barrier interposed between the conduit ID and the superconducting cable.

The meaning of the term 'propagation velocity', as applied to a growing normal zone in a conductor of this configuration, is nebulous at best, for the transition to the normal state does not progress from one location to the next along the length, so much as it progresses in time over the whole length at once. The behavior of this conductor under fault conditions in an inductive device would be characterized by a quench that dissipates the stored energy over the entire volume of the device.

Although copper was chosen as the low resistivity material for the conduit for these test samples because of the ease of fabrication and availability, for use in large-scale devices such as those built for high energy particle detection facilities, conductors of this configuration would most likely utilize a high-strength aluminum alloy. The selection of the most suitable alloy, and the development of a cost-effective manufacturing and fabrication technology for its utilization will be an important facet of the overall conductor development program.

12. Detectors II

MONOLITHIC JFET PREAMPLIFIER FOR

IONIZATION CHAMBER CALORIMETERS

Larry A. Rehn and Dan E. Roberts

InterFET Corporation
322 Gold Street
Garland, TX 75042

ABSTRACT

A prototype preamplifier circuit is presented for use in SSC ionization chamber calorimeters. It consists of a new type of silicon integrated circuit comprised of very low noise junction FET (JFET) components. Presently, monolithic preamplifier circuits for use in highly segmented detectors are made of implanted channel JFETs or MOS devices. While such circuits solve the density problems, they do not perform to the same level of low noise characteristics as found in discrete JFET components. The JFETs which comprise this new integrated circuit preserve the excellent low noise performance normally found only in discrete JFETs. JFETs also are much more radiation resistant and less prone to damage by electromagnetic discharges than MOS transistors. Two innovative fabrication processes are discussed. They solve the difficult gate-to-gate isolation problem needed to manufacture JFET integrated circuits. Both allow the use of an epitaxially formed channel and a diffused gate, as in standard discrete JFET processing. This, presumably, results in JFETs which exhibit lower noise than those made with implanted channels.

INTRODUCTION

Detector electronics for SSC ionization chamber calorimeters face a number of difficult challenges. Large volume and fine segmentation will result in up to 200,000 signal channels. Geometrical constraints and requisite fast signal processing times prohibit the use of cabling to bring these signals outside the calorimeter.[1] This means that signal processing electronics may be located inside the calorimeter, where space is at a premium and power consumption is a problem. Integrated circuits offer the obvious solution to these requirements. However, a number of other problems complicate the use of conventional integrated circuits for this application.

At the design luminosity of the SSC the p-p collisions will generate substantial levels of ionizing radiation as well as neutron flux.[2] The estimated total doses of radiation over the operating life of the SSC pose a serious constraint on the choice of electronics. Even radiation hardened MOS circuitry cannot normally withstand these radiation levels.[2] In this respect, bipolar and JFET technologies offer superior radiation tolerant properties.[3,4]

The exact type of calorimeter has not yet been decided for use in the SSC. However, those using liquid argon offer several advantages over other choices. These include fine segmentation, uniform time and spatial response, and high tolerance to radiation.[5] If the electronics are to be located near the detector electrodes this would require that they operate at around 90K.[1] This constraint limits the use of bipolar devices, since they cannot operate effectively at these temperatures.

Very low noise signal detection circuitry is a mandatory attribute of the detector electronics. This is necessary due to the summing of the many channels and the very small signal levels of interest.[6] Their inherently low 1/f and series thermal noise characteristics make JFETs a good choice.

The SSC experiments are likely to continue for many years.[7] Inability to readily access much of the electronics once installed dictates the need for very reliable devices. Susceptibility to electrostatic discharge (ESD) between closely spaced, high voltage plates within the detectors is also likely. The reliability of JFETs and their resistance to ESD damage compared to MOS devices, once again make JFETs a preferred choice.

Because of their good low-noise characteristics, radiation and ESD tolerance, and proven reliability JFETs appear to be the best choice for SSC calorimeter front-end electronics. Bipolar and MOS circuits are better suited to downstream signal conditioning and multiplexing functions where the operating environment is less harsh and servicing is convenient.

Presently, no high performance integrated JFET circuits exist for the SSC calorimeter preamplifier application.[7] This is due largely to the inherent difficulties found in attempting to integrate JFET components into a monolithic chip. The easiest approach is to use conventional integrated circuit (IC) wafer fabrication techniques.[8] Typically, this is done by providing diode isolation between transistor components by the use of implanted wells into the single crystal substrate. This is illustrated for JFETs in Fig.1. JFETs are more difficult to isolate in an IC than MOSFETs because of the nature of their construction in the silicon. A MOSFET can be made in a single diffused well into the substrate, and the gates are formed over oxide on the surface. A JFET requires three

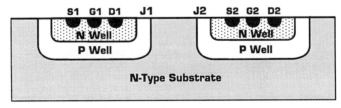

Fig. 1. JFET IC with Implanted Channels. JFET components J1 and J2 are shown using a triple-diffused process.

alternating polarity reversals in the silicon in order to define the gate and channel regions. The usual method to manufacture JFET ICs is to use ion implantation to form diffused wells for the channel and gate regions.

While such circuits solve the density problems, they do not perform to the same level of low noise characteristics as found in discrete JFET components. The primary advantage of a discrete JFET is that its noise performance is much better than those found in ICs. Standard discrete JFETs are made from silicon that has an epitaxial layer over a heavily doped substrate. The substrate is a very low resistance silicon, and the epitaxial layer can be carefully controlled for thickness uniformity and resistivity. The epitaxial layer forms a uniform JFET channel while the substrate forms the low resistance backside gate. Epitaxially formed channels do not have counter-doping and have less crystal damage than those formed by implantation. This, presumably, results in JFETs which exhibit lower noise than those made with implanted channels.

This paper investigates a new JFET integrated circuit which preserves the excellent low noise performance normally found only in discrete JFETs. Innovative fabrication processes are described which solve the difficult gate-to-gate isolation problem needed to manufacture JFET integrated circuits. Both allow the use of an epitaxially formed channel and a diffused gate, as in standard discrete JFET processing.

The attributes of the circuit are described more fully by Demicheli, et al.[9], and is typical of other published low noise circuits. Fig. 2 shows an electrical schematic of the circuit. The purpose of this circuit is to allow cryogenic operation in a liquid argon calorimeter. This allows shorter connecting electrodes to promote faster signal processing. All transistors were chosen to be n-channel JFETs to simplify processing requirements.

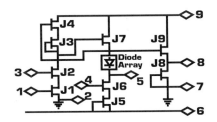

Fig. 2. Preamplifier circuit using JFET components and diode array level shifter.

DEVICE DESCRIPTION

The circuit shown in Fig. 2 consists of nine JFETs and a diode array level shifter comprised of eight diodes in series. Of the nine individual JFETs of this IC, six are unique. These range in gate width from 0.016-0.450 inch (16-450 mils). Channel length is either 0.6 or 1.0 mil. Isolation is provided for all JFETs which operate at different gate potential.

The first evaluation design consists of four cells arranged in a 2 X 2 array on the wafer. These four elements are:

1.   the basic IC,
2.   a piecemeal IC with internal test points ("kit parts IC"),
3.   discrete JFETs of each IC component, and,
4.   a diagnostic cell with miscellaneous process and test structures.

The premise of this approach was to allow a maximum amount of knowledge to be derived about each process alternative in the event that manufacturing problems arose to prevent proper operation of the complete IC.

Two basic types of silicon substrate are proposed which result in four different process variations. In all cases, the JFET channel is formed by an epitaxial layer of silicon over the substrate, rather than by diffusion following ion implantation of the channel dopant. This is believed to be important in order to preserve the superior low noise attributes of discrete JFETs. It is possible to control the doping profile to a much greater extent using an epitaxial process than for a diffused process. Since a diffusion process requires addition of the dopant source from the surface, the resulting profile into the silicon follows either a Gaussian or error function distribution.[10,11] Therefore, the epitaxially formed channel is more uniform than a diffused channel. In addition, the use of ion implantation as a channel doping source is a process that is disruptive to the silicon crystal lattice. High temperature or other annealing processes are necessary in order to attempt restoration of the integrity of the crystal lattice. Also, each polarity reversal requires greater and greater concentrations of compensating dopant material. With an epitaxially formed channel no counterdoping occurs and only one polarity reversal is necessary to form the gate and the finished JFET, compared with two or three reversals for the implanted method.

The first isolation alternative uses a double layer of epitaxial silicon over a standard silicon substrate and involves a single process option using etched V-grooves for device isolation. The second alternative is a single epitaxial layer over a dielectric isolated (D.I.) silicon substrate. Three different process options are described using the D.I. approach.

Double Epitaxy/V-Groove Method (DEVG)

The double epitaxy, V-groove method begins with an n-type silicon substrate on top of which are grown two layers of epitaxial silicon. The top layer is n-type and ultimately forms the channel for the JFET. The buried layer is p-type and forms the lower boundary or backside gate to the channel. The substrate wafer and both epitaxial layers are generally specified and ordered from merchant silicon suppliers.

A finished JFET cross section is shown in Fig. 3. Wafer processing begins with the typical p-type diffusion guard ring that surrounds each device and extends down to the buried p-epitaxial layer. Gate fingers are diffused part way down into the channel, and overlap the guard ring at the ends; thereby, electrically connecting top and backside gates.

The next step is to etch V-shaped grooves into the silicon using orientation-dependent silicon etch. This requires that the silicon be of [100] crystal orientation, and that the groove patterns be rectangular in shape, and aligned with the edges parallel or perpendicular to the primary wafer flat. The etchant removes the silicon about 100 times faster in a direction parallel to the [111] crystal planes, relative to the perpendicular direction. The groove is etched deep enough to pass through both of the epitaxial layers, thereby separating the backside gates from one another. Back-to-back diodes through the common substrate provide the only electrical path between devices.

Following the groove formation the contact openings and the metal leads are formed. These photolithography steps are more difficult than normal since they must allow for the pooling of photoresist in the grooves. This results in thin photoresist at the edges of the grooves that is subject to pinhole problems. In addition, the presence of the grooves makes step coverage of the aluminum leads a potential problem. These process related problems do require additional process steps, and do impact manufacturing yield compared to a planar process; however, it does have the advantage of a great deal of manufacturing experience at InterFET Corporation.

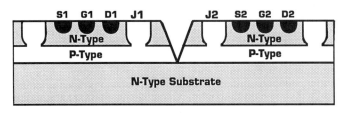

Fig. 3. JFET manufacturing process using double layer epitaxy and V-groove isolation.

Dielectric Isolation With Multiple Tubs/Planar Process

The basic dielectric isolation (D.I.) process begins with a starting wafer of single crystal silicon in which V-grooves are etched to form the pattern of isolated elements. A high quality oxide is thermally grown over all the single crystal silicon. Next, undoped polysilicon is grown over the oxide, filling the grooves and building to a thickness comparable to the original single crystal wafer. Both sides are polished flat and most of the original single crystal silicon is ground and lapped away until the tips of the V-grooves are exposed. This results in isolated single crystal silicon islands or tubs surrounded by the thick thermal oxide and separated by the polysilicon substrate.

Since the thickness of the remaining single crystal silicon is determined by mechanical processes, its tolerance is not adequate for the manufacture of the JFET channel. Instead a single layer of n-type epitaxial silicon is grown over the D.I. substrate, and provides a uniform channel for the JFET.

A cross section of the finished D.I. planar process circuit is shown in Fig. 4. Separate single crystal tubs are formed to define the electrically isolated components in the circuit. Upon completion of the D.I./epitaxy process, wafer processing proceeds very much as for a standard discrete JFET. A p-type diffusion surrounding the transistor active area is driven down through the epitaxial layer to the original p-type silicon. This electrically isolates components by back-to-back diodes outside the isolation ring. A polysilicon bridge between these diodes is formed during the epitaxial growth, since the underlying layer is polysilicon. Electrically this process is very similar to the double-epitaxy/V-groove process with an additional series resistor between diodes.

Dielectric Isolation With Multiple Tubs/Mesa Separation

The dielectric isolation with multiple tubs using a mesa separation is similar except that a redundant isolation provision is included. It consists of a groove etched into the silicon down to the underlying oxide surrounding the single crystal tubs. This leaves silicon mesa structures within each tub thereby eliminating the diode/polysilicon bridge path between components. This approach is shown in Fig. 5. The disadvantage of this approach is that it adds more processing steps, and more importantly, makes processing more difficult due to the nonplanar surface. Photoresist and metal step coverage is a problem because of the deep grooves around each mesa.

Dielectric Isolation With Single Tub/Mesa Separation

The dielectric isolation method with single tub/mesa separation is similar in concept to the previous mesa structure approach. Since free-standing mesas over oxide provide excellent electrical isolation, the multiple D.I. tubs are eliminated in favor of one large tub for the entire circuit. This ultimately would allow significant space savings over the redundant design.

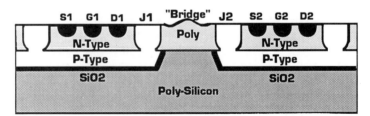

Fig. 4. Dielectric Isolation planar process.

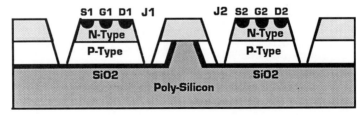

Fig. 5. Dielectric Isolation with single layer epitaxy. Mesa process, multiple tubs.

RESULTS AND DISCUSSION

Wafer Fabrication and Processing

Processing results from all manufacturing alternatives have been encouraging. The DEVG approach follows the existing InterFET process flow for standard monolithic dual JFETs. This process was well behaved as was expected and yielded excellent results. Process steps following the formation of the grooves can potentially cause a yield loss due to step coverage problems, but little loss was encountered on the initial wafers.

Processing of the D.I. approaches has also been very positive. The planar (non-mesa) approach is the simplest from a manufacturing point of view since groove steps are eliminated as well as the yield losses associated with the mesa processes. Figure 6 shows an SEM micrograph of the D.I. planar process. The area between two adjacent single crystal tubs is shown, and is defined by the L-shaped oxide layers. The polysilicon substrate is the center, between the tubs. Above the ends of the oxide is the epitaxially grown layer; single crystal above the tubs, and polysilicon between. Note the difference in rate of growth of this layer over single crystal silicon, oxide, and polysilicon. A metal lead is shown over the top surface.

The two D.I. mesa alternatives are closer in wafer fabrication to the DEVG approach. The larger grooves necessary for the mesa structures relative to the DEVG method causes somewhat greater difficulty in processing, but still is acceptable.

Preliminary Electrical Results

The completed DEVG wafers were tested for functioning JFETs within the circuit, DC parameters of the individual JFETs in the discrete component cell, and device-to-device electrical isolation within the IC. This was done using both a standard manual curve tracer and an automatic tester and wafer probe station.

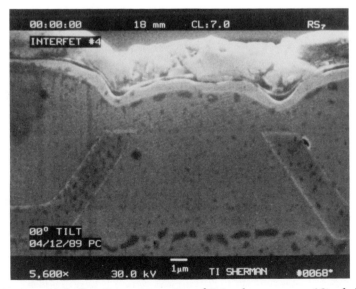

Fig. 6. SEM micrograph showing cross section of D.I., planar process/ Single layer of epitaxy is used over D.I. substrate.

The curve tracer analysis of the discrete JFETs resulted in a very high yield of functioning transistors with normal operating characteristics. The results indicated that all of the individual JFETs functioned properly and within the designed parametric limits. Device-to-device isolation demonstrated very low leakage characteristics with breakdown voltages which typically exceeded that of the individual devices.

Further testing of the individual JFETs from the discrete cell of the array was done on an automatic probe station using an automatic JFET tester. This method obtained sufficient data for some statistical presentation of the common DC electrical parameters. Figures 7 through 9 show some of these statistical test results. This data was taken from one wafer with approximately 200 data points for each parameter.

Figure 7 shows reverse bias gate leakage current, Igss, as a function of gate width of the individual JFETs while Fig. 8 shows drain saturation current, Idss. The gate leakage current is seen to increase approximately linearly with gate width. This is due to the linear increase of junction area with the increase in gate width. The indicated gate leakage currents are well within the typical limits of comparably sized discrete JFETs and are expected to decrease after the devices are separated from the wafer and tested individually. The drain saturation current of Fig. 8 is also seen to increase linearly with gate width. This is due to the linear increase in the cross sectional area of the channel with the increase in gate width. Both of these results are expected and follow the design rules of the JFETs for this integrated circuit.

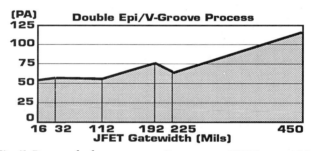

Fig. 7. Reverse leakage current, Igss, versus JFET gatewidth.

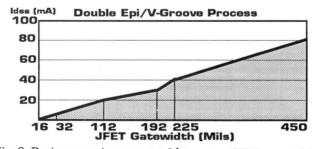

Fig. 8. Drain saturation current, Idss, versus JFET gatewidth.

The graphs of Figure 9 show the distribution of gate cut-off voltage, Vgs(off), for each of the six JFET designs across the wafer. It can be seen that this distribution is nearly identical for each JFET design across a 28-fold range in gate width. The actual median points vary only 160 mV for a change in gate widths of 0.016 inches to 0.450 inches. This uniformity in Vgs(off) occurs as a result of close epitaxial thickness tolerance and controlled fabrication processes which are common to discrete JFET fabrication. The absolute value of the Vgs(off) is controlled on a lot-to-lot basis in the same manner that is used for discrete JFET fabrication. These parameters are important to the functionality of the integrated circuit.

One observed problem with the DEVG approach is that latchup to the substrate can occur among circuit elements under certain bias conditions. This can occur because the latchup conditions are a function of the physical parameters of the parasitic npn and pnp bipolar transistors present through the common n-type substrate. It was observed that these parasitic devices typically exhibited very low gain thereby reducing the possibility of latchup. This does impose a limitation on the voltage shift possible for the diode array and could effect the maximum power supply voltages and dynamic range of the IC.

Some initial electrical evaluation results are available from the D.I. planar process devices. The individual JFETs demonstrate good DC characteristics and are well behaved. The overall yield of the process looks quite good. The integrity of the isolation diodes is excellent and the polysilicon "bridge" exhibits a resistance in excess of 40K ohms. The polysilicon bridge does not affect the operation of the JFETs but does have some effect on the forward biased diode array. This is seen as a bulk distributed resistive path in a series-parallel path with the array and introduces varying amounts of noise and voltage shift uncertainty. Generally the value of this resistance is large enough to allow proper functioning of the IC.

Some packaged devices of the various design cells have been assembled, but no significant test data is yet available. Much of this electrical evaluation is being conducted by Istituto Nazionale di Fisica Nucleare (INFN) and Brookhaven National Laboratories.

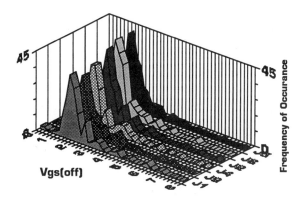

Fig. 9. Probed wafer distibution of pinchoff voltage, Vgs(OFF) for discrete JFETs used in I.C.

CONCLUSIONS

The ability to successfully fabricate simple integrated circuits containing JFET components which have characteristics similar to discrete JFETs has been demonstrated for several innovative process alternatives. These approaches use unconventional isolation methods to achieve a wafer fabrication process that is compatible with that used for the manufacture of discrete JFETs. Initial results of standard DC characteristics are encouraging and as expected. All of the approaches investigated have demonstrated acceptable manufacturing yield, but the planar D.I. method appears to be best in this regard. Other critical performance parameters including low noise measurement, operation of the circuit parts and functionality of the IC have not yet been fully investigated. Solutions to the observed problems and improvements to the weaknesses are being investigated with positive results expected.

ACKNOWLEDGEMENT

The authors wish to thank P.F. Manfredi and V. Radeka for their technical support. They also acknowledge S. Rescia for support with the manuscript and providing samples for prototype assembly. Thanks also are due R. Pennell at Texas Instruments for providing support with the dielectric isolated silicon, and A. Stephens for providing the SEM analysis.

This work is supported by INFN; contract numbers N0368/71, N0796/71, and N1403/71.

REFERENCES

1.	G. R. Kalbfleisch, "Status of Some of the Silicon Detector Development in the United States," Silicon Detectors for High Energy Physics, Proceedings of the Workshop at Fermilab, October 15-16, 1981.
2.	D. E. Groom, editor, "Radiation Levels in the SSC Interaction Regions, Task Force Report," Proceedings of the 1988 Summer Study of the Physics of the Superconducting Super Collider, Snowmass, Colorado, 1988.
3.	D. Allen, F. Coppage, G. Hash, D. Holck, and T. Wroble, "Gamma-Induced Leakage in Junction Field Effect Transistors," IEEE Transactions on Nuclear Science, Volume NS-31, Dec. 1984.
4.	A. E. Stevens, J. W. Dawson, V. Radeka, and S. Rescia, "Rad-Hard Electronics Development Program for SSC LiquidArgon Calorimeters," International Industrial Symposium on the Super Collider, Miami Beach, Florida, March 14-16, 1990.
5.	T. J. Devlin, A. Lankford, and H. H. Williams, "Electronics, Triggering and Data Acquisition for the SSC," Proceedings of the 1986 Summer Study of the Physics of the Superconducting Super Collider, Snowmass, Colorado, 1986.
6.	M. Bertaloccini, G. Padovini, D. V. Camin, P. F. Manfredi, J. A. Preston, and L. A. Rehn, "Perspectives in the Design of Transformerless, Low-Noise Front End Electronics For Large Capacitance Detectors and Calorimeters in Elementary Particle Physics," Nuclear Instruments and Methods in Physics Research, A264, 1988.
7.	V. Radeka, S. Rescia, Private Communication, Brookhaven National Laboratory, Upton, NY, December 1988.

8. D. J. Hamilton and W. G. Howard, *Basic Integrated Circuit Engineering,* McGraw-Hill, 1975.

9. M. Demicheli, P. F. Manfredi, V. Radeka, S. Rescia, and V. Speziali, "Design of Monolithic Preamplifiers Employing Diffused N-JFETs for Ionization Chamber Calorimeters", presented at 1988 Elba Conference on Frontier Detectors for Frontier Physics, Elba, Italy, to be published in Nuclear Instrumentation and Methods.

10. S. M. Sze, *Physics of Semiconductor Devices,* John Wiley & Sons, New York, 1981.

11. A. S. Grove, *Physics and Technology of Semiconductor Devices,* John Wiley & Sons, New York, 1967.

RAD-HARD ELECTRONICS DEVELOPMENT PROGRAM

FOR SSC LIQUID-ARGON CALORIMETERS

A. Stevens and J. Dawson

Argonne National Laboratory
High Energy Physics Division
Argonne, IL 60439

H. Kraner, V. Radeka, and S. Rescia

Brookhaven National Laboratory
Instrumentation Division
Upton, NY 11973

ABSTRACT

The development program for radiation-hard low-noise low-power front-end electronics for SSC calorimetry is described. Radiation doses of up to 20 MRad and neutron fluences of 10^{14} neutrons/cm^2 are expected over ten years of operation. These effects are simulated by exposing JFETs to neutrons and ionizing radiation and measuring the resulting bias, leakage current and noise variations. In the case of liquid-argon calorimeters, a large part of the front-end circuitry may be located directly within the low-temperature environment (90 K), placing additional constraints on the choice of components and on the design. This approach minimizes the noise and the response time. The radiation damage test facilities at Argonne will also be described. These include sources of neutrons, electrons, and gamma radiation.

INTRODUCTION

The high energy and luminosity at SSC will generate a severe radiation environment for the detector materials. Estimates indicate that the radiation caused by the normal beam/beam interactions and subsequent particle/detector-material interactions will be equivalent to an accidental beam loss into the detector every six days. The expected radiation environment of a generic detector has been well described in [1]. The size and complexity of the proposed detectors as well as the high event rate for the collider dictate that the instrumentation be placed near or inside the detectors. This places a particularly harsh constraint on the electronics, which must remain operational for the projected ten-year lifetime of the detector.

A block diagram of a possible SSC detector is shown in Fig. 1 [2]. In this figure, the predicted radiation doses for neutrons and ionizing radiation are shown for several positions inside the calorimeter. It should be noted that these numbers are only rough estimates, as more accurate calculations of the radiation will depend largely on the actual geometry and materials used in construction. As shown in Fig. 1, the calorimeter is split into barrel and forward sections, with the most severe doses being close to the beamline in the forward section.

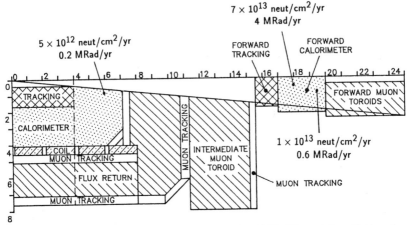

Figure 1. A possible detector implementation for SSC. The total radiation dose is shown for several locations within the calorimeter [2].

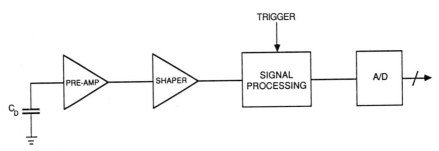

Figure 2. Detector front-end electronics.

A liquid-argon ionization chamber is one type of calorimeter which has been proposed for use in SSC detectors. It consists of plates of a heavy absorber material (typically iron, lead, or uranium) which sit in a cryostat filled with liquid argon. The designed segmentation of the detector will determine the number of plates and therefore the number of readout channels; quantities upward of 200,000 channels have been estimated for SSC. This places several constraints on the electronics associated with the calorimeter:

- They must be reliable over a large temperature range (90 K to room temperature).

- They must be low power, both due to the large number of channels, and due to thermal dissipation constraints inside the liquid-argon cryostat.

- They must be fast due to the high event rate (60 MHz).

- They must be low noise in order to maximize the energy resolution of the calorimeter.

- They must be radiation resistant to the levels described above.

A schematic of the detector front-end electronics is shown in Fig. 2. In this configuration, a charge pre-amplifier is connected directly to the detector plates in order to measure the charge generated by particles traveling through the calorimeter. A shaping amplifier is used to maximize the signal-to-noise ratio with optimal or near-optimal filtering. An analog signal processor follows the shaper and contains the waveform sampler, analog pipeline, trigger amplifiers, and event storage. Finally, an analog-to-digital converter is necessary to store the information in a computer.

The requirements and location of each section will determine the semiconductor technology to be used. In the case of the pre-amplifier, the low noise, temperature, and radiation-susceptibility requirements make JFETs the technology of choice. This is due to:

- Inherent radiation hardness due to the fact that the JFET is a majority carrier device and does not rely on oxides for charge control.

- Excellent noise performance of JFETs due to extremely low leakage currents and $1/f$ noise.

- The JFET noise performance is optimum at low temperatures, typically near 120 K.

The radiation hardness of JFETs will be discussed in this paper. We will give a qualitative account of the damage mechanisms due to radiation as well as test results on device performance at fluences up to 10^{15} neutrons/cm^2. Previous pre-amp designs have used commercially available JFETs with excellent results [3]. For example, at the HELIOS experiment at CERN, JFET pre-amps are used inside the liquid-argon cryostat and have shown 99.8% reliability. Although these pre-amps were not specifically designed for radiation immunity, we have also tested their radiation behavior with neutrons and gamma radiation. The balance of the electronics (shaper, signal processing, A/D) is currently under development. We are evaluating several rad-hard full-custom processes for this purpose, including CMOS, BiCMOS, SOS, and monolithic JFET.

Radiation damage testing will be necessary for all potential materials used in SSC detectors. Thus, the accessibility and availability of radiation sources will be key for any successful rad-hard design. Argonne National Laboratory has complete facilities for irradiation of materials with neutrons and ionizing radiation. These facilities are available to outside users and are currently used for evaluation of semiconductors and scintillators.

RADIATION DAMAGE MECHANISM IN JFETS

Radiation damage in JFETs has been well characterized with respect to DC and small-signal parameters [4,5,6]. However, to our knowledge, no comprehensive study has been made on the behavior of the device noise with respect to radiation. A JFET is pictured in Fig. 3 and consists of a conducting channel between source and drain whose dimensions are modulated by a variable width depletion region. The depletion width is controlled by varying the voltage on the gate terminal with respect to the source. Sources of noise in the JFET are as follows:

- Leakage current between the gate and source due to generation in the depletion region.

- $1/f$ noise due to generation-recombination centers in the depletion region [7].

- Thermal noise due to the finite resistance of the conducting channel between source and drain.

Upon exposure to neutrons, the JFET will experience damage to the crystal lattice due to collisions between the neutrons and the silicon atoms (Fig. 4). The resulting kinematic displacements, or vacancies, will have a finite mobility and will tend to accumulate around donor atoms. In addition, within a short range of a given collision, a cascade of silicon recoils will cause larger defect "clusters." Displacements and clusters will cause intermediate energy states, or traps, to exist in the bandgap. These mid-band states will make it easier for a carrier to jump to the conduction band. This translates into an increased leakage current ΔI_G in the device, which will be proportional to the density of the traps,

$$\Delta I_G = \alpha \phi V \tag{1}$$

where V is the volume of the depletion region, ϕ is the neutron fluence, and α is a leakage current damage constant which relates the neutron fluence to the displacement density and

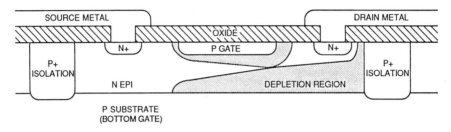

Figure 3. Cross-section of junction field-effect transistor.

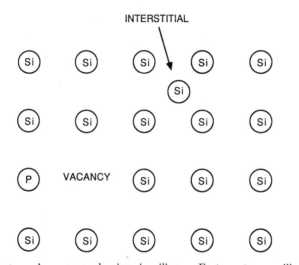

Figure 4. Neutron damage mechanism in silicon. Fast neutrons will damage the crystal and cause vacancies in the lattice, which will increase leakage currents [5].

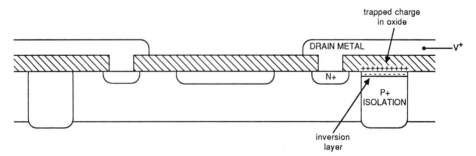

Figure 5. Ionizing radiation damage mechanism in the JFET. Charge build-up in the oxide can cause inversion in the p-type isolation region [3].

will depend on processing, temperature, and neutron energy [8]. Thus, the leakage current is strictly a volume effect in the silicon, and may be minimized by using smaller devices.

Ionizing radiation will cause carrier generation in all of the transistor materials, including the metal, silicon crystal, and silicon dioxide. As shown in Fig. 3, the oxide is used to insulate the metallization from the top of the die. In the case of the metal and silicon, the excess generated charge will be conducted away with little effect. However, in the case of the oxide, the charge will become trapped, causing a net charge to build up in the oxide. Enough of this built-up charge, together with a positive voltage on the metal, can cause a surface inversion in the p-type isolation region (Fig. 5). The mechanism is similar to that of an MOS transistor, where the threshold voltage V_t is given by

$$V_t = \Phi_{ms} + 2\Phi_f + \frac{Q_b}{C_{ox}} - \frac{Q_{ss}}{C_{ox}} \qquad (2)$$

where Φ_{ms} is the metal-silicon work potential, Φ_f is the Fermi potential, Q_b is the channel charge, C_{ox} is the oxide capacitance, and Q_{ss} is the trapped charge in the oxide. The net result of the inversion is an increased leakage current due to the increased area of the p-n junction [4]. This effect can be minimized by heavily doping the p-type isolation (increasing Q_b), which will increase the threshold voltage and make it harder to cause inversion.

Ambient temperature will have an effect on the radiation damage, both with respect to neutrons and ionizing radiation [4,6]. Annealing effects occur because at higher temperatures (e.g. room temperature), the displacements and trapped charges caused by the radiation will tend to dissipate due to thermal motion. The result will be a decrease in the radiation damage. At SSC, annealing will generally not occur because the devices will always be held at liquid-argon temperature. However, annealing is an issue in the damage testing environment where the devices may be tested at room temperature. Thus, care must be taken when comparing tests done at low temperature (where noise is lower but there is no annealing), and at room temperature (where noise is higher but may be lessened due to annealing).

NEUTRON DAMAGE TESTING

Neutron damage testing was conducted at the IPNS facility at Argonne with the neutron spectrum shown in Fig. 6. The JFETs were irradiated at room temperature while under typical bias conditions. Measurements were performed approximately one week after the actual irradiation, which allowed the neutron activation of the devices to subside. All annealing effects should stabilize during such time. Fig. 7 shows measurements of I-V curves and gate leakage current before and after the irradiation. In the case of the I-V curves, ~5% variations were observed after exposure to 10^{14} neut/cm^2 (typical 10-year dose at SSC). Measurements of transconductance showed <2% variation. Thus, these JFETs could be used successfully at SSC with little or no loss in gain or bandwidth.

The gate leakage current (Fig. 7b) increased by one order of magnitude after the irradiation at room temperature. However, between room temperature and liquid-argon temperature, the leakage can be expected to drop by about six orders of magnitude. By extrapolating from the post-irradiated room-temperature measurement, the leakage current at 90° K will be on the order of 0.1 fA, which will not adversely affect the noise of the pre-amplifier.

Fig. 8 shows the room-temperature series noise for neutron fluences ranging from 10^{13} to 10^{15} neut/cm^2. The main effect of the neutrons is to increase the $1/f$ noise due to the displacement damage. Due to the limitations of our test equipment, we have not been able to directly measure the noise at the high frequencies which are of most interest. However, we can get an estimate of the high-frequency behavior by linearly extrapolating the $1/f$ noise and assuming the thermal noise to be constant (dashed lines in Fig. 8). This assumption

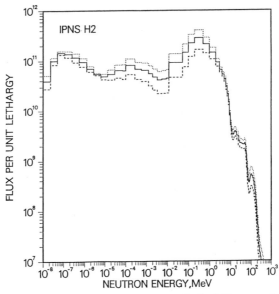

Figure 6. Neutron spectrum at Intense Pulsed Neutron Source.

should be valid because the thermal noise is proportional to the transconductance, which, as mentioned above, remains essentially unchanged by the neutrons. Assuming a bipolar shaping function of 100 ns peaking time, the noise will be bandlimited to the high-frequency region shown in the figure. After 10^{14} neut/cm^2, the noise increase, though measurable, is far less dramatic than the increase of the equivalent noise voltage at lower frequencies. This was confirmed by using one of the irradiated JFETs as the input transistor to an un-irradiated HELIOS pre-amp. An increase of 25% in the equivalent noise charge (ENC) was measured with 100 ns bipolar shaping at room temperature. This increase will be smaller at true operating conditions, i.e. at liquid-argon temperature and at faster shaping times.

IONIZING RADIATION DAMAGE TESTING

Several of the HELIOS JFET pre-amplifiers were subjected to 12 MRad(Si) of Cobalt-60 gamma radiation at Brookhaven at the rate of 5 kRad(Si)/hr. The devices were irradiated with power applied while kept at their normal operating temperature of ~120 K, and were kept cold for the duration of testing. The resulting change in the amplifier noise is shown in the noise spectrum of Fig. 9, which shows the ENC with respect to the shaping time constant for unipolar shaping. The ENC worsens, especially at longer shaping time constants. This corresponds to an increase of leakage current as the positive charges in the oxide drift to the oxide/silicon interface and cause inversion in the silicon. However, long shaping times are not of interest for SSC applications, therefore making this behavior unimportant. At the shorter shaping times (<100 ns), the noise increase is less than 20%.

RADIATION TEST FACILITIES

Because Argonne National Laboratory was for a number of years the site of one of the world's major high energy physics accelerators (ZGS), experienced groups were developed to supervise radiation safety, dosimetry, etc. After the ZGS was closed, these facilities continued to serve the needs of the various accelerators which continued to operate at Argonne. The

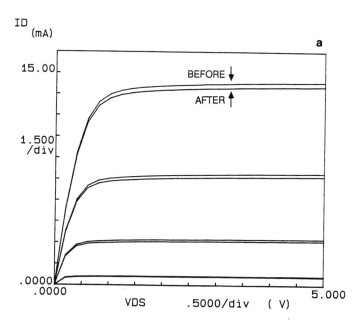

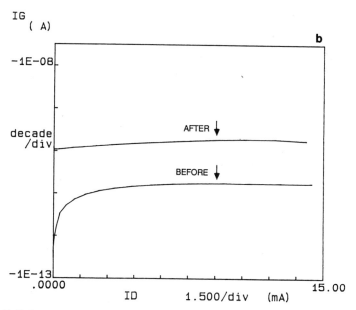

Figure 7. DC characteristics for SNJ132L JFET, before and after exposure to 10^{14} neutrons/cm^2. (a) I-V curves. (b) Gate leakage current.

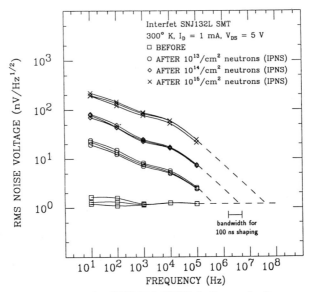

Figure 8. Input series noise for SNJ132L JFET, before and after exposure to neutrons.

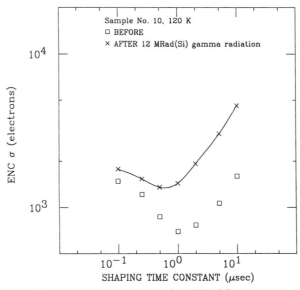

Figure 9. Equivalent noise charge spectrum for HELIOS pre-amplifier, before and after exposure to 12 MRad (Si) gamma radiation.

availability of these services makes the performance of radiation damage studies at Argonne particularly straightforward and efficient from the experimenter's point of view.

Intense Pulsed Neutron Source (IPNS)

The Intense Pulsed Neutron Source at Argonne is used routinely in an on-going program of material science research. It is a large well-staffed facility which runs a well defined schedule on average of two weeks per month throughout the year. This device is unique in that it can give in minutes or hours neutron fluences characteristic of several years of operation in the SSC environment at very forward pseudorapidities. Additionally, the IPNS offers the advantage over a reactor in that there is virtually no gamma contamination of the neutron flux. This is particularly valuable in semiconductor radiation damage testing because it allows the effects of displacement damage to be distinguished from the effects of ionization damage.

The IPNS uses a 50 MeV Linac and 500 MeV Rapid Cycling Synchrotron to produce a 500 MeV proton beam which is transported to a spallation target in a large experimental hall. The shielding in this facility has a number of ports which are used for material science experiments, and two test ports which are used parasitically for such things as radiation damage studies. One port is approximately 1 cm (3/8") diameter and the other is approximately 5 cm (2") in diameter. The neutron flux at either port is approximately 10^{12} neut/cm^2/sec, and the neutron energy is shown in the spectrum of Fig. 6. Fortunately, the neutron spectrum is roughly similar to the spectrum of the albedo neutrons in an SSC detector, peaking at approximately 1 MeV.

We have done a large number of neutron damage exposures using the test ports at fluences ranging from 10^{12} to 10^{17} neut/cm^2. These exposures have involved MOS, JFET, and bipolar semiconductors, and plastic scintillator.

Cobalt-60 Source

Within the Biology Division, Argonne has a large Co-60 source which is used routinely in life science research. This source is capable of producing up to 2 MRad(Si)/hr. The source is periodically calibrated with an ion chamber which is calibrated by NIST, and subsequently only half-life corrections are made in dose calculations. This source has been used in tests of radiation damage in CMOS devices at room and cryogenic temperatures.

22-MeV Electron Linac

The Chemistry Division at Argonne has a 22-MeV electron linac which is available for radiation damage research. The nominal beam current of the linac is 50 microamps, however this number may be reduced by as much as three orders of magnitude at the user's discretion. Typically, the beam is focused to a spot 0.6 cm (1/4") in diameter, however the user can de-focus the beam so that it fills the beampipe. Accordingly, the flux of 22 MeV electrons can be varied to fit the users requirements over a very large magnitude. This is particularly useful in semiconductor radiation damage research.

Fast Neutron Generator

The Fast Neutron Generator in the Engineering Physics Division at Argonne is is capable of producing neutrons for radiation damage research through the Be(d,n) reaction, and is used currently for neutron physics. It consists of a dynamitron, which is similar to a large Cockroft-Walton accelerator, and can accelerate 7 MeV deuterons onto a thick beryllium target to yield 2.5 MeV neutrons and produce a flux of up to 10^{10} neut/cm^2/sec. The spatial constraints at the dynamitron are much more relaxed than at IPNS, allowing much easier installation of a cryostat for neutron damage research at cryogenic temperatures.

CONCLUSIONS

The radiation environment at the SSC interaction regions will place additional restrictions on the already conflicting requirements of high speed and low power for the electronics. However, we have shown that JFET technology is particularly well suited for use in this environment. The main source of damage due to neutrons is an increase in the $1/f$ noise of the devices. Our measurements show that after exposure to 10^{14} neut/cm^2, the pre-amplifier noise will increase by 25% for a bipolar shaper with 100 ns peaking time at room temperature. This figure should improve at faster shaping times and lower temperatures. For the case of ionizing radiation, the damage will be due to increased leakage currents. For 12 MRad(Si) of gamma radiation, the pre-amp noise increased by 20% for a unipolar shaper with 100 ns shaping time constant at 120 K. However, this figure will also decrease at faster shaping times.

Radiation sources will be necessary for damage testing of all detector materials. Argonne has complete facilities, including sources and dosimetry, for damage testing of total dose and dose rate effects with neutrons, electrons, and gamma radiation.

ACKNOWLEDGMENTS

The authors would like to thank T. Scott and G. Schulke of IPNS for help with the neutron irradiations.

REFERENCES

[1] D.E. Groom, ed., "Radiation Levels in the SSC Interaction Regions," *SSC-SR-1033*, June 10, 1988.

[2] J.W. Dawson, L.J. Nodulman, "Development of Radhard VLSI Electronics for SSC Calorimeters," in *Supercollider 1,* M. McAshan, ed., Plenum Press, New York, 1989, pp. 203-216.

[3] HELIOS Collaboration, unpublished report.

[4] D.J. Allen, et al., "Gamma-Induced Leakage in Junction Field-Effect Transistors," *IEEE Trans. on Nucl. Sci.,* Vol. NS-31, No. 6, Dec. 1984, pp. 1487-1491.

[5] S.S. Naik, W.G. Oldham, "Neutron Radiation Effects in Junction Field-Effect Transistors," *IEEE Trans. on Nucl. Sci.,* Vol. NS-18, No. 5, Oct. 1971, pp. 9-17.

[6] G.C. Messenger, M.S. Ash, *The Effects of Radiation on Electronics Systems,* Van Nostrand Reinhold Co., New York, 1986.

[7] M.B. Das, J.M. Moore, "Measurements and Interpretation of Low-Frequency Noise in FET's," *IEEE Trans. on Electron Devices,* Vol. ED-21, No. 4, April 1974, pp. 247-257.

[8] H.W. Kraner, Z. Li, K.U. Posnecker, "Fast Neutron Damage in Silicon Detectors," *Nucl. Instr. and Meth.,* A279 (1989), pp. 266-271.

PERFORMANCE MEASUREMENTS OF HYBRID PIN DIODE ARRAYS[*]

J. Garrett Jernigan and John F. Arens
Space Sciences Laboratory, University of California
Berkeley, CA 94720

Gordon Kramer
Hughes Aircraft Company, El Segundo, CA 90245

Tim Collins and Jim Herring
Hughes Aircraft Company, Carlsbad, CA 92009

Stephen L. Shapiro
Stanford Linear Accelerator Center
Stanford University, Stanford, CA 94309

Colin D. Wilburn
Micron Semiconductor Inc., 126 Baywood Ave.
Longwood, FL 32750

ABSTRACT

We report on the successful effort to develop hybrid PIN diode arrays and to demonstrate their potential as components of vertex detectors. Hybrid pixel arrays have been fabricated by the Hughes Aircraft Co. by bump bonding readout chips developed by Hughes to an array of PIN diodes manufactured by Micron Semiconductor Inc. These hybrid pixel arrays were constructed in two configurations. One array format having 10×64 pixels, each 120 μm square, and the other format having 256×256 pixels, each 30 μm square. In both cases, the thickness of the PIN diode layer is 300 μm.

Measurements of detector performance show that excellent position resolution can be achieved by interpolation. By determining the centroid of the charge cloud which spreads charge into a number of neighboring pixels, a spatial resolution of a few microns has been attained. The noise has been measured to be about 300 electrons (rms) at room temperature, as expected from KTC and dark current considerations, yielding a signal-to-noise ratio of about 100 for minimum ionizing particles.

---

[*] Work supported by Department of Energy contract DE–AC03–76SF00515.

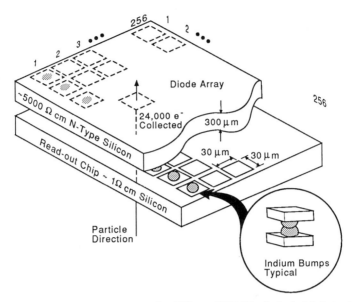

Fig. 1.   Schematic representation of a Silicon PIN Diode hybrid detector.

INTRODUCTION

An architecture well suited for charged particle detection at the SSC is that of a hybrid.[1-3] The charged particle detector and the readout electronics are constructed as two separate silicon chips, each optimized for its specific function. The two chips, indium bump bonded together, then provide the basic building block for the construction of a detector array.

The choice of the hybrid design (viz., one in which each diode of the detector array is bonded to an independent amplifier readout circuit on a mating VLSI chip via an array of aligned indium metal bumps that cold-weld under pressure to form ohmic contact), allows for additional flexibility in the selection of detector and readout electronics.[4] For instance, a change in the leakage current specification of the detector array will not affect the readout electronics, nor will a change in the VLSI chip oxide thickness to accommodate a radiation hardness specification affect the detector array. Figure 1 is a schematic representation of a silicon PIN diode array hybrid.

SILICON HYBRID ARRAYS

Development of hybrid vertex detectors has been the goal of the authors since late 1984. To this end, two hybrid arrays have been designed and fabricated. The sensor arrays were fabricated by Micron Semiconductor and the readout arrays by Hughes Aircraft. The indium bump bonding was done by Hughes with bumps measuring under 15 μm in diameter. The properties of the two arrays are described in Table 1.

Figure 2 is a schematic diagram of the MOSFET circuit of the 10 × 64 readout array. The diagram has been divided into its several functional portions. The section replicated for each pixel contains four MOSFETs. Signal charge is generated by the detector diode, and is fed to the gate of the signal MOSFET where it stays until a readout is made. The pixel selection circuit indicates how a sequence of address lines can select an individual pixel by turning on the gates of the V_{DD} bias MOSFET and

Table 1. Summary of Device Parameters

Array dimension	10×64	256×256
Pixel size	$120 \ \mu$m	$30 \ \mu$m
Detector material	Silicon	Silicon
Number of readout channels	10	2
Power during "write" cycle	0 mW	0 mW
Power during read cycle	10 mW	2 mW
Present clock speed	1 MHz	2 MHz
Theoretical clock speed	10 MHz	10 MHz
Readout mode	Random access	Random access
Processing power	20 MIPS/channel	20 MIPS/channel
Radiation hardness	1 Mrad	?
Noise at room temperature	$<300 \ e^-$ rms	$<300 \ e^-$ rms

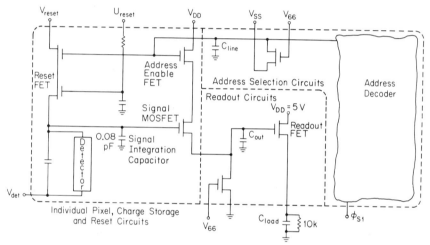

Fig. 2. Schematic drawing of the readout electronics of the 10×64 readout array.

the enable gate of the reset MOSFET. The U_{reset} signal allows the gate of the signal MOSFET to be reset to the V_{reset} level for any pixel that is enabled by the reset MOSFET. All of the signal MOSFETs in a column of the array are connected to a readout MOSFET in a source follower configuration which provides power for driving an external circuit.

The present readout chips allow random access to any pixel, which then operates as an independent detector. By virtue of its geometry alone, each pixel detector (PIN diode) provides about 3000 times less sensitivity to diode leakage current than a microstrip detector, which will increase the radiation hardness to neutrons. To complement this increase in the radiation hardness of the detector diode, the readout chips can be fabricated in a technology which is radiation hard to 1 Mrad of ^{27}Co gamma rays at cryogenic temperatures.

The two readout chips are similar, but have some differences. The 10×64 array has a random access architecture in that a unique setting of its address lines will select one and only one row of pixels. The 256×256 array, on the other hand, is random

Fig. 3.   A photograph of the 10×64 Silicon PIN Diode Hybrid in its mount in the laboratory.

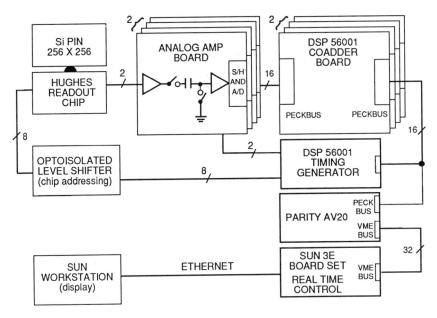

Fig. 4.   A block diagram of the High Energy Physics readout and display electronics.

access in that the pixels are addressed via row and column shift registers. This feature makes addressing a given pixel more complicated but allows easy implementation of a sparse scan algorithm. This particular readout chip has been optimized to collect electrons but is bipolar at the signal levels we expect.

DESCRIPTION OF PRESENT HARDWARE

Figure 3 is a photograph of the 10×64 array in its mount in the laboratory. Figure 4 is a block diagram of the high energy physics data acquisition system. This system, used to read out the 256×256 array, is more modern and more up-to-date than the system described in Ref. 2 on which the 10×64 data described in this paper were taken. A Sun Microsystems Sun-3/110LC-4 workstation controls a system housing a Sun 3/E CPU, a Motorola 68020 bus converter board, amplifiers, ADCs, digital

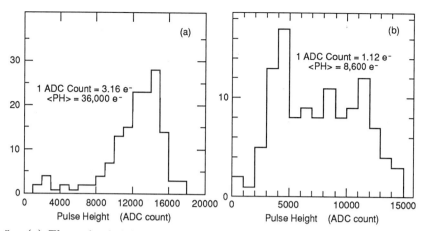

Fig. 5. (a) The pulse height spectrum of the alpha data; and (b) the pulse height spectrum of the beta data.

signal processors, and a clock generator. The digital signal processor is the Motorola DSP56001. This device acquires data at a rate of 10 MIPS, processes it, and passes it, via the MC68020 to the Sun 3/E. The bus converter board, a Parity systems AV20 dual port processor, interfaces the local analog bus (PECKBUS) to the VME bus.

Detector development funds were used to begin the fabrication of the dedicated high energy physics data acquisition system described above. The amplifier/ADC/DSP56001 boards and the clock generator in the present system, however, are circuits remaining from the infrared data acquisition system. These need to be redesigned to high energy physics criteria.

A data acquisition and display software package based on the previous system has been written. The operating system is UNIX, the DSP has been programmed in assembly language, and various control functions are written in Magic/L, an interactive language derived from Forth. The Parity AV20 dual port processor has been programmed in C, and the DSP will be reprogrammed in C.

RADIOACTIVE SOURCE TESTING OF THE ARRAYS

Testing of the 10 × 64 hybrid array started in mid-September 1989 at the Space Sciences Laboratory in Berkeley while testing of the 256 × 256 array started in December 1989 at Hughes Aircraft, Carlsbad.

A ^{106}Ru beta source and an ^{241}Am alpha source were used to irradiate the 10 × 64 array. Spectra for both alphas and betas were obtained, and the data are presented in Fig. 5. From this data, we roughly calculate both the signal in the hit pixels and the noise in the surrounding pixels. We have verified that both signal and noise agree with models of the readout electronics.

In our geometry, particles enter the device on the cathode of the PIN diode, the side farthest from the bump bonds. Thus, much of the charge collected must drift across the entire depletion distance of 300 μm. The charge cloud, spread by scattering of the initial radiation and by diffusion, will have a finite lateral size. If a particle were closer to the edge of a pixel than to the center, one would expect charge to be shared by adjacent pixels. On average, we see charge spread over 9 pixels; a 3 × 3 array.

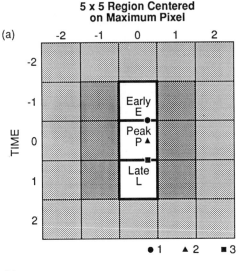

**5 x 5 Region Centered
on Maximum Pixel**

(a)

TIME

● 1   ▲ 2   ■ 3

(b)

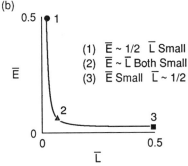

(1) $\overline{E} \sim 1/2 \ \overline{L}$ Small
(2) $\overline{E} \sim \overline{L}$ Both Small
(3) $\overline{E}$ Small $\overline{L} \sim 1/2$

Fig. 6. (a) A schematic of a 5 × 5 pixel region centered on a particle hit; and (b) a schematic plot of the $\overline{E}$ versus $\overline{L}$ planes indicating the result of various hypothetical incident positions of a particle within the central pixel.

A detailed analysis of a set of pixels near a particle hit allows the calculation of four quantities: signal-to-noise, noise, size of the charge cloud, and spatial resolution. In Fig. 6(a), a schematic of a 5 × 5 region centered on a particle hit is shown. The pixel with the maximum signal is located at $(0,0)$ is labelled "$PEAK$" and the pulse height of its signal referred to as P. Nearby pixels are located by offsets between -2 and $+2$. Each pixel in the 5 × 5 region has its zero set by analyzing other frames of data wherein there have been no particle hits in its vicinity. Further, in analyzing a particular hit, an average of the outer boarder of 16 pixels is subtracted from the inner 9 pixels to remove any systematic offset which arises because of the separation in time between the measurement of the background frame and the frame containing the particle hit. The direction of time in Fig. 6(a) refers to the sequence of the readout of the pixels. In this device, columns are read out in parallel, thus, the five pixels in the row labeled -2 at the top of the figure are readout first, with sequential readout, by row, from -2 to $+2$ on the time axis. For convenience, the pixel just above the $PEAK$ is labelled "$EARLY$," and its contents referred to as E, while the pixel just below the $PEAK$ is labelled "$LATE$," and its contents referred to as L.

E, P, and L are used to derive the $\overline{E}$ and $\overline{L}$ ratios from the relation:

$$\overline{E} = E/(E + P + L) \qquad \text{and} \qquad \overline{L} = L/(E + P + L) \quad .$$

Figure 6(b), a schematic plot in the $\overline{E}$ versus $\overline{L}$ plane of what should happen if a particle were to hit different parts of the central pixel. Possibilities 1, 2, and 3 are indicated in the central pixel, with their corresponding results shown in Fig. 6(b). The position of possibility 2 will be close to or far from the origin depending on the size of the charge cloud.

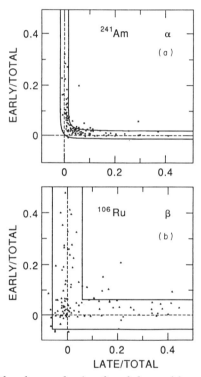

Fig. 7. Scatterplots of the charge sharing for alpha and beta radiation incident on the 10×64 array. The $\pm 2\sigma$ limits are shown on the solid lines.

Figure 7 shows the $\overline{E}$ or $EARLY/TOTAL$ versus $\overline{L}$ or $LATE/TOTAL$ scatterplots for the alpha and beta data. The trend of the data to fall near the axes, as explained schematically by Fig. 6(b), is as expected. The difference in the two cases is dominated by the somewhat larger charge cloud generated by the alphas compared to the betas. If one uses the measured signal to noise of the total charge deposited by a particle hit, one can derive the expected deviations from perfect correlation in the $\overline{E}$ versus $\overline{L}$ plane. The $\pm 2\sigma$ limit curves for the alpha and beta data are shown in Fig. 7. Note that the data scatter approximately as expected. These curves are not fits to the data shown, but are instead predicted from the signal-to-noise ratio only.

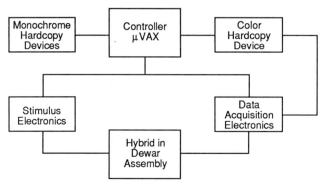

Fig. 8.   A block diagram of the Hughes/Carlsbad readout system.

In Fig. 7, the data points near the ends of the L-shaped locus of points represent data which have no charge deposition in either the *LATE* or *EARLY* pixel. Thus, the variance of these data is a measure of the fluctuation of the ratio of the signal of a typical nonhit pixel and the total signal from the three pixels in a hit column. This variance is the inverse of the average signal-to-noise ratio. For this analysis, the variance of $\overline{E}$ for data where $\overline{L} \geq 0.25$, and the variance of $\overline{L}$ for data where $\overline{E} \geq 0.25$ is used. The signal-to-noise ratio for the beta sample from this variance is about 36. The same analysis for the alpha data set yields a signal-to-noise ratio of 113. The mean signal size for the betas is about 8,600 electrons, while that for the alphas is 36,000. Thus, the two independent measurements of noise yield 239 and 319 electrons, respectively, for the beta and the alpha data. The error in making this measurement is not statistical, but is dominated by systematic errors, estimated to be at the 20% level. Thus, for both alphas and betas, the noise measurement corresponds to about 300 electrons (rms).

The noise measurements are somewhat higher than a calculation of KTC noise and dark current noise, which are the dominant sources of noise at room temperature. For our devices, these are 110 electrons (rms) and 120 electrons (rms), respectively yielding a combined noise of 165 electrons (rms). We attribute this extra noise to the external electronics, and expect that it can be removed by better calibration and filtering. We have, however, achieved our goal of 300 electrons (rms) at room temperature.

From the ratio of the number of data points which lie near the axes and the number of data points in the cluster near the origin, one can determine the average size of the charge cloud produced by an incident charged particle. For betas, this size is 19 μm (1σ) and for alphas this size is 28 μm (1σ), assuming a two-dimensional Gaussian profile.

To determine the spatial resolution, one uses the fact that the cluster of data points near the origin corresponds to events which deposit nearly all of their charge in the central pixel. Thus, within this region, no interpolation is possible. For those events which are near the edge of a pixel, one can use the variance noted earlier to estimate the error in the position within the pixel. For the betas which strike within 30 μm of an edge, the centroid of the 19 μm distribution can be located to within about 2 μm. A similar result for the alphas is 2–3 μm across the entire 120 μm pixel, due to the larger size of the alpha-induced charge cloud.

Preliminary testing of the 256 × 256 array has been started in the test laboratory of Hughes Aircraft in Carlsbad, California. Figure 8 is a block diagram of the readout

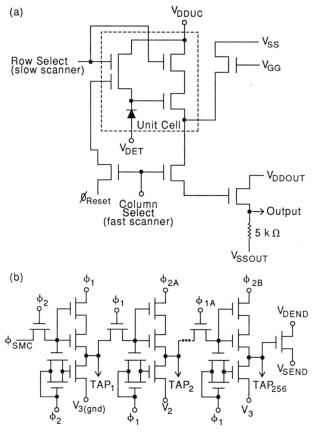

Fig. 9. (a) A schematic diagram of the MOSFET circuit of one cell of the 256 × 256 array; and (b) a schematic diagram of the six transistor slow scanner.

system used. The hybrid, mounted in a 68 pin leadless chip carrier, is mounted in a Dewar on a pc card having an amplifier gain of 2.5. DC power, and clock stimuli are provided by a general purpose test station designed and fabricated by the Hughes Technology Center. This station has emphasized flexibility over high data rate. A DEC Workstation II (μVax) is used to control the data acquisition and stimulus electronics. Data reduction, evaluation and storage codes have been developed. Data is acquired at a 60 Hz frame rate, resulting in a 500 ns pixel read time. The system employs low noise, high bandwidth drive electronics, with tunable rise and fall times for the clock pulses. Voltages are maintained with a resolution of 0.5 mV.

To acquire the data, acquisition and gain electronics is interfaced to a TRAPIX 5500 imaging system supplied by Recognition Concepts. This electronics features adjustable gains from times 1/4 to times 64, and adjustable offsets with very low noise, 8 MHz, track and hold circuitry. A 12 bit, 10 MHz, A/D converter is interfaced to the Q bus of a μVax processor. Digital data is displayed in real time on the high resolution color monitor, and is stored on hard disk. By using the array processing capabilities of the TRAPIX 5500, events with signal levels only a few mV above the read-noise level may be detected in real time.

Figure 9(a) is a schematic diagram of the MOSFET circuit of one cell of the 256 × 256 array, and Fig. 9(b) is a schematic of the six transistor slow shift register.

Fig. 10. A noise spectrum of the 256 × 256 array at 200 °K.

The fast shift register enjoys the same circuit configuration, but has transistors of slightly different geometry.

The bulk of the preliminary data on this hybrid was taken at 200 °K. The hybrid evidenced dark current at room temperature at twenty times the expected level. Rather than wait for a new hybrid, it was felt that characterizing the existing chip, cold, would be more instructive.

Figure 10 is a measurement of the noise for this hybrid at 200 °K. Sixteen frames of data were taken and a pixel by pixel average taken at the normal frame rate of 60 Hz. The data displays the average pulse height at the output of the chip in mV. The truncated mean is about 3.5 divisions at 0.188 mV/div. Using a pixel capacitance of 40 fF, one calculates the noise to be 165 electrons (rms) at this temperature. KTC noise is about 65 electrons (rms) and the dark current is negligible at these temperatures. This data was taken with no double-correlated sampling techniques used, thus, other sources of noise such as $1/f$ noise will remain in the measurement.

The array was exposed to radiation from a ^{207}Bi source. This provides a source of 1 MeV electrons, which directly give a measurement of the signal to be expected from minimum ionizing particles. Figure 11 is a spectrum of the radiation from the Bismuth source. The 1 MeV electrons will result in 24,000 electron hole pairs (channel 1000 in Fig. 11) if they transit the PIN diode detector array normally. However, due to multiple coulomb scattering they will certainly not do this, but rather will exit the bottom of the array after scattering through an angle of about 38 degrees, and going through about 380 μm of material. This will result in an average pulse height of about 30,000 electrons (channel 1200).

The explanation for the pulse height distribution which occurs below channel 1,000 is found in the myriad of gamma rays which also are produced by the Bismuth source. For every 100 Bismuth decays, one gets 9–1 MeV electrons; but one also gets: 100–574 KeV gammas, 75–1 MeV gammas, 75–(70–90) KeV gammas, and 30–(10–13) KeV gammas. When all of these gamma rays are multiplied by their respective conversion probabilities in 300 μm of silicon, one observes that for every 100 Bismuth decays one sees: 1 event from the 175 high-energy photons, 9 events from the 1 MeV electrons, and 10 events from the 70–90 KeV photons. There are no events recorded from the low-energy photons, as we set a threshold of about 6,250 electrons for this data set. Figure 12 is a three-dimensional plot of a characteristic minimum ionizing particle

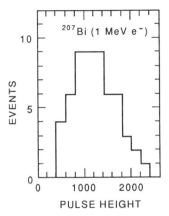

Fig. 11.   The pulse height spectrum measured using the 256 × 256 array when irradiated by a ^{207}Bi source, emitting 1 MeV electrons, and a number of gamma rays.

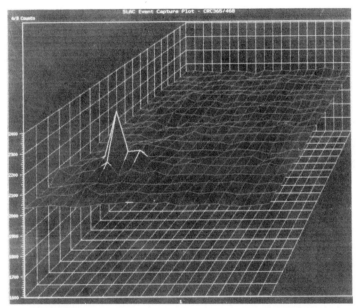

Fig. 12.   A three-dimensional plot of a 1 MeV electron being detected by the 256 × 256 array.

event. Figure 13 is a photograph of a 256 × 256 silicon PIN diode hybrid similar to the one used for these measurements.

SUMMARY OF RESULTS

Room-temperature operation of the 10 × 64 arrays has been demonstrated with noise levels less than 300 electrons (rms). Charged particles have been detected and imaged using this array with excellent signal to noise, and a preliminary measurement of 2–3 μm spatial resolution has been achieved. A measurement of the noise in the

Fig. 13. A photograph of a 256 × 256 silicon hybrid similar to the one used in these measurements.

256 × 256 array has been made at 200 °K, and found to be about 165 electrons (rms). Charged particles have been detected and imaged and pulse heights consistent with those expected have been achieved.

REFERENCES

1. S. Shapiro and T. Walker, "The Microdiode Array—A New Hybrid Detector," SLD New Detector Note No. 122 (1984).

2. S. Shapiro, W. Dunwoodie, J. Arens, J. Gernigan, and S. Gaalema, "Silicon PIN Diode Array Hybrids for Charged Particle Detection," *Nucl. Instrum. Methods* **A275**:580 (1989).

3. S. Gaalema, G. Kramer S. L. Shapiro, W. Dunwoodie, J. Arens, and J. G. Jernigan, "Silicon PIN Diode Hybrid Arrays for Charged Particle Detection: Building Blocks for Vertex Detectors at the SSC," *Proc. Int. Industrial Symposium on the Supercollider,* 1989, New Orleans, LA, p. 173.

4. S. Gaalema, "Low Noise Random-Access Readout Technique for Large PIN Detector Arrays," *IEEE Trans. on Nucl. Sci.* **NS–32**, No. 1:417 (1985).

A STRAW-TUBE TRACKING SYSTEM FOR THE SSC

C. Lu and K.T. McDonald

Joseph Henry Laboratories
Princeton University
Princeton, NY 08544

ABSTRACT

We briefly introduce the use of a straw-tube tracking system in an SSC experiment, and outline some of the many technical issues to be resolved in R&D programs.

WHAT IS A STRAW-TUBE TRACKING SYSTEM?

In experiments at the SSC we wish to infer the momentum of the various charged particles produced in the proton-proton collisions. This can be done be measuring the deflection of the particles' trajectories in a magnetic field.

A **tracking system** is a set of position-sensitive particle detectors designed to implement the momentum measurement. A **straw-tube chamber** is a gas-filled proportional counter in which the outer cylindrical wall is a spiral-wound straw made from two (or more) layers of polyester (or polycarbonate) film (see Fig. 1) The straw walls can support the tension of the central anode wire, can withstand several atmospheres of gas pressure, and are metallized on the interior surface to form the chamber cathode.

The straw tubes are glued together in bundles, or 'superlayers,' perhaps 8-12 layers thick, to provide rigidity against transverse buckling (see Fig. 2). At each end of a superlayer module are low-mass structures that provide the mechanical positioning of the tubes, distribute the chamber gas, and provide the electrical connections between the chambers and the VLSI readout chips (also mounted on the end structures).

The tracking system is then an array of superlayer modules such that each particle traverses $\sim$ 8 superlayers, yielding 64-96 samples per track. Figure 3 sketches a tracking system for the Bottom Collider Detector at the SSC; other SSC experiments will likely have tracking systems with cylindrical geometry. In a typical proton-proton collision some 50 charged particles are to be tracked, yielding about 5000 position measurements. To avoid confusions where more than one particle enters the same straw, the straws should be configured so the probability that of any given tube being struck is of order 1%; hence the total number of straws in the tracking system is of order 500,000. Figure 4 shows a computer simulation of the particle 'hits' in the tracking system for a single proton-proton collision.

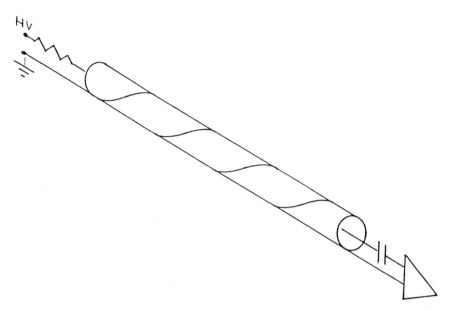

Fig. 1. Sketch of a straw-tube chamber.

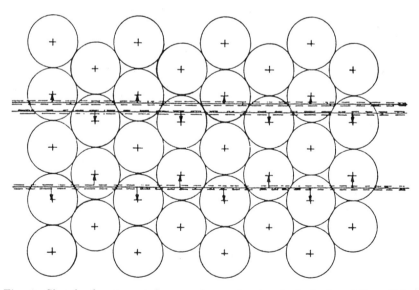

Fig. 2. Sketch of a straw-tube superlayer showing trajectories of 3 particles.

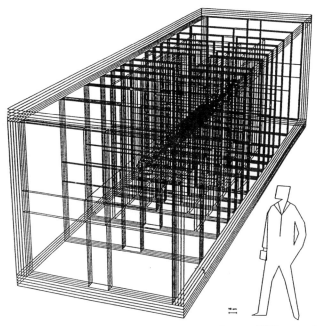

Fig. 3. Perspective view of a tracking system for an SSC experiment.

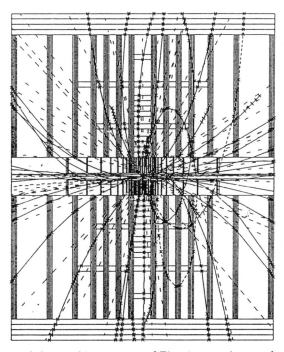

Fig. 4. Top view of the tracking system of Fig. 3, superimposed with tracks from a computer-simulated proton-proton collision. The solid tracks are from charged particles that are deflected in the magnetic field which is perpendicular to the paper. The dashed tracks are photons, which are not detected in the tracking system.

R&D ISSUES

Mechanical

- Choice of material for the 0.5-mil-thick inner ply of the straw: conductive polycarbonate *vs.* conductive Kapton.

- Thickness of the metallization of the inner straw ply (want low resistance, but heat load of metallization may destroy the film).

- Production winding of the straw tubes. Use of an adhesive that minimizes 'creep.'

- Straw-to-straw glueing procedure (want mass of the glue small compared to mass of the tubes).

- Alternative chamber construction: honeycomb (see. Fig. 5, from Oak Ridge National Lab).

- Thickness of the anode wire (low tension *vs.* low resistance).

- Supports for the anode wire against the electrostatic instability (and gravity).

- Need for an exoskelton to provide mechanical rigidity for very long straw bundles.

- Low-mass tube ends (see Fig. 6 for one scenario):
  - Provide alignment of the straws to 0.5 mil.
  - Distribute the chamber gas.
  - Provide coaxial electrical coupling of the chamber to the readout electronics.

- Cooling of the chamber against ionization heat load (~ 1 mWatt per tube) and electronics heat load (~ 10 mWatt per tube).

- Quality control during assembly for gas tightness, wire tension, electrical continuity, *etc.*

- Mounting and aligment of the superlayer modules.

Electrical

- High-voltage distribution.

- Miniature high-voltage capacitors on the tube ends.

- Possible use of active termination on non-readout end of tube.

- Front-end preamplifiers:
  - Use of Bipolar or BiCMOS technology.
  - Optimization of gain, rise time, input impedance, output shaping and discrimination, *etc.*

- Front-end time digitzers (CMOS):
  - Design of analog storage mechanism.
  - Design of digitizing circuit.

- Data-collection chips (a few per superlayer module)

- Routing and coupling of data to the outside world (via fiber optics?).

696

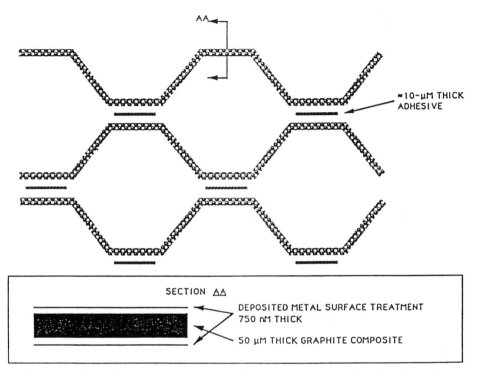

Fig. 5. Alternative superlayer construction using a carbon-fiber honeycomb.

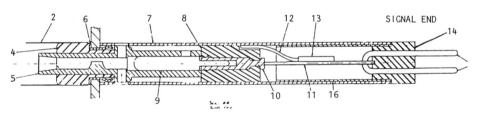

Fig. 6. Possible scenario for a straw-tube end plug: **2**: the straw tube; **4**: aluminum insert; **5**: Ultem feedthrough; **6**: collar spring; **7**: metallic sleeve; **8**: plastic socket; **9**: taper pin; **10**: pin socket; **11**: G-10 board with anode lead on bottom; **12**: cathode lead; **13**: blocking capacitor; **14**: plastic collar; **16**: Mylar sleeve. The anode- and cathode-signal pins at the right end plug into the front-end electronics board. The vertical plate at the left is made of Macor.

<u>Signal Analysis</u>

- Hardware processors:
  - Custom digital track-segment finders.
  - Analog track-segment finders (neural nets?).
  - Local digital processors (Hughes 3-D computers?).

- Software analysis – on a commercial processor farm.

UNIVERSITY COLLABORATIONS

There are four university collaborations pursuing straw-tube technology for SSC applications. The contactpersons are

Al Goshaw, Duke University (919-684-8134).

Gail Hanson, Indiana University (812-855-5942).

Kirk McDonald, Princeton University (609-258-6608).

Scott Whitaker, Boston University (617-353-8916).

A STRAW TUBE DESIGN SUITABLE FOR MASS PRODUCTION

Leonard Schieber

PCK Technology Division
Kollmorgen Corporation
Melville, NY

ABSTRACT

A straw tube drift chamber appears to be a viable candidate for a tracking system for the SSC. Several designs using straw tubes have been proposed. While straw tubes have been used in a number of smaller tracking chambers they have not been made in the quantities needed for the SSC. The problem to be addressed is not so much whether straw tubes will work but how can they be made in very large quantities and assembled into a useable detector on the scale required for the SSC.

A number of straw tube detectors have been made or are in the process of construction as illustrated in Table I. The number of straws needed for these detectors were significantly smaller than the SSC detector requirements. A design proposed for the SSC, by H. Ogren, at the Workshop on 4π Tracking Systems for the Superconducting Supercollider is also shown in Table I.[1] The proposed design estimates the number of straw tubes needed for the SSC to be around 800,000. Building a detector of this size gives rise to a number of problems concerning efficient manufacturing processes for straws and structural problems in construction of a large array. Among the engineering problems to be solved are the threading of a 12.5μ wire thru a 4 mm tube a meter or more in length; proper tensioning of the wire; the assembly of the fragile straws into an array that can be handled easily; and the interconnection of the read out electronics and power supply. This paper will present a novel method of constructing an array of straw tubes which lends itself to automated fabrication procedures.

The basic concept begins by splitting the straw tube in half longitudinally. The tubes can then be formed in large sheets. Either circular or hexagonal straws could be easily formed by vacuum forming aluminized mylar, or aluminum sheets over a suitable mold. Certain materials could also be machined or molded and metalized in a variety of ways, such as sputtering or plating, to provide a conductive coating.

Table 1. Experience in Making Straw Tube Detectors

Experiment	#Cells	Straw Dimensions	GAS	Resolution
HRS Vertex Chamber Indiana Univ. NIM254(1987)542	356	Aluminized Mylar–85μ 7mm dia.46 cm	Ar– Ethane(25%)	s=100μ
MAC Vertex Chamber 4 ATM SLAC NIMA261(1987)399	342	Aluminized Mylar–75μ 6.9mm dia.43 cm	Ar– CO_2(50%)	s=45μ
Mark II Vertex Chamber Univ. Colorado NIMA255(1987)486	552	Aluminized Mylar 75μ 8mm dia.75cm	Ar– Ethane(50%)	s=90μ
CLEO I Vertex Chamber Ohio State	192	Aluminized Poly- carbonate 12.5μ mylar 20μ 5.8mm–50cm	Ar– Ethane(50%) DME	σ=90μ σ=40μ
Mark III Vertex Chamber ATM, stereo layer NIMA276(1989) 42	640	Aluminized Mylar 75μ 8mm dia.84cm	Ar– Ethane(50%)	s=49μ
AMY Vertex Chamber Ohio State	576	Aluminized Polycarbonate 12.5μ mylar 12.5μ 5mm–55cm	Ar– Ethane(50%)	σ=85μ
CLEO II Vertex Chamber Ohio State	384	Aluminized 12.5μ mylar 12.5 4.5mm–50cm	Ar– Ethane (50%) DME	?
SSC Proposed Design	800K	Aluminized Polycarbonate + mylar–25μ total 4mm–100cm	CF_4	

The formed sheets are filled, preferably during the forming operation with a removable thermoplastic material such as polyethylene glycol, wax or a number of other suitable materials. This material has two main functions: first it provides rigidity to assist in handling the structure; second it provides an adhesive surface on which a wire can be positioned in the center of each tube. The resulting structure is shown in figure 1. A wiring machine such as the Multiwire Machine made by the Electronic Equipment Division of Kollmorgen Corporation, shown in figure 2, can be used to place wires in the center of each tube in a cost effective manner. This machine is capable of accurately positioning the wire in the tube half with a controlled amount of tension. The wire is fed at a servo controlled rate under an ultrasonic stylus while the straw is moved under the stylus at a slightly differing rate; producing the desired amount of tension. The ultrasonic stylus is heated by an ultrasonic power supply causing the thermoplastic material to melt around the wire. The thermoplastic material solidifies quickly capturing the wire in its tensioned state. When a layer is completely wired its mating layer is attached with conductive adhesive or clamping. Layers are stacked to form a complete module. A printed circuit board attached to each layer is used for connecting to the read out circuitry and providing a permanent tensioning device. The leachable material may be removed when the module is completely assembled or left until the entire detector is completed.

The wiring machine is capable of placing wires at a rate of one meter per minute with a true position accuracy of .0025 cm. Wires can be tacked in place on the circuit board at the end of the run on a strip of permanent wiring adhesive and connected to pads on the circuit board.

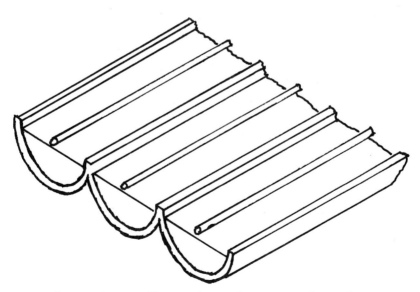

Figure 1. Configuration of Straw Tube Halve

Figure 2. Multiwire Wiring Machine

Figure 3. Wired Sample of Straw Tube Half

A variety of leachable materials are available which are compatible with the Multiwire process, and are light enough to allow easy handling of modules but strong enough to avoid damage during assembly. One material, suggested by Dave Duchane of LANL which holds promise is polyethelene glycol foam. This material has a high strength to weight ratio, is soluable in cold water and melts at a reasonable temperature. Small samples of panels have been wired using this material. (See figure 3) it can be cast in the straw tube layers, extrusion molded or pressed into them while they are being formed.

The Multiwire machine consists of a numerically controlled wiring table driven in the x direction by a digitally controlled servo system capable of positioning the table to an accuracy of 25μ and controlling the table speed to 2.5μ/sec. Wire is fed thru a wiring head, suspended over the table on a granite bridge. The rate at which the wire is fed is measured by an optical encoder and controlled by a servo motor in the feed mechanism shown in figure 4. The head is moved along the granite bridge by a servo utilizing a linear optical scale and rotated in azimuth by a third servo. Intermediate supports of a permanent adhesive may be positioned in the straw as shown in figure 5 with a dual purpose. First, they will limit sag in the wire if placed at 1 meter intervals and second, they will assist in maintaining an even tension in the wire after the leachable material is removed.

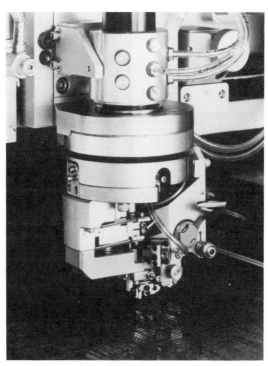

Figure 4. Wiring Machine Head

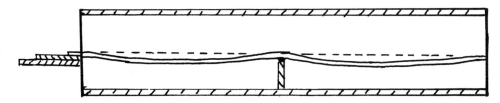

Figure 5. Tensioning Supports

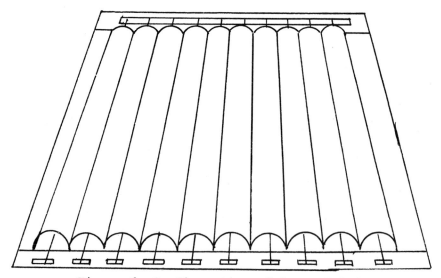

Figure 6. Configuration of Straw Halves

PROPOSED METHOD OF CONSTRUCTION

A feasible process for making straw tube detectors using the ideas presented would proceed as follows. A block of leachable material would be molded or machined in the shape shown in figure 6. A 25μ aluminized mylar sheet is vacuum formed around the leachable material to form the outer wall of the straw tube. An unfilled formed mating straw half layer is attached to the aluminized mylar sheet with conductive adhesive. Permanent adhesive strips are placed at appropriate intervals on the filled side and the layer is ready for wiring. Wires are laid in the centers of each tube half at a speed of 200 inches per minute. Wire ends are terminated on printed circuit boards attached to each end of the assembly. Individual layers are then assembled into a module of convenient size and the leachable material is removed (figure 7).

The Multiwire process may have some additional advantages beyond its ability to mass produce straw tubes in a cost effective manner. Due to the high precision of its wire positioning capability, the diameter of the straw tube could be made less than the proposed 4 mm. This would allow reduction of the anode voltage without sacrificing transit time. Since the machine is capable of laying wires within 240μ of each other it does not impose a limitation on the diameter of the straw.

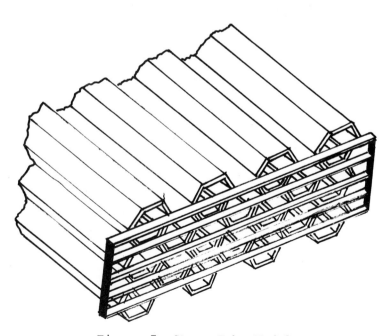

Figure 7. Straw Tube Module

SUMMARY

To summarize, the method of construction described provides a way to rapidly place wires in the straw tubes using an automated process while simultaneously tensioning them. Arrays of straw tubes are assembled into modules which are rugged and easy to handle. After electrical connections are made the modules are assembled into the detector. The leachable adhesive may be removed from each module or after assembly of the complete detector. Several samples have been made to evaluate the feasibility of wiring on leachable materials and test the critical processes in the procedure and some are available here for our examination. This proposed method of construction would turn a tedious, complicated and time consuming process of making straw tubes into a simple, rapid and cost effective procedure.

REFERENCE

1.   Harold Ogren - "Straw Tube Drift Chambers" presented at The Workshop on 4π Tracking Systems for the Superconducting Supercollider, Vancouver, Canada (July 24-28, 1989).

13. Education

BUILDING BLOCKS OF THE UNIVERSE

Ernest Malamud

Science and Technology Interactive Center
Aurora, IL 60506

Charles O'Connor

Ohio's Center for Science and Industry(COSI)
Columbus, OH 43215

Alan Cooper

The Open University, England

Abstract

COSI, a well established science center, and SciTech, an
emerging one, have formed a collaboration to develop a group of
original interactive exhibits conveying to a wide audience the
nature of the most fundamental features of the Universe, as
revealed in the fascinating world of nuclear and particle
science. These new exhibits will add to, and be supported by,
the basic science exhibits which have already attracted large
numbers of visitors to both centers. The new project, called
"Building Blocks of the Universe", aims to foster an
appreciation of the way all features of the Universe arise from
simple, basic rules and to lead the visitor from the perceived
complexities of our surroundings, to the unperceived, but
simpler features of the sub-nuclear world. It has already
become apparent from individual prototypes that these simple
but immensely far-reaching ideas can indeed be conveyed by
hands-on exhibits. These exhibits will be linked and enhanced
by an effective museum environment, using pictorial diagrams,
accurate non-technical text, and artistic displays to create an
atmosphere in which visitors can learn about phenomena beyond
the range of direct perception. This paper describes the
goals, content and organization of the exhibition. We also
outline our experience with prototype exhibits, and thereby
invite additional input into the development process.

I.   Introduction

Scientific literacy--which embraces science, mathematics, and
technology-- eludes us in the United States. A cascade of
recent studies has made it abundantly clear that our

educational institutions fail to attract and train adequate students in science and technology, despite the ever increasing importance and scope of the subjects. The solution cannot be achieved by the institutions alone. The standing of science among the general population must be improved. The same studies have shown that a better public understanding of science does lead to greater public support for science, which can bestow on students the encouragement and confidence to turn to science and technology. This also translates directly into increased public and political support for SSC.

Contemporary science centers are a valuable resource to help achieve this goal. They provide an introduction to the exciting and diverse worlds of science and technology for people of all ages, parents as well as children, teachers as well as students. Their exhibitions and programs allow the public to participate, to become involved in the thrill of scientific discovery, and to appreciate the application of scientific principles to industry and technology. Last year more than 50 million visitors enjoyed exploring science in centers across the country.

The hands-on experience is an important way in which to raise scientific literacy.

"I hear and I forget".
"I see and I remember".
"I do and I understand".

Chinese proverb

II. Project Objectives

Interactive science centers have developed from an initial emphasis on optics and perception into many areas of classical science, and are successful in both satisfying and stimulating public interest in these areas. COSI and SciTech are already playing their part in these activities. But the most fundamental patterns in our Universe are beyond direct perception -- notably those of the structure of nuclear matter and the form of the explosive creation of the Universe, which still governs its evolution. Why should we tackle these areas? First, because many aspects of physical science fall into place when these fundamental patterns are appreciated. Nuclear forces take their place alongside, and indeed are unified with, the electrical and gravitational forces currently demonstrated by exhibits in science centers. The second reason is quite different. Large amounts of public money support intense and expensive research in these areas. The SSC will, of course, become increasingly a focus of such research. Although members of the general public can have little direct participation in the work, they certainly can benefit from and enjoy the results. This is no empty claim. Such attempts as have been made to report back the findings to the public have revealed a considerable latent interest. Several books on fundamental particles and cosmology have risen, by their own intrinsic interest and merit, into the bestseller lists. It is time for interactive science centers to communicate these basically simple but immensely far-reaching ideas with hands-on exhibits to a wider audience.

This project is bringing together the complementary skills of outstanding subject expertise and science center management and craftmanship at SciTech and COSI to produce prototypes of 12 exhibits for testing and evaluation, followed by 8 exhibits which will form a permanent display in both science centers. These will break new ground in subject areas covered by science centers. The exhibits developed can be used by other science centers on an international basis. Associated teacher resources and other material will be developed on a copyright-free basis.

III. Project Design

To create a successful group of exhibits, linked and enhanced by an effective museum environment, is a big challenge. The aim of the exhibits is to lead the visitor from the perceived complexities of our surroundings to the unperceived, but deeper and simpler features of the sub-nuclear world. This will be done by a series of "gateways": from the molecular to the atomic to the nuclear to the sub-nuclear worlds. Pictorial diagrams and artistic displays will create an atmosphere in which visitors will be receptive to learning phenomena beyond the range of their perception. Careful attention will be paid to the development of accurate and effective, but non-technical text.

On the next page is a possible arrangement of the exhibition, covering 1000 sq ft. In the future it will be augmented and completed by exhibits on cosmology - the large scale patterns in the Universe - and this is indicated by the area on the left. There are intimate links between the large scale and small scale patterns in the universe, which make it appropriate to envisage the combined exhibits from the outset.

There are only three forces in the new, sub-nuclear world and all are actually different aspects of a single force. (The effects of gravity, the fourth known force, is too small to be of any consequence.) There are only a few basic particles, the quarks, leptons, and messenger particles, from which all other particles are made. Interactions between the particles, within the quota of energy bestowed by the creation, shaped the future of the whole Universe. The unification of forces, and the linking of particle physics and cosmology, represent the most far-reaching synthesis achieved by the human mind. Now that the underlying simplicity of the ideas has been revealed, after an intense international research effort, they are accessible to a wide audience. The basic ideas, being so fundamental, are unlikely to change over the next decade. Detailed features, on the other hand, such as the search for the top quark at Fermilab, or the Higgs boson at SSC are the subject of current research, and the exhibits will form a framework in which future advances can be incorporated.

IV. Project Members

COSI, Ohio's Center Of Science & Industry, is a science center celebrating its 27th Anniversary. This year approximately

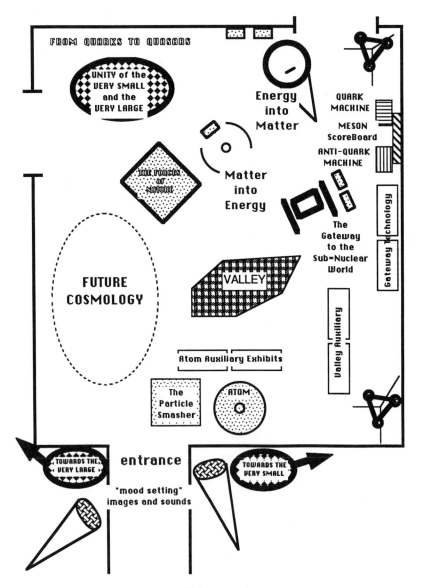

Figure 1

700,000 visitors will enjoy the over 100,000 sq ft of exhibits
at COSI. The staff at COSI have extensive experience in
exhibit development, evaluation, design and production. In the
last five years they have been responsible for creating over $7
million of interactive permanent and traveling exhibits for
COSI and other science centers.

The Science and Technology Interactive Center, SciTech, located
in the fast-growing western suburbs of Chicago, is a brand-new
science center. What SciTech lacks in experience and staff is
made up with an energetic core of volunteer scientists who are
committed to communicating the excitement of modern science and
technology to the public. The exhibit builders are comprised
of a wealth of professional talent from Fermilab, AT&T Bell
Laboratories. The Amoco Research Center, Argonne National
Laboratory and many outstanding local high school physics
classrooms. In 1989 SciTech opened a temporary 7,000 sq ft
exhibit development facility in Naperville and served over
20,000 visitors and 100 field trips. At the beginning of 1990
SciTech signed a 10 year lease for the former U.S. Post Office
on the Fox River in downtown Aurora. Renovation of this
historic, architecturally beautiful 37,000 sq ft building is
underway. The opening to the public is planned for late spring
or early summer, 1990.

V. Project Organization

The work of the two museums will be overseen and guided by an
Advisory Panel comprising recognized experts in the field of
particle physics, educators, and museum professionals. The
Panel will review the content of the proposed exhibits with
special attention to scientific accuracy and public
understanding and will assist with the evaluation of the
exhibits. Because of the particular nature of the topic, there
will in addition be a "lay person" focus group, meeting
monthly, to give detailed advice on how to make the exhibits
attractive and meaningful to the general public.

The national impact of this project was confirmed when a survey
was conducted to assess the interest other museums may have in
either purchasing a copy of the exhibition or hosting a
traveling version of the exhibition. We were overwhelmed with
the positive responses. Virtually everyone we contacted
responded affirmatively. There is no doubt that the growing
public interest in the subject has been, and will continue to
be, stimulated by the excitement of the SSC project.

Design and construction drawings, and the educational material
developed in conjunction with the exhibits, will be made
available to all interested institutions through the
Association of Science and Technology Centers (ASTC). We will
make the designs modular, for ease of transport, in view of the
results of our survey.

We have budgeted $483,158 for development and construction of
the exhibition. We expect to receive a grant from the National
Science Foundation of $249,482 towards this effort. The two
centers have committed an additional $65,000. We are planning
to form partnerships with corporations active in the field of
particle physics and accelerator construction to provide the
balance of the necessary funding.

VI.   Exhibit Elements

It has already become apparent from individual SciTech prototype exhibits that these simple but immensely far-reaching ideas can be conveyed by hands-on exhibits.   Following is a short description of each of the exhibit elements.

1.   The Atom

The atom is at the lower limit of objects which can be made visible by magnification.   In passing inside the atom we enter a new world. The idea of passing into a new world, where special behavior and structures can be identified but never seen is an important one.   The aim of this exhibit is to convey by simulation some of the features of that quantum world.   The model atom, when stimulated will react as a real atom is known to do.

2.   The Atom Auxiliary Exhibits

Different atoms emit different wavelengths of electromagnetic radiation when they de-excite.   Participatory exhibits and charts will expand on the basic idea shown by The Atom.

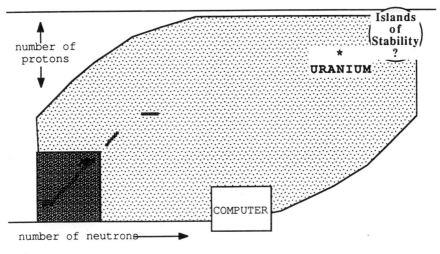

Figure 2

3.   The Valley of the Isotopes

The limited number of arrangements of the nucleons in a nucleus impose strong patterns which show up in the behavior of nuclei − their lifetime and their mode or modes of decay.   In the model each nucleus is represented by a column whose position represents its constituents, with the number of neutrons, N, measured off along one direction and the number of protons, Z, measured off along a direction aat right angles. The height of the column represents the   (negative of) the binding energy/nucleon which is a major determinant of the lifetime of the nucleus.   Each isotope is topped by an independently addressable indicator light.

A prototype containing the first 250 isotopes (up to iron) has been built and operated successfully.   The program which drives the exhibit already produces different types of display, according to the topic selected by the visitor: chemistry, radioactive decay, history, stellar evolution etc.

4. Valley auxiliary exhibits

Exhibit goals are to explain in a very basic way the meaning of the word "isotope". For the visitor who wants to dig deeper we attempt to explain the reason why there are stable and unstable nuclei and why certain combinations of neutrons and protons do not exist at all.

Prototypes of models of nuclei have been made of rubber balls joined by springs. The balls represent the constituents of the nucleus and the springs represent the forces between them, and the consequent oscillations of within the nucleus, called the Fermi energy. It will be possible to make some of the model nuclei fuse together, or undergo fission. More complex reactions can be illustrated by computer animation, and we have, as a prototype, made an animation of the production of deuterium.

5. Half-Life

This exhibit(s) combines the important concepts of "random", "chaos" and "half life" in a visually attractive way. This playful exhibit will show how radioactive decay can be modeled by an ensemble of random events. Many ideas are being explored; some will be prototyped before choosing the best one.

6. Gateway to the Subnuclear World & Auxiliary Exhibits

On the far side of this gateway we pass into the subatomic world, where particles cannot be seen but their effects can be recorded. The aim is to symbolize this step in the exhibit component sequence by an imposing gateway made up of spark chambers detecting sub-nuclear particles resulting from cosmic rays generated in distant parts of our Galaxy, or in some cases in other galaxies.

Figure 3

7. Gateway Technology

There are many interesting aspects to the technology: the color of the sparks relates to atomic excitation and ties to the Atom exhibit; the speed of propagation of signals and electromagnetic waves in space can also be illustrated.

8. Particle Smasher

This exhibit is a central part of The Quark Room whose theme is "Diversity out of Simplicity", or how the entire universe is built from a few simple building blocks: quarks, gluons, leptons, and messenger particles. This exhibits shows the relationship of organic structures, chemical compounds, elements, molecules, atoms, and sub-atomic particles.

It takes the form of a highly interactive computer game. A working version has been used, and has proved very popular with visitors. Further refinements will be added.

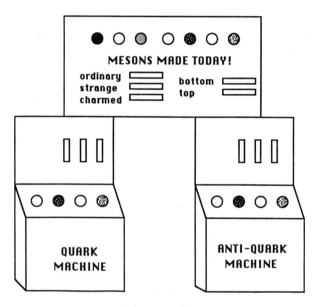

Figure 4

9. The Quark Machines

The visitor encounters the quark machines after passing through the Gateway. The visitor is now within the world of a single nucleon, an extremely simple world, containing only quarks and the forces between them. The goal is for the visitor to discover the recipe for making quarks into nucleons. With nucleons (and leptons and messenger particles) one can make a Universe!

One machine has been built, and been successfully used both at SciTech and at the Chicago Academy of Sciences. It stood up to vigorous use at both locations.

10. Energy into Matter

The aim of these two side-by-side exhibits is to show how matter and energy are so intimately linked as to be interconvertible. In the early universe space was laden with energy and matter could precipitate out of it as it cooled. Such concentrations of energy can only be produced nowadays where particles collide or decay. At big accelerators energy is converted to matter. The visitor can watch a small, repeat performance. Such processes are random and cannot be made to occur at will. However, the visitor will be able to control the working of the detectors, and the types of matter involved.

11. Matter into Energy

Antimatter meets matter. Making pure energy is easy if you have antimatter to start with. We will use antielectrons (positrons) as the antimatter, and allow them to annihilate with electrons. This process produces two gamma rays with very characteristic back-to-back trajectories, and specific energies, so the annihilation process can be easily recognised.

12. The Forces of Nature and their Unification

Gravity, magnetism and nuclear forces seem to be very different and they act over completely different distances. One thing unifies them: they are all carried by messenger particles. The most familiar force, gravity, has the most elusive messenger. On the other hand it is easy to find the nuclear messengers, but we never feel their short range force. This exhibit(s), still in the conceptual design stage, will create an opportunity for visitors to understand the differences and similarities between different forces.

VII. Conclusions

The project "Building Blocks of the Universe" harnesses the combined and complementary expertise and facilities of two interactive science centers: COSI in Columbus, Ohio and SciTech in Aurora, Illinois. It aims to bring the ideas of the fundamental patterns in nature into the scope of interactive science centers. There are strong indications that many other science centers will exploit the results of the project.

We believe that this project is important because the exciting discoveries of the simplicity of nature at the most fundamental level now form part of general scientific literacy, and the raising of the level of public appreciation of science is important in national terms. We believe that it is timely because the construction and operation of the SSC, an international focus of basic research, will provide a continuing stimulus to public interest, which our science centers can foster and satisfy.

EDUCATIONAL OPPORTUNITIES FROM THE SSC

George J. Doddy and Alan H. Rider

Daniel, Mann, Johnson, and Mendenhall
Washington, D.C.

Albert H. Halff

Albert H. Halff Associates
Dallas, Texas

High energy physics and education are very closely interwoven. Most
physics laboratories are located at universities or are operated by
consortiums of universities. Fermilab and the SSC are operated by the
Universities Research Association, URA, a consortium of 69 major research
universities and 3 associate members. Another example of this laboratory
and universities relationship is the Continuous Electron Beam Accelerator
Facility, CEBAF, which is operated by the Southeastern Universities
Research Association, SURA, another consortium of 39 major universities.

The educational potential inherent in the planning, construction and
operation of the SSC is immense. The SSC, as the world's largest
scientific instrument and, as the most powerful accelerator, will have a
natural attraction as the preeminent institution in the scientific
community. In addition to the primary objective of probing the
fundamental composition of matter, the SSC will appeal to a broad segment
of the population and will create the opportunity for both passive and
active educational experiences on the part of staff, students and
visitors.

On the esoteric level, the SSC will be a magnet for the scientific
community and will attract from around the world the finest minds in the
field of high energy physics. On the human level, the laboratory will
become an integral part of the community and will be an object of great
interest to local residents and visitors. The SSC planners should
recognize the opportunity to be a contributing institution to both the
local and the world community.

The SSC can make its greatest contribution in the primary and
secondary education levels. The success of the SSC will ultimately
depend upon on a continuous flow of skilled manpower, some of whom are
now in elementary and high schools. The project can become an

inspiration to an entire generation of Americans in much the same way as the space program was a generation ago in the 1960's.

As the first step, a public relations campaign could be utilized to inform the public about the educational potential of the SSC. The public relations efforts will build and maintain community support while dispelling potentially negative perceptions about the project. As part of this program, tours can be provided of the site during construction and ultimately during operation. These tours will generate interest and maintain enthusiasm for the work of scientists and engineers as well as instill a sense of anticipation in the students who should be encouraged to participate at this stage so that they can grow with the SSC from its very beginning.

As an adjunct to the public relations program, the SSC could develop a system of exhibits to make the laboratory a focus for a general introduction to the world of science. Through the use of exhibits, resource centers, classrooms, and observation galleries, the SSC in itself will become a part not only of the local community but also of the entire scientific community as well.

The avenues for an educational program are many and varied and will require consideration during the conceptual phase of the project to allow the facilities to be included either as part of the initial construction or as part of later construction depending on funding availability.

Within the planned headquarters building, space can be set aside for exhibits which will be viewed by students as well as the general public. These exhibits will provide information on the fundamentals of high-energy physics, an introduction to the SSC and how it works, a description of the research programs, and technology transfer benefits in industry and medicine - both of whom could be approached to assist in the development and funding of these spaces.

In addition to an education center, a science museum adjacent to the headquarters building could expand upon the exhibits and provide hands-on interaction with these exhibits. "Walk-through" models of major components of the SSC complete with animated mock-ups of equipment, audio visual displays, and video tapes of the history of science depicting the major discoveries along with their related contemporary benefits - all this could serve as an exciting introduction to the magic of science and how it affects our daily lives. One way to achieve this is to identify a principle of science, illustrate the method by which this principle is adapted to industry and then display products that result from this interaction.

Another ingredient to enhance the educational value as well as to stimulate public interest would be to organize self-guided tours of the entire laboratory facility. These tours would include stops at significant components of the SSC and would allow observation of that specific facility in operation. Controlled access galleries could be planned as a part of the initial construction. This would allow the public to observe operations in areas of interest such as the collision halls, refrigeration plants, service and access structures, tunnel section, and the industrial shops without interfering with those operations. Displays and video tapes of these components in action could serve to highlight the importance of the separate parts of the SSC and how they relate to the whole of the laboratory.

In addition to attracting visitors who have a general interest in the SSC, the facility will attract educators as well as students who have a specific interest in science. The SSC campus could be planned to include an education center with classrooms, lecture halls, teaching labs and resource centers. Students for these classrooms would come from the nearby schools, community colleges and area universities as well as national and international schools and universities.

One level higher in the ultimate plan of the education center could include a science-specific Community College under the auspices (and perhaps funding) of the various universities in Texas. This college would serve as the next logical step for those students who have science inclinations but are not certain of the direction they wish to take. In addition to the educational center, this campus could include a national science museum as well as cultural and recreational facilities which could be utilized by the students along with the SSC staff and their families. Including dormitories could provide economical and convenient facilities for long term transient students and scientists.

A further means of expanding the educational and scientific viability of the SSC will be the establishment of a world-wide communications network which would link the SSC Laboratory with educational institutions, appropriate government agencies, other physics laboratories, and supercomputer centers. To achieve these goals, the campus could include satellite communication system and TV-production facilities. An integral component of this global communications network could be a teleconferencing center. This center would serve not only as a global network but also would be part of an-in-laboratory network which could unite the far flung components of the laboratory unto itself with the minimum of lost travel time to distant meetings.

The ultimate success of the SSC will be determined by its ability to maintain a staff of skilled scientists and the support of the public and its elected representatives. To assure a continued supply of skilled technicians, the SSC must actively participate in the educational process. In addition to the establishment of educational facilities on the SSC campus, the development of an outreach program would allow laboratory staff to make personal visits to schools through all grade levels to supplement the educational facilities and to act as personal ambassadors for the SSC.

Other laboratories have had educational programs in place for many years. One example of the success of this collaboration is that of Fermilab where the following formal education programs are in place:

o   DOE High School Honors Research Program which brings students to Fermilab from each state, Puerto Rico, the District of Columbia and six Economic Summit countries for a two-week program of seminars with Fermilab scientists as well as to work on a Fermilab experiment.

o   DOE Teacher Research Association Program which gives some 20 teachers selected from both regional and national schools an opportunity to work with a scientist or engineer on a research project during the summer at Fermilab.

o   Summer Institute for Science and Mathematics Teachers Program which provides the opportunity for 45 high school biology,

chemistry and physics teachers as well as 15 mathematics
teachers to spend 4 weeks at the laboratory attending lectures
by research scientists and mathematicians.

o   Chemistry and Physics teacher networks that allow high school
     teachers to share skills, teaching strategies and materials.

o   <u>Topics in Modern Physics</u> is a teacher resource manual
     containing curriculum materials.

o   Kindergarten through 8th Grade Programs that include Beauty and
     Charm at Fermilab are currently sponsored by DOE. This is a
     hands-on curriculum that has trained over 90 teachers and has
     brought over 5000 students to tour Fermilab.

The success of education at Fermilab has led to the construction of
the new Science Education Center which is the result of collaboration
between Fermilab and the education community. The importance of this
facility was underscored by the presence of Secretary Watkins who gave
the keynote address at the ground breaking in October of 1989.

The key to establishing the SSC as a viable educational tool is the
ability to incorporate any or all of the educational awareness concepts
into the laboratory. This can be accomplished by developing a master
plan that not only recognizes importance of education but also provides
the flexibility to adapt contingent requirements in the present or in the
future as funds become available.

Construction and operation of the SSC presents an outstanding
opportunity to enhance our scientific educational resources along with
the capability to increase our industrial and scientific base.

By its very scope, the SSC has excited our imagination. Let the SSC
provide therefor the opportunity to excite the imagination of today's
young people who will be the thinkers and leaders of the future.

14. Magnets II

AN AUTOMATED COIL WINDING MACHINE FOR THE SSC DIPOLE MAGNETS

S.Kamiya,T.Iwase,I.Inoue,I.Fukui,K.Ishida,S.Kashiwagi,
Y.Sato,T.Yoshihara,S.Yamamoto,E,Johnson* and C.Gibson*

Kawasaki Heavy Industries, minami-suna,koto-ku,Tokyo,Japan
*General Atomics, San Diego, California

ABSTRACT

We have finished the preliminary design of a fully automated coil winding machine that can be used to manufacture the large number of SSC dipole magnets. The machine aims to perform all coil winding operations including coil parts inserting without human operators at a high productive rate. The machine is composed of five industrial robots.
In order to verify the design, we built a small winding machine using an industrial robot and succesfully wound a 1 meter long coil using SSC dipole magnet wire. The basic design for the full length coil and the robot winding technique are described in this paper.
A fully automated coil winding machine using standard industrial components would be very useful if duplicate production lines are used.

INTRODUCTION

The SSC dipole magnets require high field quality, identical performance, and reliability, and they must be built at low cost and large quantity in a real industrial facility. Specifically regarding the coil winding machine,the following functions should be required :
· Accurate and repeatable winding · High production rate winding
· Minimum operator skill · Minimum number of operators
Initially, the coil winding operation and the associated requirements were investigated. The result was that industrial robots were found to be very effective. We then designed the fully automated machine using industrial robots, which satisfies the above requirements. We also verified that the robot winding technique is feasible by building a small winding machine which incorporates a robot.

COIL WINDING OPERATION

The coil is assembled from superconducting cable, copper wedges and various spacers. To build the coil,the coil winding steps in the following chart (Fig.1) are necessary, based on information from the Phase I Technology Orientation.
Coil winding operations shown in Fig.1 are mainly divided into two categories ; cable winding and coil parts (spacers, wedges) inserting. In each operation, the following operations are required :

Cable winding

· Put cable correctly on the mandrel at all position
· Adjust cable tension at all winding locations
· Clamp cable to maintain tension during winding

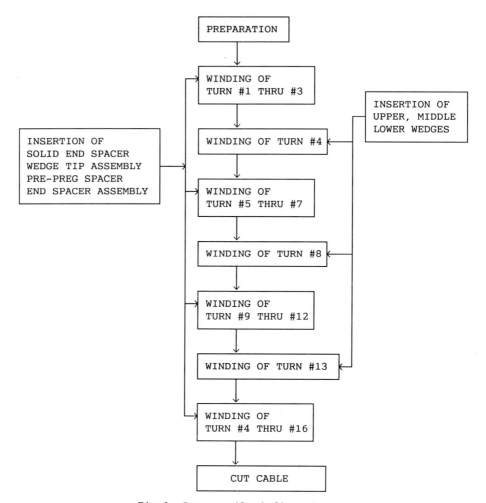

Fig.1 Inner coil winding steps

Coil part (spacers, wedges) inserting

- Insert wedge with 27" length all along one side of the coil with proper of the orientation. Space between wedges is within 0.015".
- Insert various spacers such as soild end spacers wedge tip, laminated end spacers, etc.
- Clamp coil parts so they do not fall out.

BASIC DESIGN OF FULLY AUTOMATED COIL WINDING MACHINE

      The coil winding machine for manufacturing the large number of SSC dipole magnets must not only perform the coil winding operations mentioned above, but also have the following functions :
- Accurate and repeatable winding
- High production winding
- Fully automated operation
      We have finished the preliminary design of a fully automated coil winding machine(called Robot coil winding machine) which can perform all the coil winding tasks (excluding cable preparation and cutting cable) without human operators. A schematic view is shown is Fig.2. The design major specification are as follows.

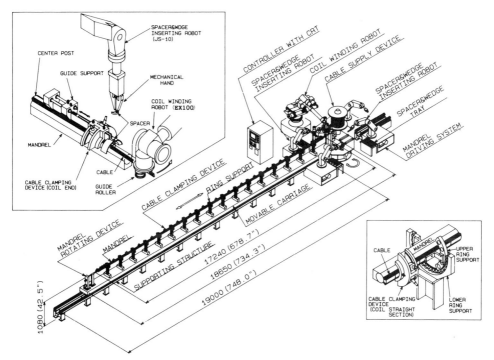

Fig.2 Schematic view of fully automated coil winding machine

Type   : Shuttle type using industrial robots
Applicable coil : SSC dipole coil
                coil dia 1.57"(40mm) ∼2.35"(60mm)
                length 690"(∼17m)
Winding speed :
    straight side   0∼75 feet/min (0∼23m/min)
    coil end   ∼15 sec
Control   : Personal Computer and Programable controller

Components

The fully automated coil winding machine is composed of Kawasaki
industrial Robots for coil winding and spacer & wedge inserting, a movable carriage which translates the mandrel on axis, a cable supply device
which feeds the mandrel with cable and other components. The major
specification of the components are shown in Table 1. Since EX-100 and
JS-10 robot are normally used for spot welding handling parts and
small assembly work, they are designed for reliability and long life.
The tools for guiding cable and assembling coil parts are similar in
function to a human hand and are located on the end of the robot arm.
Working flow of the winding machine.

The flow diagram of the winding machine operation is shown in Fig.3.
The first step includes mounting the mandrel and cable spool, setting the
robots and the movable carriage to their inital positions, and program
-ing the winding conditions. Next the automated coil winding machine is
started. When the winding machine comes to a coil end, the robot winds
the cable three-dimentionally on the mandrel without rotating the mandrel.
If a spacer needs to be inserted, the spacer robot picks up the correct
spacer and places it on the mandrel.

727

Table 1. Major specifications of components

Components	Type Description	
Robot for Coil winding	Type Degree of motion Pay load Repeatability	Kawasaki EX100 6 100 kg ± 0.5 mm
Robot for Coil parts inserting	Type Degree of motion Pay load Repeatability	Kawasaki JS-10 6 10 kg ± 0.1 mm
Movable carriage	Variable speed Drive method Position accuracy	0 ~ 75 feet/min(0 ~ 23m/min) Pinion & Rack Two Servo motor ± 0.2 mm (at low speed)
Mondrel rotation device	Drive method Rotation	Two Servo motors 360° CW & CCW
Cable supply device	Rotation Drive method Tension control	360° CW & CCW Pinion & Rack, Two Servo motors 0 ~ 50 ℓb (0~22.7 kg)
Cable clamping device	Actuating method	Air Actuating (or Robot Actuating)

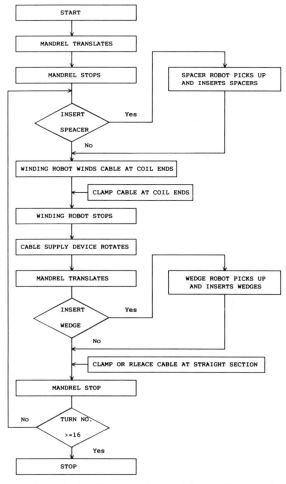

Fig.3 Block diagram for coil winding work

After coil end winding, the cable supply device rotates 180° counter clockwise in the case of inner coil. Coil winding on the straight section is performed by translating the mandrel. Related to cable clamping, there are two methods : the air actuating and robot actuating. The robot actuating method is preferred because the robot controls the clamping device instead of an operator and has a much simpler support structure since no actuating power is required.

WINDING SHORT COIL USING ROBOT

In order to verify the feasibilty of a robot winding machine, we wound a 1 meter length short coil with SSC superconducting cable. The coil winding machine used here was modified from an existing coil winding machine, and an industrial robot was applied to it. In this prototype, insertion of coil parts was done manually.

Winding machine

The winding machine shown in Fig.4 is composed of a Kawasaki EX-100 robot, a cable supply device, and a rotating table. The EX-100 is the same type which will be used in the full length winding machine. The cable supply device with tension controller is installed at a stationary position and cannot rotate. This is different from the full length winding machine design. However the robot winding m otion is the same as that of the full length winding machine.

The rotating table has a mechanism which translates and rotates the mandrel. To develop the tool for guiding the cable, both of two roller and three roller manipulators were manufactured.

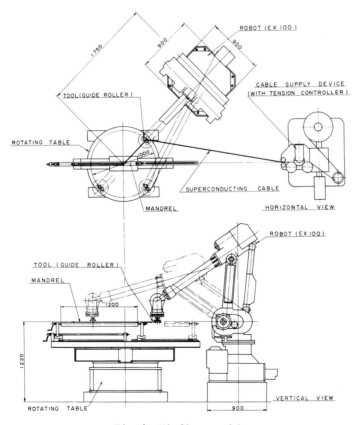

Fig.4 Winding machine

Coil winding Process using robot

The cable laying profile of each turn is programmed by using a "Teaching Box". There are seven programmed points per turn at the coil end. The mandrel does not move during the winding of the coil ends. The manipulator on the end of the robot arm guides the cable into place around the stationary mandrel. This winding method is equivalent to one with rotatings cable supply as in the basic design. Winding at the straight sections is conducted by translating the mandrel. Spacer and the wedge inserting and cable clamping are done by hand. The coil winding is shown in Fig.5.

Results

We obtained the following results :
· The robot can place cable on the mandrel in the correct position
· Winding speed at coil ends is faster than the conventional winding machine
· A two roller tool mounted to the tip of robot arm was preferred
· Robot cable winding is easily adaptable to the full length coil.

PROBLEMS TO BE SOLVED

In order to complete the fully automated winding machine, there are still the following problems to be solved, related to coil parts inserting work.
· How to pick up the correct spacer from various spacers and insert it in the right position
· How to automatically clamp the cable if it rises from the mandrel at the coil ends
· How to insert a wedge between turns precisely

The planned JS-10 robot, which has high-speed performance (max. 5 m/sec) and high precision movement, will be able to perform coil inserting work, in concert with the other components' (movable carriage, cable clamp device) movement.

These problems require more developments, but are not thought to be insurmountable in building an effective automated coil winding machine for the SSC depole magnets.

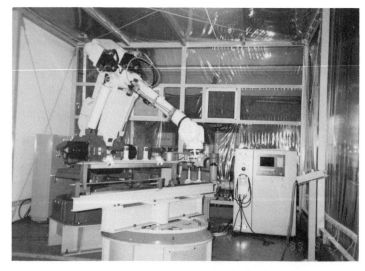

Fig.5   Photograph of coil winding work

EXPERIMENTAL EVALUATION OF VERTICALLY VERSUS

HORIZONTALLY SPLIT YOKES FOR SSC DIPOLE MAGNETS

J. Strait, K. Coulter, T. Jaffery, J. Kerby, W. Koska, and
M. J. Lamm

Fermi National Accelerator Laboratory
P.O. Box 500
Batavia, IL 60510

ABSTRACT

The yoke in SSC dipole magnets provides mechanical support to the collared coil as well as serving as a magnetic element. The yoke and skin are used to increase the coil prestress and reduce collar deflections under excitation. Yokes split on the vertical or horizontal mid-plane offer different advantages in meeting these objectives. To evaluate the relative merits of the two configuration a 1.8 m model dipole was assembled and tested first with horizontally split and then with vertically split yoke laminations. The magnet was extensively instrumented to measure azimuthal and axial stresses in the coil and the cold mass skin resulting from cooldown and excitation. Mechanical behavior of this magnet with each configuration is compared with that of other long and short models and with calculations.

INTRODUCTION

The tests described in this paper were carried out on a 40 mm aperture dipole magnet[1] of the design originally proposed for the Super Conducting Super Collider (SSC). This magnet has a "cos θ" type coil which is clamped by interlocking stainless steel collars. A circular iron yoke of inner diameter 111 mm and outer diameter 267 mm increases the magnetic field by approximately 20%. It is surrounded by a 4.8 mm thick cylindrical stainless steel skin which clamps the yoke about the collared coil, supports the magnet between the cryostat support posts and serves as a helium containment vessel. In the original design the yoke served as a purely magnetic element and was mechanically decoupled from the collared coil except for positioning tabs at the vertical and horizontal mid-planes. Azimuthal and radial constraint to the coil was provided entirely by the collars and axial restraint was provided by end plates and the ill-defined friction between the collar positioning tabs and the yoke. Full length dipoles of this design exhibited excessive quench training.[2,3,4] In more recent magnets the yoke and collars are closely coupled mechanically. The yoke, when clamped by the skin, serves not only to enhance the magnetic field but also to provide additional preload to the coil, to decrease collar deflection under excitation and to transfer the axial component of the Lorentz force to the skin via coil-collar-yoke-skin friction.

Dipoles of this design have showed dramatically reduced training, typically reaching plateau in three or fewer quenches.[4,5,6]

For assembly reasons the yoke must be split along a diameter. If the yoke is mechanically decoupled from the collars, the direction of this split is unimportant. However, with the yoke acting as a mechanical element, the yoke split direction may effect the magnet behavior. Most SSC magnets have used horizontally split yokes. Prototype magnets[7] for the proposed Large Hadron Collider (LHC) at CERN, which also use the yoke to provide additional mechanical support to the collars, use a vertically split yoke.

Because the Lorentz force is mainly horizontal in the body of the magnet, collar deflections are minimized if the yoke supports the collars near the horizontal mid-plane. The ability to provide this support depends on the relative thermal contractions of the yoke and collar materials, the choice of yoke split direction and the details of the relative shapes of the collar outer surface and the yoke inner surface. Two different steel alloys have been used for the collars in SSC model magnets: Armco Nitronic 40 with an integrated thermal contraction to 4 K of -2.9×10^{-3} and a Kawasaki high-manganese alloy with an integrated thermal contraction of -1.7×10^{-3}. For comparison, the integrated thermal contraction of the low-carbon steel used for the yoke is -2.1×10^{-3}.

The details of the collar's design and its interaction with the yoke depend on the thermal contraction of the collar material. Figure 1 shows the yoke-collar configurations for a horizontally split yoke with both collar materials and for a vertically split yoke with Nitronic 40 collars. To be inserted easily into the yoke at room temperature the collar diameter in the direction of the yoke split must be no larger than the yoke inner diameter. Nitronic 40 collars shrink away from the yoke during cooldown and may lose contact with the yoke along the split direction. If Nitronic 40 is used with a horizontally split yoke, the collared coil is designed to be sufficiently vertically oversized that when clamped in the yoke it deflects horizontally to contact the yoke. (See Figure 1a.) High-manganese collars (Figure 1b) are designed so that after deflection due to the coil prestress the collared coil is circular with a diameter equal to the inner diameter of the yoke. With cooldown the collars shrink less than the yoke and firm 360° contact is achieved.

Each of these collar materials has its own advantages and disadvantages. The lower thermal contraction material allows a simpler design to achieve close yoke-collar contact and results in better horizontal support against the Lorentz force. However, the low thermal contraction coeficient results in a larger prestress loss with cooldown. The larger contraction material requires a less simple design to achieve horizontal yoke-collar contact at operating temperature, but suffers less cooldown prestress loss.

Both horizontally split yoke designs are sensitive to the magnet-to-magnet variation in collared coil vertical diameter due to prestress variation. The measured rate of deflection is about 4 μm/MPa resulting in an expected range of 100-150 μm in vertical diameter for a preload range of ±20 MPa. Thus there is the possibility that the yoke mid-plane gap will not close under all conditions for higher prestress coils. This is particularly likely for the Nitronic 40 collars in which vertical contact with the yoke is required by the design. Finite element calculations[8] indicate that if there is sufficient vertical interference between the yoke and collars to ensure horizontal contact for the lowest preload coils, the mid-plane will be open by 50-100 μm at liquid helium temperature for the highest preload coils. At room temperature a mid-plane gap is likely to exist for all coils. With Kawasaki steel collars the mid-plane gap is likely to be closed at room temperature but will tend to be open for the higher prestress coils at 4 K.

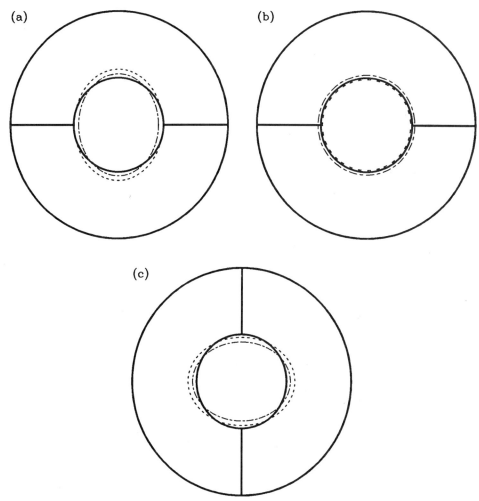

Fig. 1. Horizontally split (a and b) and vertically split (c) yoke configurations with high (a and c) and low (b) thermal contraction collars. The solid lines represent the yoke laminations, the dotted lines represent the free collared coil at T=300 K and the dot-dashed lines represent the free collared coil at T=4 K. (The collared coil distortions relative to the yoke are greatly exaggerated.)

In contrast, the collared coil horizontal diameter is relatively insensitive to coil prestress. (The vertical force from the coil prestress is applied at a smaller radius than the opposing force for the keys. This couple causes an inward bending of the sides of the collars that almost perfectly cancels the outward deflection due to the internal pressure of the coils.)[9] The rate of horizontal deflection is less than 20% of that in the vertical direction; horizontal diameter varies by < 30 μm for the full range of expected coil preloads. If a vertically split yoke design is adopted it is relatively easy to ensure both good horizontal support to the collared coil and a closed mid-plane gap independent of the coil preload. This can be achieved with the larger thermal contraction collar material by appropriately choosing the horizontal and vertical diameters of the collars. In this case (Figure 1c) the collared coil is oversize in the horizontal direction and undersize in the vertical direction to allow easy insertion into the yoke for even the highest preload coils. Under cooldown the collars move away from the yoke at the vertical radius but positive contact is maintained at the horizontal radius, guaranteeing both horizontal support and transfer of the axial force to the skin. At zero field, the skin tension is balanced primarily by a pressure between the mating surfaces of the yoke halves. As the field increases, the mid-plane progressively unloads as the horizontal Lorentz force is transferred to the yoke. As long as the skin tension exceeds the Lorentz force, the mid-plane gap remains closed and the yoke behaves as a rigid solid structure. The mechanics of this design are described in more detail in Reference 7.

TEST MAGNET DESIGN

To compare the relative merits of the two yoke split directions a 1.8 m model magnet (DSS012), built and previously tested[10] at Brookhaven National Laboratory (BNL), was partially reassembled twice at Fermilab with the two yoke types. In the original assembly at BNL horizontally split yoke laminations which did not contact the collars were used. Similarly to long magnets DD0012 and DD0014[4] shims were placed between the yoke and the collars to fill the gap. This magnet was not the ideal vehicle for these tests because it is built with the low thermal contraction Kawasaki steel collars which work well with a horizontally split yoke, but, unlike the design discussed above and shown in Figure 1b, the collared coil is significantly oval in the vertical direction.

In the first reassembly the horizontally split yoke laminations used were the same as those used in other recent 40 mm SSC dipoles.[5] These are designed so that at room temperature there is "line-to-line" contact between the inner surface of the yoke and the outer surface of the undeflected collar and they are expected to result in mechanical behavior of the collar-yoke system equivalent to the original assembly. Because of deflections of the collar due to coil prestress there was approximately 170 μm radial interference in the vertical direction and 0-25 μm radial interference in the horizontal direction between the yoke and the free collared coil. When clamped in the yoke the collars contact the yoke around the full circumference and the yoke mid-plane gap is open. Because the collars shrink less than the yoke, similar conditions exist at 4 K. This configuration is shown in Figure 2a.

Vertically split yoke laminations were used in the second assembly. Because the collared coil is vertically oval and shrinks less than the yoke with cooldown, the design described in Figure 1c could not be used. To achieve the proper mechanical interface to the collars the inner surface of the yoke lamination deviated from that required by the magnetic design. The inner surface was made elliptical with the horizontal and vertical radii larger than the "line-to-line" contact yoke by 25 μm and 280 μm respectively. At room temperature the there was clearance between the yoke and the collars of

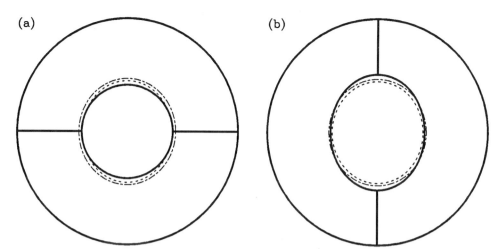

Fig. 2. Yoke-collar configurations used for the tests of DSS012:
(a)horizontally split yoke with yoke-collar shims and "line-to-
line" contact yoke and (b) vertically split yoke. The solid lines
represent the yoke laminations, the dotted lines represent the
free collared coil at T=300 K, and the dot-dashed lines
represent the free collared coil at T=4 K. (The collared coil
distortions relative to the yoke are greatly exaggerated.)

0-25 μm and approximately 110 μm in horizontal and vertical directions
respectively. At 4 K the vertical clearance was reduced to about 80 μm and
there was 0-25 μm of interference along the horizontal radius.   This
configuration is shown in Figure 2b.

The goal of this experiment was to compare the mechanical behavior of
the horizontally and vertically split designs with each other and with
expectations from calculations.   In both configurations the magnet was
extensively instrumented to measure coil azimuthal stress, force between the
end of the coil and the end plate, and circumferential and axial stresses in the
cold mass skin.   Instrumentation used for coil stress and end force
measurements are described in Reference 11.   Strain gages were mounted
directly on the cold mass skin to measure circumferential stress as a function
of azimuth at two points approximately 8 cm apart near the center of the
magnet and axial stress as a function of longitudinal position at an azimuthal
position approximately 50° from the yoke parting plane.

MAGNET TESTS

In the horizontally split yoke case the skin gages were mounted before
the skin was welded around the magnet and the strain-free resistance of all
gages were measured at room temperature and 4.2 K.   This allowed
measurement of absolute strain and of absolute stress where no yielding
occurred.   Near the mid-plane weld the skin was observed to yield
significantly. Far from the weld the azimuthal stress is 150-200 MPa at room
temperature and grows to 300-350 MPa with cooldown due to the larger
thermal contraction of the stainless steel skin than of the yoke. The clamping
force of the skin at operating temperature is approximately 3×10^6 N/m or
about twice the horizontal Lorentz force.

Tests were performed in pool boiling liquid helium in a vertical dewar at
the Fermilab Superconducting Magnet R&D test facility at Lab 2.   The data

acquisition system is similar to that used at the Fermilab Magnet Test Facility.[12] Strain gage, dewar pressure and temperature were recorded at 10 minute intervals through cooldown, the cold testing period and warm-up to room temperature. Care was taken to continue data recording on the warmup cycle until the magnet had fully reached room temperature. (Because of the large difference between the thermal expansion coefficients of the coil and the high-manganese steel collars the coil prestress changes at a rate of approximately 140 kPa/K.) The magnet was cooled to approximately 80 K by flowing liquid nitrogen through copper tubes band-clamped to the outer surface of the skin. Final cooldown was achieved by filling the dewar with liquid helium. The magnet was warmed by blowing room temperature helium and then nitrogen gas through the dewar and the copper tubes and by powering the magnet coils with 10 A.

Cold tests emphasized strain gage data to compare the mechanical behavior of the different configurations. Strain gage data were recorded at roughly equal intervals of IxB in cycles to increasing peak current until a quench occurred. The magnet was then trained to the quench plateau at 4.2 K, and strain gage data were recorded up to a current just below the plateau. Recent 17 m model dipoles have been build with several different horizontally split yoke configurations. These magnets are instrumented similarly to DSS012, and a similar test procedure has been followed allowing comparisons among them and the various DSS012 configurations.

Figure 3 shows the quench history for the three tests. The first test, with yoke-collar shims, is a retest in the same configuration as the earlier tests at BNL[10] with no intervening disassembly and reassembly. In the tests at BNL no "retraining" was observed after a thermal cycle to room temperature. Here, in tests roughly a year after the tests at BNL, two retraining quenches occurred. In the next two tests the yoke was replaced but the collared coil was not changed. A similar number of training quenches occurred each time.

Coil stress is a function of the current squared (proportional to IxB) is plotted in Figure 4a. The prestress and stress versus I^2 slope are almost identical for the two horizontally split yoke cases. In both cases there is sufficient vertical interference between the yoke and the collars that the yoke mid-plane gap is open under all circumstances. The force from the skin tension is transferred to the collared coil increasing the coil prestress by about 15 MPa. Full circumferential yoke-collar contact occurs in both cases. In the vertically split case there is a small yoke-collar horizontal interference and vertical clearance. The coil prestress is essentially that of the free collared coil and, with the large (35 MPa) cooldown loss, is quite small at the operating temperature. As a consequence the coil unloads at the pole at 5.5 kA (5.6 T). The prestress change with excitation is displayed in Figure 4b along with data from two 17 m magnets.[5] DD0017 and DD0015 both use Nitronic 40 collars. DD0017 uses horizontally split yoke laminations identical to those used for the second assembly of DSS012. DD0015 uses yoke laminations similar to those used for the first assembly of DSS012; however, no shims are placed between yokes and collars so the collars are unsupported by the yoke. The smallest collar deflection and hence the smallest prestress change with excitation occurs with the horizontally split yoke and Kawasaki steel collars. The prestress decrease is somewhat more with the vertically split yoke because the collars contact the yoke over less of the circumference. With Nitronic 40 collars and a horizontally split yoke (DD0017) there is also less than complete circumferential contact, but because the yoke-collar gap is now at the horizontal radius, the collar deflection is larger than in the vertically split case. With free-standing collars (DD0015) the deflection and prestress decrease are the largest.

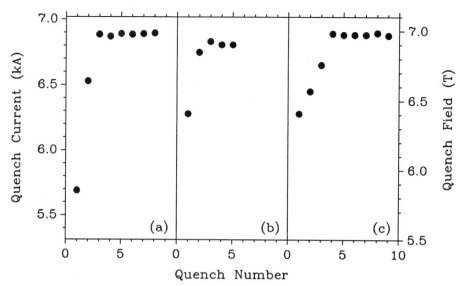

Fig. 3.   Quench histories for DSS012 on three assemblies: a) original
assembly with a horizontally split yoke and yoke-collar shims,
b) second assembly with a horizontally split "line-to-line"
contact yoke, and c) third assembly with vertically split yokes.

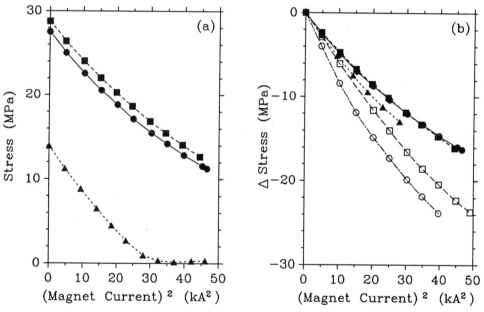

Fig. 4.   Inner coil stress (a) and stress change (b) at the pole as a
function of magnet current squared for the three assemblies of
DSS012 and two 17 m magnets.   The filled circles are DSS012
with a horizontally split yoke and yoke-collar shims, the filled
squares are DSS012 with a horizontally split "line-to-line"
contact yoke, the filled triangles are DSS012 with a vertically
split yoke, the open squares are DD0017, and the open circles
are DD0015.

The skin stress change with excitation is plotted in Figure 5. Azimuthal stress changes may result from bending of the yoke, which causes the radius of curvature of the skin to change, or from changes in the yoke mid-plane gap, which cause the skin to stretch. The former will have its largest effect 90 degrees from the yoke parting plane and the latter will have its largest effect near the parting plane. Longitudinal stress changes result from the axial Lorentz force transferred from the coil to the skin through the end plates and through coil-collar-yoke-skin friction. Because there are active stress changes in both directions, account must be taken of the Poisson effect in converting the measured strains into stresses. This is done assuming that the longitudinal stress is independent of azimuth. In the vertically split yoke case longitudinal strain is measured at the same point as one of the azimuthal gages. The

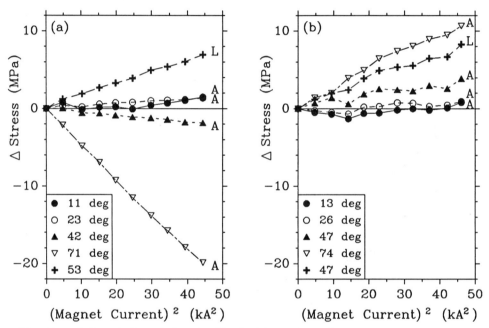

Fig. 5.   Azimuthal and longitudinal skin stress changes with excitation near the center of DSS012 with a) a horizontally split "line-to-line" yoke and b) a vertically split yoke. The angles listed in the legend represent the distance from the yoke parting plane. Data labelled "A" and "L" are azimuthal and longitudinal stresses respectively.

Poisson resolution is done at this point and the resultant longitudinal stress is used to correct the other azimuthal measurements. In the horizontally split yoke case the longitudinal strain was measured between two of the azimuthal gages. The azimuthal strain at the location of the longitudinal gage was gotten from a quadratic interpolation among the three nearest azimuthal gages. Because of the uncertainty of the interpolation, there are uncertainties in the stresses that are roughly proportional to I^2 and reach approximately 0.7 MPa and 0.2 MPa for longitudinal and azimuthal stress respectively.

In the horizontally split case (Figure 5a) the azimuthal stress becomes progressively more compressive with increasing distance from the horizontal mid-plane. This is the behavior expected if the yoke bends outward at the horizontal mid-plane under the Lorentz load. The stress near the yoke parting plane (gages at 11 and 23 degrees) appears to be slightly tensile. This may result from the yoke mid-plane gap opening slightly at the outer radius as the yoke bends outward. In the vertically split case (Figure 5b) the azimuthal stress becomes more tensile with increasing distance from the vertical mid-plane. Here yoke bending tends to decrease the skin radius of curvature in the horizontal direction. The azimuthal stress near the vertical mid-plane is essentially zero even at the highest current, demonstrating that the yoke mid-plane gap remains closed. With the gap closed, the yoke bending, as measured by the azimuthal stress far from the parting plane, is much smaller than in the horizontally split case in which the mid-plane gap is open. In either case, however, the stress change is small compared with the total azimuthal tension of 300-350 MPa and with the stress change of 150 MPa that would result if the entire Lorentz force were balanced by the skin.

DISCUSSION

Successful magnets can and have been built with both horizontally and vertically split yokes. In this sense the choice of yoke split direction may not be crucial. However, each system offers different advantages and disadvantages in meeting the requirements of the collar-yoke system:

1)   The yoke and skin can be used to increase the coil prestress.

2)   At T = 4 K, B = 0, there should be close contact between the yoke the collars near the horizontal mid-plane to minimize collar deflection under excitation and to provide axial restraint.

3)   Yoke mid-plane gap should be closed under all circumstances: at T = 4 K to well above the operating field and at room temperature. A closed gap improves the field quality and eliminates the possibility of non-reproducible behavior with thermal cycling if the gap opens and closes.

A horizontally split yoke is more efficient in increasing the coil prestress because the yoke and skin load the collars vertically. It is more straightforward to achieve close horizontal contact with vertically split yokes, particularly with collars whose thermal expansion coefficient is larger than that of the yoke. It is easier to guarantee that the yoke gap is closed at assembly with the a vertically split yoke for two reasons. First, the collared coil radius varies from magnet to magnet less in the horizontal than the vertical direction. Second, the collared coil is significantly more compressible if it is free to expand in the direction orthogonal to the direction it is being clamped. In the vertically split yoke case there is no reason that the collars must make contact near the vertical radius, but in the horizontally split yoke case the collars must contact the yoke at the horizontal radius (requirement 2). Once such contact is made, it is difficult to compress the collars further. The mid-plane gap is less likely to open with excitation with a horizontally split yoke because the horizontal Lorentz force is opposed by the yoke acting as a C-frame. However, at 4 K the skin tension supplies a clamping force about twice the Lorentz force at full field, so the mid-plane is likely to stay closed with a vertically split yoke as well. In addition, finite element calculations[13] indicate that even with the mid-plane gap closed, there is enough bending of the yoke and hence the collars that about half the Lorentz load is taken by the collars. This adds an additional factor of two mechanical margin and ensures that the mid-plane gap will stay closed.

The effect on quench training behavior of coil prestress and collar deflection (or any other mechanical property) is not well demonstrated at a detailed level. Indeed short (1-4 m) model SSC magnets have been built that unloaded at the pole well below 5 T yet reach their critical currents with little training. On the other hand, early 17 m SSC dipoles with low preload trained very poorly and several never reached their critical currents. Quench performance improved dramatically when the collars were firmly clamped in the yoke. This clamping serves to increase the prestress, decrease collar deflections with excitation and improve the axial restraint. Which of these has the dominant effect or whether all three are necessary for good performance is unknown. However, it seems likely that a structure in which the coil prestress is maximized and all deflections - radial, azimuthal, and longitudinal - are minimized is likely to perform well. Minimizing radial collar deflections requires close yoke-collar contact at the horizontal radius. This is difficult to achieve with a horizontally split yoke and collars made of material with a higher coefficient of thermal expansion than the yoke. This is demonstrated by the larger coil stress decrease with excitation in magnets with Nitronic 40 collars than those with Kawasaki steel collars. (See Figure 3). On the other hand, use of low thermal expansion collars result in a larger prestress loss with cooldown. With a vertically split yoke, close horizontal yoke-collar contact at 4 K can be achieved using collar material of any thermal expansion coefficient, allowing both prestress loss with cooldown and collar deflection with excitation to be minimized. In the vertically split yoke test discussed here the collars and yoke made contact only near the horizontal mid-plane, as they would with collars with a larger thermal expansion coefficient. The collar deflection, inferred from the coil stress decreased in Figure 3, is less than with Nitronic 40 collared magnets and a horizontally split yoke. Recent finite element calculation,[13] done since this experiment was designed, suggest that the horizontal yoke-collar interference can be increased somewhat and still maintain a closeed yoke mid-plane under all conditions. This should improve the collar support and allow it to approach that of the Kawasaki steel collar, horizontally split yoke case. Collar and yoke laminations for this improved design have been ordered for several 1 m long, 40 mm aperture models to be built later this year. A similar design will be used for both short and long 50 mm aperture SSC dipoles to be built at Fermilab.

ACKNOWLEDGEMENTS

We would like to thank the Magnet Division of Brookhaven National Laboratory who provided us with the model magnet on which these tests were done. We would also like to thank the staff of the Fermilab Superconducting Magnet Fabrication Group who did the disassembly and reassembly work and the staff of Lab 2 who carried out the tests.

REFERENCES

1. Conceptual Design of the Superconducting Super Collider, SSC-SR-1020, March 1986, revised September 1988, J. D. Jackson, ed.

2. J. Strait, et al., Tests of Prototype SSC Magnets, Proc. of the 12th Particle Accelerator Conf., 1540 (1987), E. R. Lindstrom, and L. S. Taylor, eds.

3. J. Strait, et al., Tests of Prototype SSC Magnets, IEEE Trans. Magn. 24:730 (1988).

4. J. Strait, et al., Tests of Full Scale SSC R&D Dipole Magnets, IEEE Trans. Magn. 25:1455 (1989).

5. J. Strait, et al., Full Length SSC R&D Dipole Magnet Test Results, Proc. of the 1989 IEEE Particle Accelerator Conf., 530 (1989), F. Bennet and J. Kopta, eds.

6. J. Tompkins, et al., Performance of Full-Length SSC Model Dipoles: Results from 1988 Tests, Supercollider 1, 33 (1989), M. McAshan, ed.

7. D. Leroy, et al., Design of a High Field Twin Aperture Superconducting Dipole Model, IEEE Trans. Magn. 24:1373 (1988).

8. J. Cortella, private communication.

9. C. L. Goodzeit and P. Wanderer, "Summary of Construction Details and Test Performance of Recent Series of 1.8 Meter SSC Dipoles at BNL," presented at the 1990 International Industrial Symposium of the Super Collider, Miami Beach, FL, March 14-16, 1990.

10. P. Wanderer et al., Test results from recent 1.8-m SSC model dipoles, IEEE Trans. Magn. 25:1451 (1989).

11. C. L. Goodzeit, et al., Measurements of Internal Forces in Superconducting Accelerator Magnets with Strain Gauge Transducers, IEEE Trans. Mag. 25:1463 (1989).

12. J. Strait, et al., Fermilab R&D Test Facility for SSC Magnets, Supercollider 1, 561 (1989), M. McAshan, ed.

13. J. R. Turner, private communication.

SUMMARY OF CONSTRUCTION DETAILS AND TEST PERFORMANCE OF

RECENT SERIES OF 1.8 METER SSC DIPOLES AT BNL

C. Goodzeit and P. Wanderer

Brookhaven National Laboratory
Upton, New York 11973

INTRODUCTION

Certain design features of the SSC dipole magnets are evaluated in short (1.8 meter) 40mm aperture magnet tests before being incorporated into the design of the full length magnets. This report summarizes the results of recently tested short magnets at Brookhaven National Laboratory.

DISCUSSION

Description of Magnets Tested

The five magnets that are described in this report represent the SSC Baseline Design for the 40mm aperture dipoles. The coil cross-section is known as C358D and has been described in detail in other publications[1]. The cross-section of the cold mass is shown in Figure 1. The five magnets all have several common features which include the following:

    a.   Nitronic 40 stainless steel collars (90 Kpsi minimum yield strength).

    b.   Collars spot welded in pairs, alternating L/R to produce a twist free collared coil assembly.

    c.   Zero clearance nominal fit between collars and yoke.

    d.   One piece, 1½-inch thick end plates.

    e.   Ends of coil loaded by means on set screws mounted in end plates.

    f.   Epoxy bonded stainless steel yoke blocks on ends for axial rigidity and lowering field at end.

    g.   A beam type strain gauge transducer mounted at the minimum coil size section to measure inner and outer coil polar stress.

    h.   Forty one voltage taps to locate the origin of quenches.

---

[1] BNL -43775, "Status Report on SSC Dipole R&D"

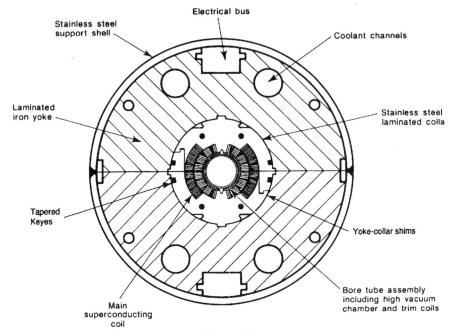

Stainless steel
support shell

Electrical bus

Coolant channels

Laminated
iron yoke

Stainless steel
laminated colla

Tapered
Keyes

Yoke-collar shims

Main
superconducting
coil

Bore tube assembly
including high vacuum
chamber and trim coils

Figure 1.

In addition to these common features, two of the tests were with magnets that had a design variation of the collar to reduce the vertical ovality of the collared coil assembly (DSV016 and DSS019). This type of collar has been called "anti–ovalizing". The reduction in vertical ovality has been achieved by relocating the slot for the tapered key by .010 inches closer to the midplane. This results in an unloaded collar assembly that is slightly flat (nominally .010) but stretches to a round shape (still with some vertical ovality) when containing the assembled coils.

The superconducting cables used in the inner and outer coils were of two slightly different types whose characteristics are shown in Table I. This table includes the measured value of the current density in the superconductor and the critical current carrying capacities for the inner and outer cables.

Assembly Characteristics

Starting with the collaring operation, the inner and outer coil stresses are measured with the beam type strain gauge collar pack[2]. A typical stress history for the collaring and subsequent operations is show in Figure 2 for DSS017. Some significant stress values are as follows for the inner coils:

---

[2] C.L. Goodzeit, et al., "Measurement of Internal Forces In Superconducting Accelerator Magnets with Strain Gauge Transducers", IEEE Trans. Magn. 25, No. 2 (1989):1451

Table I

Cable Characteristics of Magnets

A. Inner Coils

Magnet	Wire Mfg.	Cable Jc (5T) A/sq mm	Cable Ic (7T)	JCu @ Ic	Cu:SC	Filament Dia. (μ)
DSS016	IGC	2775	7741	1104	1.47	6.0
DSS017	IGC	2775	7741	1104	1.47	6.0
DSS018	OST	2682	7020	985	1.53	6.0
DSV016	IGC	2775	7741	1104	1.47	6.0
DSS019	OST	2682	7020	985	1.53	6.0

B. Inner Coils

Magnet	Wire Mfg.	Cable Jc (5T)	Cable Ic (5.6T)	JCu @ Ic	Cu:SC	Filament Dia. (μ)
DSS016	IGC	2549	8130	1300	1.72	4.8
DSS017	IGC	2549	8130	1300	1.72	4.8
DSS018	IGC	2549	8130	1300	1.72	4.8
DSV016	IGC	2549	8130	1300	1.72	4.8
DSS019	SCN	2715	8586	1365	1.75	6.0

Manufacturers: ICG - Intermagnetics General
OST - Oxford Superconducting Technology
SCN - Supercon

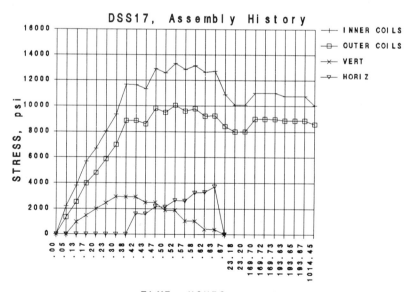

Figure 2

The maximum stress during the collaring operation occurs usually when the tapered keys are just starting to be inserted and in this case reaches a value of about 13,000 psi. After the keys are inserted (.67 hours), the hydraulic pressure is relieved and the stress immediately drops to ~11,000 psi. Note that there is a loss of stress due to creep in the first day after collaring of about 1000 psi (~10%). After about a week the shell has been welded around the yoke and the weld shrinkage causes an increase in the coil stress by about 1000 psi as the ovality of the collared coil is reduced by compression in the yoke.(Note that this is a line to line fit collar–yoke assembly). The coil continues to creep, but at a slower rate now and after about 33 days has relaxed another 700 psi. A summary of these significant stress values for the five magnets is shown in Table II.

Collared Coil Ovality Measurements

The structural response of the collars to the load applied by the prestressed coils is to produce a vertical deflection or ovality with little or no horizontal deflection. This action can be understood by referring to the force diagram of the collars as shown in Figure 3. The reaction of the keys on the collars produces a moment which tends to bend the collars inward at the midplane while the tensile hoop stress tends to stretch the collar vertically. The measured vertical and horizontal deflections along the length of a collared coil is shown in Figure 4 for DSS017. Since the ends of the coils are rather compressible compared with the straight section, the reduced loads on the collars produces a smaller distortion in those sections. In order to partially compensate for the ~.012 inch vertical deflection, the location of the key slots in the standard collars have been moved .010 inch closer to the midplane for

Table II

Summary of Significant Coil Stresses During Assembly
(stress in psi)

A: Inner Coils

Magnet	Maximum Collaring	After Collaring	Delta	Before Welding	After Welding	Delta	Comment
DSS016	11315	8876	-2439	8165	7772	-393	Note 1
DSS017	13300	10888	-2412	10102	11015	913	
DSS018	11839	9316	-2523	7875	8879	1004	
DSV016	12034	11964	-70	11280	13350	2070	Note 2
DSS019	11356	9532	-1824	8200	8520	320	Note 2

B: Outer Coils

Magnet	Maximum Collaring	After Collaring	Delta	Before Welding	After Welding	Delta	Comment
DSS016	7660	5984	-1676	5587	6091	504	
DSS017	10047	8422	-1625	8007	8868	861	
DSS018	8010	6158	-1852	5292	6221	929	
DSV016	8259	8049	-210	7717	9826	2109	Note 2
DSS019	6653	5510	-1143	5070	5316	246	Note 2

Notes:
1. Stress loss caused by overheating of magnet due to abnormal shell welding condition.
2. Anti-ovalizing collars used in this assembly.

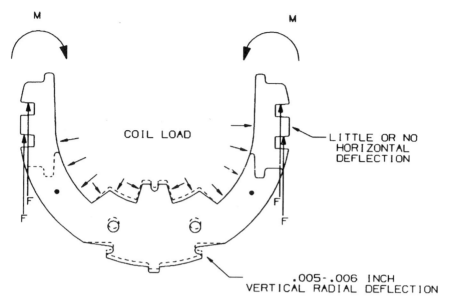

REACTION FORCES (F) OF KEYS AND
MOMENTS (M) ON COLLAR PAIRS

Figure 3

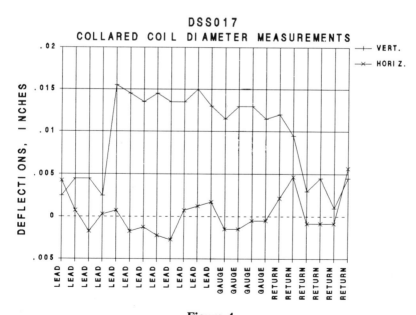

Figure 4

the anti-ovalizing effect. In the case of the five magnets that are reported, three of them had regular collars and two (DSV016 and DSS019) were assembled with anti-ovalizing collars. The vertical deflections of the collars (at the strain gauge transducer location) for these case is shown in Figure 5 along with the calculated value of the ovality based on the structural model of the collar. According to these measurements, the sensitivity to coil stress level tracks well with the calculation. It is also seen that for the anti-ovalizing collars, the vertical deflection has been reduced form typically .012 inch to the .004-.007 inch range.

Quench Performance The complete quench-test history of one of the magnets, DSS018, is shown in Fig. (PW1). The test sequence is typical of all the magnets discussed in this paper. After initial studies at the SSC design operating temperature, 4.35 K, the mechanical reserve of the magnet is tested as the temperature is reduced in 0.5 K steps. Following further testing at 4.35 K the magnet is taken to room temperature and then recooled to 4.35 K (thermal cycle) to determine whether it retains its initial training. At each temperature, most of the quenches occur during a ramp at 16 A/sec., but during the last run to quench the current is increased stepwise so that the strain gauge readings can be taken at constant current. Also at each temperature, the magnet's performance at the limit of the conductor is compared to an estimate based on the current-carrying capacity of a short (1 m) sample of cable taken from the same spools used to wind the magnet. In the case of DSS018, the estimate is very close at SSC conditions. (The temperature dependence of the estimate is an average over a number of cables.)

The quench histories of the five magnets reported here are given in summary form in Fig. (PW2). The initial training is limited to one or two quenches, the magnets reach the limit of the conductor at 3.35 K at central fields of approximately 8 T with little additional training, and retraining is negligible. (For DSS017, after the thermal cycle, the magnet was initially tested at 3.35 K, as indicated.)

Multipole Measurements. The 1.8 m models are not "field quality" magnets (made as identically as possible) since they are still being used as test beds for refinements in production. Thus, the focus is on understanding the relation between the measured field uniformity and magnet construction rather than on detailed comparison with SSC tolerances. For this discussion, it is most interesting to ask whether there are differences between the three magnets made with standard collars (DSS016, DSS017, DSS018) and the two with anti-ovalized collars (DSV016, DSS019).

The standard expression for the multipole representation of the fields is:

$$B_y + iB_x - B_o \sum_{n-0}^{\infty} (b_n + ia_n) (xtiy)^n$$

The multipole coefficients are evaluated at a radius of 1 cm, in dimensionless units of 10^{-4} of the dipole field.

The measured allowed and unallowed geometric coefficients at 4.35 K are given in Tables (PW1) and (PW2), respectively. Apart from differences in collars, two factors are known to significantly affect the multipoles. First is magnet-to-magnet differences in the sizes of the shim placed between the pole turn of the coil and the collar. The shim size is changed to make the prestress as uniform as possible while the development of coil curing procedures is underway. Second is out-of-tolerance distortions in the stainless steel shell which is welded around the yoke. These magnets have used such shells, while a satisfactory vendor was being identified. Both of these factors are expected to improve in production. For these magnets, no differences in either the allowed or unallowed multipoes can be traced to the type of collar.

Magnet Testing Measurments

The coil stress measurements made by the beam type strain gauge transducers are compensated for thermal and magnetic effects and therefore can accurately measure the coil stress of the magnet under operating conditions. However, stress measurements between ambient and operating temperature are not regarded as reliable because of the large variation in strain gauge grid resistance between 15K and 100K. Thus, we can only obtain reliable measurements near ambient temperature or at operting temperature in the range 3.5K to 4.5K. The change in coil stress in cooling and with magnet excitation are discussed below:

Thermal Stress Loss

A summary of the stress changes of the five magnets tested for the inner and outer coils are listed in Table III. Except for the case of DSS019, the stress loss is significantly larger than what would have been calculated from the intrinsic thermal contraction properties of the yoke, collar and coil materials. This effect has been

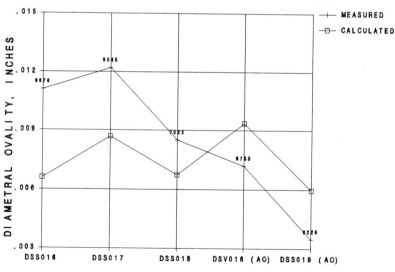

Figure 5

Table III

Coil Stress Changes During Testing
(σ is stress in psi)

A: Inner Coils

Magnet	σ Start Warm	σ Start Cold	Delta σ	σ End Warm	Global Delta σ	Comment
DSS016	7651	3143	-4508	7669	18	
DSS017	10039	5540	-4499	9393	-646	
DSS018	6914	3218	-3696	6043	-871	
DSV016	12282	8594	-3688	10607	-1675	Note 1
DSS019	7364	4991	-2373		-7364	Note 1

B: Outer Coils

Magnet	σ Start Warm	σ Start Cold	Delta σ	σ End Warm	Global Delta σ	Comment
DSS016	6063	3027	-3036	5570	-493	
DSS017	8586	6642	-1944	8119	-467	
DSS018	5223	3213	-2010	4787	-436	
DSV016	9418	6946	-2472	8674	-744	Note 1
DSS019	4349	3408	-941		-4349	Note 1

Notes:
1. Anti-ovalizing collars used in this assembly.

attributed to the effect of coil ovality on the fit between the collared coil and the yoke and has been analyzed by Chapman, et. al.[3] This analysis and measurements, as well, show that the coil ovality produces and extra force between the collars and yoke which appears as compressive stress in the coils. When the magnet is cooled, the extra force decreases as the fit between the collars and yoke change. Collars with the greatest amount of ovality show the largest thermal stress loss. In the magnets that are reported here, the least oval collared coil assembly was that from magnet DSS019. Figure 5 shows the diametral ovality to be about .0035 inches compared to .010–.012 for the other magnets. Note that the thermal stress loss for this case was also the lowest as shown in Table III.

Stress Change During Magnet Excitation

When the magnet is energized, the Lorentz forces tend to pull the coils away from the poles and compress it more at the midplane. Since the coil stress transducers measure the polar stress, one can see this effect form the measurements od the coil stress as a function of magnet current. Figure 6 shows the measurements of the polar coil stress for DSS017 up to the ~8000A. At the operating current of 6500A (for 6.6T), the inner coil stress has dropped from about 4200 psi to 1200 psi for a 3000 psi change which is about 90% of the value predicted by analysis without

---

[3] M. Chapman, et. al. "Mechanical Analysis of Different Yoke Configurations for the SSC Dipole", From the SSC Central Design Group (No report number)

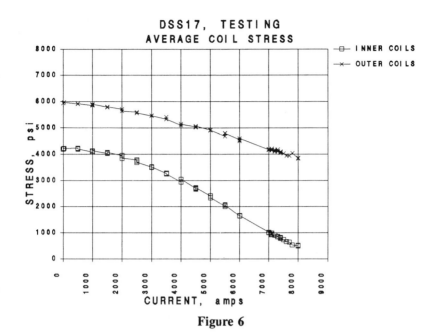

Figure 6

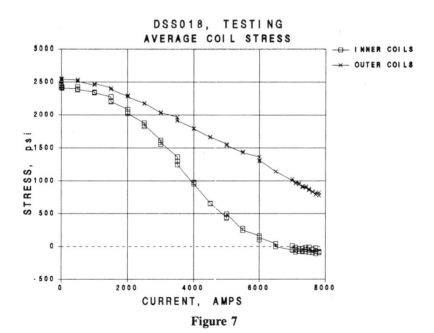

Figure 7

friction. The Lorentz forces on the outer coil are less than those on the inner coil and this effect is also seen by the drop of about 1500 psi for the outer coil stress at operating current.

A specified operating condition for the magnet is to maintain positive contract between the coil and the pole at operating current. This requires that the initial prestress of the collared coil must be high enough to ensure that this condition is met. The coil stress measurements indicate that a minimum cold coil stress (for the inner coil) is sufficient to keep the coil in contact with the pole. However, during the R&D phase of the program this condition was not always met. In this case the polar turn of the inner coil would show no compressive stress and such a case is seen for DSS018 in Figure 7. In this case it is seen that the stress does not appear to decrease quadratically with current but shows an inflection and drops to zero at about 6000 A.

INVESTIGATION OF THE MECHANICAL

PROPERTIES OF SUPERCONDUCTING COILS

F.W. Markley and J.S. Kerby

Fermi National Accelerator Laboratory*
P.O. Box 500
Batavia, Illinois   60510

ABSTRACT

This paper presents data on 3 of the important mechanical properties of SSC type superconducting coils. The measured properties are: 1) The azimuthal elastic modulus of the coil samples made for the stress relaxation tests. 2) The rate of stress - relaxation of collared SSC outer coils molded to different sizes and 3) The pressures that various insulations can withstand during molding or collaring before turn-to-turn shorts develop. Additional data on these and other properties are available but omitted here because of space limitations.

SAMPLE PREPARATION

Coil samples for stress relaxation and modulus testing were molded in 3" long molds. During the cure, a shim was placed between the top of the fixture and the hydraulic ram loading the coil. This shim determined the cured size of the coil. The hydraulic pressure was continuously adjusted by hand so that this shim could just be moved back and forth between the fixture and the ram. Thus we were able to monitor the force required to just keep the coil at the desired size during the cure. When an additional shim is added to this shim to make an oversize sample, it is identified as being made with a negative shim. When an additional shim is placed on top of the coil during cure to make an undersized sample, it is identified as being made with a positive shim.

COIL MODULUS

We measured the modulus of all the samples and found the stress strain curve to become reasonably linear near the high stress end. All samples gave a modulus of $1.4 \times 10^{+6} \pm .4 \times 10^{+6}$ psi, regardless of the size to which they were molded.

STRESS RELAXATION

Figure 1 is a drawing of the stress relaxation fixtures used (there are three). The solid steel blocks on either side of the load cells are spacers to allow the use of several different load cells with different dimensions. The curved block supporting the coil has a slightly larger radius than the block used to cure the coil because now the ground insulation has been added. The notch just above the open space holds a steel spacer with the coil I.D. that keeps the coil in place during assembly of the coil into the fixture.

---

*Operated by Universities Research Association under contract with the U.S. Department of Energy

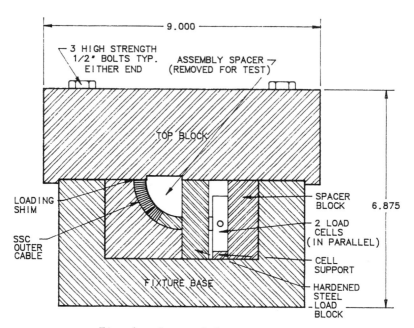

Fig. 1. Stress Relaxation Fixture

Fig. 2. Stress Relaxation Setup

We decided to build the fixtures to have the same total compliance that is expected for the SSC collars. Robert Wands had done a finite element analysis for the collars that predicted a compliance of .45x10^{-6} inches per pound in the azimuthal direction. One of the load cell and fixture combinations turned out to have a larger compliance than planned. The compliances of the load cell-fixture combinations were measured by substituting a solid steel block and a series of shims for the coil and measuring the load cell output when the upper half of the fixture was tightened down. The compliance was then taken as the slope of the line on the plot of shim thickness versus load. Two fixtures were used to take the data reported here and one had a compliance of 1.2 10^{-6} inches per pound and was used for samples numbered 4, 7, 11, and 13, 13R. The other had a compliance of .5 10^{-6} inches per pound and was used for samples numbered 8, 12, 14, 15, 16, and 19. These compliances are only used in the fixture corrected stress relaxation calculation.

Figure 2 is a photo of the entire creep apparatus. The fixture is enclosed in a plastic box with an electric heater and a fan. The heater is controlled by a PID controller and a thermocouple in the air downstream from the fan. The air temperature is controlled to 90° F, ±.2°. The coil temperature is measured with a thermocouple in the fixture near the top of the coil.

The two load cells in each fixture are powered in series by a very stable HP6186C constant current power supply and their output voltages are measured with an HP3457A 6 1/2 digit voltmeter. The voltmeter has an IEEE488 buss and is computer controlled to measure the voltages every 4 minutes and record them on magnetic discs where they are later converted to load.

Figure 3 is the raw load versus time for 9 coil samples. The shims used to mold these samples are as follows:

Coil Sample	Mold Shim Thickness	Measuring Shim Thickness	Initial PSI Load
#13	-.008"	-.020"	3020
#13R	-.008"	+.007"	12718
#14	-.008"	+.018"	8511
# 4	.000"	.000"	2469
# 7	.000"	.000"	6313
# 8	.000"	.000"	10105
#11	+.008"	?	5790
#12	+.008"	+.029"	12100
#15	+.016"	+.025"	9250
#16	+.024"	+.026"	10032
#19	+.040"	+.026"	8918

The loads given for samples 13R and 15 are the first measured points, the rest are extrapolated to 0 times. At first we tried to mold and measure a coil using the same shim thickness, but the loads were too low for accurate measurement so we began using larger measuring shims (similar to collaring to a smaller size) and trying to start at the same load. The data in Figure 3 are very hard to compare because of the large load range and the small variation. Obviously some kind of normalization is needed.

We can justify the normalization in the following way. The stress relaxation modulus is $G(t) = \sigma(t)/\varepsilon$ and the unrelaxed modulus (the value at t=0) is $G_0 = (\sigma_0)/\varepsilon_0$. Remembering that strain is nearly constant; i.e. $\varepsilon_0 = \varepsilon$, we can combine these equations to get $G(t) = G_0*(\sigma(t)/\sigma_0)$, and divide by the area for $G(t) = G_0*(1/1_0)$. Dividing our loads by the initial load will nicely normalize our data to a value of G_0. Since the actual value of G_0 is not important to comparing data, we choose to use for it's value the ordinary modulus we have previously measured. Figure 4 is a graph of the data normalized in this way, using E = 1.1 x 10^{+6} psi.

The most unusual feature of the data is that samples 12 and 13 do not agree with the rest. In fact, sample 11 is also unusual in that it seems to be decreasing at a nearly linear rate. These differences are not a function of the shim size used in the molding of the samples since 12 was made with a +.008 inch shim and 13 was made with a -.008 inch shim. We have remeasured

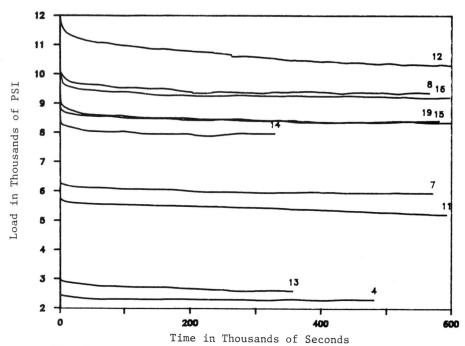

Fig. 3. SSC outer coil test load in psi as a function of time for samples molded to various final sizes.

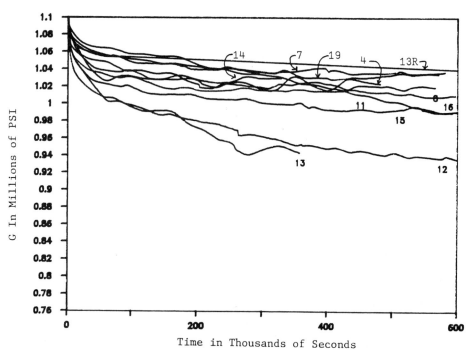

Fig. 4. Stress relaxation modulus as a function of time (assuming an unrelaxed modulus of $1.1 \times 10^{+6}$).

sample 13 after a long relaxation time and got very different results that agree with the majority of the other samples. This data is shown on the graph as sample 13R. The straight line portion of 13R is due to a data acquisition failure and the line connects to additional points off the scale of the graph. We would like to remeasure sample 12 also.

The second obvious feature of the data is that most of the samples relax at about the same rate regardless of the shim thickness used in their molding. It has been suggested that the Kapton part of the insulation system would undergo stress relaxation at a rapid rate during the high temperature molding cycle, and would recover from that history at a lower temperature and therefore, a slower rate. This stress history might then affect the relaxation rate after collaring. No such effect is apparent. There are too many experimental variables to conclude much beyond the obvious that it is not easy to change collared coil visoelastic behavior by modifying coil molding parameters.

We were concerned that the 11 different relaxation curves were obtained from two different fixtures with different compliances and one was somewhat different from the predicted compliance of the SSC collars. We have, therefore, computed a correction to the data that will approximately remove the effect of compliance. The stress relaxation modulus $G = \sigma(t)/\varepsilon_0$ which we can rewrite as $G = (1(t)/A)\ (L/\Delta L)$ where 1 (t) is the load, A the area, L the length, and ΔL the change in length. Now as the load on the sample decreases with time, the load on the fixture decreases accordingly and the fixture opening decreases imposing an additional deformation on the sample. This additional deformation can be written as $\Delta L = \Delta L_0 + c(1_0-1)$, where c is the compliance of the fixture and ΔL_0 is the deformation at time zero, and 1_0 is the load at time zero. Substituting this value of ΔL into the equation for G gives us $G = (1/A)\ (L/(\Delta L_0 + c(1_0-1)))$. If we note that at time zero G is the unrelaxed modulus $G_0 = (1_0/A)/\ (L/\Delta L_0)$, we can solve for ΔL_0 and substitute into the equation for G. Rearranging terms $G = (1/1_0)*G_0*(1/(1+(c*A*G_0/L)\ (1-1/1_0))$. Note that when the fixture is perfectly rigid, the compliance is 0 and the expression for G reduces to the same expression we used to normalize our data. Also note that for c>0, the fractional term is less than 1, which implies that for any given value of G the load 1 must be larger than it would be for the c = 0 case. This is to say that stress relaxation in any real fixture occurs more slowly than for a hypothetical case of constant strain. Figure 5 shows the data modified in this way to correct for the fixture compliance. The effect is seen to be an increase in the spread of the data which is unfortunate even though it does make a sample 12 look less far off the median.

Figure 6 shows some very long term data. Texts on visoelastic theory[1,2] state that crosslinked plastics like epoxies should relax to some constant value of stress while thermoplastics like Kapton may continue to relax indefinitely due to an actual irreversible flow of the material. The long term data shows a continuing relaxation. It can easily be seen that if the data had stopped at some unfortunate earlier period, it might have been interpreted as having reached a limit. This illustrates just one of the difficulties in these measurements.

We can look back at an earlier paper[3] where we measured relaxation rates on straight stacks of Tevatron cable and find a remarkable agreement considering the difference between the two experiments. Since the primary stress relaxing element in both experiments is assumed to be the Kapton, we are encouraged to use the time-temperature shift factor measured in the earlier case as the best available data until it can be remeasured with actual SSC coils.

INSULATION BREAKDOWN

The maximum pressure that insulated SSC cable can withstand before electrical breakdown occurs turns out to be a surprisingly complicated subject. It is most important because of the pressure that must be applied both in molding a coil and in collaring the final coils. We have not considered the insulating value of the helium, but instead have applied a turn-to-turn voltage of 2 kilovolts which is sufficient to cause breakdown whenever the plastic insulation has ruptured.

We have found three different modes of failure of the Kapton insulation. The first mode is found whenever there is a flaw in the cable construction. Under the microscope we have found cables with strands of varying diameters which cause irregular decreases in the flat area on the surface of each strand which area actually supports the applied load. We have found strands with distinct bumps on the flat surface especially near the cable edges. These bumps can pierce

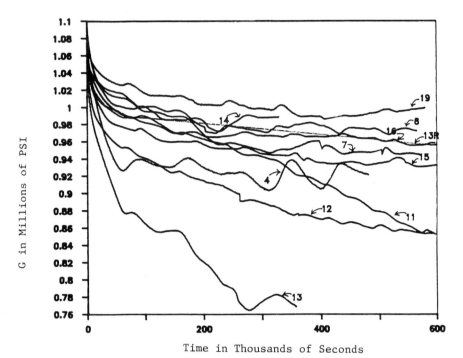

Fig. 5. Stress relaxation modulus corrected for
fixture compliance.

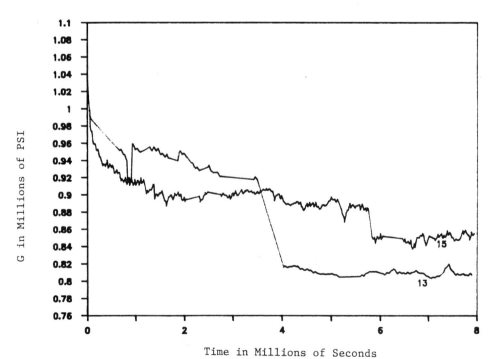

Fig. 6. Long term stress relaxation modulus.

the insulation much like a foreign inclusion would. We have also found cables where some of the strands have badly pitted surfaces. These pits look as if the copper had been torn out of the flattened cable surface by a galling contact with the cable flattening dye. The Kapton appears to extrude into these holes and in doing so, thins at the hole edges where ruptures then occur. We have seen many instances of cable strands with sharp protrusions on the edges of the flattened regions. Others have reported seeing these protrusions so bad that they break off the strand edges as long copper hairs. All of these flaws can cause failure of the insulation at low values of pressure. It will be vital to have a good inspection system for incoming cable to detect flaws and reject cable. We have tried to correct flawed cable by sanding and/or polishing the cable surfaces, but have not been successful. The complex behavior of these kinds of flaws is illustrated by the tests in which we took good cable and created flaws by sanding and scratching the surface. When insulated and tested, only surprisingly small decreases in the pressure to breakdown were found. Figure 7 shows the breakdown pressures of 4 different cables. Of the two SSC cables, the one with the lower average breakdown pressure was found to have strands with the pitted surfaces mentioned previously.

The second mode of failure is a series of cuts through the Kapton where the cuts are found along and aligned with the edges of the strand flat surfaces. They were found on both cables at a mating surface. Several such cuts are usually found even though the charred indications of an electrical arc are only found at a single spot. Obviously, a failure must be located at the intersection of an upper and lower strand flat surface edge. We consider these failures to be the predominate type on high quality cable.

The third failure type is found whenever extra layers of Kapton are used or the insulation system otherwise modified to give rise to failures at a much higher value of pressure. In these cases, the higher pressure causes the cable to spread in the width direction which tears the Kapton apart. If the cable is restrained from this spreading, much larger pressures can be reached before insulation failure. In the present design, the outer coil might be so restrained by the presence of the inner coil, but the inner coil is not restrained. It is conceivable to design a magnet where the inner coil is also restrained by filling the space between the beam tube and the inner coil with a material capable of transferring stress from the coil to the beam tube. We have measured the residual increase in width of several types of cable as a function of the pressure to which we have exposed them. There is a small residual width increase at very low pressures which does not change much as the pressure is increased. At a pressure of around 50 Kpsi, the residual width begins to increase in a nearly linear manner with pressure. This pressure is the transition point to insulation failures of the third type.

In Figure 7, the Staybright cable is not only much better than the other cable types, but it is visually different in having a much larger flat area on the top of each strand. The cable must have been flattened much more than usual in the final rolling. This prompted us to make a measurement under the microscope of the total flat area found in the Staybright cable and the two SSC cables. The flat area of a single strand was measured very carefully with a microscope and video camera measuring system. For ease of calculation, this area was used to find the width of a strand having the same area, but a uniform width. We then calculated the area of the intersections of the upper and lower cable strands of this width. This gives a good approximation of the true area of the cable actually supporting the applied load. Using the calculated areas, the pressure at breakdown was found to be 113 Kpsi for Staybright, 156 Kpsi for SSC22-0006, and 125 Kpsi for SSC23-384b. Thus the apparent superiority of the Staybright cable can be completely explained by it's increased area, but the superior cable SSC22-0006 compared to SSC23-384b must still be explained by the damaged surface of the latter cable. The data of Figure 8 were taken using a single 1/2-lapped layer of Kapton and two cables stacked narrow edge next to wide edge and loaded between parallel steel plates (the Fermi fixture). No epoxy fiberglass was used.

In Figure 8, we compare the breakdown pressure of the Kapton only insulation with data obtained when epoxy fiberglass is added in both the uncured and cured condition. Also shown are data from both the Fermi style fixture and a fixture originated by Brookhaven. The Brookhaven fixture used a shorter length of cable in a round die shaped to take two cables stacked narrow edge to narrow edge and constraining those cables on their edges. We have built fixtures of this type to take both inner cable pairs and outer cable pairs. Unless otherwise noted, our data was taken with outer cable.

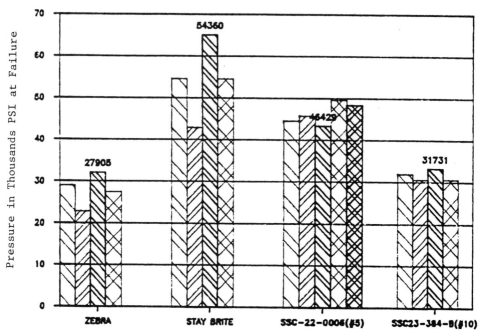

Fig. 7. Pressure to cause electrical breakdown of 1
         layer of 1/2 lapped Kapton insulation on
         various superconducting cables.

SSC23 - 384b Cable

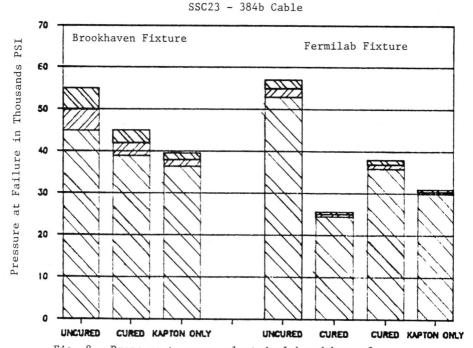

Fig. 8. Pressure to cause electrical breakdown of
         insulation consisting of Kapton only, or Kapton
         plus epoxy fiberglass, in both the cured and
         uncured state.

The Brookhaven and Fermi fixtures are seen to give comparable results at pressure below the point where appreciable width expansion occurs. The cured epoxy-fiberglass was just slightly better than the Kapton only, which is to be expected since the epoxy does not form an integral film. The uncured epoxy-fiberglass samples withstood much higher pressures. We think that this is due to the lubricating effect of the uncured epoxy. The lubrication allows the layers of Kapton to slide past each other locally reducing the stresses in these places.

In order to be quite sure that our tests were giving the same pressure to breakdown that would be found in an actual coil, we molded a 3 inch long section of SSC outer coil. We loaded it in a curved fixture to simulate the collar and noted the pressures at which turn-to-turn shorts occurred. The results were that, with the exception of the one unusual short that was found right away (we did not test until 4 kpsi), the shorts began at 31 kpsi and when the test stopped at 37 kpsi, 10 of the 19 turn-to-turn gaps were still unshorted. This is in very good agreement with the cable pair test data.

Figure 9 shows data from 8 different tests of Kapton only breakdown done with a section of SSC23-384b cable taken from the cable used to wind 4 different short coils. Each bar labeled F is the average of 10 samples tested in the Fermi fixture, and the bars labeled B are the averages of 5 or 6 samples tested in the Brookhaven type fixture. The boxes at the top show the +/- standard deviations. The numbers are coil numbers. Also shown are 3 tests of 5 samples each of new Kapton from DuPont in the same configuration of 1 layer of 1/2 lapped Kapton only. The insulation was cut into 1/4" wide strips from sheets supplied by DuPont, and wrapped on the SSC23-384b cable by hand. The 100HA is .001" film of amorphous Kapton (the regular Kapton is crystalline), and the 100MT is the same film but filled with a powdered aluminum oxide, and the 130MT is the same filled film, but in a thickness of .130". The results of new and old film are very much the same.

Figure 10 shows the data from 6 different combinations of 2 layers of new Kapton each 1/2 lapped. Also shown are data from two tests of old Kapton in the same two layer 1/2 lapped configuration. One of these is labeled Kapt.epo.cured., and was made from film which had been coated with a .0001" layer of a 3M's adhesive. This sample was cured before testing. This epoxy coated Kapton is the same that is being used to make our low beta quads. As in preceding figures, this one also shows the average of 5 tests as a single bar. All of these tests give averages in the range of 45 kpsi to 55 kpsi which is considerably higher than the 30 kpsi to 38 kpsi found for the single thickness of Kapton, but they do not show much difference from one film type to the other. All of the data of Figures 10 and 11 were obtained in the Brookhaven type fixture which restrains the cable from widthwise expansion.

DuPont has made the new Kapton films coated with a thermoplastic polymide adhesive and Brookhaven has wound some 3 foot coils from these films. We have cut and polished a section from one of these coils supplied to us by Brookhaven. The adhesive was a .0002" layer on either side of the film and the cables were wound with 2 layers 1/2 lapped. Thus there are 8 layers of adhesive between each cable for a total thickness of .0016". With that much adhesive, the 5000 kpsi pressure and 225° C temperature caused the adhesive to flow into the void space between cable strands pushing the Kapton layers with it.   Photomicrographs of cross-sections of the Brookhaven coil and an SSC coil and one of our low beta quads have been made. They show the varying degrees to which the insulation systems fill the void between cable strands. The Brookhaven-new-Kapton-thermoplastic adhesive coil shows the coil completely filled, the SSC coil shows a void only partly filled, and the low beta quad shows a completely empty void space. It is presently uncertain what effect this has on coil performance. It should be noted that future Brookhaven coils may use less adhesive.

It has been noted that filling in the void between cable strands on the cable surface might increase the breakdown pressure by better distributing the loads. This effect might be seen in Skaritka's study of these new DuPont adhesive coated films. We have looked at this effect in two ways. For the first way, we have filled in the surface spaces between strands with a cured epoxy, and then wrapped the cable with Kapton with the usual 1 layer 1/2 lapped and tested it in the Brookhaven type fixture which restrains widthwise expansion. On our second attempt to completely fill the surface with epoxy, three replications gave breakdown values ranging from 65 kpsi to 91 kpsi, or 2 to 3 times the breakdown with the same cable (SSC23-384b)! For the second

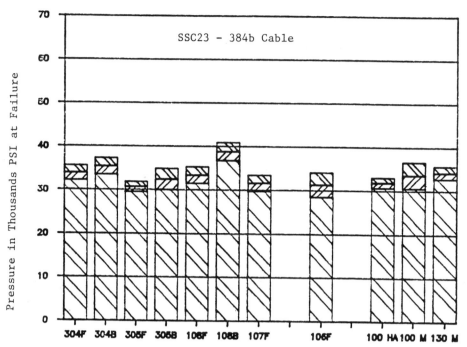

Fig. 9.   Pressure to cause breakdown in a complete quality
control series of SSC cable. Tests of new Kapton
are also included.

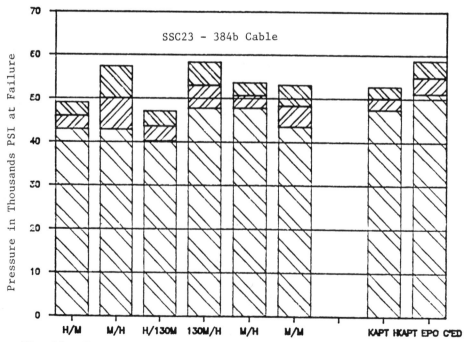

Fig. 10.   Pressure to cause electrical breakdown in insulation
consisting of 2 layers 1/2 lapped of new Kapton, and
also of old Kapton with and without cured epoxy adhesive.

way, we took solid bars of copper of about the same cross-section as SSC cable except they were not tapered. We then had to test them in the Fermi type fixture with its parallel plates. Of course, we used the same insulation system as the epoxy filled cable. These 5 samples broke down at at an average value of 66 ±7 kpsi (using the original unloaded area to calculate psi) and the failures were type 3; i.e., they failed when one of the copper bars expanded widthwise and tore the Kapton. We have not attempted to measure such pairs of copper bars in the Brookhaven fixture for fear of destroying our moderately hardened steel fixtures.

ACKNOWLEDGEMENTS

We are pleased to acknowledge the assistance of Choudet Khuon, Selles Morris, Daniel Rogers, Barbara Sizemore, Laurent Stadler, and Jim Cahill in taking and reducing the data.

REFERENCES

1. Ritchie, P.D., "Physics of Plastics," D. Van Nostrand Co., Princeton, NJ

2. Ward, I.M., "Mechanical Properties of Solid Polymers," John Wiley & Sons Ltd., New York

3. Carson, J.A., and Markley, F.W., "Mechanical Properties of Superconducting Coils", IEEE Transactions on Magnetics, March, 1985, Volume MAG-21, #2. A publication of the IEEE Magnetic Society, 345 East 47th Street, New York

DEVELOPMENT OF HIGH MANGANESE NON-MAGNETIC STEEL

FOR SSC DIPOLE SUPERCONDUCTING MAGNET COLLAR

K. Nohara, Y. Habu, S. Sato, K. Okumura and H. Sasaki

Kawasaki Steel Corporation
1, Kawasaki-cho, Chiba-shi, 260, Japan

ABSTRACT

The SSC dipole magnet collar requires the material that is of high strength and sufficient stability in magnetism along with adequate thermal contraction at liquid helium temperature as well as at room temperature. The newly developed high Mn non-magnetic steel is shown to meet such requirements better than other materials like stainless steel. The new steel is featurized by a marked magnetic stability, temper cold rolling without any rise in magnetic permeability, and a little smaller thermal expansivity. High Mn steel demonstrates a satisfactory stampability and spot weldability similar to those of stainless steel. It is subjected to the fabrication of both short and long model magnets as collar material of a cold mass to exhibit favored results in their fabricability and magnet performance.

1. Introduction

The SSC dipole magnet is designed to realize the tremendously large magnetic field reaching 6.6 tesla and its uniformity to the order of 10^{-4} within a bore tube and from one bore tube to another. This means the necessity of non-magnetic material as a collar whose magnetic permeability at crogenic temperature shall be low and its fluctuation in each coil or coil to coil shall be small as well. Quantitatively, specific magnetic permeability, μ, at room temperature and 4K is to be less than 1.002, and its alteration, $\hat{I}\mu$, is to be less than 0.0005, so that magnetic field fluctuation within a bore tube, $\hat{I}H/H$, maintains smaller value than 3×10^{-4} or so at 6.6 tesla.

Table 1 shows the necessities of such magnetic property and other requirements for a non-magnetic collar material. Particularly, yield stress, $\hat{A}_y$, at room and cryogenic temperatures must be essentially high to overcome local micro-yield due to the electro-magnetic force induced by 6.6 tesla magnetic field and to withstand local stress concentration due to the pre-stress applied for keeping a magnet system configuration correctly during cooling-down. It is noted that the current target of $\hat{A}_y$ at room temperature is $\hat{A}_y$ A90,000PSI.

Other characteristics of importance are (1) thermal coefficient which is desired to be close to those of iron yoke and /or superconducting coil and its insulator (Kapton), (2)usabilities which are made up of stamping performance as one of magnet fabrication processes and spot welding for laminating each collar, and (3) also the production fabricability like hot workability as well as (4) the economical aspect or the cost performance.

2. High Mn Steel for SSC Dipole Magnet Collar

2-1 Chemical composition and structure

To cope with the requirements described above for the collar material, several candidates have been listed: austenitic stainless steel, aluminum alloy, titanium (alloy), copper alloy, etc. Lately, high Mn austenitic steel as structural material has been developed to meet the need from the application under the strong magnetic field in the course of superconductivity technology[1] considering the fact that each candidate material poses each own drawback for practical use.

We have performed the research work to develop a new high Mn steel suited to the SSC dipole superconducting magnet collar in terms of chemical composition that is essential. Figure 1 shows the Fe-Mn-C ternary phase diagram at 1100°C that is a basis of high Mn steel, and magnetic permeability, μ, at 4K. This presents a newly compositional area illustrated by a hatched mark that is free from certain problems in manufacturing or characterization -(1) unfavorable magnetic property to give higher values of μ at lower Mn content outside the hatched zone, (2) less usability such as welding, machining, and stamping at higher C content outside the hatched zone, and (3) less fabricability of hot rolling at higher Mn content outside the hatched zone again.

To be more exact, high Mn steel developed here has chemical composition as shown in Table 2 (an example of two charges) that is marked by low C-high Mn-Fe compositional balance added with about 0.1%N and small amount of V plus Ni, Cr and Ca. For reference, Table 2 offers N-bearing austenitic stainless steel (N-SS) which is another candidate for the collar materials. The optical micrographs of high Mn final product specimen that is fully annealed and then slightly temper cold rolled are presented in Photo.1. Morphologically, the micro-structures are of typical austenitic grains with annealing twins and deformed bands induced during a light temper rolling.

2-2 Magnetic and Mechanical Properties

The major concern with the collar structural material is compatibility between magnetic stability and high strength. Figure 2 denotes the magnetic permeability, μ, at 10kOe of field applied to high Mn, N-SS and AISI 316LN as a function of temperature between room temperature and 4K. The provisional target value of μ for the SSC is below 1.002 through the whole temperature as stated in Table 1. AISI 316 owns the Néel temperature, T_N, around 30K to show excessively large μ than the target, and N-SS has T_N around 100K to demonstrate rather smaller value of μ at low temperature than the conventional AISI 316LN, still exceeding the target at cryogenic temperature. While high Mn steel shows very lower values of μ than the target, 1.002, with extremely small temperature dependence in the whole range, which implies a remarkable stability in magnetism.

With regard to strength, yield strength at room temperature, $\hat{A}_{y(RT)}$ of high Mn, is almost 35kg/mm^2 in the as -annealed state, whereas normal N-SS' $\hat{A}_{y(RT)}$ is a little larger than 60,000PSI, that is coped with by high Mn steel after conventional employment of skin-pass process. It is notable that the

Table 1. Proposed specifications of collar material for the SSC dipole magnet.

Item		R T		4 K	N o t e
Mechanical property	Yield stress (σy)	σy ≥ 60.000PSI (42kg/mm²) ⇒	σy ≥ 90.000PSI (63kg/mm²)	σy ≥ 170.000PSI (120kg/mm²)	① Local micro-yield (EMF due to 6.6T magnetic field) ② Local stress concentration (Pre-stress)
	Tensile strength	—		O	
	Toughness	—		O	
	Fatigue	O		O	
Physical property	Magnetic permeability (μ)	μ≤1.0020 . $\Delta\mu$≤0.0005			Magnetic field fluctuation $\Delta H/H \le 3.0 \times 10^{-4}$ (at 6.6T)
	Thermal coefficient (β)	$(8\sim15)\times10^{-6}$/deg (RT~4K)			Thermal contraction 0.05~0.4% (RT~4K)
	Specific heat	—		O	
Usability	Spot welding	O		—	
	Stamping	O		—	(Shipping)
	Corrosivity	O		O	
Others	Fabricability				Hot workability
	Economy				Cost performance

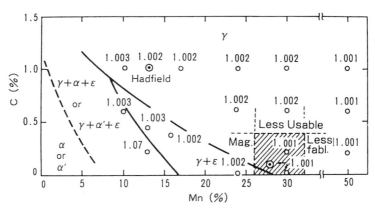

Fig. 1. Phase diagram of Fe-Mn-C of at 1100°C and magnetic permeability at 4K of the alloys with various composition balance.

Table 2. Chemical composition of newly developed high Mn steel (N-SS: N-bearing austenitic stainless steel).

(wt%)

	Item	C	Si	Mn	P	S	Al	Ni	Cr	V	Ca	O₂	N	Mo
High Mn	M0005 (01-1109)	0.107	0.71	27.75	0.032	0.003	0.02	1.01	7.13	0.063	0.0043	0.050	0.1037	—
	M0006 (01-1110)	0.106	0.61	28.4	0.033	0.002	0.022	1.04	7.10	0.063	0.0047	0.056	0.0998	—
Ref.	N-SS	0.03	—	8.9	—	—	—	7.3	20.2	—	—	—	0.3	0.16

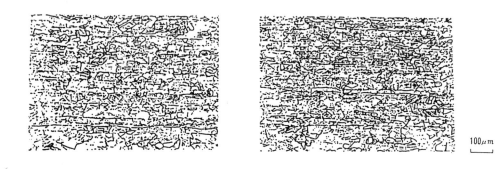

M0005   M0006

Photo 1. Optical micrographs of high Mn steel.

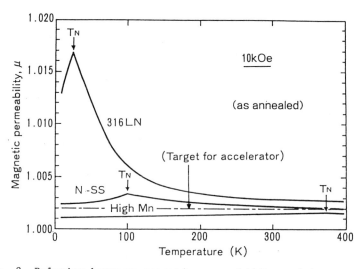

Fig. 2. Relation between magnetic permeability and temperature.

required strength of over 90,000PSI can be obtained when the temper cold rolling of approximately 14% or more is given to the as-annealed high Mn steel as shown in <u>Fig.3</u>. Metallurgy-wise this treatment corresponds to "work hardening" that is technically reliable and more economical process to give rise to higher strength than "precipitation hardening" or "solution hardening" process. The problem in the present case is the generation of magnetization usually accompanied with plastic deformation of austenitic steel. <u>Figure 4</u> is a result of measuring μ at 4K of each material including Hadfield manganese steel (1C-13Mn-Fe). The data are very suggestive in the light of the plastic deformation dependences of permeability according to different materials. Namely, magnetization of high Mn steel is almost independent upon deformation to lead to the constant low permeability below 1.002. Then the proposed specifications, $\mu < 1.002$ and $\hat{A}_{y(RT)}$ A 90,000PSI, are feasible in case of high Mn steel. Such compatibility is hardly realized in other materials such as AIS 316LN, N-SS, Hadfield Mn steel etc.

The thermal contraction during cooling-down process of a magnet is of another concern in view of the available materials combination among (1) supper conducting coil with insulator, (2) supporting collar, and (3) shielding yoke. <u>Figure 5</u> shows the relation between thermal coefficient, B, between 4K and RT and the Néel temperature, T_N, shown in Fig.2 in terms of austenitic steels which involve high Mn steel, 316LN, N-SS, etc. It is found that the inverse semi-linear correlation exists between β and T_N, and that β of high Mn and iron (yoke)is close each other, while β of N-SS is close to that of coil, respectively. The assembled structure or cold mass composed of three different components (namely, materials) will be discussed later in view of thermal contraction behavior. Additionally, β of manganese steel is found to be controlled by the chemical composition balance and expressed by the following empirical equation,

$$\beta \ (\times 10^{-6}/\text{deg}) = 51.6 + 7.60[C] - 0.92[Mn] - 0.52[Cr] - 6.4[N] - 0.40[Ni]$$

where the unit of each element is wt%. Hence, basically β and also T_N of high Mn steel can be controllable mainly by Mn, C, N, etc, where there exists inter-correlation of thermal behavior with magnetic and mechanical properties as well as usability and steel production fabricability.

2-3 Stamping and welding

As the SSC magnet collar is supposed to be composed of a laminated structure of stamped thin sheets. This is why the stampability of collar material is of significance in the construction of magnets.

High Mn steel ($\hat{A}_{y(RT)}$: 60,000 and 90,000PSI grade) and N-SS specimens with thickness of 1.5mm are subjected to a laboratory level stamping test on a 10 ton HP hydraulic press machine. The stamping conditions (clearance and blank diameter) are given in <u>Table 3</u> together with the experimental results of burr height and maximum hydraulic pressure. An example of burr height measurement on a surface roughness meter for high Mn specimen with the conditions of 60,000 PSI, 0.08mm clearance and 24mmd blank is exhibited in <u>Fig.6</u>, and its image on scanning electron microscopy is presented in <u>Photo.2</u>. Basically, the stamping performance of high Mn steek in the way of burr height is similar to that of N-SS (austenitic stainless steel) and independent upon yield strength (of 60,000 or 90,000 PSI). Its stampability in the way of maximum hydraulic pressure has a sign of being low compared with N-SS. The degree of clearance and blank sizes does not affect much the stamping behavior. Concerning the die life (the die material:SKD11), it is obtained that 60,000 PSI class high Mn steel gives a life of about 5000 times for re-grindinga die, that is compared with about 3000 cycles in N-SS, and that there is no particular difference in the influence of stampimg between

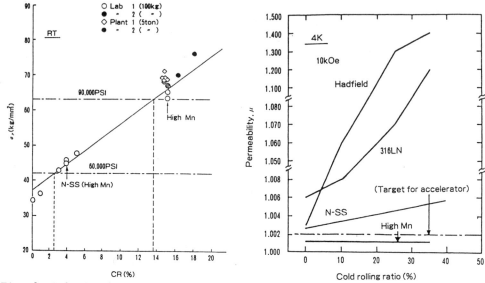

Fig. 3. Relation between yield stress σ_y and temper cold rolling ratio.

Fig. 4. Relation between permeability and temper cold rolling ratio.

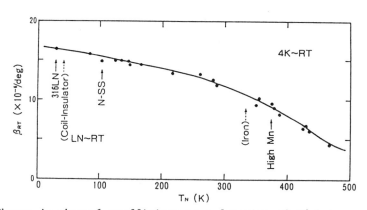

Fig. 5. Change in thermal coefficient as a function of Néel temperature T_N.

Table 3. Measurement of burr height and maximum hydraulic pressure.

σy	Clearance Blank	High Mn		N-SS	
		0.08mm	0.14mm	0.08mm	0.14mm
60,000PSI	12mmφ	90μ	89μ	98μ	94μ
	18mmφ	78	76	82	76
	24mmφ	72 ¦200**	62	78 ¦240**	81
90,000 PSI		80 ¦200**	—	—	—

Ⅰ Average at four directions **Max. kydraulic pressure(ky/cm²)

770

90,000PSI class high Mn steel and AISI304 stainless steel without giving any
need for re-grinding as far as approximately 3000 stamping cycles are
concerned.

The stamped collars are followed by spot welding process for lamination.
Then spot weldability should be guaranteed at least to the degree of
austenitic stainless steel, N-SS. A comparative test of high Mn steel with
N-SS is carried out under a spot welding condition as follows:

    1) Voltage   0.5-1.3V
    2) Current   1.6-7.3KA
    3) Load   600kg
    4) Welding cycle   5, 10, and 20
    5) Electrode (chip)   Cu; dome, d=5/t

The results are illustrated in Fig.7. The experimental data say that (1) the
weld lobe area where spot welding can be available without a trouble of either
expulsion or unwelding is roughly the same in both materials, showing somewhat
wider region in high Mn steel; (2) The nugget width does not show any
difference in both specimens regardless of the change in the current applied;
and (3) maximum load for separation on a cross type removing tensile test
produces the similar results in high Mn steel and N-SS with a tendency of
presenting an optimum region around 5kA of current which falls in the weld
lobe area close to the boundary to expulsion zone. As a whole, newly
developed high Mn steel has a spot weldability equivalent to that of
austenitic stainless steel. In addition, it is noted that the new steel does
not show any change in magnetic permeability of welded molten portion due to a
strong magnetic stability, which is in quite contrast with stainless steel
after welding.

3. Application of High Mn Steel to SSC R/D Dipole Magnet

3-1 Assembly and field test

High Mn steel is employed to the SSC model magnets collar of small scale
(1.8m long) and large scale (17m long). The steel is manufactured to be
1.52mm long x 120.7mm wide temper rolled cold strip through the processes of
melting on induction furnace, casting to 5 ton ingot, slabbing, hot rolling,
cold rolling, annealing, pickling, and finally temper cold rolling. Then it
is shipped to BNL to be subjected to assebly of cold mass after stamping and
spot welding. In case of a small magnet, it is built up at BNL and the magnet
field test is also conducted at the same place. In case of a large magnet, it
2is finally completed at FNAL, where the magnet field test is carried out. As
far as the new steel is concerned, no particular technological problems have
happened in the course of the entire, building up processes.

Figure 8 shows the quench test result on 1.8m short magnet with high Mn
and N-SS collars installed. DSS12 (High Mn) requests the training twice that
is just followed by saturation of fully critical current corresponding to 20
TeV. This can be compared with DSS11 (N-SS) and somewhat superior to DSS10
(N-SS). In addition, pre-stress loss is measured on these short magnets. The
results are given in Table 4. The data are obtained from high Mn and N-SS,
inner coil and outer coil, cooling-down process and current charging process,
and before and after the processes and the differences. Inner coil pre-stress
is larger than outer coil in any case. The substantial difference is not
seen. However high Mn steel can adopt greater pre-stress to maintain larger
value even after cooling-down and the imposition of critical current, as seen
in high Mn inner coil.

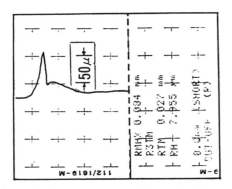

Fig. 6. Measurement of burr height on surface roughness meter.

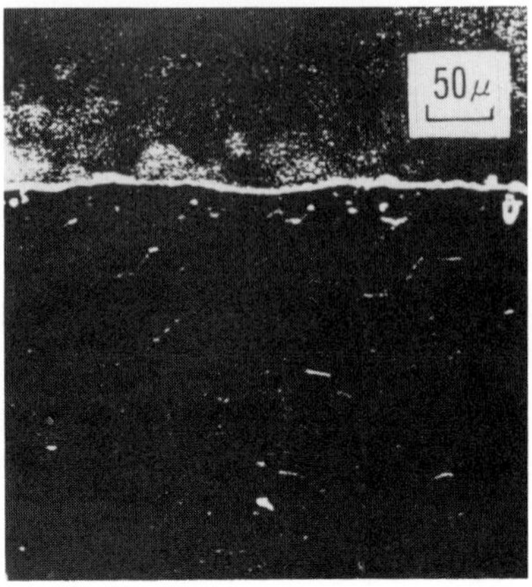

Photo 2. Electron micrograph of burr on SEM.

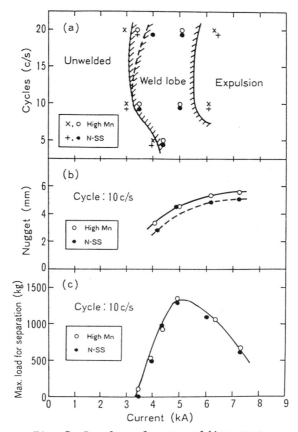

Fig. 7. Results of spot welding test.

Photograph 3 shows an appearance of the 17m long magnet having high Mn collar which was assebled at BNL and built up at FNAL. This long magnet was completed with no trouble. The magnet was subjected to the field evaluation test. The result is demonstrated in Fig.9. DD0019 magnet containing high Mn collar attains the satulated current (6.5 kA) without any trouble and any training. This is a remarkable result when the magnet is actually operated. Thermal cycle was tried twice. In this case, the magnetic field is maintained beyond the specifications. Thus, the first one of trial long magnets was successful in the field test. It sounds natural that the magnet characteristics is determined by the whole factors governing the respective performance, therefore a role of supporting collar is said to be not easy to abstract independently.

Related to the thermal contraction during cooling-down process, it was measured in various materials including inner and outer coils, while the data were relative to Ti silicate. The measurement was made at the temperature range of room temperature and 4.4K (Fig.10). Needless to say, invar alloy is lowest in thermal contraction, or 0.48 mils/inch. The value from high Mn steel is 1.7mils/inch, a little lower than that from yoke iron, or 2.1 mils/inch, while N-SS displays the value of 2.6mils/inch, comparatively large value compared with yoke iron. Thermal contraction of each coil changes with the direction and with neither inner coil nor outer coil. Generally, these absolute values are referential, and the important thing is a combination of coil, collar and yoke which pose the respective thermal contraction characteristics in terms of coherency at each boundary.

3-2 Uniformity of magnetic field and occurrence of thermal contraction

As has been stated, the uniformity of magnetic field within a bore tube is of great significance. This poses the two aspects in view of the configurative way of thinking:

    (1) Uniformity within each magnet based on position...($\dot{I}H/H$)m
        (H: magnetic field, $\dot{I}H$: its deviation)
    (2) Uniformity in one magnet to another at particular position...($\dot{I}H/H$)c

According to Shintomi and the work at BNL, ($\dot{I}H/H$)m is a function of magnetic permeability, μ, of the material. Figure 11 shows the change in ($\dot{I}H/H$)m within each position of a bore tube with a diameter of 50mm as a parameter of μ. The calculated value of ($\dot{I}H/H$)m when μ is 1.001 varies a bit in a circular region 15mm radius from the center. This is in quite contrast to consider that ($\dot{I}H/H$)m when μ is 1.01 fluctuates largely even in the same circular zone. Thus, the smaller μ is, the better the magnetic uniformity is. On the other hand, it is considered that ($\dot{I}H/H$)c is dependent on $\dot{I}\mu$ that is a deviation of μ of the same type of collar material used in each magnet, in other word the deviation of μ due to different charges or coil.

Figure 12 shows the relation between ($\dot{I}H/H$)m or ($\dot{I}H/H$)c and μ or $\dot{I}\mu$ of high Mn steels, N-SS and AISI304N with the respective yield stress at room temperature. The two double circles are calculated values from N-SS (μ: 1.0025) and AISI304N (μ: 1.0082) (a bore tube is 40mm dia.), and they are connected with a straight line to be extrapolated to smaller value zone at ($\dot{I}H/H$)m or ($\dot{I}H/H$)c which corresponds to high Mn steel. It is seen that ($\dot{I}H/H$)m changes with μ, and that the value of μ less than 1.002 or 1.001 is very critical, judging from the target value of ($\dot{I}H/H$)m (3×10^{-4} or 5×10^{-4}). It is also seen that ($\dot{I}H/H$)c is 3.1×10^{-4} when $\pm \dot{I}\mu$ is 0.0005, and that such value of 3.1×10^{-4} or 0.0005 is another critical thing which can be a target.

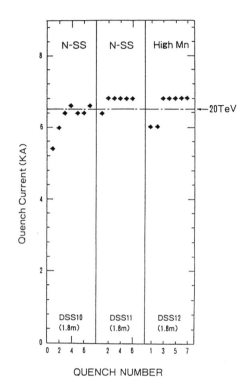

Fig. 8. Quenching test of 1.8m short magnet with high Mn collar.

Table 4. Result of prestress loss measurement.

I t e m		Cool down			Current 0 → 6800A(=6.6T)			Total
		Before	After	Dif.	Before	After	Dif.	
N—SS	Inner coil	PSI 9400	PSI 5000	PSI 4400	PSI 5000	PSI 3700	PSI 1300	5700
	Outer coil	6500	4800	1700	4800	2400	2400	4100
High Mn	Inner coil	11600	6800	4800	6800	5200	1600	6400
	Outer coil	5300	4500	800	4500	2200	2300	3100

Photo 3. Appearance of 17m long magnet with high Mn collar.

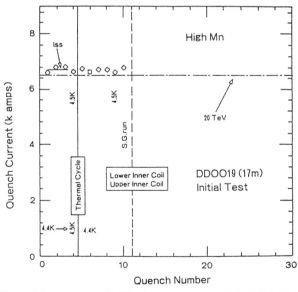

Fig. 9. Quenching test of 17m long magnet with high Mn collar.

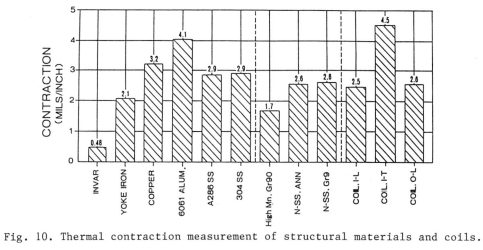

Fig. 10. Thermal contraction measurement of structural materials and coils.

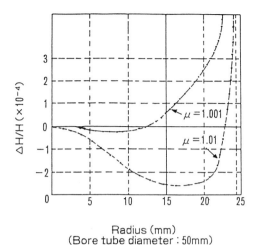

Radius (mm)
(Bore tube diameter : 50mm)

Fig. 11. Change in magnetic field deviation ΔH/H in 50 mmφ bore tube [7].

Regarding the thermal contraction problem, the materials combination seems important among tube, collar and yoke as was stated earlier. A schematic illustration is a given in __Fig.13__ along with the observed values of thermal contraction, β', cited from Fig. 10. If β'of coil and yoke are 4.5 and 2.1 mils/inch, respectively, the relation of β's of those three components is expressed by Fig. 13, depending on the choice of high Mn steel (β'=1.7mils/inch) and N-SS (β'=2.6mils/inch) as a supporting collar. (1) High Mn case: A collar is fastened by yoke because β'(collar) < β'(yoke), and a gap is liable to be formed between coil and collar because β'(coil) >> β'(collar). However the gap can be compensated by the application of prestress. (2) N-SS case: Gaps are induced at boundaries between collar and yoke, and coil and collar, because β'(collar) > β'(yoke) and β' (coil) >> β'(collar). The latter gap can be avoided with prestress applied, but the former gap is hardly be lessened. As a result, the combination would be more favored on high Mn steel collar than N-SS collar in view of thermal contraction.

4. Conclusion

To keep a uniformity of magnetic field within a bore tube and to withstand both local prestress applied plus electromagnetic force, high Mn steel has been newly developed for the SSC superconducting magnet collar use. The results are summarized as follows:

(1) High Mn steel containing as much as 28%Mn, 0.1%C, 0.1%N, 7%Cr, 1%Ni, 0.05%V and so on is developed to meet various requirements.
(2) The new steel is characterized by prominent non-magnetic behavior and its stability. Magnetic permeability, μ, less than 1.002 offers a magnetic uniformity in each bore tube. The deviation of permeability, Îμ, guarantees a magnet-to-magnet uniformity.
(3) Elevated yield strength over 90,000PSI at room temperature and about 170,000PSI at 4.2K was obtained to bear local prestress and electromagnetic force.

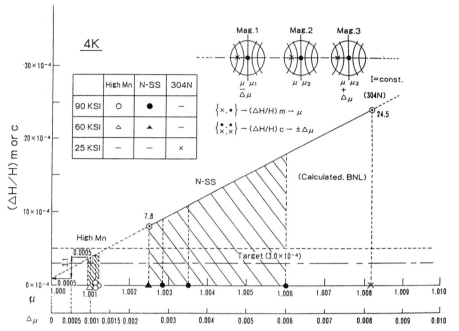

Fig. 12. Change in magnetic field deviation (ΔH/H)m (each magnet) and (ΔH/H)c
(magnet-to-magnet) with permeability μ and its deviation Δμ.

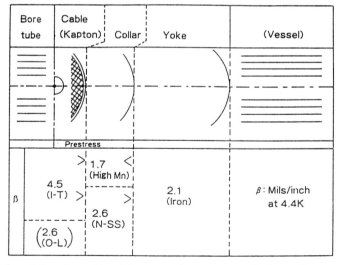

Fig. 13. Schematic view of cold mass structure and thermal contraction of
each material.

(4) Favored usability such as stamping performance and spot weldability becomes reality.
(5) The steel is desired to preserve dimensional tolerances when thermal contraction between collar and yoke, and the application of prestress to coil and collar are taken into consideration.
(6) The new material has been used in both 1.8m short and 17m long tial magnets without any fabrication trouble. The quenching test carried out shows good performance and the prestress loss measurement is acceptable.

REFERENCES

1) T. Sasaki and K. Nohara : Trans. ISIJ, 22 (1982), 1010
2) K. Nohara and K. Kato : ICEC9-ICMC (1982), Kobe
3) J. W. Morris Jr. and S. K. Hwang : Prenum Press, N.Y. (1978)
4) K. Nohara and K. Kato : "Advances in Cryogenic Engineering", 30(1984), 193, Plenum Pub. Corp.
5) H. Froschel and F. Stangler : Zeit.für Metallk., 66(1975), 311
6) K. Nohara and Y. Habu : ASM International Conference on Mn Containing Stainless Steels, Proceeding, (1988), 33
7) T. Shintomi : Conference on High Energy Physics Accelerator, Tsukuba, (1989)

Please note: The new steel was used as supporting collar to fabricate and test superconducting model magnets at BNL and FNAL.

15. Technology Transfer

INDUSTRIAL PARTICIPATION IN TRISTAN PROJECT AT KEK, JAPAN

Satoshi Ozaki

Brookhaven National Laboratory
Upton, L. I., N. Y. 11973

ABSTRACT

Industry-Laboratory collaborations played a very important role in
the construction of TRISTAN electron-positron colliding beam facility,
and brought this construction project to a successful completion in a
scheduled time. What had motivated the collaborations, what were the
important elements in the successful collaborations and how the
collaborations worked will be given based on the authors experience as
the TRISTAN Project Director.

It is my pleasure to participate in this meeting of IISSC, and to
present a talk on my experience with industry-laboratory cooperations in
the construction of a major high energy accelerator facility in Japan;
namely the TRISTAN electron-positron colliding beam facility project at
KEK, (National Laboratory for High Energy Physics) in Tsukuba, Japan.

TRISTAN PROJECT

Let me, first, give you a very brief description of the TRISTAN
facility which was commissioned in 1986 after five years of
construction. It is an electron-positron colliding beam facility at the
nominal collision energy of 60 GeV, the world's highest energy from the
time of commissioning until 1989 when it was exceeded in energy by SLC
at SLAC and LEP at CERN with successful operation of respective collider
at close to 100 GeV. The main collider ring of this facility is a
storage accelerator of 3 kilometers in circumference. It consists of
2.2 kilometer of four arc-quadrants where dipole-, quadrupole- and
sextupole magnets are installed to guide the bunches of particles around
a closed loop orbit, and four straight sections each 200 meters long
connecting the arc quadrants. In order to compenciate a large energy
loss of electrons at high energy due to emission of intense synchrotron
radiation, approximately 330 meters of radio frequency (acceleration)
cavities are installed in these straight sections, providing a total of
570 MV of acceleration potential per turn supplied by 30 units of
500 MHz klystrons with 1.2 MW (CW) rating. Of these cavities, 48 meter
length are of superconducting type made of Niobium metal operated at the
liquid Helium temperature. These cavities are more efficient than
common room temperature copper cavities and provide greater than 3 times
acceleration potential per meter. An ultra high vacuum beam tube goes
around the ring. A highly sophisticated computer system controls magnet

and radio frequency power supplies and other devices to maintain a stable operating condition of the accelerator. Having two bunches of electrons and positrons counter rotating around the ring, they collide at the center of four straight sections. In order to capture the details of what happens at the moment of particle collisions, a large and sophisticated particle detector (typically measuring 10 m cube and weighing 2000 tons) are built at these collision points.

The scope of the TRISTAN project as expressed in the cost of construction was: approximately $270 million for the civil construction of the tunnel and experimental facilities, $300 million for accelerator technical components including an addition of superconducting cavities, and about $100 million dollars for three major detectors. TRISTAN was the first major facility in Japan of this scope dedicated to the research of the vary basic science such as high energy physics. In spite of its scope the project was carried out successfully, namely on time and within budget, with the extensive cooperation of Japan's top industries. A list of major technical components of the TRISTAN facility and the names of industrial firms involved in the development and fabrication of these components is shown Table 1.

MOTIVATIONS FOR THE INDUSTRY-LABORATORY COLLABORATION

KEK was established in 1976 as the inter-university center for high energy physics research in Japan under a direct jurisdiction of MONBUSHO, the Ministry of Education, Science and Culture of the government. The idea behind this is to build big and expensive tools of high energy physics research such as large particle accelerator facilities at this center for use by university scientists from all over Japan (and lately from all over the world). This is a departure from the traditional way the basic science research had been carried out there, i.e. at national universities. Although the laboratory's institutional nature and missions, thus the way it should have been operated, is different from those of academic universities, the organization of KEK followed that of traditional style. The resulting organizational structure based on Professor, Associate Professor and Assistant system did not induce high level engineers and highly skilled technician to join the laboratory. In addition, the ceiling imposed by the government on the staff level of the laboratory, all of whom are permanent government employees, did not allow a rapid increase of needed manpower at the beginning of the project. Also, like in most academic universities, the laboratory's machine shop capability was quite limited. In other words, the laboratory had been compelled to depend on the engineering and fabrication capabilities elsewhere for much of engineering development work for the project.

Japanese industries have a high level of engineering skills and quality fabrication capability and are very willing to engage in a frontier science project. This willingness must have been based on a number of benefits the industry may obtain in such an involvement. There, the industry can receive a transfer of forefront technology as well as intellectual stimulation which sharpens their engineer's way of thinking and enhances their engineering capability. Such involvement will provide them with demonstrable accomplishment in the advanced technology product line nationally as well as internationally. This is of great prestige or PR value; and with that there will be some profit, or at least some merit of keeping their factory running. Namely, I have had a distinct feeling that, in many cases, the industries involved in the TRISTAN project have pursued the technological advances which would promise long term growth of the company, rather than a short term profit which comes from the

Table 1. List of Major Technical Components for TRISTAN Colliding Beam
Accelerator and Detectors and Name of Companies which were Involved in
the Development and Fabrication

TRISTAN MAIN RING:

Dipole Magnets	HITACHI LTD.
Quadrupole Magnets	HITACHI LTD.
Magnet Instal'n & Align't	HITACHI LTD
RF Cavities, Room Temp.	MITSUBISHI Heavy Ind.
RF Cavities, Superconducting	MITSUBISHI Heavy Ind.
500MHz, 1MW RF Krystrons	TOSHIBA
	VOLVO of West Germany
Vacuum System	ISHIKAWAJIMA-HARIMA Heavy Ind.
Control Computer System	HITACHI LTD.

DETECTORS:

Superconducting Solenoid
    AMY: 3.0 T, Thick Coil   HITACHI LTD
    TOPAZ:1.0 T, Thin Coil   FURUKAWA Electric Works
    VENUS:0.75T, Thin Coil   MITSUBISHI Electric Inc.

Magnetic Flux Return and Detector Structure
    AMY: 700 tons   MITSUI Ship Building Co.
    TOPAZ: 2000 tons   ISHIKAWAJIMA-HARIMA Heavy Ind.
    VENUS: 2000 tons   KAWASAKI Heavy Ind.

Particle Track Detector
    AMY: Cyl. Drift Ch.   KEK Shop & Research Group
    TOPAZ: Time Proj. Ch.   TOSHIBA
    VENUS: Cyl. Drift Ch.   KAWASAKI Heavy Ind & Research
                                     Group

EM Calorimeter
    AMY: Pb-Drift Tube   The US University Laboratory
    TOPAZ: Pb-Glass Blocks   OHARA Glass Work & IHI
    VENUS: Pb-Glass Blocks   NIKON and MITSUBISHI Heavy Ind.

---

manufacturing and sales of a particular equipment to the project. In addition, I have noticed that this willingness is also based on the corporate commitment and pride in their engagement in the pioneering scientific research program.

When I assumed the position of directing the project in 1981, after being accustomed with the American way of mounting a project with a strong in-house engineering support, I had a very serious concern in the lack of engineering and technical man-power in launching a new major project on the scale of TRISTAN. I was assured by my colleagues, then, that there will be a strong industrial support coming for the project. I realized very quickly that a close industry-laboratory collaboration can be established, and was able to accomplish the task quite well without having a large number of in-house engineers and technicians on the staff for the project.

MY WORKING EXPERIENCES

In the course of the project, I have learnt that the laboratory-industrial cooperation does not work well without a strong technological capability at the laboratory. Namely, a cognizant scientist in the project must know exactly what he wants and have knowledge of technologies to be used in the development of the product. The knowledge must be sufficient to convince the industry engineers that they can depend on the laboratory's scientific and technological leadership. This is also a way to open a real communications channel with them. Industry engineers then, can develop the fabrication technology and technical know-how for the product development which is suitable to their manufacturing facilities and carry out the tasks effectively.

One of many successful examples was the fabrication of superconducting solenoids for three detectors at TRISTAN, namely, AMY, TOPAZ and VENUS. Owing to the fact that the TRISTAN project as it was originally envisioned had included a construction of a superconducting magnet ring to accelerate and store protons in a multi-hundred GeV energy, there had been an extensive R&D effort on superconducting magnet technology at KEK. With this basic training in hands, KEK had among its staff members a number of scientists who became expert in this superconducting technology. These people became cores of successful development of very advanced superconducting solenoids for these detectors in collaboration with industrial engineers. Another example is in the development of such standard accelerator components as magnet, radio-frequency acceleration cavities, vacuum and control system. Here again, from the experiences in building proton and electron linear accelerators and 12 GeV Proton Synchrotron at KEK built prior to the TRISTAN project, we had a good accumulation of technology within KEK. Thus the collaboration went quite well, developing the world frontier devices. On the other hand, we have experienced some difficulties in the area of a high power radio-frequency sources, klystron, where we could not provide our expertise at the beginning.

HOW THE INDUSTRY-LABORATORY COLLABORATIONS WERE STARTED

In most cases, this type of cooperation begins with technical discussions between scientists on the project and engineers from one or more industries. These discussions sometimes led to pre-contract industrial studies at no cost to the project. There usually are two ways these discussions can get started:

i)   Through sales representatives of industries who visit the
        laboratory frequently and who are knowledgeable on the project
        as well as their industrial capabilities

ii)   Through a request by a cognizant scientist of the project to
        specific industry that is known to have special technology and
        know-how needed to solve certain technical problems.

These discussions help the project scientists in formulating technical
and engineering ideas to accomplish his mission, and help write more
comprehensive specifications.

    The job is, then, put on a competitive bid to qualified manufacturers
without prejudice on the pre-contract studies. In this case, the
qualification is judged only on the basis of the capital size (or
financial resources) of the firm relative to the scope of the task.
Any task which is estimated to cost more than a few hundred thousand
dollars (depending upon the foreign exchange rate) has to be opened for
international bidding.

    The most important part of the collaboration comes right after the
contract is awarded. Namely, this is the time when a full fledged
technical discussion will take place to establish a clear and detailed
understanding of the tasks ahead. This also is the time when the
industry engineers and the project scientists get to know each other
well, forging a close partnership needed to accomplish the task.

    Thereafter, a close association based on timely plant visits by the
project scientists for inspection and guidance and laboratory visits by
industry engineers for consultations keep the task rolling on the
correct road. In the process, we sometimes encountered changes in the
design or changes in the approach. In these cases, a clear
understanding of what is to be accomplished for the task established at
the beginning helped produce an amicable solution on how the additional
cost are to be shared without placing a blame to one side or the other.

CONCLUSION

    As a whole, the industry-laboratory collaboration worked very well in
the case of the TRISTAN project. The accelerator and detector facility
was successfully built in time and within budget thanks to the
corporate commitment to the project. The availability of industrial
engineering and fabrication capability to us played an important role
in construction of this most advanced facility. In many cases, the
most advanced technology available in major industries in the area of
materials, machining and material treatment, electronics and computer
software were made available to the project. Also, from the
laboratory's point of view, the industrial help has made it possible to
accomplish the task of building a major accelerator and detectors
without an enormous bulge in the engineering and technical staff level
which would have been necessary if we were to carry out the task of the
construction without the help of the industry.

    At the same time, I trust that the industries involved have gained
something from this association. I hope that the next speaker, Dr.
Takao Suzuki will have something to say about it.

POSSIBLE SSC DIPOLE DESIGN/MANUFACTURING

IMPROVEMENTS BASED ON HERA EXPERIENCE

Daniel Bresson, Thierry Evrard, Pierre-Jean Ferry,
Gérard Grunblatt, and Christophe Koch-Mathian

GEC ALSTHOM
90018 Belfort France

ABSTRACT

For the HERA collider GEC ALSTHOM (formerly ALSTHOM) manufactured in
its Belfort factory in France and supplied to DESY 6 preseries plus
120 complete quadrupoles (coils and cryostats).

Design of these quadrupoles was developed by CEA (French Atomic
Energy Commission) under direct contract to DESY. CEA built also two
prototypes to test this design.

Contract received by GEC ALSTHOM, following competitive bidding,
provided for DESY to supply a complete set of manufacturing drawings as
well as two complete lines of toolings used by CEA and some materials
(superconducting cable, bore tube, etc...).

Starting from these drawings and toolings GEC ALSTHOM manufactured
6 preseries coils and 3 complete quadrupoles. During this phase a lot of
improvements were made to the design, the tooling or the manufacturing
procedures. These improvements which were possible to be made only at
that phase (i.e. after initial prototype of CEA) were achieved using GEC
ALSTHOM large resources and long experience in the field of superconducting magnets for high energy physics. These improvements can be
listed in various categories :

- Change in basic design tolerances based on real material/component
  received or purchased - Modification of basic design to make an
  overall cheaper product.

- Change of manufacturing process to adapt to mass produced mate-
  rials/components - Modification, improvement of tooling.

Changes listed above were thereafter used during manufacture which
is also described (plant and human resources management). The 120 quadrupoles which were delivered to DESY on time were all accepted by DESY
upon cold testing so far.

INTRODUCTION

GEC ALSTHOM experience in the field of superconducting magnets goes back more than 20 years before the manufacture of quadrupoles for HERA.

The major achievements are as follows :

1968 : Development and manufacture of coils and cryostat for the "Big European Bubble Chamber" at CERN.

1972 : Manufacture of winding and cryostat for ALEC dipole for CEA.

1976 : Manufacture of a VERTEX detector for CERN.

1977 : Manufacture of 9 quadrupoles for the LEP-CERN. For these magnets, GEC ALSTHOM has cold-tested all the windings in its own cryogenic test facilities.

Concerning HERA, GEC ALSTHOM's work began during the manufacturing of the prototypes at CEA SACLAY. As a matter of fact, long before the order was placed, the development of these quadrupoles was followed by engineers from BELFORT. Industrialization was thus accelerated by this collaboration.

Together with the order, DESY supplied :

- All manufacturing drawings for windings and cryostats (drawings made by CEA).

- All components for preseries quadrupoles (non-insulated cable – Kapton – wedges – stamped collaring laminations – stamped cold iron laminations).

- Non-insulated cable for the series production.

- Steel for magnetic core (non cut-out raw material).

- AISI 316 LN steel materials required for the enclosure and the helium pipe tubulures.

- Beam tubes

- Corrective dipoles

- Diodes

- Superconductor for bus-bars

- Cables for the beam monitor

On the other hand, the contract binding GEC ALSTHOM and DESY stipulated that the toolings used for manufacturing the prototypes would be transferred to GEC ALSTHOM.

Among others, these toolings comprised :

- 2 winding lines

- 2 polymerization presses

- 1 measuring bench for the module of each coil

- 1 magnetic measuring bench for room temperature measurements

- 1 collaring press

- 2 clamping systems for cold iron sheets

- 2 platforms for cryostat assembly

- etc...

For manufacturing these magnets, 2 separate workshops have been installed in the BELFORT Plant : one was for the production of collared coils and the other for the production of completed magnets.

The workshops were thus smaller, more specialized and favorable to a team spirit, therefore to higher productivity and quality.

It is mainly during the production of the pre-series magnets that most of the modifications on the design, toolings and manufacturing procedures have been carried out.

The table below shows the result that these modifications have brought about on the pre-series and on the series. Please note the very good improvement in the result for the series.

Table 1

	Pre-series 10		Series 120	
	number	%	number	%
Fault-short circuit between turns	1	10 %	1	0.8 %
Fault-earth breakdown	2	20 %	–	
Fault-high dielectric current	1	10 %	1	0.8 %
Fault-self	3	30 %	1	0.8 %
Fault-short circuit between coils	1	10 %	–	
Number of magnets having had 1 or 2 faults	6	60 %	3	2.4 %

IMPROVEMENTS CARRIED OUT WITH RESPECT TO THE INITIAL CEA PROJECT

We will enumerate after the main modifications which have been carried out. It is very clear that apart from these modifications listed below, we have adapted, modified or reconceived an important number of small toolings, which, although unitary of a low importance, have contributed to the good manufacture of this project.

DESIGN MODIFICATION

The main modifications have given rise to the points below :

- Initially, the cable insulation was made up of two half-lapped layers of a 25 microns thick kapton tape, all be it a total thickness of 0.1 mm. Cable being at its maximum tolerance, it has been decided without changing the trial voltage values, to lessen the total insulation thickness to 50 microns. Instead of keeping the same 25 micron tape with a single half-lapped layers we have recommended the use of a 13 micron kapton tape in two half-lapped layers, the use of two half-lapped layers appears to bring about a higher reliability than just one.

- The insulation in the heads has been modified. A kapton layer has been replaced by a Nomex layer which is more resistant to compression. The head wedge forms have also been modified to limit the local compression constraints.

- An insulating washer has been added in the coils connection zone between them.

- Then helium pipeworks have been improved considerably :

  . the welded elbows have been replaced by successive bent elbows in a single tube.

  . certain ends and tolerances have been modified on the inertia and tightness tubes.

- The clamps have been modified to facilitate the assembly and to improve the functioning.

- The coupling parts between the different shells have been replaced.

- In a general manner these parts are submitted to upstream supplementary controls in order to detect possible faults before assembly in the cryostats.

TOOLING MODIFICATION

The main modifications have given rise to the points below :

- Massive steel coil winding inner form for ease in taping down, a saving in time and a tooling longevity.

- Improvements in the cable clamping systems - insulation protection - saving in time.

- Fitting out of the four coils assembling post allowing work for one person rather than two.

- Bus conformation of tooling production

- Manufacture of several gauges for assembly of cryostat elements.

MANUFACTURE PROCEDURE MODIFICATIONS

The main modifications have given rise to the poins below :

- Simplification of the installation of the kapton strips, made up of the insulation mass of each coil, by the strips of the single length instead of 3 necessary strips.

- Use of key of same length as total magnet length. It was anticipated in the beginning 14 parts, 150 mm in length.

REMARKS

A careful preparation of all the constituants is essential and allows an appreciable saving in time and an optimal security.

Presently, more than 100 magnets have been cold tested at DESY and the first ones have been installed in the ring. Only one quadrupole is posing problems and for several months DESY has been doing research to find out the reason or reasons for the fault (conductor fault or manufacture fault ?).

Taking this affair as a whole, the industrialization has been carried out in a relatively short contractual time-period.

Order : 12.05.1986

Delivery of the first series magnet : 10.13.1988

Average rate of production : 4 magnets per week

CONCLUSION

It seems essential for the success of such products, that the production of a pre-serie is foreseen in the general schedule of the project with a sufficient number of pre-serie magnets to be manufactured in a time-period allowing the best industrialization possible.

TECHNOLOGY TRANSFER - PAST, PRESENT AND FUTURE

George J. Doddy

Daniel, Mann, Johnson, and Mendenhall
Washington, D.C.

The present and the future were never more a product of the past
than in Science, where one thing generally leads to another. Science has
continuously produced discoveries and events that have benefited society.
In reviewing some of the benefits which we so readily enjoy today, I
could not help but wonder when they occurred and to what they owed their
inceptions. I have subsequently found that the course of many of these
benefits had their beginnings during the period of time spanning the turn
of the century and as such are early examples of technology transfer from
pure research. This segment of science history is stated lucidly by
Close, Martin and Sutton in "The Particle Explosion."

THE PAST

In 1879, William Crooke's learned that when he passed an electric
current through a vacuum tube, an eerie cold glow appeared. A stream of
rays emanating from a cathode within this tube caused the tube to glow
where it was struck by these rays. The Crooke's tubes were subsequently
called cathode ray tubes. Examples of applied technology from this
scientific discovery are the extensive use of cathode ray tubes in
laboratories, the electron microscope and, of course, the most familiar
form of cathode ray tube - the television picture tube.

In 1897, J.J. Thompson, using a better vacuum than those in Crooke's
tubes, measured the motion of rays generated by electric as well as
magnetic fields. He came to the conclusion that the negatively-charged
particles were approximately 2,000 times lighter than hydrogen and that
these rays really were matter in a new state. These particles were
called electrons and J.J. Thompson was awarded the Nobel prize for its
discovery in 1906. The electron made possible the revolution in
computing and electronics, including television, and is perhaps the prime
example of pure research in basic physics leading to fundamental and
dramatic changes in society.

In 1895, Wilhelm Rontgen was one of the many people who investigated
the strange lights in Crooke's tubes and he found that photographic
plates were fogged when exposed to the rays emanating from these tubes.

One night, not remembering whether he turned out the lights, Rontgen returned to a darkened laboratory and saw that a coated paper was glowing despite the fact that there was no light in the room. Since he had covered the Crooke's tube, he concluded that the glow was caused by the same invisible rays that had fogged the photographic plates. Not knowing what to call this phenomena, he called them x-rays. Rontgen also discovered that these same x-rays would pass through many objects, such as skin, and his first x-ray photograph was the hand of his wife. We are all indebted to the many industrial and medical benefits that have been transferred to us through the discovery of x-rays.

In 1897, Henri Becquerel noticed that uranium salt crystals would glow for some time after they had been removed from sunlight and decided to use the same salt in his x-ray experiments. As luck would have it, when he was ready to expose his uranium salt experiment to sunlight, clouds had blocked out the sun. Consequently, he put everything away in a drawer thinking that nothing would happen to his experiment without sunlight. It wasn't until three days later that the sun came out. When Becquerel prepared to begin the experiment, he found a very clear imprint on the photographic plate. He then concluded that the uranium compounds had emitted rays continuously and thus was discovered radioactivity for which Becquerel was awarded the Nobel Prize in 1903. He shared this prize with the Curies who by this time had isolated radium.

In 1897 and 1898, Ernest Rutherford discovered two components for radioactivity - alpha, which is easily absorbed radiation, and beta, which is low intensity and more penetrating radiation. In 1900, Rutherford suggested that the latent energy that lies within the atom might be released if a rearrangement of its constituents could occur. Rutherford also worked with a young physicist from Germany, Hans Geiger, in the use of prototype counters for their experiments. Geiger would later go on to develop a more sophisticated counter that bears his name. In 1920, Rutherford was alone in speculating about the existence of electrically neutral particles within the nuclei. Although the Curies later observed also that neutral particles were at work, they failed to recognize the signs of this particle. It wasn't until later in 1932 that James Chadwick, who also worked with Hans Geiger, was impressed by the Curies' paper in 1932. He conducted numerous experiments until it became clear that nuclei consisted not only of positive charged protons but of neutral particles as well and in the process he discovered the neutron.

Beginning in 1928, an important paper by Russian theorist George Gamow initiated a new chain of events in the world of physics. John Cockroff saw the importance of this theory and realized that artificial nuclear disintegration was possible by applying millions of volts of electricity to the particles. He was joined by Ernest Walton and together they developed an apparatus that revealed for the first time, in 1932, the artificial disintegration of the nucleus of an atom. One thing leading to another, Otto Hahn and Fritz Strassman shortly thereafter in Germany surprised everyone by firing the newly discovered neutrons at uranium and they succeeded in splitting the nuclei in two. This process of fission releases energy and frees more neutrons which then triggers the fission process of adjoining nuclei which frees even more neutrons and so forth, and so forth, and so forth. If this process is harnessed and controlled, we have nuclear power. If this process is uncontrolled, we have a chain reaction that will continue to multiply until we have an explosion - as happens in an atomic bomb. The technology transfer from this discovery not only has provided us with means of defense and electric power, but also has enabled us to develop the tools with which to diagnose illness and to treat cancers.

THE PRESENT

And then came the Accelerators.

In 1930, Ernest Lawrence built the first successful cyclotron and thereby changed the course of particle physics. Much has happened in the intervening years from that small 80 KeV accelerator to our present energy level of 1 TeV. Significant technological developments have begun to accrue as a result of the application of the discoveries generated by this science and we look forward to what research with the SSC holds in store.

Many of us have witnessed and have even personally experienced the numerous present day benefits in both industry and medicine.

IN INDUSTRY

X-rays

We have had a continual benefit of x-rays in industry and adding super conductivity to them will enable us to develop x-ray lithography. With this super precise process, we will have the capability to make unprecedentedly compact integrated circuits, microchips, and thus take miniaturization down to another level.

Super Conductivity

Until recently superconductivity was a laboratory curiosity. Successful operation of the Tevatron at Fermilab has demonstrated the feasibility of building the SSC. Superconductivity has moved from the laboratory through the cooperation of industry, universities and national laboratories. It has made possible medical diagnostics and treatment techniques and high speed magnetically levitated trains (maglev).

Synchrotron Radiation

The synchrotron radiation emitted from accelerated electrons is finding many industrial uses. In addition to x-ray lithography, the use of intense beams of various wave lengths (x-rays, ultra violet and infrared beams) enables the investigation of complex surface mechanisms and ultra fast photo chemical reactions.

Cathode Ray Tubes

Sired by the Crooke's tube, these cathode ray tubes are used extensively in laboratories, in television picture tubes and in a very advanced state - the electron microscope.

Super Computers and High Speed Data Communications

The need to assemble large amounts of data quickly during experiments has spawned the development of super computers with programs that can do just about anything in translating data into spectacular graphics that have become common place in industry and medicine.

Cryogenics

Superconducting requirements for the magnets has created markets for magnet construction techniques, wire fabrication, refrigeration systems, transfer lines, storage systems and cryogenics controls.

Superconducting Radio Frequency - SRF

Large scale use of super conducting radio frequency cavities such as those at CEBAF has benefited the suppliers of niobium and cryogenics materials in addition to providing the technology to design a continuous electron beam.

High Vacuum

The need for high and clean vacuum in beam tubes has generated a requirement for sophisticated pumps and systems and has spawned a wide variety of industries to develop and produce this equipment.

Nuclear Power

Despite its controversial nature, nuclear power has enabled us to expand our economy by providing the required electricity. Even obsolete nuclear power plants can be converted to gas cogenerating plants thus spawning an entirely new technology transfer within a previous technology transfer, as is the case in Midland, Michigan.

IN MEDICINE

The January 1987 issue of National Geographic has a lengthy and dramatic treatise entitled "Medicine's New Vision," which illustrates how incredible new machines can peer into the human body.

X-ray Treatment of Cancer

Our old friend the x-ray not only has the ability to produce internal images, but also has well-defined electron particles that have the ability to destroy malignant tissues. Since x-rays are not always successful in destroying some cancerous tissue, oncologists have discovered that radiation from other subatomic particles are more successful for certain types of tumors.

Neutron Therapy of Cancer

Because neutrons are bulkier than electrons and can be more devastating than electrons, they are very effective at treating radio resistant tumors of salivary glands, sarcomas, carcinomas, use pancreas, and melanomas. An example of this type of facility is the Midwest Institute for Neutron Therapy at Fermilab where they have treated over 2000 patients since 1976. While I was with DMJM at Fermilab in the early 1970's, I had the privilege of attending the first meetings to discuss the modifications of the Linac to accommodate the new particle treatment of cancer.

Proton Therapy of Cancer

Radiation can control cancer by killing cancer cells beyond their capacity to repair their own damage provided that the absorbed dose is sufficient. The problem is that normal tissues receive a similar dose as that of the cancerous tissue and are consequently damaged in the process. Therefore a less than optimal dose is used in order to minimize damage to normal tissue. However, protons of a specific energy have a definite range in matter and the cell-killing effects of the radiation can be localized on the diseased cells while generally sparing the adjacent healthy tissue. The Loma Linda Proton Therapy facility is a true partnership between medicine and high energy physics in that the proton synchrotron was developed and built at Fermilab, then disassembled to be shipped to Loma Linda and reassembled in place. The facility will begin operation in the summer of 1990 and expects to treat 100 patients per day.

Computed Tomography - CT

CT scanners convert x-ray pictures into digital codes to make high resolution video images using computer graphics similar to those that are used by NASA to reassemble pictures sent back from outer space probes. The body is penetrated with a thin fanshaped beam and a CT scanner produces cross sections of tissues. These pictures depict bone structures in fine detail and show small differences between normal and abnormal tissues in the brain, lungs and other vital organs.

Magnetic Resonance Imaging - MRI

MRI relies on the principle that hydrogen atoms, when exposed to a magnetic field, line up precisely and spin like tops. This is accomplished in medicine by using cylindrical superconducting magnets to align the hydrogen nuclei, the protons, and then stimulating these protons with weak radio frequency waves which makes the protons move out of alignment. When the signal stops, the protons move back to their original positions. The energy released in this process is measured and a computer uses this information to construct an image on a cathode ray tube (TV) where it is recorded for further instant replay and analysis. MRI is very helpful in diagnosing brain and nervous system disorders; cardiovascular disease; cancer of the uterus, ovaries, prostrate liver, pancreas, lymph nodes, bladder, kidneys, vocal cords and spinal cord.

Already there are over 1400 MRI units in the United States with approximately another 500 units throughout the rest of the world. Not only is there industrial benefit in the manufacture of these units at about $2,000,000/unit but also in the regular supply of helium and nitrogen to cool the superconducting magnets.

Sonography - Sono

An outgrowth of the sonar technology which was developed in World War II, Sono uses ultra sonic high frequency sound waves to look within the body. The sound waves are reflected back to a piezoelectric crystal which then reconverts these sound waves to electric signals which are transmitted to a computer that translates the signals to images. This diagnostic tool is recommended for pregnant women and is also well suited for examination of breasts, heart, liver and gall bladder.

Digital Subtraction Angiography - DSA

Using a digital x-ray scanner, a picture of the heart is made. Next, a contrast agent is injected into the coronary arteries. This is followed by a second x-ray image which shows the agent flowing through the heart's vessels. A computer subtracts the first image from the second leaving only that which has changed and thereby allowing the physician to diagnose the problem. DSA is an extremely effective tool in diagnosing heart problems.

Positron Emission Tomography - PET

Positron Emission Tomography utilizes a small low-energy cyclotron. The radioactive solution is injected into the body and emits positrons which collide with electrons releasing a burst of gamma rays which shoot out in opposite directions and strike a ring of detectors that transmit the information to computer that in turn translates the data into an image. By tracing the radioactive substance, the doctor can pin-point areas of abnormal activity.

The Particle Detective

A different but nonetheless interesting application of radioactivity is in the art world. Forgeries or overpaintings can be detected by exposing a painting to a flux of neutrons for about an hour which generates a low level of radioactivity within the paints. Since the nuclei within the different paints have different half-lives, some paints will finish emitting radiation before others. Therefore, photographs taken at regular intervals will reveal different images because of the different half-lives of the paints. Because of this characteristic, it is possible to determine if a painting has been altered or forged completely simply by knowing the half-lives of the paints that the masters were using at the time they were painting.

Practical applications of discoveries in scientific research usually have a long gestation cycle. Neutron therapy for cancer is in its childhood almost 60 years after the discovery of the neutron by James Chadwick in 1932. Proton therapy for cancer is in its infancy almost 50 years after the first prediction of its application by Robert Wilson in 1946, and the transistor is in its adulthood almost 90 years after the discovery of the electron by J.J. Thompson. The wait was definitely worth the while. Just as past scientific research has given benefits that we enjoy in the present, so must we have confidence that our present research will produce benefits in the future.

THE FUTURE

What can we expect in the future? No one really knows, but having looked at the past and the present will certainly help us look to the future. Aside from discovering the fundamental laws that govern nature, we can learn how these basic forces behave and we may even learn to harness those forces for our own benefit. It is important, then, that basic research continue aggressively so that we can provide a technological legacy to future generations.

What can we expect from science in the future? Some non-guaranteed predictions have been elicited from various sources:

o   **High speed railroads**

Economical superconducting materials will make magnetic levitation more economical and high speed trains will be developed. Japan and Germany already have prototypes which have been tested to speeds up to 400 Kmh. Germany is considering the construction of a maglev system between the Cologne/Bonn and Dusseldolf airports.

o   **Holography**

Developed in visual particle detectors, holography can lead to industrial applications in pest control, fog characteristics, vaporization, optimum fuel droplet size, spark and ignition mechanisms, and exhaust particles.

o   <u>X-ray lithography</u>

Continued development will lead to unprecedentedly compact
integrated circuits and further miniaturization of products.

o   <u>Laminar tooling</u>

This technology was developed for the manufacture of
accelerator magnets and could be applied for the manufacture of
very long objects that require precision machining.

o   <u>Superconductivity</u>

In addition to continual refinement in its present laboratory,
industrial and medical applications, high temperature
superconductors will increase the practicality of magnetically
levitated transport, low cost electric transmission lines,
cost-effective energy storage systems and superminiature
microchips.

o   <u>Free electron lasers - FEL</u>

Continued experiment and development of equipment will make FEL
technology suitable for the study of surgical applications
where its high power and short pulses will result in smaller
scars and faster healing compared to present surgical lasers.
In addition, FEL has benefits in physics and chemistry research
as well as potential industrial applications in communications,
radar and plasma heating.

o   <u>Previously owned cyclotrons</u>

As particle therapy increases in use for the treatment of
cancer, so will the need for additional treatment centers
increase. Cyclotrons that have become obsolete for physics
could be converted to use for medical purposes as has the ones
at Harvard, and Lawrence Berkley Laboratory as well as in Japan
and the Soviet Union. Old cyclotrons should never die - they
can be revived to help people live.

o   <u>New cyclotrons</u>

The Loma Linda Proton Treatment Facility is preparing to open
this summer. If its anticipated success of treating 100
patients a day is realized, more of these cyclotrons that are
expressly designed for medicine could become a reality.
Perhaps provisions could be made in the SSC to permit its
eventual adaptability for both neutron and proton treatment of
cancers.

o   <u>Personnel resources</u>

One little discussed technology transfer is the roll over from
science into industry of many high energy physicists who, in
this way, are making an important pay-back to society on its
investment in science.

o   <u>Electrical generators</u>

British scientists claimed that they have built the world's
first electrical generator using new types of superconducting

materials recently discovered. This experimental project is the result of a partnership between two industries and a university and utilizes a product developed by a third industry.

o Old nuclear power plants

By converting abandoned nuclear power plants into gas cogeneration plants, new life may be fused into these facilities. There are approximately 20 of these abandoned nuclear power plants around the country at this time.

o Superconducting Magnetic Energy Storage - SMES

This potential technology would utilize superconducting coil windings to store electricity in an underground enclosure about 1000 meters in diameter. The SMES unit can be developed to store electricity when demand is low and then release its energy into the power grid when needed.

TECHNOLOGY TRANSFER PARTNERSHIPS

Physics laboratories throughout the U.S. and the world have a history of partnership with industry. Some examples of this relationship can be seen in the following as described in the November 1988 issue of the Cern Courier.

Fermilab

o Fermilab has its "Industry Affiliates" and since 1980 has received 11 awards for excellence in technology transfer.

o The Midwest Institute for Neutron Therapy (MINT) is located at Fermilab.

o Fermilab has designed and built cyclotron for the Loma Linda University Medical Center for the use of protons for treatment of cancer.

Brookhaven

o The synchrotron radiation facility has been serving industry for years and the staff has received special awards for excellence in technology transfer.

o Brookhaven hopes to build Superconducting x-ray lithography sources to manufacture high speed computer chips for unprecedentedly compact integrated circuits.

Los Alamos

o "Industrial Applications Office" insures that the laboratory's science and technology base is used to produce significant industry applications.

o Research is directed at developing new materials for new classes of relatively high temperature superconducting materials.

Cornell/CEBAF

    o   Both have relied heavily on Industrial participation in the development of superconducting radio frequency cavities.

CERN

    o   The "Committee for Industrial Relations" has been instrumental in working with industry to develop high speed electronics and computers, electrical equipment, vacuum and cryogenics technology, special welding processes and precision medicines.

CONCLUSION

    When we look into the past we can see into the future. While we are in the present we enjoy and take for granted the benefits of the discoveries of those few in the past who had the vision and determination to pursue their goals. We owe them a great debt of gratitude. We can only hope that we in the present will pursue our goals with equal vigor so that our legacy to future generations can be as good or better. There is a great desire among scientists to discover fundamental laws that govern nature which, in the end, should take us back to the beginning. The better that we understand our beginnings, the greater will be the opportunity we will have to control our future. One instrument with which we can continue to pursue the goals of our future is the SSC.

REFERENCES

Bete, "What You Should Know About Magnetic Resonance Imaging."
Boslow, "Worlds Within the Atom" in the May 1985 issue of National Geographic.
Close, Martin and Sutton, "The Particle Explosion."
Cobb, "Living with Radiation" in the April, 1989 issue of National Geographic.
Cohen, MD; Hendrickson, MD; "Neutron Therapy for Nonresectable Radioresistance Tumors."
Lennox, "Neutron Therapy - A Progress Report" in the November/December 1989 issue of Fermilab Report.
Slate, MD; Miller, Ph.D.; Archanbeau, MD; Livdahl; and Cole; "Proton Therapy and the Control of Cancer."
Sochurek, "Medicine's New Vision" in the January 1987 issue of National Geographic.
SSC 1989, "To the Heart of the Matter - The Superconducting SuperCollider."
"What Particle Physics Gives to Technology" in the November 1988 issue of Cern Courier.

Attendees

Debbie Abrahamson
U. S. Department of Energy

Frank S. Adams
TSA

Gerald Adams
Ceramaseal

S. Carl Ahmed
Technology Transfer Specialists Inc.

R. K. Ahuja
Union Carbide Linde

Delton Ake
Lone Star Gas Co.

Gary Albert
General Dynamics Space Systems

James Allen
ITEN Industries

Mike Allen
SSC Laboratory

Ed Altgilbers
Inland Steel Co.

E. Anamateros
Italcompositi

Bruce Andersen
H & J Tool & Die

Charles E. Anderson
Air Products & Chemicals Inc.

Kathy Anderson
SSC Laboratory

Catherine M. Anderson
Universities Research Assoc.

Mike Anderson
City of Mesquite

Rich Andrews
Fermi National Accelerator Laboratory

Owen W. Anglum
ARMCO Inc.

Thomas Ankerman
Koch Process Systems, Inc.

Michael Antochow
Howden Wirth Inc.

William Appleton
CVI Inc.

Yasuo Arai
KEK, National Laboratory for HEP

Richard Araujo
C. S. Draper Laboratory

H. Harold Aronson
Andersen Consulting

Stanislaw D. Augustynowicz
SSC Laboratory

Clicerio Avilez
Instituto de Fisica

M. Baggett
Brookhaven National Laboratory

Neil V. Baggett
SSC Laboratory

Rich Bailey
General Dynamics Space Systems Division

David Bailey
SSC Laboratory

Robert W. Baldi
General Dynamics Space Systems Division

J. Ballam
SSC Laboratory

Bob Balliett
NRC, Inc.

Oscar Barbalat
CERN

Gentry Barden
Texas National Research Laboratory
Commission

Victoria Bardos
SSC Laboratory

Joseph A. Barresi
American Superconductor Corp.

Joseph Bartling
Oracle Corp.

Representative Joe Barton
(R) Texas

Suzanne Bass
Office of Rep. Jim Chapman

Kurt E. Bassett
Martin Marietta Strategic Systems

D. E. Baynham
SFRC Rutherford Appleton Laboratory

M. Begg
Tesla Engineering Ltd.

W. F. Bensiek
Babcock & Wilcox

David Berley
National Science Foundation

Alfred Bertsche
Brookhaven National Laboratory

F. P. Bevc
Westinghouse Electric Corp.

Daryl Bever
General Dynamics Space Systems

Representative Tom Bevill
(D) Alabama

R. Bharat
Rockwell International

T. S. Bhatia
Los Alamos National Laboratory

Jack P. Biegalski
Inland Steel Co.

Jeffrey R. Bilton
Advanced Cryo Magnetics

Ed Bingler
Texas National Research Laboratory
Commission

Paul Bish
SSC Laboratory

Charles Bizilj
Howden Compressors

James L. Black
IBM

George Blanar
Le Croy Corp.

R. J. Blanken
IBM Corp.

Jeffrey Bodkins
Worthington Industries

Dipl. Ing. Boeer
Interatom

Mike Boivin
New England Electric Wire Corp.

Peter Bonanos
Princeton University

John W. Bonn
CVI Inc.

Franz Boos
MPI International, Inc.

Regina Borchard
Martin Marietta

William N. Boroski
Fermi National Accelerator Laboratory

Rodger Bossert
Fermi National Accelerator Laboratory

Robert Botwin
Grumman

Weldon Brackett
Dielectric

William C. Breen
Fluor Daniel

Daniel Bresson
GEC Alsthom

Richard J. Briggs
SSC Laboratory

B. E. Briley
AT&T Bell Labs

Paul Brindza
CEBAF

Donald P. Brown
Brookhaven National Laboratory

John C. Bruno
J & L Specialty Products

Larry D. Buhl
Copper & Brass Sales

Jack W. Burke
Southwestern Labs Inc.

Sibley C. Burnett
Advanced Cryo Magnetics

Catherine Burns
Texas National Research Laboratory
Commission

Thomas O. Bush
SSC Laboratory

David Butler
Air Products & Chemicals Inc.

Rod Byrns
Lawrence Berkeley Laboratory

Brian Byrwa
MKS Instrument Inc.

Robin Caldwell
IBM

H. Glenn Campbell
Babcock & Wilcox

Jan Campbell
Outokumpu Copper

John Campbell
J. R. Campbell & Associates

Rod C. Camper
The M. W. Kellogg Co.

Cyrus D. Cantrell
The University of Texas at Dallas

William M. Caracciolo
SSC Laboratory

J. W. Carey
Triumf/Kaon

Ronnie P. Carleton

Pete Carlson
Hutchinson Technology

Susan Carlson
IBM

Lee Carlson
Superconductor Industry

Jim Carney
U.S. Department of Energy

Ruben Carragno
SSC Laboratory

Richard A. Carrigan Jr.
Fermi National Accelerator Laboratory

John A. Carson
Fermi National Accelerator Laboratory

John Carusiello
MTM Cryotech Laboratory

Shlomo Caspi
Lawrence Berkeley Laboratory

Annette Caudiano
Mitsui & Co. (USA) Inc.

Sergio Ceresara
Europa Metalli-LMI SpA

Representative Jim Chapman
(D) Texas

Anthony K. Chargin
Lawrence Livermore Laboratory

Coby Chase
Texas National Research Laboratory
Commission

Wendell Chen
University of Texas at Arlington

Allen Cheng
d'Escoto, Inc.

Denis Christopherson
SSC Laboratory

W. Gilbert Clark
U C L A / Physics

John Clarke
Belding Corp.

Herschel W. Clay
SSC Laboratory

Pat Clemens
Sverdrup Technology, Inc.

Richard Cole
The Associated Press

E. W. Collings
Battelle

Tim Collins
Hughes Aircraft Co.

James A. Collins
H. B. Fuller Co.

Eugene P. Colton
Los Alamos National Laboratory

Catherine Connor
Parsons Brinckerhoff Quade & Douglas Inc.

Jerry Conville
Gardner Cryogenics

Roger W. Coombes
SSC Laboratory

Germenia Corrado
Italcompositi

Walter Correa
Instituto de Fisica

William J. Courtney
IBM

Scott Covington
Cryenco

Dennis Cox
SSC Laboratory

Jeffrey J. Cox
SSC Laboratory, URA Inc.

John Cozart
Cozart Communications Co.

J. Cozzolino
Brookhaven National Laboratory

George Craig
Lawrence Livermore National Laboratory

David Creighton
Westinghouse I&CSD

Robert L. Cummins
Asea Brown Boveri Elec. Comp.

Paul Cunningham
PMCA

Frank Cunningham
Swagelok Co.

Richard L. Curl
Morrison-Knudsen Co.

Michael W. Curtin
Booz Allen & Hamilton

Dan D' Armond
U. S. Department of Energy

Rodrigo D' Escoto
d'Escoto, Inc.

Per F. Dahl
SSC Laboratory

John W. Daiber
Hughes Aircraft Co.

Gary Damiano
SSC Laboratory

John Dateo
Babcock & Wilcox

Robert A. Davis
IBM Corp.

Tim Davis
Digital Equipment Corp

William J. Davison
Baltimore Specialty Steels

Alberta M. Dawson
MIT/Plasma Fusion Center

George Day
SSC Laboratory

Tim Day
Office of Congressman Joe Barton

Christopher Day
Lawrence Berkeley Laboratory

David Deacon
Deacon Research

David Dealy
Cray Research Inc.

James F. Decker
U. S. Department of Energy

Carl Dickey
SSC Laboratory

Robert E. Diebold
U. S. Department of Energy

George Diehl
Air Products and Chemicals

Reinhart Dietrich
Vacuumschmelze

Nicholas Digiacomo
Martin Marietta

Hollye C. Doane
National SSC Coalition

George J. Doddy
DMJM

Philip G. Dolan
Oberg Industries Inc.

Tom Dombeck
SSC Laboratory

August E. Doskey
Shimizu America Corp.

Rens L. Dubbeldam
Holec Ridderkerk

Tom Dudley
Flexonics Inc.

William L. Dunn
Quantum Research Services, Inc.

Charles Durr
M. W. Kellogg

Steve Dwyer
SSC Laboratory

Phill Eckels
Westinghouse Electric Corp.

Henry Edelson
IMO / Varo Electron Devices

Don Edwards
SSC Laboratory

Edmund C. Eglinton
Associated Power Systems

Yuri Elisman
Brookhaven National Laboratory

Randy Erben
Texas National Research Laboratory
Commission

Timo Erkolahti
'Outokumpu Copper USA, Inc.

Jean Ernwein
CEN - SACLAY

Bob Esmeiser
Philips Components

Homer Faidas
University of Tennessee

John Farraro
Phelps Dodge

Roger A. Farrell
Intermagnetics General Corp.

Ronald W. Fast
Fermi National Accelerator Laboratory

Anthony Favale
Grumman

Representative Vic Fazio
(D) California

Tracy Fields
Associated Press

Robert W. Fieseler
Neuman, Williams, Anderson & Olson

William A. Fietz
U. S. Department of Energy

Ron Fincher
AT&T

Terry Fleener
Ball Aerospace

Esso Flyckt
Philips International

D. G. Fong
Energy, Mines & Resources

Harold K. Forsen
Bechtel Group Inc.

Eugene Foster
UTD Inc.

Gilbert Fox Jr.
Plainfield Stamping-Illinois

Jim Fraivillig
E. I. Du Pont

Arthur W. Francis
Union Carbide Industrial Gases

Gordon Fraser
CERN Courier

Nell W. Fraser
I C F Kaiser Engineers, Inc.

Jim Freeman
Fermi National Accelerator Laboratory

Jim Freim
Furukawa Electric Technology

Richard Fremont
Dour Metal

Jim French
SSC Laboratory

Douglas Fritz
SSC Laboratory

David Frost
Supercon Inc.

Robert K. Fry
Micron Metals, Inc.

Bob Fuller
Ellis County Appraisal Dist.

Henry M. Gandy
Americans for the SSC

Theodore Garavaglia
SSC Laboratory

Marjorie Gardner
University of California at Berkeley

Donald Gasch
Emerson Electric Co.

Marum Gettner
Northeastern University

Chuck Gibson
General Atomics

Gary O. Gigg
General Dynamics Space Systems Division

William Gilbert
Lawrence Berkeley Laboratory

Paul H. Gilbert
Parsons Brinckerhoff

Frederick J. Gilman
SSC Laboratory

Leonard M. Goldman
Bechtel Corp.

Carl Goodzeit
Brookhaven National Laboratory

Peter J. Gossens
General Dynamics

Stephen R. Gottesman
Grumman Corp. Research Center

Jim Gray
SSC Laboratory

Michael A. Green
Lawrence Berkeley Laboratory

David Green
SAES Getters/USA Inc

Arthur F. Greene
Brookhaven National Laboratory

Eric Gregory
IGC Advanced Superconductors

John A. Gruver
Westinghouse Electric Corp.

Glenn Guenterberg
Martin Marietta

Donati Guglielmo
Europa Metalli-LMI SpA

Ed Gurbuz
SSC Laboratory

Ahmet Gursoy
Parsons Brinckerhoff

Gregory M. Haas
U. S. Department of Energy

Toshihiko Hakushi
Kawasaki Steel America

Robert W. Hamm
AccSys Technology Inc.

Ken Hammer
IBM

Charles Roy Hannaford
SSC Laboratory

William E. Harrison
Brookhaven National Laboratory

Tom Hart
Hewlett Packard

Bill Hawkins
Iron Workers Local 481

Representative Jimmy Hayes
(D) Louisiana

Tricia Heger
Fermi National Accelerator Laboratory

Carl Henning
Lawrence Livermore Laboratory

Steve L. Hensley
CVI Inc.

Ron Henson
Texas Instruments

Wilmot N. Hess
U. S. Department of Energy

Stuart Hesselson
EEV Inc.

Steve Hicks
IBM Corp.

Gale E. Hill
MK-Ferguson

John Hill
UTD Inc.

D. James Hindle
Inland Steel Co.

Frank W. Hintze
Canada Wire and Cable Ltd

Masao Hisada
Hitachi Ltd.

Brian Hoang
Mitsui & Co. (USA) Inc.

Frayne Hobbs
SSC Laboratory

Lisa Hoffman
Dallas Times Herald

Vernon A. Holcomb

Seung Hong
Oxford Superconducting Tech.

Charles B. Hood
CVI Inc.

Frank Horak
SSC II

Nobuyuki Hosomi
Hitachi Ltd.

Dennis Howland
CryoLaboratory

Richard Hubbard
CEN - SACLAY

Robert E. Hughes
Associated Universities, Inc.

Robert Hume
Fiber Resin Corp.

Masaru Ikeda
The Furukawa Electric Co., Ltd.

Toru Tony Ikio
Mitsui & Co. (USA) Inc.

Kiyoshi Inoue
National Research Inst. for Metals

Toshio Inoue
Mitsui & Co. Ltd.

Ellwood Irish
Unistrut Corporation

Hisashi Ishida
Kawasaki Steel America

Tsutomu Isobe
The Furukawa Electric Co. Ltd.

H. J. Israel
Holec Ridderkerk

Genzo Iwaki
Hitachi Cable Ltd.

Cezary Jach
SSC Laboratory

Robert A. Jake
American Magnetics, Inc.

Andrew J. Jarabak
Westinghouse Electric Corp.

Garrett Jernigan
UC Berkeley

Wendell Jesseman
New England Electric Wire Corp.

Peter Paul Jodoin
U. S. Department of Energy

David Johnson
Howden Compressors

Robert A. Johnson
General Dynamics Space Systems Division

David E. Johnson
SSC Laboratory

Dave Johnson
Standard Mfg. Co. Inc.

Evan Johnson
General Atomics

Eric C. Johnson
MIT

Thornell T. Jones
IBM

Guilford Jones
Boston University

Proctor Jones
U. S. Senate App.Cte.Subcte. Energy &
Water Dev.

N. B. Buck Jordan
Waxahachie Chamber of Commerce

Robert Jozwiak
Allen-Bradley Co.

Terry Kabel
CVI Inc.

Ralph Kalkbrenner
Westinghouse PQC

Jill Kallsen
Teledyne SC

John T. Kamino
AT&T

Shoji Kamiya
Kawasaki Heavy Industries Ltd

Hem Kanithi
IGC/Advanced Superconductors Inc.

Alvin Kanofsky
Lehigh University

Joe Karpinski
U. S. Department of Energy

Raphael G. Kasper
SSC Laboratory

Steven Kenneth Kauffmann
SSC Laboratory

Juris E. Kaugerts
SSC Laboratory

Vincent J. Kavlick
Fluor Daniel

T. Kawai
Nippon Steel USA, Inc.

Osamu Kawamata
Hitachi Cable Ltd.

Lewis Keller
CEBAF

E. Kelly
Brookhaven National Laboratory

Anne Kelly
Inland Steel Co.

Jim Kerby
Fermi National Accelerator Laboratory

Ken Kikuchi
National Laboratory for High Energy Physics

E. R. Mike Kimmy
General Dynamics Space Systems

Glenn E. Kinard
Air Products & Chemicals, Inc.

Nancy King
SSC Laboratory

Francois Kircher
CEN - SACLAY - D Ph Pe/ STCM

Thomas B. W. Kirk
Argonne National Laboratory

Gerald Kirschner
Cray Research Inc.

P. A. Kitchin
Tesla Engineering Ltd.

Brenda A. Klafter
AT&T

Robert Kleinhans
Oracle Corp.

Quintin W. Kneen
CBI NA-CON, Inc.

Theodore A. Kobel
Koch Process Systems Inc.

T. Koizumi
Furukawa Electric Tech., Inc.

David Koopman
MPI

Wayne Koska
Fermi National Accelerator Laboratory

Jackie Koszczuk
Fort Worth Star-Telegram

Theodore A. Kozman
SSC Laboratory

Gordon Kramer
Hughes Aircraft Co.

David Kramer
Inside Energy

Robert F. Krause
Inland Steel Co.

Helmut Krauth
Vacuumschmelze GmbH

Detlef Krischel
Interatom GmbH

Rick Kubisch
Allen-Bradley Co.

Ken Kurek
Flexonics Inc.

Jim Kurowski
Standard Mfg. Co. Inc.

Robert R. Kurth
MPI International Inc.

Tamio Kuzuhara
Kawasaki Steel America

David S. La Fleur
MKS Instruments Inc.

Elizabeth J. La Rosa
Digital Equipment Corp.

James L. Lammie
Parsons Brinckerhoff Quade & Douglas

Horace N. Lander
Nippon Steel USA, Inc.

Andy Lankford
SLAC

Robert J. Lari
Vector Fields

Eric Larson
Fermi National Accelerator Laboratory

Warren L. Larson
Supercon Inc.

F. C. Larvie
MK-Ferguson

Barry A. Lauer
LTV Steel

Giuliano Laurenti
CERN

Charles Laverick
Consultant

Maury Lawson
IBM Corp.

George Leakey
Atomic Energy Canada Ltd

Peter J. Lee
University of Wisconsin-Madison

Daniel R. Lehman
U. S. Department of Energy

Thomas A. Leiser
Trammell Crow Company

Scott Lemans
Spaulding Composites Co.

Herbert Lesage
Kloeckner Wilhelmsburger

James Leslie I I
ACPT Inc.

Louis J. Lestochi
Air Products & Chemicals, Inc.

Jim Lieberenz
IMO / Varo Electron Devices

Neil C. Lien
Baker Manufacturing Co.

Qitang Lin
CERN

Goran Lindholm
Technology & Business

Arie Lipski
Fermi National Accelerator Laboratory

John Littlejohn
CRSS Inc.

Licio Loche
Ansaldo

Nigel Lockyer
Univ. of Pennsylvania

Stewart C. Loken
Lawrence Berkeley Laboratory

Susan Lord
Particle World

Roy J. Loring
Babcock & Wilcox

Marcello Losasso
Ansaldo

Robert Lowry
Alliant Computer Systems Co.

Peter W. Lucas
Fermi National Accelerator Laboratory

Alfredo Luccio
Ansaldo

H. B. Lyon
SSC II

Robert B. Mack
IBM

John Mackay
SSC Laboratory

Michael Mahoney
IBM

Bob Maitlin
House Science, Space & Technology
Committee

Ernest Malamud
Sci Tech

G. T. Mallick Jr.
Westinghouse STC

Robert J. Malnar
SSC Laboratory

Jorge Mandler
Air Products & Chemicals, Inc.

Paul Mantsch
Fermi National Accelerator Laboratory

Gerhard Mara
Elin Union

John Markas
IBM

Finley Markley
Fermi National Accelerator Laboratory

Paul E. Marshall
Stone & Webster Engineering

Peter G. Marston
MIT/Plasma Fusion Center

David M. Martin
SSC Laboratory

Ronald L. Martin
ACCTEK Assoc.

Kent Martin
U. S. Department of Energy

Warren Marton
U. S. Department of Energy

Michael D. Marx
University of New York at Stony Brook

T. Masuda
Mitsui & Co. Ltd.

Tom Mathers
Balzers Corp.

William R. Matlach
Grumman Space Systems

Akira Matsui
Kawasho International

Robert Matyas
SSC Laboratory

Mike May
Fermi National Accelerator Laboratory

Michael May
Phelps Dodge

Peter Mazur
Fermi National Accelerator Laboratory

Michael Mc Ashan
SSC Laboratory

Jim Mc Auley
Best Southwest

Jack Mc Carthy
Dynapower Corp.

Glenn Mc Comas
IBM Corp.

Gary Mc Curley
Trammell Crow Company

John R. Mc Donald
Parsons Brinckerhoff Quade & Douglas

N. S. Mc Gladdery
Westinghouse Electric Corp.

Arthur Mc Guigan
Westinghouse

Glen E. Mc Intosh
Cryogenic Technical Services

James Mc Kinnell
University of Wisconsin-Madison

Del Mc Lane
County of Ellis, Texas

Kevin J. Mc Laren
IBM

Murray Mc Mahon
Westinghouse Electric Corp.

David D. Mc Phail
S & W Technical Services, Inc.

Anthony P. Meade
Brookhaven National Laboratory

Marilyn F. Meigs
Morrison Knudsen

Stan Mendelsohn
Grumman Space Systems

Hugh Menown
EEV

Robert Meserve
New England Electric Wire Corp.

Paul Messina
Westinghouse PQC

John E. Metzler
U. S. Department of Energy

Howard Meyer
Digital Equipment Corp

Morton H. Meyerson
Texas National Research Laboratory
Commission

Michael S. Milillo
E - Systems, Inc.

James R. Miller
Sheet Metal Local 68

M. Minami
Mitsui & Co. (USA) Inc.

Michael Minot
American Superconductor Corp.

John E. Mitchell Jr.
Martin Marietta

Tatsuro Miyatake
Furukawa Electric

Chiaki Chris Mizuno
Mitsui & Co. (USA) Inc.

T. Mizuuchi
Nippon Steel USA, Inc.

Martin W. Moffat
Teledyne Metal Forming

James E. Monsees
Parsons Brinckerhoff

Suhas Mookerjee
ABB Technology Co.

W. Henson Moore
U. S. Department of Energy

Billy Moore
Office of Rep. Jim Chapman

John Morena
Brookhaven National Laboratory

Stefano Moretti
Ansaldo N. A., Inc.

Alan Morgillo
Brookhaven National Laboratory

Vicki Lynn Morris
Structural Composites Industries

S. Mulhall
Brookhaven National Laboratory

Pam Mundo
Midlothian Chamber of Commerce

Naseem A. Munshi
CTD, Inc.

Fumio Murase
Kawasaki Steel America

H. Murayama
Nippon Steel USA, Inc.

Michael Murphy
Oxford Superconducting

Francis X. Murray
Energy R&D

Donald E. Naftzger
RMI Company Extrusion Plant

David Nahmias
Air Products & Chemicals, Inc.

K. Naruse
Mitsui & Co. (USA) Inc.

Kelly Nay
Rea Eng. Wire Prod.

Priscilla Nelson
SSC Laboratory

Bill Newman
City of Mesquite

Thomas H. Nicol
Fermi National Accelerator Laboratory

Masaru Nishikawa
University of Tokyo

Kiyohiko Nohara
Kawasaki Steel Corp.

John Nonte
SSC Laboratory

Mary Ann Novak
Parsons Brinckerhoff Quade & Douglas, Inc.

Phil O' Larey
Teledyne Wah Chang Albany

Mark O' Mahony
ARMCO Advanced Materials Corp.

Jim Ochsner
E. I. Du Pont

Tsuguo Ohkuma
IHI

Isamu Ohno
Ishikawajima-Harima Heavy Ind. Co. Ltd

Kenneth Olsen
Martin Marietta Corp.

Paul F. Oreffice
The Dow Chemical Co.

Maurice Osborn
City of Midlothian

Satoshi Ozaki
Brookhaven National Laboratory

Mike Packer
General Dynamics Space Systems Division

Bob Palmer
ISC Subcommittee

Peter A. Panfill
Lake Shore Cryotronics Inc.

C. Dino Pappas
PS Consultants

Adam Para
Fermi National Accelerator Laboratory

Hal Parish
CVI Inc.

Joe Parish
E. I. Du Pont

Dhiru Patel
Hammond Manufacturing

Pamela S. Patrick
ALCOA

Lee Patterson
General Dynamics Space Systems Division

Mark Patterson
EBCO Industries

Steve Paxton
Hewlett-Packard

James B. Peeples Jr.
CVI Inc.

Arlin Pennington
SSC Laboratory

John Peoples
Fermi National Accelerator Laboratory

Yvette Perez
Particle World Magazine

Marc Pesetsky
Associated Press

Karl Pfister
MPI International, Inc.

Steve Pidcoe
General Dynamics Space Systems

James G. Pierce
CVI Inc.

Thomas H. Piquette
EG&G Pressure Science

Joe Pohlen
Martin Marietta

Elizabeth Polk
IMO/Varo Electron Devices

Ilene M. Pollack
U. S. General Accounting Office

N. Porschek
P.E.C. Engineering

Thomas F. Porter
Westinghouse STC

John Scott Poucher
AT&T Bell Laboratoryoratories

Robert L. Powell
Koch Process Systems, Inc.

John Powell
CRSS Inc.

Robert J. Powers
Powers Assoc. Inc.

Ronald P. Pratt
Weldaloy Products Co.

Hartmut Preissner
DESY

Peter E. Price
Industrial Materials Tech

Mark Price
Fiber Resin Corp.

Neil Prosser
Union Carbide Corp.

Representative Carl D. Pursell
(R) Michigan

Joseph Pusateri
Grumman Space Systems

Fred Putnam
IBM/Labtech

Hans H. Quack
Sulzer

Donald J. Quigg
Neuman, Williams, Anderson & Olson

Al Raftis
Hammond Manufacturing

Leo J. Rahal
Advanced Sciences, Inc.

R. Rajaram
Bharat Heavy Elect. Ltd

Myrna Ramos
Hewlett-Packard

Robert N. Randall
Supercon Inc.

M. G. Rao
Aniga Enterprises

Joseph Rasson
SSC Laboratory

George A. Ratz
Niobium Products Company, Inc.

Paul Reardon
Science Applications International

Penny Redington
Ellis County Judge

Tom Reed
SSC Laboratory

Lou Reginato
SSC Laboratory

Margareta Rehak
Brookhaven National Laboratory

Larry A. Rehn
InterFET Corp.

Anne Reifenberg
Dallas Morning News

Robert Remsbottom
SSC Laboratory

Mark Rennich
Engineering

Regenia Richardson
Fermi National Accelerator Laboratory

Mack Riddle
I C F Kaiser Engineers

Harold J. Rietveld
General Dynamics

Verrill Rinehart
Cray Research, Inc.

Ernst Ringle
Noell GmbH

Representative Robert A. Roe
(D) New Jersey

E. Parke Rohrer
Brookhaven National Laboratory

L. David Roper
Virginia Poly. Inst. & State Univ.

Carl H. Rosner
Intermagnetics General Corp.

Robert Rowlands
Hewlett Packard

John M. Royet
Lawrence Berkeley Laboratory

William G. Rueb
MK-Ferguson Company

C. Rupprecht
Leybold Vacuum Products Inc.

C. Jim Russell
Superconductivity News

Bob Russell
Ant Bosch Telecom

Lori Ryan
VAT Inc.

Tim Ryan
KDFW-TV

Hans Rykaczewski
ETH Zurich

Richard Sah
Brobeck Div. of Maxwell Labs Inc.

Shigeo Saito
Sumitomo Electric Industries Ltd

Tetsuo Saito
Shimizu Corp.

Chris Saltmarsh
SSC Laboratory

Michael Saltzman
U. S. Department of Energy

N. P. Samios
Brookhaven National Laboratory

Keith Sams
IBM

Phil Sanger
SSC Laboratory

Tom Sarao
Day Associates

Shinji Sato
Kawasaki Steel Corp.

Sharon Saupp
Union Carbide

John Scango
U. S. Department of Energy

Ron Scanlan
Lawrence Berkeley Laboratory

Giuseppe Scarfi
Ansaldo

Wolfgang Schellmann
Kloeckner Wilhelmsburger

Charles Schiano
Grumman Data

Leonard Schieber
PCK Technology Div

William Schiesser
SSC Laboratory

Marty Schmalhorst
IBM Corp.

Stanley O. Schriber
Los Alamos National Laboratory

Steve Schuermann
Morrison-Knudsen Co.

Pete Schumacher
Westinghouse

Bill Schumacher
ARMCO

Roy F. Schwitters
SSC Laboratory

James Self
City of Waxahachie

Bill Shannon
Hovair Systems Inc.

Kenneth Shearer
Martin Marietta

Bob Sheldon
SSC Laboratory

Philip E. Shelley
SSC Laboratory

Noboru Shibata
Hitachi Ltd.

Jeng Shih
SSC Laboratory

Tohru Shimizu
Kuraray Co. Ltd

Jack Shultz
Tempel Steel

Charles W. Simpson
Lipsen, Whitten & Diamond

S. K. Singh
Westinghouse STC

George Sintchak
Brookhaven National Laboratory

John R. Skaritka
SSC Laboratory

Patrick Skubic
University of Oklahoma

Donald Slanina
IBM Corp.

David B. Smathers
Teledyne Wah Chang

Robert Smellie
SSC Laboratory

Sally Anne Smith
SSCL/EGG

Volker Soergel
DESY

Bob Sokoll
City of Waxahachie

John Sondericker
Brookhaven National Laboratory

Torben Sonderskov
Danfysik A/S

Frank Spinos
Grumman Space Systems

Vic Stack
Cray Research, Inc.

Stuart Stampke
SSC Laboratory

Michael J. Stanko
Westinghouse Science & Technology Center

Piotr M. Starewicz
Resonance Research, Inc.

Ray Stefanski
SSC Laboratory

Elizabeth Stefanski
Argonne National Laboratory

John Stekly
Intermagnetics General Corp.

Christine Stelter
A T & T

Andrew E. Stevens
Argonne National Laboratory

R. Stiening
SSC Laboratory

Gary Still
EG&G

William Stokes
Brookhaven National Laboratory

Becky Stokes
InterFET Corp.

Jay Stone
U. S. Department of Energy

Jack Story
SSC Laboratory

Richard H. Strader
Trammell Crow Company

James Strait
Fermi National Accelerator Laboratory

Bruce P. Strauss
Powers Associates, Inc.

H. D. Stringer
E-Systems Inc.

Gael M. Sullivan
The LTV Corp.

Takeshi Suzaki
Kawasaki Heavy Industries Ltd

Takao Suzuki
Hitachi Ltd.

Sven B. Svendsen
DMJM

John E. Swindle
Best Equipment Co.

Vito P. Sylvester
GTE Products Corp.

S. Takemura
Nippon Steel USA, Inc.

Richard Talman
SSC Laboratory

Claude D. Tapley
Kabelmetal America

Clyde E. Taylor
Lawrence Berkeley Laboratory

Zachary Taylor
Structural Composites Industries

David W. Taylor
ARMCO Inc.

L. Edward Temple Jr.
U. S. Department of Energy

Robert K. Tener
SSC Laboratory

Roger E. Tetrault
Babcock & Wilcox

Juhani Teuho
Outokumpu Copper

Jay C. Theilacker
Fermi National Accelerator Laboratory

Robert L. Thews
University of Arizona

Kenneth M. Thomas
Brobeck Div. of Maxwell Labs Inc.

Don Thomas
UCSD-La Jolla

Tommy Thompson
SSC Laboratory

Tim Thurston
SSC Laboratory

Kuniyasu Toga
Hitachi Ltd.

Yoshi Tokunaga
Nippon Steel USA

John S. Toll
Universities Research Assoc.

Timothy E. Toohig
SSC Laboratory

Gerry Tool
SSC Laboratory

George Trilling
Lawrence Berkeley Laboratory

M. Tsuji
Nippon Steel USA, Inc.

Gary Tucker
Fiber Resin Corp.

Bill Tucker
Concurrent Computer Corp.

Michael Twerdochlib
Westinghouse

Wendell W. Uldrich
IBM Corp.

George J. Urich
Everson Electric Co.

Jim Van Dine
Elano Corp.

Peter Van Duyne
Koch Process Systems Inc.

Voitto Vanhatalo
Outokumpu Copper

Alex Varghese
Gardner Cryogenics

Wally Wade
EG&G

Masayoshi Wake
KEK National Laboratory

Ian J. Walker
GMW Associates

John Walker
Texas Instruments

Roy M. Wallace
Remmele Engineering

William Wallenmeyer
SURA

M. Gully Walter
Tempel Steel Co.

Bert Wang
Wang NMR Inc.

Robert E. Warren
Hughes Aircraft Co.

Yoshiyuki Watase
KEK National Laboratory

Jerry M. Watson
SSC Laboratory

Myron Wecker
IBM Corp.

De' Ann Weimer
United Press International

Paul B. Weiss
United Tech Optical System

Siegfried Wenger
ELIN Co.

William W. Wesley
Neuman, Williams, Anderson & Olson

James E. West
Cryogenic Consultants, Inc.

Mark West
SSC Laboratory

Ed Whiting
SSC Laboratory

Ben C. Wiant
Westinghouse Electric

Joe W. Wildmon Sr.
LTV Steel Corp.

Leonard Wilk
C. S. Draper Laboratory

Clark Wilkinson
The PB/MK Team

Erich Willen
Brookhaven National Laboratory

D. R. Willis
Tesla Engineering Ltd.

Kenneth F. Wilson
CVI Inc.

Kathy Wilson
North Texas Commission

Russell A. Winje
SAIC

Stefan L. Wipf
DESY

Mark S. Wissinger
Westinghouse Electric

Dan Wolff
Fermi National Accelerator Laboratory

Peter Wolochow
Intel Scientific Corp.

Jack Woltz
SSC Laboratory

John Womersley
Florida State University

James Wong
Supercon Inc.

William A. Worstell
Boston University

Jim Worth
Oxford Superconducting Tech.

Russell L. Wylie
SSC Laboratory

Steve C. Yakub
KDFW-TV

Hiroyuki Yamaguchi
Teledyne Japan K.K.

Tateki Yamamura
Teledyne Japan K.K.

Meir Yogev
Embassy of Israel

Rodney Young
Hutchinson Technology

Jon Zbasnik
SSC Laboratory

Bruce Zeitlin
IGC/ASI

Ren-yuan Zhu
CalTech

Art Zinszer
General Dynamics Space Systems Division

Author Index

Subject Index

ABB, *see* Asea Brown Boveri
Accelerator Controls Network, *see* ACNET
Accelerators, *see also* specific types
  accumulator, 418
  aperature evaluation and, 68-73
  cold diodes and, 28
  fixed target, *see* Fixed target accelerators
  highly corrected, *see* Highly corrected accelerators
  introduction of, 797
  moderately corrected, 68
  test schedules for, 177-178
  US/Japan cooperation in, 3, 6
  US projects in, 9-21
AccSys Technology Inc., 269, 280, 282-283
Accumulator accelerators, 418
ACNET, 420, 422, 425, 426
AFA Industries, 365
AGS accelerator, 19, 21
Air core toroids, 311-317, 437-443
Air Products, Inc., 247, 254, 501
ALEC dipole, 790
ALEPH, 632, 641
Allowed multipoles, 63, 65, 749
Amoco Research Center, 713
AMY, 5, 700, 785, 786
Ansaldo ABB Componeti, 445-450, 451-455
ANSYS program, 452, 458
Antiprotons, 10, 19
APC, *see* Artificial pinning center
Application Specific Integrated Circuits, *see* ASICs
Argonne National Laboratory, 673, 675, 676-679, 713

Artificial pinning center (APC), 341-348
Asea Brown Boveri (ABB) Ltd., 319, 330, 558
ASICs, 138-139, 142, 143
ASPEN/SP simulation, 245-254
Association of Science and Technology Centers (ASTC), 713
ASTC, *see* Association of Science and Technology Centers
Astrophysics detectors, 142
Atomic Energy Commission, France (CEA), 789, 790
AT&T Bell Laboratories, 713

Barium fluoride calorimeters, 269-283
  features of, 273-281
  photon and electron physics in, 270-272
  R&D program for, 281-283
BCD, *see* Bottom collider detector
Beam line magnets, 631, 632-639
Beam position monitoring (BPM) system, 109-112
Beam Sync systems, 112, 113, 117
Beam tests, 305
Beam transfer synchronization, 109, 120-124
Beam tubes, 220, 375, 406, 783, 790, 798
  in insulation tests, 399
  laser measurement and, 470, 471
  SPITS and, 382, 385, 386
Beijing Glass Research Institute (BGRI), 282
BGRI, *see* Beijing Glass Research Institute

825

Central Design Group (CDG), 7, 26, 146, 418, 547
Central Helium Liquifier plant, 418
Central Tracking Detector (CTD), 445
CERN, 8, 109, 117, 319, 326, 732, 783, 790
  accelerator projects at, 13
  CAD simulation at, 266
  calorimeters and, 278, 281
  conductor development and, 631, 632, 641
  data acquisition and, 135, 137, 138
  JFETs and, 673
  scintillator detectors and, 267
  technology transfer and, 803
  timing system of, 112
  twin aperture dipole and, 451, 452
CESR, 13, 17, 21, 641
CESR PLUS, 19
Charles Stark Draper Laboratory, 324
Charm quarks, 13
Chlorophyll, 288
Chromaticity, 77, 79, 98, 100, 102, 103
Chromaticity sextupoles, 41, 68
Chromophores, 288, 290, 291, 292
CICC, see Cable-in-conduit conductor
CLEO, 13, 21
CLEO I, 700
CLEO II, 632, 641, 700
C358 magnets, 494
CMM, see Cable Measuring Machine
Coil curing, 753
  in 1.8 meter dipoles, 748
  in short model dipoles, 483, 484, 487, 488, 491
Coil end construction, 483, 484-487
Coils, 85, 89, 731, 734, 735, 736, 738, 740
  for air core toroids, 438
  for bubble chamber, 790
  bulk modulus load cells and, 519
  for CICC, 645-648, 651
  cryogenics and, 220
  for HEB, 371, 372, 457
  for HERA, 559, 560, 564, 789, 791, 792
  high Mn collars and, 765, 769, 773, 777, 779
  insulation breakdown in, 757-763
  insulation tests and, 397-398, 399, 402, 403, 404
  for L* Detector, 322, 324
  MagCom and, 145, 150, 152

Coils (continued)
  MCPMR and, 411, see also 8-coil system; 32-coil system
  mechanical properties of, 753-763
  mechanical support of, 457-467
  for 1.8 meter dipoles, 743, 744, 746-748, 749, 750, 751, 752
  multipoles and, 61, 63
  NbTi cable and, 349, 353
  for short model dipoles, see under Short model dipoles
  transverse cooling and, 210, 211, 213, 217
  for ZEUS, 446, 447, 448, 449
Coil-to-coil insulation, 397
Coil-to-ground insulation, 397, 483, 494-495
Coil winding, 106, 483, 484
Coil winding machine, 725-730
Cold bore dipoles, 533-536, see also Cold magnets
Cold diodes, 25-37
Cold magnets, 10, 12, 457, see also Cold bore dipoles
Cold mass, 220
  high Mn collars and, 765, 771, 778
  installation of, 159, 160
  laser measurement of, 469, 470, 471-473, 474
  of 1.8 meter dipoles, 743
  suspension system for, 378
  yokes and, 731, 735
Collars
  bulk modulus load cells and, 519
  coil mechanical properties and, 757
  coil mechanical support and, 457-467
  cryogenic system and, 220
  for HEB, 376-377
  for HERA, 559
  high Mn type, 765-779
  insulation tests and, 399
  for 1.8 meter dipoles, 743, 744, 746-748, 749, 750, 752
  multipoles and, 60
  quenches and, 505
  for short model dipoles, 483, 495-498, 499, 500
  yokes and, 731, 732, 733, 734, 736, 737, 738, 739, 740, 765, 769, 773, 777
Collider rings
  aperture evaluation and, 59, 60, 61, 63, 65, 67, 69, 70, 71, 72, 73-74, 75-76

Collider rings (continued)
    correction magnet system for, 97-106
    egress spacing and, 185
    HEB and, 371, 378
    particle loss tracking in, 77-84
    plausibly corrected, see Plausibly corrected machines
    refrigeration and, 200, 202
    of TRISTAN, 783
Computed tomography (CT), 799
Computer simulation
    ASPEN/SP and, see ASPEN/SP
    CAD system and, see CAD system
    for helium core heat exchangers, see under Helium core heat exchangers
    Monte Carlo, see Monte Carlo simulation
    of radiation effects, 33-35
Conductors, 601-608, see also Cable-in-conduit conductor; Cables; Nb3Sn conductors; NbTi conductors
Construction schedule, 177-183
Continuous Electron Beam Accelerator Facility (CEBAF), 10, 19-21, 719, 798, 803
Cornell University, 803
    accelerator projects and, 9, 10, 13, 16, 19, 21
    conductor development and, 632
COSI, see Center for Science & Industry
Critical current density
    in conductors, 601, 603, 604, 608, see also under specific conductors below
    in Cu/NbTi cable, 621, 622, 625, 626, 627, 628, 630
    in HERA cable, 558, 560-562, 566-570, 578
    in 1.8 meter dipoles, 744
    in Model Dipole Magnet Program, 611, 612
    in multifilamentary strand, 581
    in Nb3Sn conductors, 632, 633, 639, 642
    in NbTi cable, 350, 351, 352, 353, 593, 595, 598, 599
    in NbTi conductors, 631, 632, 633, 634, 637
    in NbTi superconductors, 342, 343, 345, 348
    in passive superconductors, 391
Crooke's tubes, 795-796, 797
Cryogenic systems, see also Refrigeration systems

Cryogenic systems (continued)
    applications of, 797
    beam monitoring systems in, 109
    in electronics and photonics, 427-432
    helium core heat exchangers for, see Helium core heat exchangers
    for MTL, 219-225
    redundancy in, 223
Cryostats, 260, 790
    for cold diodes, 29
    for HEB, 371, 377-379
    for HERA, 789, 792
    quench protection and, 25, 26
CT, see Computed tomography
CTD, see Central Tracking Detector
Cu/NbTi cable, 621-630
CUSB detectors, 13
Cyclotrons, 797, 801, 802

Data acquisition
    for BCD, 299, 307
    instrumentation in, 135-144
    real-time, 163-173
Data General, 137
D-16-B1 dipoles, 390, 391-394
D-15-C2 dipoles, 390, 391, 392
DD0012 magnets, 736
DD0015 magnets, 736, 737
DD0017 magnets, 736, 737
DD0019 magnets, 773
Decapoles
    aperature evaluation and, 63
    passive superconductors and, 389
    random, 103
    in ring correction system, 97, 98, 99, 103, 104, 105, 106
DecNet, 146, 147
DEC systems, 509, 647, 689
Defocusing quadrupoles, 97, 98, 99
Department of Energy (DOE), U.S., 3, 5, 7, 177, 178, 721, 722
DESY, see Deutsches Elektronen-Synchrotron
Detector magnets, 631, 639-642
    CICC for, 645-655
Detectors, see also specific types
    air core toroids and, 443
    CAD system for design of, 257-266
    Swiss R&D program for, see Swiss R&D program
    for TRISTAN, 785, 786
    US/Japan cooperation in, 3, 6
Deutsches Elektronen-Synchrotron (DESY), 8, 36, 91, 789, 790, 793
    cable development and, 558

828

Deutsches Elektronen-Synchrotron
(DESY) (continued)
conductor development and, 632,
641
field integral/field direction
testing at, 357-362
ZEUS tests and, 446, 447, 450
DEVG method, *see* Double epitaxy/V-
groove method
Digital Equipment Corporation, 137,
170, 426
Diode arrays, 112, 661, *see also*
PIN diode arrays
Diodes
cold, 25-37
injection laser, 428-430, 432
light-emitting, *see* Light-
emitting diode
Dipoles, 19, 85, 86, 87, 89, 365,
557, 558, 560, 561, 565,
576, 586, 790
aperture evaluation in, 59-76
in BCD, 300
coil winding machine for,
725-730
conductors for, 601-608, 632,
637, 638, 639, 642
correction of magnetization
sextupole in, 389-395
cryogenics for, 219, 220, 223
Cu/NbTi cable for, 626-627, 628,
630
development schedules for, 549
field integral/field direction
in, 357-362
for HEB, 371, 374, 375, 377, 378,
457
high Mn collar in, 765-779
improvements in, 789-793
insulation requirements for,
397-404
laser measurement and alignment
of, 469-481
long model, 483, 488
MCPMR and, 405-413
1.8 meter, 743-752
NbTi cable for, 349, 353
particle loss tracking and,
77-84
R&D in, 3, 7
in ring correction system, 97,
98, 99, 100, 101, 102,
103-104, 105, 106
short model, *see* Short model
dipoles
test schedules for, 177-178
transverse cooling and, 210, 214
for TRISTAN, 783
vertically vs. horizontally split
yokes for, 731-740
DMA system, 143

DOE, *see* Department of Energy,
U.S.
D0 High Energy Physics Experiment,
10, 420
Double epitaxy/V-groove (DEVG)
method, 662-663, 665
Dour Metal, 365, 368, 369, 604
DSA1508 diodes, 29, 30
DS6000 diodes, 26, 29, 30, 31, 35,
36, 37
DS0307 magnets, 494, 495
DS0308 magnets, 492, 494, 495, 498,
499
DS0309 magnets, 494, 495, 498
DS0310 magnets, 495
DSS012 magnets, 734, 737
DSS016 magnets, 745, 748, 750
DSS017 magnets, 744, 745, 746, 747,
748, 750, 751
DSS018 magnets, 745, 748, 750, 751
DSS019 magnets, 745, 748, 749, 750
DSV016 magnets, 744, 745, 748, 750
DSV019 magnets, 744
Duodecapole errors, 41
DuPont, 761
Dye-polymer conjugates, 287-296
Dynamic aperature, 40, 41, 81, 83,
84, 97
D-Zero liquid argon calorimeter,
259, 260, 262

EB Norsk Kabel, 319
Egress spacing, 185-194
Eidgenossische Technische
Hochschule (ETH) Zurich,
324
8-coil system, 405, 407-409, 411,
412, 413
Electromagnetic calorimeters, 269,
272, 280, 281, 300, 785,
see also specific types
Electron accelerators, 13
Electron beams, 21
Electron colliders, 137
Electronic noise, 279, 335
Electronics, 319, 325, 330-337,
427-432
Electron linac, 13, 679
Electron linear accelerators, 786
Electron machines, 10
Electron-positron colliders, 5, 13,
see also specific types
Electrons, 717, 795
accelerator technology and, 9, 13
calorimeters and, 270-272, 280
JFET exposure to, 671
PIN diode arrays and, 690
trigger systems and, 167
TRISTAN and, 784
Electron spin resonance (ESR), 534,
535, 536

New England Electric Wire Co., 365, 368, 604, 611
NIM systems, 137, 139, 140, 143
NINA, 91
NIST, 679
Noise, *see also* specific types
  JFETs and, 659, 661, 662, 671, 673, 675-676, 678, 680
  PIN diode arrays and, 681, 685, 686, 688, 690, 691-692
Nonmagnetic scattering, 584
Norland Products Co., 300
Norsk Data, 137
Novosibilsk, 8
NSC, *see* Texas A&M University Nuclear Science Center
NSF, *see* National Science Foundation
NS method, *see* Neuffer-Simpson method
Nuclear Magnetic Resonance (NMR), 94, 357, 359, 360, 405, 407, 410, 412, 533, 534, 535, 536

Oak Ridge National Laboratory (ORNL), 305, 324
Octupoles
  aperature evaluation and, 73
  in ring correction system, 97, 98, 99, 101, 102, 103, 106
ODE system, 238, 239, 240
Ohio State University, 700
OPAL project, 8
OPERA system, 425
Optical data links, 163, 166-168, 171
Opto-electrical conversion, 337
ORNL, *see* Oak Ridge National Laboratory
Oxazoles, 290

Parallel optical data transfer, 319, 330-337
Parameter Page program, 424
Particle loss tracking, 77-84
Passive quench protection, 25-28, 37
Passive superconductors, 389-395
PCOS III system, 140
PECKBUS, 685
PEP collider, 13
PET, *see* Positron Emission Tomography
PFLOW program, 509
Photon beams, 538
Photonics, 427-432
Photons, 429, 430
  calorimeters and, 270-272, 278, 280, 282, 283

Photons (continued)
  liquid scintillators and, 288
  PIN diode array exposure to, 690
  trigger systems and, 167
Photosynthesis, 288
PIN diode arrays, 305
  hybrid, 681-692
Pions, 19
Pixel detectors, 299, 305, 307, 683
Plausibly corrected machines, 72, 73-74, 75-76, 77, 79, 81, 83, 84
Plot Package program, 424-425
PMAA, *see* Polymethacrylic acid
Polycarbonate, 693, 696
Polyester, 693
Polyethylene glycol, 701, 703
Polymers, 291
Polymethacrylic acid (PMAA), 287, 292-293, 294, 295
Polyphenyls, 290, *see also* specific types
Positron Emission Tomography (PET), 799-800
Positrons, 9, 717, 784
pp Collider, 21
Preamplifiers, 428, 431, 527-531, *see also* specific types
  very low noise junction FET, *see* JFET preamplifiers
Precision electronics, 319, 325, 330-337
Precision timing systems, 109, 112-120
Pre-production cable, 615-620
Princeton University, 305, 324
Proton beams, 679
Proton linear accelerators, 786
Proton-proton collisions, 693
Proton rings, 67, 557
Protons
  accelerator technology and, 9-10, 19
  in cancer therapy, 798-799, 800, 802
  long-term tracking of, *see* Long term tracking
  in timing systems, 120
  TRISTAN and, 786
Proton synchrotrons, 3, 13, 786
Psi particle, 13
PSPICE program, 277
P-terphenyl (PTP), 287, 291-292, 293, 294, 295
PTP, *see* P-terphenyl
Purdue University, 324
PYTHIA program, 266, 270, 272

Quadrupoles, 85, 86, 87, 89, 783, 789, 790, 793

Quadrupoles (continued)
  aperature evaluation and, 61, 65,
      65-66, 74
  in BPM system, 109
  cables for, 365, 557, 558, 560
  conductors for, 601-608, 631,
      632, 637, 638, 642
  cryogenics in, 219, 223
  long term tracking and, 41
  magnetic field errors and, 65-66
  magnetization sextupole in, 389
  particle loss tracking and, 81
  quench protection and, 26
  regular, 41
  ring correction system and, 97,
      98, 99, 100, 102, 103,
      104, 106
  skew, see Skew quadrupoles
  transverse cooling and, 210
Quark-gluon plasma, 19, 21
Quarks, 9, 10, 13, 272, 711,
      716-717
Quench analysis, 501-517
  computer codes used in, 509
  conservation in, 505, 506-507
  electromagnetic model of, 504,
      505, 517
  iteration methods in, 507, 508
  thermohydraulic model of, 504,
      505, 517
Quenches, 19, 20, 60, 155, 200,
      223, 228, see also Quench
      analysis; Quench
      protection
  adiabatic, 509, 510, 511, 514
  CICC and, 647, 648, 654
  cold diodes and, 25, 28, 30, 31,
      35-36, 37
  defined, 25
  in HEB, 378
  HERA cable and, 560, 563
  high Mn collars and, 771, 774,
      775, 779
  insulation tests and, 399, 401
  MagCom and, 145, 151
  in 1.8 meter dipoles, 743, 748
  NbTi cable and, 353
  non-adiabatic, 509, 514
  ring correction systems and, 106
  in short model dipoles, 492
  in twin aperture dipole, 455
  yokes and, 505, 731, 732, 736,
      737, 740
  in ZEUS, 447, 448, 449
Quench protection, 25-35, 418, 422
  active, 26
  passive, 25-28, 37

RA20-A diodes, 29
RA20-D diodes, 29, 30

Radiation, 155, see also Radiation
      hardness
  effects of on preamplifiers,
      527-531
  ionizing, see Ionizing radiation
  PIN diode arrays and, 685-691
  simulation of, 33-35
  synchrotron, see Synchrotron
      radiation
Radiation hardness, 143
  of barium fluoride calorimeters,
      269, 275, 277-278
  of BCD, 305
  of cold diodes, 25, 26, 28-29,
      33-35, 36
  of JFETs, 659-660, 671-680
  of liquid scintillators, 287,
      288
  of muon spectrometers, 325, 335,
      336
  of PIN diode arrays, 683
Radiation tests, 29-33
Radiation transparency, 632, 639
Radio Frequency Quadrupole (RFQ)
      accelerator, 269, 280, 282,
      283
Random decapoles, 103
Random multipoles, 58
  aperture evaluation and, 60-61,
      62, 63, 65, 70, 74
  particle loss tracking and, 77,
      78, 81, 82
Random quadrupoles, 61
Random sextupoles, 61, 70, 73, 77,
      82
Random wound magnets, 106
Ratiometer test, 397, 404
R&D, see Research and Development
RDBMS, see SYBASE relational
      database management system
Real-time data acquisition,
      163-173
Reduced instruction set computer
      (RISC), 170
Refrigeration systems, 199-208,
      553, 797, see also
      Cryogenic systems
  ASPEN/SP simulation of, see
      ASPEN/SP
  quench analysis and, see Quench
      analysis
  redundancy in, 199, 202-204
  refrigeration/liquefiers in, see
      Helium
      refrigeration/liquefiers
Regular multipoles, 60
Regular quadrupoles, 41
Regular sextupoles, 41, 60, 70, 82
Relative Intensity Noise (RIN),
      432

Relativistic Heavy Iron Collider
(RHIC), 7, 19, 20, 582,
585, 587, 588
Research and development (R&D)
in barium fluoride calorimeter,
281-283
in Switzerland, see Swiss R&D
program
U.S./Japan cooperation in, 3, 6-7
Resistance tests, 397-398, 401
Resonant magnet network, 85-87, 88,
90, 91, 92
RFQ, see Radio Frequency
Quadrupole
RHIC, see Relativistic Heavy Ion
Collider
RICH detector, 300
RIN, see Relative Intensity Noise
RISC, see Reduced instruction set
computer
Robotics, 725-730
Rotating coils, 406
RSX-M11+ operating system, 509
Rutherford cables, 593, 596
in-house facility for, 365-369
NbTi type, 349-355

Saddle magnets, 581-591
Sarclay, 557
Saturation sextupoles, 374
SBIR, see Small Business
Innovative Research
SCDR, 417, 418
Science and Technology Interactive
Center (SciTech), 709-717
Scintillation fibers, 267, 268,
275
Scintillation light, 276, 277, 278,
283
Scintillator detectors, 163, 164,
168, 267-268
Scintillators, liquid, see Liquid
Scintillators
SciTech, see Science and Technology
Interactive Center
SDRC, 260
SEDAN III, 33
Self-propelled in-tube shuttle
(SPITS), 381-387
Semiconductor lasers, 166
Semiconductors, 170, 171, 428, 538,
see also specific types
Sextupoles
aperature evaluation and, 59, 60,
61, 63, 68, 70, 73
chromaticity, 41, 68
HEB and, 374-375
in long term tracking, 41
NbTi cable and, 350
particle loss tracking and, 77,
82

Sextupoles (continued)
passive superconductors and,
389-395
regular, see Regular sextupoles
in ring correction system, 97,
98, 99, 100, 101, 102,
103, 105, 106
saturation, 374
in TRISTAN, 783
Shanghai Institute of Ceramics
(SIC), 269, 281-282
SHIP, 5
Short model dipoles, 483-500
coil curing in, 483, 484, 487,
488, 491
coil end construction in, 483,
484-487
coil size and, 489-492
coil-to-ground insulation in,
483, 494-495
coil winding in, 483, 484
collars for, 483, 495-498, 499,
500
instrumentation for, 483,
499-500
skins for, 498-499
yokes for, 483, 496-499, 500
SIC, see Shanghai Institute of
Ceramics
Signal-to-noise ratio
in BPM system, 110
in fiber optic transmission
systems, 429, 430, 431,
432
in JFETs, 672
MCPMR and, 410
PIN diode arrays and, 681, 686,
687, 688, 691
Silicon
in JFETs, 661, 662-663
in PIN diode arrays, 682-684
radiation and, 673, 674, 675
Silicon microstrip detectors, 299,
300, 305, 307
Skew multipoles, 60, 103
Skew quadrupoles, 41, 61, 65, 74,
103
Skins
aluminum, 457, 461, 465-466
coil mechanical support and,
457-467
in insulation tests, 399
for short model dipoles, 498-499
stainless steel, see Stainless
steel skins
yokes and, 731, 734, 735, 738, 739
SLAC, see Stanford Linear
Accelerator Center
SLC, 13, 21, 112, 783
Small Business Innovative Research
(SBIR), 282-283, 365, 369